Business and Environmental Sustainability

Environmental sustainability is increasingly important to organisations, whether for regulatory, financial or ethical reasons. *Business and Environmental Sustainability* looks at the environmental aspect of sustainability for all organisations pursuing competitive advantage. The book provides theoretical foundations from science, economics, policy and strategy, introduces three environmental challenges (climate change, pollution and waste) and looks at how corporate functions can address these.

This textbook provides a thorough foundation by introducing readers to the science, reasoning and theory behind environmental sustainability and then delves into how these ideas translate into principles and business models for organisations to use. Next, it covers environmental challenges from climate change, pollution and waste, and then goes on to examine the different corporate functions (from supply chain management to human resources) to illustrate how environmental sustainability is managed and put into practice in organisations. Finally, a set of integrative case studies draws everything together and enables the reader to apply various analytical tools, with the aim of understanding how companies can not only reduce their environmental footprint but can positively contribute to environmental sustainability.

Written by an award-winning lecturer, *Business and Environmental Sustainability* boasts a wealth of pedagogical features, including examples from a range of industries and countries, plus a companion website with slides, quiz questions and instructor material. This will be a valuable text for students of business, management and environmental sustainability and will also be suitable for broader courses on corporate responsibility and sustainability across environmental studies, political science and engineering.

Sigrun M. Wagner is Senior Lecturer in International Business and Sustainability at Royal Holloway, University of London, UK, and Head of the Department for Strategy, International Business and Entrepreneurship in the School of Business and Management.

Business and Environmental Sustainability

Foundations, Challenges and Corporate Functions

SIGRUN M. WAGNER

Routledge
Taylor & Francis Group
LONDON AND NEW YORK

First published 2021
by Routledge
2 Park Square, Milton Park, Abingdon, Oxon OX14 4RN

and by Routledge
52 Vanderbilt Avenue, New York, NY 10017

Routledge is an imprint of the Taylor & Francis Group, an informa business

British Library Cataloguing-in-Publication Data
A catalogue record for this book is available from the British Library

Library of Congress Cataloging-in-Publication Data
Names: Wagner, Sigrun M., 1978– author.
Title: Business and environmental sustainability : foundations, challenges and corporate functions / Sigrun M. Wagner.
Description: Abingdon, Oxon ; New York, NY : Routledge, 2020. | Includes bibliographical references and index.
Identifiers: LCCN 2020012342 (print) | LCCN 2020012343 (ebook) | ISBN 9781138632387 (hbk) | ISBN 9781138632424 (pbk) | ISBN 9781315208275 (ebk)
Subjects: LCSH: Sustainable development. | Business logistics | Economic development—Environmental aspects.
Classification: LCC HC79.E5 W2645 2020 (print) | LCC HC79.E5 (ebook) | DDC 658.4/083—dc23
LC record available at https://lccn.loc.gov/2020012342
LC ebook record available at https://lccn.loc.gov/2020012343

ISBN: 978-1-138-63238-7 (hbk)
ISBN: 978-1-138-63242-4 (pbk)
ISBN: 978-1-315-20827-5 (ebk)

Typeset in Sabon
by Apex CoVantage, LLC

Visit the companion website: www.routledge.com/cw/Wagner

The Future
To our nephews and nieces, Bjarne, Lizzie, James, and Thea – may you grow up to live and work in a world where business takes its responsibility towards sustainability more seriously.

The Present
To my students who have helped shape this book by choosing to study business and the natural environment.

The Past
To my late father who passed away half-way through my writing of this book and who implicitly and sometimes unknowingly lived out sustainability principles throughout his life.

CONTENTS

FIGURES

TABLES

BOXES

ACKNOWLEDGEMENTS

This book has been a result of teaching the course Management and Environmental Sustainability (MN2805), now Business and the Natural Environment. Without students choosing to take this course, the book would not have come about, so I am very grateful to each and every one of you who chose and chooses to study this important topic.

I have enjoyed writing this book, though it has taken a lot longer than initially planned. In the process, a number of people provided feedback, support, and above all, encouragement to keep going. I would like to thank them all, in particular the following.

Four anonymous reviewers provided valuable suggestions for which I am grateful and which improved the end result. I am thankful to the visiting academics at Royal Holloway who gave of their time to offer feedback and advice: Prof Jen Babbington, Dr Michelle Greenwood, Prof Joseph Sarkis.

The staff at Routledge – Isobel Fitzharris, Judith Lorton, Rebecca Marsh, Sophie Peoples, Natalie Tomlinson and Natalie Tomlinson – who answered my questions without fault and were kind and patient enough to let me take the time needed to write. Thank you!

I'm very grateful to Mary Corfield and the Greenbelt Festival for agreeing to be part of one of the case studies. Keep up the good work!

All current and former colleagues in the Centre for Research into Sustainability and in the School of Business and Management at Royal Holloway, University of London. In particular, thank you to Prof Gloria Agyemang, Dr Stephanos Anastasiadis, Dr Donna Brown, Dr Lauren McCarthy, Mr Justin O'Brien, Prof Bob O'Keefe, Prof Laura Spence, Prof Helen Tregidga, Prof Jeffrey Unerman and Prof Diego Vazquez-Brust for their feedback, support and trust as well as words of encouragement. All the staff in the Department for Strategy, International Business and Entrepreneurship, who I am privileged to lead as Head of Department, for their patience, trust and support – thank you!

All the other writers whose work has fed into the book and who have inspired me. Where I have used their work, every effort has been made to trace copyright holders and to obtain permission to reproduce copyright material, which I gratefully acknowledge – if anything is amiss, please get in touch. All remaining errors are, no doubt, of my own making.

Friends and family, near and far, and the community in Northwood, for encouragement and acceptance as I have not been around much during the writing of the book. Especially my partner's family and my awesome siblings and their lovely families, who have also seen less of me as a result of this project. My loving parents, Solveig and Manfred, who instilled in me the values of the pursuit of knowledge and hard work, product stewardship and a love of the outdoors. In particular, my late father ('Papa') who saw me start the book but didn't get to see me complete it. He embodied the values of sustainability without using its vocabulary. He bought things that were made to last (often at a price!), reused what was left over, repaired what was broken and admired good craftsmanship. He still showed me how to sharpen kitchen knives with his father's grind stone so they would last us longer in the month before he died. May I honour his memory with what I write, teach and do.

Last but certainly not least, my wonderful partner Ann – the first plans for the book were drawn before we met, but you have been with me throughout the writing process and without you, your untiring encouragement and support in action and in word, this would not have been possible. Here's to our next adventure and finally getting that electric car we've been talking about for years.

SDG

ABBREVIATIONS

μg	microgram
μm	micrometre
3BL	triple bottom line
3D	three-dimensional
3PL	third-party logistics
3P	pollution prevention pays (3M programme)
3Ps	people, planet and profit
3Rs	reduce, reuse and recycle
A4S	The Prince's Accounting for Sustainability Project
AA	AccountAbility
AASHE	Association for the Advancement of Sustainability in Higher Education
ABC	activity-based costing
AFOLU	Agriculture, Forestry, Other Land Use
AI	artificial intelligence
AIP	American Institute of Physics
AR5	IPCC Fifth Assessment Report
BAT	best available technology
BATNEEC	best available technology/techniques not entailing excessive costs
BAU	business as usual
BCG	Boston Consulting Group
BIS	Department for Business, Innovation and Skills
BoP	base of the pyramid
BPEO	best practicable environmental option
BPM	best practicable means
BPR	Business Process Re-engineering
BREEAM	Building Research Establishment Environmental Assessment Method
BSC	balanced scorecard
BSI	British Standards Institutions
CBA	cost-benefit analysis
C2C	Cradle to Cradle (certification)
CCS	carbon capture and storage (or sequestration)
CDSB	Climate Disclosure Standards Board
CDP	Carbon Disclosure Project
CDR	carbon dioxide removal
CEO	chief executive officer
CERES	Coalition for Environmentally Responsible Economies
CFCs	chlorofluorocarbons
CH_4	methane
CHP	combined heat and power
CLRTAP	Convention on Long-range Transboundary Air Pollution
CMI	Carbon Mitigation Initiative

CO	carbon monoxide
CO_2	carbon dioxide
CO_2e	CO_2 equivalent
COP	Conference of Parties
CRM	customer relationship management
CRT	cathode ray tube
CSR	corporate social responsibility
CSS	Center for Sustainable Systems, University of Michigan
DDT	dichloro-diphenyl-trichloroethane
Defra	Department for Environment, Food and Rural Affairs
DfE	Design for Environment
DG	Directorate General (of European Commission)
DIN	Deutsche Industrie Norm
DMAIC	define (the problem), measure, analyse, improve and control
EEA	European Environment Agency
EIA	environmental impact assessment
EKC	environmental Kuznets curve
ELV	end-of-life vehicles
EMAS	Eco-Management and Audit Scheme
EMF	Ellen MacArthur Foundation
EMIS	Environmental Management Information Systems
EMS(s)	Environmental Management System(s)
EPA	United States Environmental Protection Agency
EP&L	environmental profit and loss
EPP	environmental(ly) preferable purchasing
EPR	extended producer responsibility
EPS	expanded polystyrene (Styrofoam)
EQOs	environmental quality objectives
ERP	enterprise resource planning
ESG	environmental, social and governance
EV	electric vehicles
e-waste	electronic waste
FAO	Food and Agriculture Organization of the United Nations
FLW	food loss and waste
FOG	fat, oil and grease
FSC	Forest Stewardship Council
FT	*Financial Times*
GDP	Gross domestic product
GEMI	global environmental management initiative
GEO	Global Environmental Outlook
GGP	Greenhouse gas protocol
GHG(s)	greenhouse gas(es)
GICS	Global Industry Classification Standard
GLP	Green Loan Principles
GPI	Genuine progress indicator
GRI	Global Reporting Initiative
GSSB	Global Sustainability Standards Board

GWP	global warming potential
HAPs	hazardous air pollutants
HCs	hydrocarbons
HFCs	hydrofluorocarbons
HP	Hewlett Packard
HR	human resource(s)
HRM	human resource management
HSE	health, safety and environment
HVAC	heating, ventilation, and air conditioning
IAASB	International Auditing and Assurance Standards Board
IASS	Institute for Advanced Sustainability Studies, Potsdam
ICT	information and communication technology
IEA	International Energy Agency
IFAC	International Federation of Accountants
IIRC	International Integrated Reporting Council
IOGP	International Association of Oil & Gas producers
IPAT	IPAT equation where I equals environmental impact, P population, A affluence and T technology
IPBES	Intergovernmental Science-Policy Platform on Biodiversity and Ecosystem Services
IPCC	Intergovernmental Panel on Climate Change
IPIECA	global oil and gas industry association for environmental and social issues
IS	Information Systems
ISAE	International Standard on Assurance Engagements
ISEA	Institute of Social and Ethical Accountability
ISEW	Index of Sustainable Economic Welfare
ISIC	International Standard Industrial Classification of All Economic Activities
ISO	International Standards Organisation
IT	information technology
ITPS	Intergovernmental Technical Panel on Soils
JIT	just-in-time
KPI	key performance indicator(s)
LCA	life cycle assessment (or analysis)
LCD	liquid-crystal display
LED	light-emitting diode
LEED	Leadership in Energy and Environment Design
LNG	liquefied natural gas
LOHAS	Lifestyles of Health and Sustainability
m^3	cubic metre
MARPOL	International Convention for the Prevention of Pollution from Ships
MEA	Millennium Ecosystem Assessment
MIS	management information system
MNE	multinational enterprise
MOOC	Massive Online Open Course
MRP	micronised rubber powder
MSCI	Marine Stewardship Council
MSCI	Morgan Stanley Capital International
MSW	municipal solid waste

N_2O	nitrous oxide
NACE	Nomenclature statistique des activités économiques dans la Communauté européenne (Nomenclature of Economic Activities in the European Communities)
NAICS	The North American Industry Classification System
NASA	National Aeronautics and Space Administration
NBS	Network for Business Sustainability
NCC	Natural Capital Coalition
NDC	nationally determined contributions
NGO	non-governmental organisation
NH_3	ammonia
NO	nitric oxide
N_2O	nitrous oxide
NO_2	nitrogen dioxide
NO_3	nitrate
NOAA	National Oceanic and Atmospheric Administration
NOX/NO_X	nitrous oxides
NPP	net primary product
O_3	ozone
OECD	Organisation for Economic Cooperation and Development
OED	Oxford English Dictionary
OSPAR	Oslo Convention for the Prevention of Marine Polluting by Dumping from Ships and Aircraft and the Paris Convention for the Prevention of Marine Pollution from Land-based Sources
PAHs	polyaromatic hydrocarbons
Pb	lead
PCBs	polychlorinated biphenyls
PDCA	Plan-Do-Check-Act
PDSA	Plan, Do, Study, Act
PES	payment for ecosystem services
PET	polyethylene terephthalate (form of plastic)
PFCs	perfluorocarbons
pH	measure of acidity and alkalinity in water-based fluids
PHA	Planetary Health Alliance
PM	particulate matter
PM_{10}, $PM_{2.5}$	particulate matter (less than 10 μm (micrometres) and 2.5 μm)
POPs	persistent organic pollutants
ppb	parts per billion
ppm	parts per million
ppt	parts per trillion
PRI	Principles for Responsible Investment
PRM	product recovery management
PRME	(United Nations) Principles for Responsible Management Education
R&D	research and development
RCP	representative concentration pathways
RoHS	restriction of hazardous substances
RORO	roll-on roll-off
SA	Standards Australia

SAM	Sustainability Assessment Model
SASB	Sustainability Accounting Standards Board
SBSC	Sustainability Balanced Scorecard
SCAP	Sustainable Clothing Action Plan
SCM	supply chain management
SCU	Science Communication Unit
SDGs	(United Nations) Sustainable Development Goals
SEA	strategic environmental assessment
SEHN	Science and Environmental Health Network
SF6	sulphur hexafluoride
SIC	Standard Industrial Classification
SIGMA	Sustainability Integrated Guidelines for Management
SMEs	small and medium enterprises
SO_2	sulphur dioxide
SO_2 or "sox"	sulphur oxides
SO_3	sulphur trioxide
SRM	solar radiation management
SSCM	sustainable supply chain management
SSE	Sustainable Stock Exchanges initiative
STP	segmentation, targeting and positioning
SWSR	Status of the World's Soil Resources
TBL	triple bottom line
TEEB	The Economics of Ecosystems and Biodiversity
TCFD	Task Force on Climate-related Financial Disclosures
TNS	The Natural Step
TPM	total preventive maintenance
TPS	Toyota production system
TQEM	total quality environmental management
TQM	total quality management
TRBC	Thomson Reuters Business Classification
TRI	Toxic Release Inventory
UEL	uniform emission limits
UKCIP	UK Climate Impacts Programme
UKSIC	United Kingdom Standards Industrial Classification of Economic Activities
UN	United Nations
UNCED	United Nations Conference on Environment and Development (Earth Summit)
UNECE	United Nations Economic Commission for Europe
UNEP	United Nations Environment Programme
UNFCC	UN Framework Convention on Climate Change
UNGC	UN Global Compact
UNPRI	UN Principles for Responsible Investment
VOCs	volatile organic compounds
VRIN	valuable, rare, inimitable, non-substitutable (resource-based framework)
WBCSD	World Business Council for Sustainable Development
WCED	World Commission on Environment and Development
WEEE	Waste of Electrical and Electronic Equipment
WEF	World Economic Forum

WHO	World Health Organization
WMO	World Meteorological Organization
WNA	World Nuclear Association
WRAP	Waste Reduction Always Pays (Dow Chemical programme)
WRAP	The Waste and Resources Action Programme (UK initiative)
WRI	World Resources Institute
WWF	World Wide Fund for Nature
ZEV	Zero Emission Vehicle

AUTHOR BIOGRAPHY

Dr **Sigrun M. Wagner** is the Head of Department for Strategy, International Business and Entrepreneurship in the School of Business and Management at Royal Holloway, University of London, UK. She is an award-winning Senior Lecturer in International Business and Sustainability, and a Senior Fellow of the UK's Higher Education Academy. Her interest in business and sustainability stems from her different experiences of environmental perspectives when living, working and studying in the UK, Germany and Norway as well as a high school year in the United States as a 17-year-old. She gained her PhD from Loughborough University, which was on corporate political activities by the automotive industry regarding environmental regulations in the European Union – research that involved fieldwork in Brussels with a number of EU institutions, NGOs, companies and industry bodies. At Royal Holloway, she was instrumental in the introduction of a sustainability and corporate responsibility pathway for business and management students on the undergraduate BSc programmes. She can be found on Twitter @DrWagner_RHUL.

for business and environmental sustainability. The section here introduces this context, looking at the history of corporate environmental disasters and responses to them from a corporate perspective, environmental policy perspective and the societal perspective. All this has been accompanied and underpinned by academic and non-academic publications on the environment, many of which have become "classics", as highlighted by Vaz (2012) and Visser (2009).

Following the destruction by war in the 1940s and the rebuilding of societies in the 1950s, it was the 1960s that marked the beginning of the environmental movement, with Rachel Carson's *Silent Spring* published in 1962. Both Vaz (2012) and Visser (2009) add Aldo Leopold's *A Sand County Almanac* from 1949 to the list of classics to read, in which Leopold calls for a new "land ethic", and *Walden or Life in the Woods* by Henry Thoreau from 1854 is considered an even earlier precursor (Vaz 2012), alongside George Perkins Marsh's *Man and Nature* (1864, Robertson 2017). It is in the US, in the latter half of the 19th century that the first environmental organisation (The Sierra Club) and the first national park (Yellowstone in 1872) were founded (Buckingham 2008b).

The first use of the term sustainability appears in its German version (*Nachhaltigkeit*) in 18th-century forest management in Germany, which sought to avoid deforestation damages of the Middle Ages (Robertson 2017).[1] In the same century, Thomas Malthus was concerned about humanity's ability to feed geometrically increasing populations in light of arithmetically increasing food stocks (Kedia et al. 2011). This is reflected in the definition Kedia and his colleagues provide of sustainability (2011: 73):

> Sustainability implies Malthusian concern regarding the carrying capacity of nature in relation to human consumption patterns.

Going back even further, the origins of sustainability in agriculture can be traced back to Mesopotamia where the irrigation and subsequent salinisation of arable land led to changes in land use, including deforestation, and, possibly, ultimately to the collapse of the Babylonian empire (Roorda et al. 2012). Similar problems were found on Easter Island where a lack of ability to shape a social-economic-political system that would avoid the exhaustion of forest resources essential for survival led to the demise of the entire island population (Nemetz 2013).

As a full discussion of the historical context is not possible here, Table 1.1 illustrates the parallel nature of developments within the societal triangle as underpinned by literature over the last 60 years. Often environmental policies at the national level have been the result of twin pressure from international agreements at the regional (e.g. EU) or global level (e.g. UN) and from grassroots (NGOs) and local governments (Buckingham 2008b). The previous image of business as part of the societal triangle illustrates the linkages.

The development of environmental issues in terms of perception, recognition, activism and policy in the 1960s was accompanied by a substantial expansion of higher education in Europe, North America and Australasia (Buckingham 2008b). Concrete developments in business and management education regarding environmental issues are more recent, where MBA programmes and business schools have leant on geography, politics, economics and law, amongst others, to reflect a changing business environment.

The historical context of corporate environmental disasters and of the public policy environment provides an overview of the development of sustainability thinking and practice, allowing us to frame the book as offering a sustainable alternative to conventional, more unsustainable, ways of managing organisations across complex layers and levels, as discussed further in the following.

The most common usage of sustainability is in relation to sustainable development. Although the emphasis of the book is on business, environment and sustainability, it is important to

TABLE 1.1

Overview of selected historical developments in environmental sustainability

Timeperiod	State	Market	Civil society & research	Writings
Pre-1960s			1859 John Tyndall discovers how gases change the climate	1854 Walden
			1892 The Sierra Club	1864 Man and Nature
		1942 DDT first marketed	1939 DDT discovered	1949 A Sand County Almanac
		1953 Love Canal	1948 World Conservation Union	
1960s		1950s "Green Revolution"	1960s environmental movement	1962 Silent Spring
			1961 WWF	
	1969 Environmental Protection Agency (US)		1969 Friends of the Earth	1968 The Population Bomb
1970s	1972 UN Conference on the Human Environment and UNEP	1976 Seveso, Italy, dioxin industrial accident	1970 first "Earth Day"	1976 Limits to Growth
	1972 DDT banned	1978 Amoco Cadiz shipwreck oil spill	1971 Greenpeace	
	1973 First Environmental Action Programme (EU)	1979 Three Mile Island, nuclear reactor	1975 CFCs identified as threat to ozone layer	1973 Small is Beautiful
	1976 OECD MNE Guidelines			1979 Gaia
1980s	1980 Word Conservation Strategy	1984 Bhopal, India, Union Carbide disaster	1980s environmental justice movement	1987 Our Common Future
	1983 WCED			
	1987 Montreal Protocol to protect ozone layer			
		1986 Chernobyl, Nuclear Power Plant		
		1989 Exxon Valdez oil spill		
1990s	1992 "Earth Summit" Rio (UNCED), Agenda 21	1990s Corporate Environmental Reporting	1990 first assessment report of IPCC (FAR)	1993 Ecology of Commerce

	1997 Kyoto Protocol	1992 (World) Business Council for Sustainable Development, eco-efficiency	1991 Biosphere 2	1997 Cannibals with Forks
			1993 Forest Stewardship Council	
	1999 UN Global Compact	1995 Shell, Brent Spar oil platform	1995 IPCC second assessment report (SAR)	
		1999 Dow Jones Sustainability Indices	1997 Marine Stewardship Council	1999 Natural Capitalism
2000s	2001 Millennium Ecosystem Assessment	2000 Baia Mare, Romania, cyanide spill	2001 IPCC third assessment report (TAR)	2001 The Sceptical Environmentalist
	2002 Johannesburg Summit	2002 Prestige, tanker		2002 Cradle to Cradle
				2006 An Inconvenient Truth
			2007 IPCC AR4	2007 The Economics of Climate Change
2010s	2012 "Rio+20"	2010 Deepwater Horizon, BP	2013/2014 IPCC AR5	2013 What has nature ever done for us?
		2011 Fukushima Nuclear Plant		2014 This Changes Everything
	2015 COP21 Paris Climate Accord	2015 Samarco dam disaster, Rio Doce, Brazil, BHP Billiton		2014 Farmageddon: The true cost of cheap meat
	2016 Montreal Protocol extended to HFCs		2017 Blue Planet II (BBC)	2017 Doughnut Economics
	2018 Circular Economy Action Plan, European Union	2019 Brumadinho dam disaster, Brazil	2018 Greta Thunberg's climate strike	2019 Financial Times Moral Money newsletter launched
			2018/2019 Extinction Rebellion	

Sources: compiled by author on the basis of Elkington (1997), Buckingham (2008a), Visser (2009), Vaz (2012), Nemetz (2013), Robertson (2017), Whitfield and McNett (2014), AIP (2017), and news sources.

acknowledge the developments behind it. The World Commission on Environment and Development (WCED) whose work culminated in the Brundtland report of 1987 (WCED 1987) was tasked to re-examine environment and development problems found in conventional growth and development models and to come up with proposals for solutions. It is here we find an acknowledgement that economic growth brings risk of environmental damage (WCED 1987: 40).

Sustainable development is defined in the report as development that "seeks to meet the needs and aspirations of the present without compromising the ability to meet those of the future" (WCED 1987: 40). This carries within it the concept of "needs" and the idea of "limitations" both from technology and society on how able the environment is to meet present and future needs (ibid. 42). The commission's work gave rise to new definitions and concepts in a variety of disciplines, informing systems thinking (see Elliott 2013 for a further discussion). It provided an alternative response to conventional economic development and growth focused on gross domestic product (GDP) and per capita income, where many developing countries had experienced large debt service requirements to such an extent that essential investment in human capital (health and education) had to be reduced or curtailed (Hackett 2011). The goal of economic development to improve the well-being of members of society could not be met, and out of this failure the sustainable development movement rose. Daly (2005) calls this uneconomic growth – when increases in production come at the expense of resources and well-being.

The WCED report (1987: 40) acknowledges that there is "no single blueprint of sustainability" to be found as the systems for economies and societies, and their ecological circumstances differ widely. The book here recognises this, and whilst not delivering exact roadmaps for corporate environmental sustainability, provides tools companies can use, adapting it to the context of their industry, size, location and service or product offering.

The WCED provided a focal point following the emergence of the environmental movement and of policy responses to corporate environmental disasters. It led the way for the Rio Earth Summit in 1992 where the Business Council for Sustainable Development was founded and where we see the first serious corporate moves towards environmental sustainability. Sustainable development as a whole can be seen to emphasise sustainable society, which can be illustrated as a pyramid of environmental value, social value and economic value (or ecology, equity and economy), whereas sustainable organisations can be viewed through the three dimensions of profit, people and planet, as advocated by the triple bottom line concept of John Elkington in his landmark book, *Cannibals with Forks* (1997). The individual level is rarely discussed (Cavagnaro and Curiel 2012) though could arguably be the most significant, where leadership for sustainability is situated with *care for me* reflecting society's concern for economic value and organisations' concern for profit, while *care for me and you* parallels society's interest in social value and organisations' interest in people and *care for all* the environmental value for society and the planet for organisations. This synthesis of levels and dimensions is summarised in Figure 1.2.

As business organisations are made up of individuals, it is important to acknowledge how interdependent and interlinked these three levels of sustainability are – the societal, organisational and individual form the complex context in which corporate environmental sustainability is situated and aimed for, often alongside social and economic sustainability, which Kleine and von Hauff (2009) illustrate in an integrative sustainability triangle (see also Lüdeke-Freund et al. 2018). The focus of the book is mostly on the organisational level, but acknowledging this complexity and the context is important. To the facets of people, planet and profit, Roorda et al. (2012) add time and space. This complex context is thus embodied by a number of characteristics emphasising the dynamic environment in which sustainability works.

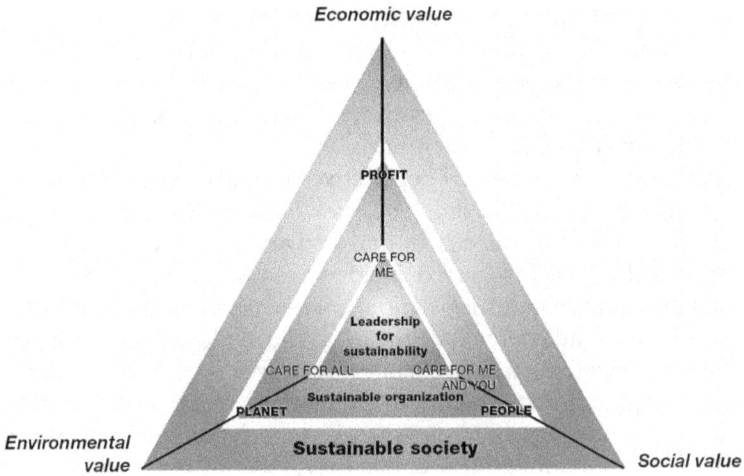

FIGURE 1.2
Three levels of sustainability

Source: Cavagnaro & Curiel, 2012, Greenleaf Publishing, used with permission.

Dynamic complexity

The context of sustainability is simultaneously global and local in terms of its scope – effects of climate change, for example, are felt locally and action to address it can be executed locally, but such a fragmented approach is only partially effective – a global approach is needed (Whitfield and McNett 2014: 14). Managing for environmental sustainability in a global context is a complex task, characterised by four attributes: multiplicity, ambiguity, interdependence and flux which are further considered by Lane et al. (2004):

- Multiplicity refers to the increasing number of stakeholders, perspectives and constraints.
- Interdependence refers to the increasing levels of interconnectedness of these stakeholders.
- Ambiguity refers to the lack of clarity concerning cause-effect relationships, and information and facts being subject to multiple interpretations.
- Flux is the result of the dynamic complexity and movements between the first three attributes.

As such, this book cannot provide the ultimate or definite answers or solutions, but it will introduce students and readers to the complexity of the task and provide tools and concepts with which to manage environmental sustainability across different challenges and corporate functions. What the book covers in particular is outlined in the following.

OVERVIEW OF BOOK

Part I: Foundations

Following this introductory chapter, part one provides the foundations for the remainder of the book, presenting both theoretical and practical underpinnings. Chapter 2 will look behind the scenes, introducing both science and economics in order to make the case for why business should be concerned about environmental sustainability. Concepts covered in this chapter include externalities, earth system and ecosystem services, the (five) capitals model and how this can be conceptualised into strong and weak sustainability. The distinction between natural and man-made capital can help measure ecosystem services through footprints and thereby carrying capacity, showing the limits of conventional growth models.

Chapter 3 translates these theoretical perspectives into principles and foundations for environmentally sustainable businesses. In doing so, we consider the tensions between various perspectives, such as inside out and outside in and between competitiveness and environmental objectives, asking how the political context handles issues such as externalities and ecosystem services (e.g. through environmental legislation) and what principles and philosophies guide companies and regulators fundamentally. Principles and philosophies underlying environmentally sustainable management, such as technocentric versus ecocentric modes of thoughts and standard-setting philosophies for environmental regulations, such as the precautionary principle, cost-benefit/risk analysis and the polluter-pays-principle, are introduced as well as the triple bottom line and life cycle analysis alongside eco-efficiency, mission zero and The Natural Step.

Chapter 4 looks at how the theoretical and philosophical foundations of the previous two chapters translate into practice – what models guide businesses (and other organisations) practically and strategically? We start with strategic responses to sustainability as value creation rather than value destruction before looking at models that put sustainability into practice, whether it is conventional strategy frameworks (e.g. generic strategies, resource-based view) or sustainability focused ones such as The Natural Step framework and natural capitalism, which includes biomimicry. We provide further context and guidance by looking at how companies can be analysed and then managed according to which phase or stage of sustainability they are at. Sustainability initiatives in businesses are covered by way of examples and by way of process – how are goals defined, how is progress measured and how is change implemented – in an incremental or transformational manner and by whom (e.g. change agents).

Part II: Environmental Challenges

Part two highlights the major (but not all) environmental challenges society as a whole and businesses in particular face. A number of avenues could have been taken here, particularly in light of the systemic nature of most environmental challenges. One possible classification could be to look at food, energy and water. Unsustainable use of resources could be another avenue. The challenges selected explicitly or implicitly cover these topics in the respective chapters and are taken to be the ones most pertinent to business, albeit other interpretations are, of course, possible. A separate introduction to the part provides further context here.

Chapter 5 covers climate change, which has implications for business from a variety of perspectives, in particular with relation to energy use. The chapter offers a brief explanation and reminder of what climate change is and how it affects businesses. The focus then shifts to the impact of business on climate change and how businesses can reduce this impact by managing their greenhouse gas emissions and by providing opportunities for consumers and other businesses. This is set within the context of general and regulatory responses to climate change – mitigation, adaptation and geoengineering, but also provides insights into the transition movement as an alternative model that builds up resilience of society and its businesses in the face of climate change.

Chapter 6 draws a link to the previous one by outlining the distinction between climate change and air pollution before setting out the foundations of pollution in terms of definition, types, sources and categories. It then investigates the three major kinds of pollution of air, water and soil, identifying the role of businesses in causing pollution and in remedying it internally and externally. Links will be drawn to the related issues of water scarcity in the context of the tragedy of the commons and the effect of climate change on water, demonstrating the interconnected nature of environmental challenges.

In Chapter 7, the final environmental challenge for businesses is waste, which is closely related to how resources are used. Environmental impacts are identified on the basis of the different types of waste and waste streams. The options for waste treatment are discussed within the context of waste hierarchies (e.g. reduce, reuse, recycle), where the principle of *re-think* can be a game changer for businesses. Concepts such as *zero waste* and *waste as a business model* are explored where products are designed with the end of life in mind and a move away from planned obsolescence is encouraged. The international nature of waste (e.g. transboundary movements) is highlighted as well.

Throughout this part, examples of how companies have addressed these challenges will draw the link between sometimes seemingly abstract challenges and theories, and concrete actions on the ground.

Part III: Corporate functions

Part three moves on to cover different corporate functions to illustrate how environmental sustainability is managed and put into practice in organisations, by looking at primary activities in the value chain in the first three chapters and then at supporting activities in the second three chapters. The introduction to part three provides more background to management and corporate functions in general.

Chapter 8 covers supply chain management (SCM). This serves as a valuable general starting point as supply chains cross company boundaries beyond the focal firm, and as their management incorporates many corporate activities, thus illustrating the complexity of sustainability in practice. Following a brief overview of supply chain management and its environmental impact, network design, logistics and transportation, sourcing and procurement are covered and the challenges and opportunities for sustainability identified. Emphasis is put on collaborations and partnerships with suppliers. Alongside tools and concepts such as the circular economy, closing the loop, cradle-to-cradle, life cycle analysis and reverse logistics, the role of labels and certifications in SCM is explored as well – a topic that re-occurs from different perspectives in later chapters (e.g. marketing and environmental management systems).

In Chapter 9, sustainable production and operations are a natural follow-up to supply chains. The chapter starts with the concepts of extended producer responsibility and product stewardship, indicating what kind of re-thinks and shifts are necessary (e.g. in research and development (R&D)) for cleaner and greener production such as for example shifting from cradle-to-grave to cradle-to-cradle in designing products (design for environment, linear to circular), shifting from providing a product to provision of a service, and shifting technology from part of the problem to part of the solution. Life cycle analysis is discussed as a core concept for this shift to happen, which is set within the context of industrial ecology that allows interdependencies to be identified where waste for one organisation becomes the input for another, as inspired by biomimicry – mimicking nature.

Chapter 10 looks at the final link of the supply chain, how environmentally sustainable products and services are marketed to consumers. The chapter highlights the limits to corporate sustainability by drawing the link to the wider context of sustainable (or ethical) consumption, discussing the tension between consumers' willingness to go green and their willingness to pay as well as the tension of sustainability's goal of reduced use of resources and marketing's goal of encouraging consumption. The role of credibility with regard to labels and greenwashing is explored after giving some thought to the distinction between marketing sustainable products or services and using sustainable marketing techniques. The chapter also includes material on communication, choice editing as well as the use of the 4P marketing mix for sustainability and its alternative, the 4Cs.

Chapter 11 addresses the question of how businesses can manage environmental sustainability systematically by using information technology and management systems. Environmental management is set in contrast to conventional economic management. The Deming cycle of continuous improvement provides a starting point for looking at particular environmental management systems such as the ISO 14000 series and the Eco-Management and Audit Scheme (EMAS). The theme of labels and certifications is re-visited in this context. The link between information systems and sustainability reporting and accounting is then made, which is covered in the following chapter.

In Chapter 12 on accounting, the focus is on how environmental accounting and reporting along with environmental auditing facilitates advanced management for sustainability as it provides, structures and organises information in a way that allows organisations to understand the true costs and impacts of their products and processes on the environment. It enables the measurement of environmental sustainability and therefore facilitates the management of progress in this regard. The chapter will offer a brief overview of environmental/sustainability accounting before outlining some principles and frameworks, such as integrated reporting and the global reporting initiative. As finance and costs are an important influence in business decisions regarding sustainability, some space is dedicated to this as well.

Chapter 13 on human resource management draws together the third part as without the commitment of an organisation's workforce, environmental sustainability is unattainable. The role of leadership is explored, looking particularly at change agents and the importance of pioneers and innovators in finding entrepreneurial solutions to sustainability challenges in organisations (intrapreneurship, environmental entrepreneurship). As the sustainability profession is still an evolving one, the role of training and development in light of the need for multiple skills (e.g. project management) and their accreditation as well as the ability to understand systems is emphasised. Some pointers regarding corporate governance and the role of incentives such as for example linking the achievement of sustainability objectives to pay and bonuses in encouraging sustainability are also included in the chapter.

In part three examples play an important role as well as exercises, putting students and other readers in the position of designing and organising strategies and functions for environmental sustainability. International aspects of functions are highlighted in "spotlight on" sections and examples (e.g. international logistics and spotlights on examples from selected countries across the world).

CONCLUSION

Chapter 14 pulls the foundations, challenges and functions together, allowing students to encounter the management of environmental sustainability in a comprehensive way. A brief overview of sectors and their classifications opens this chapter before the integrative nature of the case studies encourages readers of the book to think, plan and act holistically, systematically and strategically when putting sustainability into practice. The first case study is a longer, three-part case study on a start-up manufacturing reusable coffee cup, whilst the second case study is of medium length and on a festival illustrative of services industries. The third case study is a short scenario set in the primary sector of soya farming. The case studies are set out in a way that allows the use of the individual cases and components (where applicable) after each part of the book (e.g. case study 1 part A after Chapter 4, case study 1 part B after Chapter 7 and so on) or the use of the whole case.

Chapter 15 is the final chapter and provides an outlook into the future for the environmental challenges and the corporate functions covered, drawing together the first chapter on the past and present with food for thought about what the future will hold for environmental sustainability in organisations. Are the current frameworks and regulations as covered in part one sufficient to tackle the complexities of environmental sustainability, or are new models and changes in culture, leadership and governance required in order to foster sustainable business behaviour? How are sustainability managers being prepared for this? A focus on the importance of an interdisciplinary approach and collaborations within and across functions and corporate levels (local vs global) as well as across organisations concludes the final chapter of the book.

SUGGESTED USES

The book at hand aims to offer a thorough preparation or introduction to students and interested professionals for an employment context in which environmental sustainability is increasingly important for commercial and other organisations, whether for regulatory, financial or ethical reasons. It can be used in a variety of ways:

- as an introductory book to environmental sustainability for business and management students or for non-management students,
- as a core textbook to be used in a course focusing on the subject matter,
- as a supplementary or complementary textbook in a course addressing all aspects of sustainability (social, economic, environmental), or
- as additional reading in an integrative or capstone course covering organisational sustainability challenges across corporate functions and disciplines.

The structure is designed with flexibility in mind for how the parts and chapters can be used. Whilst the book can be used and read from start to finish, the parts and chapters could be used individually as well. The foundations set in part one serve as a general context and basis for the remainder of the book, where several of the concepts covered re-occur in more detailed or more applied form. The chapters of part two and three can therefore be used interchangeably in terms of order, and chapters could be used alternatingly (e.g. Chapter 8 followed by Chapter 5; succeeded by Chapters 9 and 6; or Chapters 8 and 9 followed by Chapter 5; then Chapters 10 and 11 before Chapter 6).

DIFFERENTIATIONS AND LIMITATIONS

There are several ways in which this book is different from others in this area. It is not simply another book on sustainability, but a book on business and environmental sustainability, firmly placing it within a corporate and/or management context (though it can be used in a non-corporate context as all organisations, whether for-profit or non-profit, have to be managed and should be managed in an environmentally sustainable way).

It is not just a book on environmental management – of which there are a few[2] – it is broader than that. Using the term environmental sustainability puts (corporate) environmental management, which involves organisational and technical activities to reduce a company's environmental impact (Cramer 1998), in a wider context, thus making it clear it is one of several sustainability aspects. An integrated approach is taken, where environmental management is not taken as a separate function but as something that needs to happen across individual corporate functions.

Inevitably, a book aiming to be comprehensive and integrated has limitations as to what it can cover and what it has to offer. An integrated approach faces its challenges as several disciplines need to be consulted.[3] In order to cover environmental sustainability for corporations comprehensively and holistically, in-depth analyses of specific aspects of sustainability are not provided. Instead, the book points to other sources that can provide more specialist knowledge in these areas.

It is not a book on sustainable development as it focuses on organisations rather than societies and as it focuses on one particular element of sustainability, that of environmental sustainability in the context of commercial, profit-seeking organisations (though non-profit organisations can also apply the concepts and tools introduced in the book in managing their environmental impacts). Such an approach does not diminish the importance of the social element but acknowledges the intersections and dependencies within the triple bottom line while focusing on the environment. With regard to legislation, for example Friedman (2011) argues that "environmental law is essentially social legislation", underlining the linkages.

Terminology may also bring limitations as sustainability is understood positively by some people and more controversially by others, making the term itself a barrier to sustainability (Robertson 2017: 333).

Beyond the limitations mentioned here, there are also external limits to corporate sustainability, such as institutional and market limits, as Paul Hawken wrote in his seminal work *Ecology of Commerce* (1993: xiv): "As hard as we may try to become sustainable on a company-by-company level, we cannot fully succeed until the institutions surrounding commerce are redesigned". Although this book operates on a company level, it is important to acknowledge that businesses are not isolated in their endeavours to become more sustainable and may find allies as well as adversaries in their surroundings, which may expand or limit those efforts. As the book is focused on business and environmental sustainability, it is outside the scope of the book to discuss these external limitations in greater detail, though some attention is paid to what enables or supports those efforts (e.g. in the form of environmental regulations).

Some readers might question the selection of environmental challenges for not explicitly including energy or biodiversity or resource use, for example. The challenges covered can be seen as headline ones chosen for their relevance to businesses. Implicitly, energy is covered within climate change and in corporate functions whilst for example resource use runs through all of the challenges. It is perhaps the most important one and therefore needs to be broken down into more manageable topics. As Blowfield (2013: 31) puts it, whatever challenge we are considering, we are confronted with a global "resource-constrained economy" that looks very different from the "resource-abundant economy" created by the industrial revolution.

It seems apposite here to underline how important it is to remember that despite all the possible issues and pressures, whether from stakeholders, managerial conflicts or senior management (see e.g. Stoner and Wankel 2008), there is reason for optimism and determination. Examples of companies that have moved towards environmental sustainability, transitioned from environmental laggard to environmental leader or incorporated environmental considerations from the start, include carpet manufacturer Interface, ice-cream makers Ben & Jerry's, clothing manufacturer Patagonia, office furniture maker Herman Miller, manufacturing conglomerate 3M, consumer goods producer Unilever, and retailer Marks & Spencer, to name but a few examples that illustrate that business is not just part of the problem but is part of the solution to addressing environmental sustainability and therefore not necessarily "the villain in the sustainability story" (Blowfield 2013: 59).

McDonough and Braungart (2002) also provide reasons to be optimistic despite the challenges society and industry are faced with in moving from a cradle-to-grave to a cradle-to-cradle approach. Their book *Cradle to Cradle: Remaking the Way We Make Things* was produced as a waterproof DuraBook™ by Melcher Media, using revolutionary technology to make an extremely durable book of synthetic paper. They see a world of abundance, not limits (2002: 15), where human creativity and prosperity combined with nature's effective design principles offer an alternative, a new industrial revolution.

To conclude this section, it goes without saying that no book can provide all of the answers in an area so vast and significant as corporate environmental sustainability. It is only one of many books that have been and will be published on similar topics, and it is hoped that it will lead to some of many steps that companies, their managers and employees take in moving towards sustainability, encouraged and informed by students and other readers educated in and called to address this important topic.

SUMMARY

The chapter has introduced the book by making the case for (another) book on business and environmental sustainability, setting its topic in the wider political and historical context, particularly drawing on the origin of sustainable development. The overall context for sustainability is characterised by multiplicity, ambiguity, interdependence and flux. The dimensions of sustainability are mirrored in the triple bottom line of people, planet and profit, or equity, ecology and economy, which dynamically interact with time and space. Despite the challenges this creates and limitations this sets, there is cause for optimism as we embark on a journey to uncover what businesses and other organisations can do for environmental sustainability.

SUGGESTED QUESTIONS AND ACTIVITIES

- Discuss what has been more important for the development of corporate environmental sustainability – environmental disasters caused by companies, environmental legislation, or societal pressure?
- Investigate how one company has moved towards sustainability and share your findings in class. How long did it take and what challenges did the organisation have to overcome?
- *If it were my home* was a website used to visualise how the scale of BP's Deepwater Horizon disaster compared to people's home town or home country. It is no longer updated, but two blogs illustrate it still. Check out http://wordpress.mrreid.org/2010/07/05/deepwater-horizon-to-scale/ and https://goodheartextremescience.wordpress.com/2010/07/06/the-size-and-scale-of-the-bp-deepwater-horizon-oil-spill/.
- Look up the video "25 Biggest Man Made Environmental Disasters In History" on YouTube.

NOTES

1 A fuller discussion of the origin of sustainability in agriculture and forestry can be found in Warde's *The Invention of Sustainability* (2011).

2 The interested reader is referred to the Institute of Environmental Management and Assessment's (IEMA) Handbook *Environmental*

Management in Organizations (2011, edited by Brady, Ebbage and Lunn), to Friedman's *Practical Guide to Environmental Management* (first published in 1988, 11th edition in 2011) as well as to Russo's *Environmental Management: Readings and Cases* (2008) and to *An Introduction to Corporate Environmental Management* by Schaltegger and colleagues (2003).

3 Cavagnaro and Curiel (2012: 5) list the following disciplines they consulted for their work on sustainability: "economics, environmental studies, human ecology, chemistry, biology, sociology, psychology, philosophy, corporate social responsibility, stakeholder analysis, leadership, management, human development, ethics, and so on". One could add to this list earth sciences, geography, political sciences and physics, which have all contributed to the field of sustainability, which truly is multidisciplinary.

SOURCES AND FURTHER READING

AIP (2017) The Discovery of Global Warming, American Institute of Physics, available at http://history.aip.org/climate/index.htm (Retrieved 02.03.2017).

Blowfield, M. (2013) *Business and Sustainability*, Oxford: Oxford University Press.

Brady, J., Ebbage, A. and Lunn, R. (2011) *Environmental Management in Organizations, The IEMA Handbook*, London: Earthscan/Routledge, 2nd ed.

Buckingham, S. (2008a) Approaching Environmental Issues, in: Buckingham, S. and Turner, M. (eds.) *Understanding Environmental Issues*, London: Sage, pp. 11–32.

Buckingham, S. (2008b) The Rise of Environmental Politics and the Environmental Movement, in: Buckingham, S. and Turner, M. (eds.) *Understanding Environmental Issues*, London: Sage, pp. 33–55.

Cavagnaro, E. and Curiel, G. (2012) *The Three Levels of Sustainability*, Sheffield: Greenleaf Publishing.

Cramer, J. (1998) Environmental Management: From 'Fit' to 'Stretch', in: *Business Strategy and the Environment*, 7(3): 162–172.

Daly, H. E. (2005) Economics in a Full World, in: *Scientific American*, 293(3): 100–107.

Elkington, J. (1997) *Cannibals with Forks: The Triple Bottom Line of 21st-Century Business*, Oxford: Capstone.

Elkington, J. (2012) *The Zeronauts: Breaking the Sustainability Barrier*, London: Routledge.

Elliott, J. A. (2013) *An Introduction to Sustainable Development*, London: Routledge, 4th ed.

Friedman, F. B. (2011) *Practical Guide to Environmental Management*, Washington, DC: Environmental Law Institute, 11th ed.

Hackett, C. (2011) *Environmental and Natural Resource Economics*, London: Sharpe, 4th ed.

Hawken, P. (1993) *Ecology of Commerce: A Declaration of Sustainability*, New York: Harper Business.

Kedia, B. L., Clampit, J. and Gaffney, N. (2011) Institutions, MNEs, and Sustainable Development, in: Jain, S. C. and Kedia, B. L. (eds.) *Enhancing Global Competitiveness Through Sustainable Environmental Stewardship*, Cheltenham: Edward Elgar, pp. 70–108.

Kleine, A. and von Hauff, M. (2009) Sustainability-Driven Implementation of Corporate Social Responsibility: Application of the Integrative Sustainability Triangle, in: *Journal of Business Ethics*, 85(3): 517–533.

Lane, H. W., Maznevski, M. and Mendenhall, M. (2004) Globalization: Hercules Meets Buddha, in: Lane, H. W., Maznevski, M., Mendenhall, M. E. and McNett, J. (eds.) *The Blackwell Handbook of Global Management: A Guide to Managing Complexity*, Malden, MA: Blackwell Publishers pp. 3–25.

Lüdeke-Freund, F., Carroux, S., Joyce, A., Massad, L. and Breuer, H. (2018) The Sustainable Business Model Pattern Taxonomy – 45 Patterns to Support Sustainability-Oriented Business Model Innovation, in: *Sustainable Production and Consumption*, 15: 145–162.

McDonough, W. and Braungart, M. (2002) *Cradle to Cradle: Remaking the Way We Make Things*, New York: North Point Press.

Nemetz, P. N. (2013) *Business and the Sustainability Challenge: An Integrated Perspective*, London: Routledge.

Piasecki, B. W., Fletcher, K. A. and Mendelson, F. J. (1999) *Environmental Management and Business Strategy: Leadership Skills for the 21st Century*, New York: John Wiley & Sons.

Robertson, M. (2017) *Sustainability Principles and Practice*, London: Routledge, 2nd ed.

Roorda, N., Corcoran, P. B. and Weakland, J. P. (2012) *Fundamentals of Sustainable Development*, London: Routledge.

Russo, M. V. (ed.) (2008) *Environmental Management: Readings and Cases*, Los Angeles: Sage, 2nd ed.

Schaltegger, S., Burritt, R. and Peterson, H. (2003) *An Introduction to Corporate Environmental Management*, Sheffield: Greenleaf Publishing.

Stoner, J. A. F. and Wankel, C. (2008) Introduction: Exploring New Frameworks, Practices, and Initiatives for a Sustainable World, in Wankel, C. and Stoner, J. A. F. (eds.) *Innovative Approaches to Global Sustainability*, New York: Palgrave Macmillan, pp. 3–10.

van Tulder, R. and van der Zwart, A. (2006) *International Business-Society Management: Linking Corporate Responsibility and Globalization*, London: Routledge.

Vaz, S. G. (2012) *Environment – Why Read the Classics?* Sheffield: Greenleaf Publishing.

Visser, W. (2009) *The Top 50 Sustainability Books*, Sheffield: Greenleaf Publishing.

Warde, P. (2011) The Invention of Sustainability, in: *Modern Intellectual History*, 8(1): 153–170.

WCED (1987) *Our Common Future*, World Commission on Environment and Development, Oxford: Oxford University Press.

Whitfield, R. and McNett, J. (2014) *A Primer on Sustainability: In the Business Environment*, New York: Business Expert Press.

PART I

Foundations

INTRODUCTION

Part one lays the groundwork, providing a robust and extensive basis for the remainder of the book. Chapter 2 introduces scientific and economic foundations for sustainability as a rationale for why businesses should be concerned with the environment. Chapter 3 translates these theoretical perspectives into principles and philosophies for corporate environmental sustainability by looking at the tensions inherent in sustainability and how the legislative framework addresses these through different types of regulations and standards, providing the context for business. The chapter also includes general tools and guidance for regulators and companies. Chapter 4 builds on this, translating the foundations of Chapters 2 and 3 into practice – what strategies and models can guide businesses with regard to environmental sustainability, and how a company can start its journey towards it, guided by change agents and leaders.

CHAPTER 2

Theoretical perspectives or "behind the scenes" of economics and science

Why business should know and care about environmental sustainability

INTRODUCTION

Whatever society, politics, history, ethics or common sense seems to command businesses to do, not everyone will be convinced by arguments found in those realms. It is therefore important to establish scientific and economic considerations as a foundation for businesses to engage with environmental sustainability. We hope to have established a sound conceptual basis for then looking into a more applied side for engaging with this, how this impacts on businesses, what consequences this has for them. For this purpose we draw here on established studies and some more recent writings on the topic, starting with the science basis, simplified for the non-scientific reader (i.e. no chemical formulae), followed by the economics basis, again simplified for the non-economic reader (i.e. no equations). The following chapter then translates this foundation into principles, philosophies and policies for corporate environmental sustainability, before Chapter 4 translates this into more applied and practical models and strategies for businesses to be environmentally sustainable. Although the chapter is structured in a certain, logical sequence, we want to emphasise that the topics are all interlinked and a different order would have equally been justified.

In short, we will show the earth as a self-regulatory system that provides services through the natural capital of the ecosystem that businesses use for free. These services are not accounted for in economic terms, leaving many goods underpriced and oversupplied. Such so-called externalities are not internalised in the price and therefore not reflected in the supply and demand of goods. Through the use of ecosystem services and goods, businesses as part of humanity leave behind ecological footprints which go beyond the carrying capacity of the earth. Whether the running down of natural capital in this way can be compensated for through human-created capital is a question the concepts of strong and weak sustainability attempt to answer.

SCIENCE AND NATURE

Let us start with the basics, the laws of thermodynamics. Thermodynamics is the study of the transformation of energy, stemming from the Greek words for heat and power. These laws, in simplified form, state that

> the Earth is a finite system where all matter that will exist is already present now, where quality is dissipated (or, in other words, disorder increases), and where the only source of new energy is the sun.

(Cavagnaro and Curiel 2012: 171, drawing on Robèrt's work, 2002). Unpacking this further, the first law states that energy cannot be created or destroyed, which means that energy that flows in must equal the energy that flows out. In particular, the second law of thermodynamics, also called the law of entropy, stipulates that disorder will increase until a state of thermal equilibrium is reached (Robertson 2017).

It is energy that is required by all living things as it represents the ability to do work (Robertson 2017). Energy drives the movement of matter, the term given to all physical substances. Matter is comprised of elements, the most important ones of which are carbon, hydrogen, nitrogen and oxygen, together with phosphorus and sulphur (ibid., Lovelock 1979). These can combine and interact in myriad ways to make life on earth possible. For matter, earth is a closed system (Robertson 2017), i.e. matter does not simply appear or disappear (Everard 2011a), mirroring the first law of thermodynamics. Energy and matter constantly move in cyclical patterns in systems. The cycles of nature are "biogeochemical" as they include biological, geological and chemical processes, such as for example photosynthesis (Robertson 2017: 28). Nature is thus not static and is constantly changing through flows of energy and materials in cyclical patterns that are regenerative.[1]

The principle of matter conservation and the laws of thermodynamics underpin our understanding of Earth as a sustainable system (Everard 2011a). The concept of systems is central to sustainability (Robertson 2017). The OED (Oxford English Dictionary) defines a system as "a set of things working together as a mechanism or network". In a system, parts make up an integrated whole. When it comes to planet earth, earth itself is a system, within which four spheres can be defined: rock (or soil), air, water and organism. Ecologists call these four the lithosphere, atmosphere, hydrosphere and biosphere, the latter including the former three (Robertson 2017). It is worth developing our understanding of earth as a system further in the next section.

EARTH AS A SYSTEM – INSIGHTS FROM EARTH SCIENCE

The scientist James Lovelock worked as an atmospheric chemist for NASA, the US National Aeronautics and Space Administration, when the organisation was looking for life on Mars and tasked scientists to find out ways to do this. He developed the concept of the planet as a self-regulating system, first known as the earth system hypothesis, now known as the Gaia theory. In a nutshell, as Visser (2012) describes it, the theory is based on observing the self-regulatory system of the earth where the climate, the atmosphere's and ocean's composition, and surface soils stay approximately in a condition that favours habitability, making it optimal for life (Lovelock 1979). At the heart of it is the concept of the biosphere as a contiguous whole (Everard 2011a).

Lovelock (1979) focused on atmospheric gases and how global temperature is affected by them. The composition of the atmosphere had previously been assumed to be given, whereas Lovelock argued that the living systems on the planet generate and maintain the atmosphere (Daly and Cobb 1989), thus making the atmosphere a dynamic extension of the biosphere rather than separating the two (Lovelock 1979).

Whilst Gaia has been elevated from hypothesis to theory (Elkington 2012), it has also had its limitations exposed (e.g. Daly and Cobb 1989) as well as its value disproven (e.g. Tyrrell 2013). Nevertheless, Lovelock's encouragement to engage with how nature and earth work is an important one. He underlines that it is our ignorance of planetary control systems that gives rise to our uncertainties about pollution and how this affects the planet (Lovelock 1979).

Looking at earth as a system is not a recent development. Ecosystem as a term was first used by Arthur Tansley in 1935 (Callicott 2012; Robertson 2017), where he accounts for how the components of an ecosystem – organisms and inorganic factors – interact to bring about a relatively stable dynamic equilibrium (Tansley 1935). To be more precise, an ecosystem is "the complexity of living organisms, their physical environment, and their entire interrelationships within a defined unit of space, through which matter and energy flow" (Green 2013: 47), or "systems created by the dynamic interrelationship of living and non-living things" (Blowfield 2013: 11), with each part playing an important role (Robertson 2017).

A more comprehensive definition of the earth system is provided by Rockström and colleagues (2009a: footnote 1)

> the integrated biophysical and socioeconomic processes and interactions (cycles) among the atmosphere, hydrosphere, cryosphere, biosphere, geosphere and anthroposphere (human enterprise) in both spatial – from local to global – and temporal scales, which determine the environmental state of the planet within its current position in the universe.

As such, humans and their activities are an integral part of this, interacting with other parts of the earth system.

Viewing earth as a system – an ecosystem – we are reminded that ecosystems change constantly, and an ecosystem's capacity to accommodate such change while retaining its properties (function, structure, identity), is called resilience (Robertson 2017), a term that some authors (see e.g. Zolli 2012) have started using in lieu of sustainability, drawing a distinction between sustainability's aims to put the world back into balance and resilience's efforts to manage in an imbalanced world. The greater an ecosystem's biodiversity, the more resilient and stable it will be (Robertson 2017). Biodiversity – biological diversity – is the variety of animal and plant life in the world (OED) and the source of much wealth (Hawken 1993). The difference between a system's stability and resilience is that stable systems are not easily moved from their current state or path by shocks, whereas resilient systems return to their previous state or path if disturbed by shocks (Nemetz 2013, drawing on Holling 1973).[2]

Systems thinking is linked to the integrated and holistic approach to sustainability encompassing the environment, ecology and economy (see Figure 1.2) – as Robertson (2017: 41) puts it, "the study of sustainability is the study of systems". Systems thinking allows the understanding of a system's properties as a whole as well as the relationships of its components (Everard 2011a), thus enabling better knowledge of complex systems, which are more than the sum of their parts.

Within this complex system we focus on businesses and the natural environment. An interdisciplinary approach is needed for such complexity – covering both science (nature) and economics (capital). Sustainability science does exactly that, aiming to understanding dynamic interactions between ecological and social systems (Robertson 2017). An interdisciplinary approach is also reflected in the earth and space sciences which draw on biological, chemical, geological and engineering fields, to name but a few (see e.g. Earth & Space Science News on eos.org, drawing on a number of peer-reviewed journals in the area).

One reason why businesses should have a rudimentary knowledge or at least an awareness of ecosystems is based on organisations' reliance on both services and goods that ecosystems provide. Hawken (1993) even likens a business to an organism as it takes food and energy from the environment, with the difference that natural creatures only consume renewable resources such as seeds, leaves and insects, whereas companies also consume non-renewable resources such as coal and oil. The services and goods that ecosystems provide are also collectively known as ecosystem services, to which we now turn.

ECOSYSTEM SERVICES

Ecosystem goods and services (together generally referred to as services) represent benefits for the human population (Costanza, d'Arge et al. 1997) and therefore, by extension, also to the business world. The concept provides a valuable framework for conceptualising the ways in which society and industry benefit from the supportive capacities of ecosystems (Everard 2011c).

The economics of ecosystem services is a relatively new area of study as these services are often taken for granted and occur outside of commercial markets (Hackett 2011). Costanza and his colleagues (Costanza, d'Arge et al. 1997) were the first to use a synthesis of over 100 published studies to estimate the economic value of ecosystem services and natural capital stocks. It represented a major and comprehensive effort to attach a monetary value to the ecological services sustaining the planet (Nemetz 2013). The authors estimated the value of ecosystem services for the entire biosphere to be in the range of US$16–54 trillion annually, with an average of US$33 trillion per year (Costanza, d'Arge et al. 1997). The economic importance is illustrated by their following quote (Costanza, d'Arge et al. 1997: 259):

> if ecosystem services were actually paid for, in terms of their value contribution to the global economy, the global price system would be very different from what it is today.

Breaking this down into manageable and understandable parts, Tony Juniper's book *What has nature ever done for us?* provides a very readable account for the lay reader of "how money really does grow on trees" (subtitle, Juniper 2013).

Although the human population is buffered against environmental changes by technology and culture, it is essentially dependent on the flow of ecosystem services, while at the same time influencing and changing these services through human and industrial activities and interactions with the ecosystem. The consequences of these changes were the focus for the Millennium Ecosystem Assessment (MEA 2005a) which assessed them and established the scientific basis for actions that could sustain ecosystems.

The MEA harmonised previously disparate classifications (Everards 2011c) and categorised ecosystem services from which people obtain benefits into provisioning services, regulating services, cultural services and supporting services (for a full list, see Box 2.1 in MEA 2005a: 40, and Figure 2.1 below):

- Provisioning services include products obtained such as food, fibre, fuel, genetic resources, biochemical, natural medicines and pharmaceuticals, ornamental resources (e.g. shells and flowers), freshwater.
- Regulating services include regulation of air quality, climate, water and erosion as well as water purification and waste treatment, disease and pest regulation, pollination and natural hazard regulation.
- Cultural services include nonmaterial benefits that provide recreational, social, aesthetic, educational and spiritual benefits.
- Supporting services include soil formation, photosynthesis, primary production, nutrient cycling, water cycling.

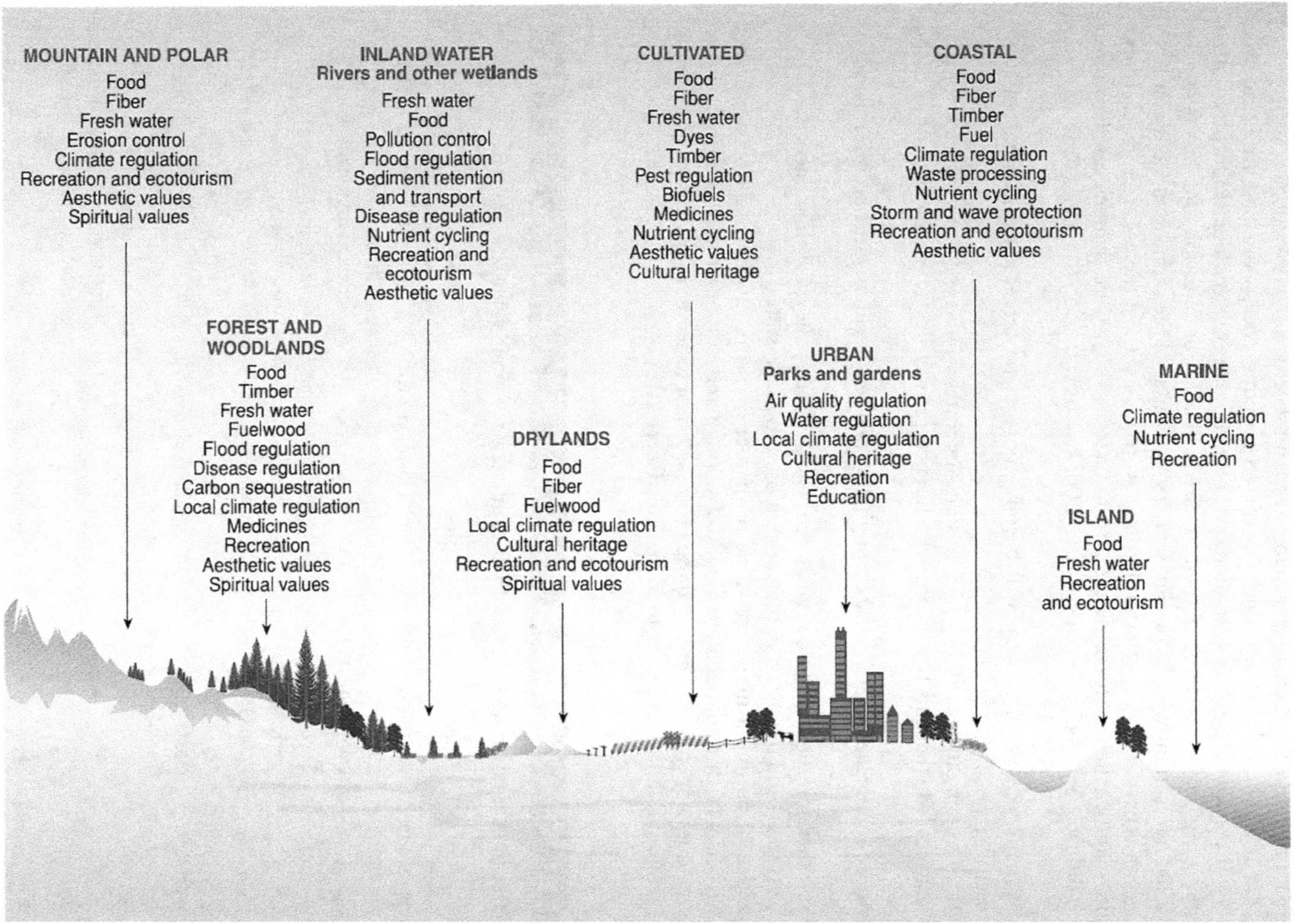

FIGURE 2.1
Ecosystem services

Source: Millennium Ecosystem Assessment, WRI, used with permission.

Supporting services differ from others as they make the production of the other services possible and are more indirect and long-term than direct and short-term (MEA 2005a).

The assessment revealed that two-thirds of the ecosystem services examined are being used unsustainably or degraded (MEA 2005b), demonstrating humanity's "ecological footprint", that is, its pressure on global ecosystems (Blowfield 2013: 11). The implications for business arise from its twofold interactions with ecosystem services – as a user of the services and as a contributor to changes in ecosystems. The relevant findings were published in the MEA synthesis report "Opportunities and challenges for Business and Industry" which were demonstrated to affect business and industry in three major ways (MEA 2005b):

1 Freely available ecosystem services will become more costly or cease to become available, thus presenting a risk.
2 The framework conditions within which business operates will be affected by the loss of ecosystem services, whether this be customer preferences, regulatory action or the cost of doing business.
3 New business opportunities will arise through new markets, technologies and incentives.

Businesses therefore need to know what ecosystem services they use or influence – which Table 2.1 illustrates – and to what risk they are exposed to as a result.

The practice of putting a monetary value on nature's services has not only been received positively. Criticism is for example offered by McDonough and Braungart (2002: 80) who point out that nature does not engage in any of these processes "just to serve people", thus questioning whether an economic value can be put to it (see also Curry 2011).[3]

TABLE 2.1

Examples of ecosystem services and their estimated economic value

Soil	Pollinators	Water	Others
>90% of world's food grown in soil 1/3 of farmed soil has degraded since mid-20th century 5.5 billion additional tons of carbon capture if soil management was changed	$1 trillion annual sales depend on animal pollination $190 billion annual services to farming by animal pollination 2/3 of major crop plants relying on animal pollination	$274 billion global GDP from fishing, fish processing and sales $50 billion extra value that could be gained from well-managed stocks $16 billion subsidies spent in ways damaging to fish stocks 20,000 jobs lost in Newfoundland as cod disappeared Oceans are worth about $21 trillion per year	$34 billion costs associated with loss of vultures in India (waste management) Link between green spaces and health in England – increased green spaces, improved health – £105 billion costs caused by mental illness, £630 million cost of maintaining 27,000 parks and green spaces in the UK

Source: Juniper 2013.

BOX 2.1 SPOTLIGHT ON PIONEERING PAYMENT FOR ECOSYSTEM SERVICES

Costa Rica and Mexico have pioneered payment for ecosystem services (PES) in an attempt to change the economics of the trade-off that comes from trees normally being worth more dead than alive, where land used as pasture or croplands is worth more than virgin forest even though these capture carbon emissions and clean up water supplies, amongst other ecosystem services. Whilst sound in principle, putting it into practice faces obstacles, such as identifying who should pay how much, how to create a market for clean water, who should receive subsidies if any, how to avoid wastage of public moneys and how to avoid leakage of efforts (e.g. where deforestation moves from a protected area to another area not covered by PES contracts). Nevertheless, studies of PES efforts in Mexico and Costa Rica suggest that deforestation is halved. Other efforts to put a monetary value on ecosystem services can be found across the globe e.g. Brazil and the Amazonian rainforest, Vietnam's planting of mangrove forests which save on dyke upkeep, helped by software InVEST (Integrated Valuation of Ecosystem Services and Trade-offs), developed by American scientists to calculate and predict economic and environmental fallouts from planned land-use change through colour-coded maps, which China also is using.

Source: The Economist (www.economist.com/node/17062651, www.economist.com/news/international/21613327-new-ideas-what-speeds-up-deforestation-and-what-slows-it-down-clearing-trees, www.economist.com/blogs/freeexchange/2015/11/economics-deforestation).

Following on from the Millennium Ecosystem Assessment, The Economics of Ecosystems and Biodiversity (TEEB) is an international initiative set up by the European Commission, hosted by the UN Environment Programme, tasked to study the economics of biodiversity loss (Blowfield 2013). It is based on the principle that ecosystems can only really be appreciated if the economic value can be shown (ibid). It is one avenue to make sustainability comprehensible to *homo oeconomicus*, who understands the value of (financial) capital.

Here it is useful to remind ourselves that ecosystem services comprise the flows of materials, energy, and information, arising from the stock of natural capital, which then combine with manufactured and human capital to generate human welfare (Costanza, d'Arge et al. 1997). To call the stock from which ecosystem services flow *natural capital* is only logical as capital is considered to be a stock of materials that generates flows of services that can be used to increase human welfare.

Other forms capital can take are manufactured capital as well as human and financial capital. With regard to natural capital, Costanza and Daly (1992) point out that this used to be (and still is in many cases) a free good due to its abundance, whereas it was only manufactured and human capital that could limit economic development. As natural capital stocks are essential for manufactured and human capital, it is worth devoting some more attention to capital in general and natural capital in particular, as used by industry.

CAPITAL – NATURAL AND MANMADE

In classical economic thinking, capital is traditionally thought of as one of the three factors of production, alongside land and labour, or as wealth used for the production of further wealth (Green 2013). No distinction is made between what kind of capital or what kind of goods are produced with its help. It is thus inherent in economics and its methodology that the dependence of mankind[4] on the natural world is ignored (Schumacher 1973), when in fact this could be considered the true capital (Kumar 2012, drawing on Schumacher 1973). This marks a shift in thinking not normally applied to natural capital – whereas in economics, stock is considered capital and dividends are taken as income, nature is regarded as a set of flows rather than as capital stock (Juniper 2013). In other words, natural capital is being "liquidated to make profits, and the process is being treated as a stream of dividends rather than the spending of capital" (Juniper 2013: 272), which is what is actually occurring. As Nemetz (2013: 79) puts it, "if any corporation were to deplete its capital and claim this as income, the management would probably be fired".

So what is natural capital then? As briefly touched upon in the previous section, it consists of the stock of resources, the elements and relationships in ecosystems, underpinning life on earth (Hackett 2011). More concretely, it comprises familiar resources used by humans – water, minerals, oil, fish, trees, soil, air and so on – as well as living systems, such as grasslands, savannas, wetlands, oceans, estuaries, coral reefs, tundras, riparian corridors and rainforests (Hawken et al. 1999). The difference of how capital is traditionally conceived – the accumulated wealth in the form of factories, investments and equipment – is that it cannot be produced by human activity (ibid.).

Further distinction can be seen in that man-made capital is normally controlled and owned by a single individual or organisation, whereas natural capital is generally a societal resource (Whitfield and McNett 2014). The terminology often varies, but other forms of human-made capital include manufactured, social, cultural, constructed and financial capital within the realms of human or human-created capital.

Natural capital itself can be further broken down into renewable and non-renewable natural resources, with some also using the term replenishable resources (Wackernagel and Rees 1996). The overall stock of non-renewable resources is not replenishable within a human time frame, whereas renewable resources are capable of replenishment over time. Here it is important to keep in mind even renewable resources can become non-renewable if depleted or over-consumed (Hawken 1993; Hackett 2011). Within non-renewable resources we can distinguish between those resources that are recyclable (such as metals) and those that are non-recyclable (such as fossil fuels) (Hackett 2011).

Having defined and explained natural capital and its forms, it is useful to characterise the other kinds of capital briefly here (Viederman 1996; Hawken et al. 1999; Hackett 2011):

- Human capital – labour, intelligence, culture and organisation.
- Constructed, created or human-created capital: technologies, productive facilities (factories, infrastructure) and inventories. This form of capital is also called
- Manufactured capital – infrastructure, machines, tools and factories.
- Social capital – stock of "civic virtues" and networks of civic engagement, trust and reciprocity.
- Cultural capital – body of knowledge, stories, myths, visions and languages shared by people.
- Financial capital – cash, investments, monetary instruments.

The relationship between the concept of different kinds of capital and sustainability's three pillars (economy, ecology, equity) is evident in Viederman's (1996: 46) definition:

> Sustainability is a community's control and prudent use of all forms of capital – nature's capital, human capital, human-created capital, social capital, and cultural capital – to ensure, to

> the degree possible, that present and future generations can attain a high degree of economic security and achieve democracy while maintaining the integrity of the ecological systems upon which all life and all production depends.

The five capitals model espoused by Viederman captures the concept that sustainability is not just about an ecological issue but comprises social, economic and environmental dimensions. It sets out the different types of capital any organisation or business needs to function. In line with the definition presented here, sustainable organisations maintain, or even enhance, the total stock of capital (Blowfield 2013).

Whatever terms are used, it is clear that a conceptual shift has taken place where instead of the traditional form of capital incorporating natural and other forms of capital, natural capital is now the encompassing form, with other forms of capital relying on natural capital and its functional integrity (Hackett 2011), which is more reflective of reality. This is illustrated in Figure 2.2.

Hawken and colleagues (1999) emphasise that an economy needs all types of capital (in their view four: human, financial, manufactured and natural) in order to function properly.

The concept of natural capital and its incorporations into notions of wealth and the monetisation of ecosystem services as a consequence has implications for how wealth and economic growth are measured. Traditional measures of gross domestic product (GDP) include e.g. the costs of cleaning up the environment after corporate environmental disasters as economic growth, when in fact all that is done is an attempt to restore the environment to its previous state with no actual wealth added. Likewise, Nemetz (2013) points out that if a country were to deplete its natural capital through mining of mineral resources (e.g. petroleum), pollution of its aquifers, logging of its forests and hunting of its wildlife and fisheries, its income as measured by conventional measures would not be affected by the disappearance of its natural resources (see also Repetto 1989). This therefore opens up questions of how the economy, how economic growth, how wealth and income are measured. This is not the place here to expand on this, but the interested reader is encouraged to look up e.g. Daly and Cobb (1989) and their Index of Sustainable Economic Welfare (ISEW) or Bhutan's Gross National Happiness Index (see also Hackett 2011;

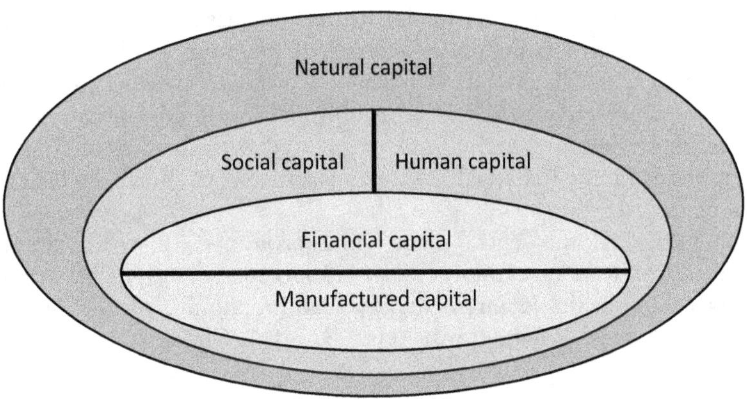

FIGURE 2.2
The five capitals model

Nemetz 2013 for other, alternative measures of economic welfare and development, e.g. the UN human development indices). The fields of environmental and ecological economics have also grappled with these questions – to which we return later.

It is the past non-scarcity of natural capital that Daly and Cobb (1989) find behind its being ignored by neoclassical economic theory and the teaching that humanly created capital can substitute for natural resources in order to maintain the overall capital stock. However, according to von Weizsäcker and colleagues (1998), the assumption of infinite substitutability of artificial capital for natural capital is a misuse of otherwise useful ("correct", ibid. 296) economic theories to which we now briefly come. We will turn our attention to economics before we return to the question of substitutability of the various kinds of capital in the last section of the chapter. The shortfall of conventional economics in recognising the role of natural capital is highlighted by Hawken and his colleagues (1999: 61):

> Economics cannot function as a reliable guide until natural capital is placed on the balance sheets of companies, countries, and the world.

ECONOMICS

Economics is traditionally concerned with the allocation of scarce resources (goods, services) amongst competing uses. The three fundamental questions that societies faced with resource scarcity have to resolve are *what to produce, how to produce, and for whom* (Hackett 2011: 31). In capitalist societies, these three questions are answered with what consumers demand, with least-cost production technology and for those consumers willing to pay. A functioning market is central in such a socioeconomic system (in contrast to e.g. planned economies, ibid.).

As a result, economics (literally the management of the household, derived from the Greek word *oikos*) deals with goods according to their market value and not according to what they really are; as Schumacher highlights (1973: 50), it is in economics' nature to "equalise and quantify everything by means of a money price". As a corollary, it excludes things that have no price or things that are priceless, or in other words includes only what can be counted, not "what really counts" (von Weizsäcker et al. 1998: 297). It also focuses on the short term rather than the long term (Schumacher 1973).

The centrality of the market is a particular feature of classical and neoclassical economics – the conventional economics most people associate with economics. These are, however, not the only kinds of economics, as Chang (2014) articulately argues, stressing that economics is not a science as there are no objective truths to be established in a field that is essentially a political argument.[5] The market, therefore, should not be equated with the economy, as this is only one way of organising the economy (Chang 2014). This is a matter of choice – which is what economics is about (Pearce et al. 1999).

Although the market takes a central place within economics, there are circumstances where it cannot allocate all resources or cannot allocate resources efficiently. This occurs when market failures happen as a result of the requirements of a well-functioning market not being met. These market failures can take the following forms (Hackett 2011; Chang 2014):

- Imperfect information
- Imperfect competition (e.g. monopoly, cartels)
- Externalities (positive or negative side effects)
- Public goods and common-pool resources and (collectively produced and/or consumed goods).

Accounting for the use of ecosystem services and natural capital as externalities – a form of market failure – could also be seen as a failure of conventional economics in not making this part of the cycle of nature and of the market as natural resources and environments clearly serve economic functions and exhibit positive economic value (Pearce et al. 1999).

Shortcomings are thus demonstrated for neoclassical economics whose essence can be represented in a simple cyclical model. Environmental economics addresses neoclassical economics' ignorance of natural capital by including the environment (Daly and Cobb 1989) while ecological economics goes one step further by embedding the economic system within the ecological system without which it cannot function (Nemetz 2013).[6] Figure 2.3 visually illustrates these three types of economics.

Visualising the economic system as situated within the ecosystem makes it clear that there are physical and biological, i.e. natural, limits to conventional economic growth, though not to development (Daly and Farley 2011). Whilst the former is quantitative, the latter is qualitative and therefore not confined by limits (Robertson 2017). Daly (2005) highlights the shift between the different kinds of economics by calling neoclassical economics – developed in a time of abundant natural resources – "empty world" economics, whilst ecological economics marks a "full

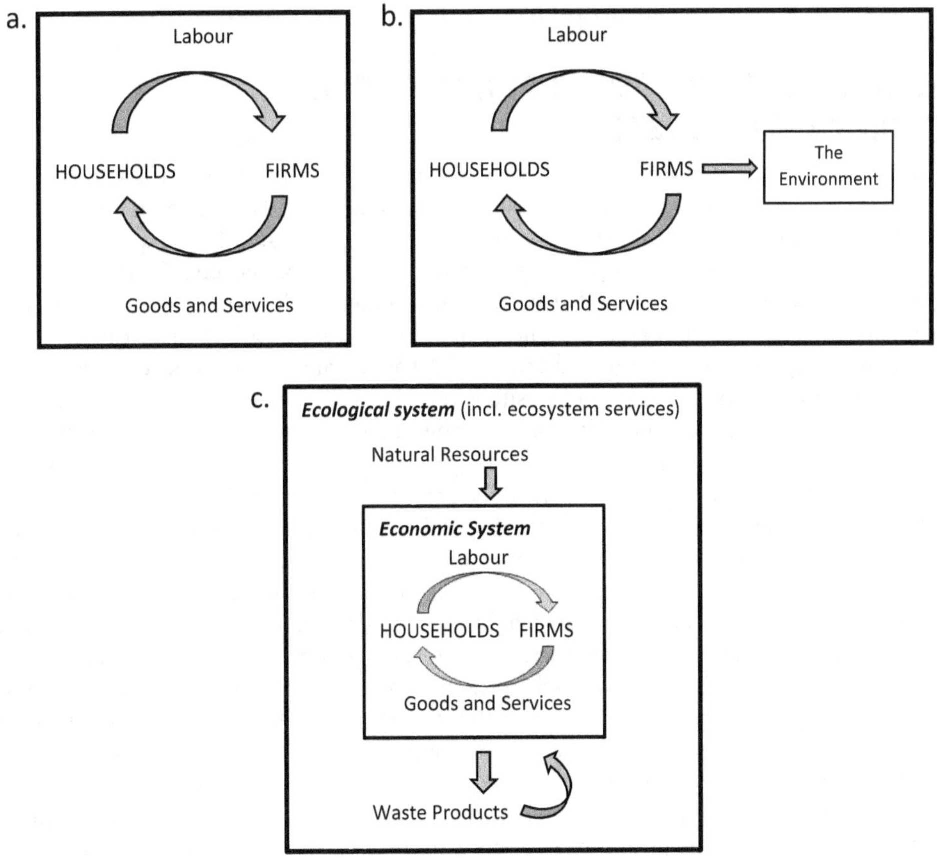

FIGURE 2.3
Comparison of three economics models

world" economics, which brings insights from ecology to the study of economics, combining the biosphere and the human sphere (Robertson 2017). This echoes the five capitals model introduced earlier, which brings together natural capital and human-made capital in its various forms.

For Hackett (2011) environmental economics provides the economic basis for addressing pollution problems, whilst ecological economics focuses on natural capital and the ecosystem goods and services flowing from it. Related to environmental economics is natural resource economics which studies the supply, demand and allocation of natural resources, distinguishing between renewable and non-renewable resources, as well as common-pool resources that are characterised by non-excludability as well as subtractability – meaning it is not possible to exclude people from using resources such as rivers or marine fisheries, and once resources are used they are no longer available to others (ibid). Here economics, as the study of allocation of scarce resources, gives us different models for different types of resources, such that for example renewable resources are allocated within the limits of maximum sustainable yield, whereas non-renewable resources are subject to dynamic efficiencies in the market where the exhaustion of a limited stock leads to future prices rises[7] (Hackett 2011).

Let us return to what links the three types of economics introduced and how they deal with the failure of the market to include the environment – externalities for neoclassical economics and environmental (or social) costs in environmental and ecological economics.

EXTERNALITIES OR HOW TO ACCOUNT FOR ENVIRONMENTAL COSTS

Externalities are benefits that are unpaid for or uncompensated costs that occur as a by-product of production or market exchanges and impact society (Hackett 2011). They are *external* as they are not represented in the classical model of market demand and supply. As a consequence these unintentional side effects of consumption and production are not reflected in the price of goods (Nemetz 2013). A distinction is made between positive externalities and negative externalities. As the (marginal[8]) private cost or benefit does not equal the (marginal) social cost or benefit, products or activities with negative externalities tend to be underpriced and overproduced while for goods or services with positive externalities, the opposite is the case – they tend to be overpriced and underproduced (Nemetz 2013). Social benefits comprise the sum of private and external benefits, whereas social costs are equal to the sum of private costs (paid by producers) and external costs, normally born by society (Hackett 2011). Social costs therefore include all costs that are associated with producing one good or service, including externalities. The relevance of externalities to corporations should therefore be obvious – it is their decisions that generate much of these costs (Nemetz 2013).

As externalities are a form of market failure, they need to be addressed by so-called *internalising*. As Anderson (2014) puts it – since externalities are felt and experienced beyond the people (or organisations) causing the effect, their internalisation can help people (or in the case of business, companies), experience (or feel) for themselves the effects of their behaviour, whether it be through taxes, subsidies or property rights, as further explored in the next chapter. As externalities often take the form of environmental pollution (or environmental disruption, Kapp 1970), and are often not voluntarily internalised, addressing them has implications for policy, in particular environmental regulations, which form part of the following chapter. Such regulations are often based on the *polluter pays principle*, which is how internalisation of costs in the case of pollution is translated into policy.

Internalisation for negative externalities provides an incentive to reduce the activity associated with the external cost (Daly and Cobb 1989). In turn, for positive externalities, products which are normally underproduced, internalisation should incentivise increased provision.

BOX 2.2 EXTERNALITIES

A prime example of negative externalities is pollution – an uncompensated harm to members of society that results from production and consumption. A factory that pollutes a river with its effluents spoils the river for fishing downstream (Daly and Cobb 1989). The price of fuel in the marketplace is cheap but the real (not yet internalised) cost is high as the future costs of climate change are not part of the price of petroleum (Robertson 2017).

Positive externalities include for example vaccinations – an immunised child reduces the likelihood of disease spreading to others, or university education benefitting not only the students in question but also society in the form of informed voters and resourceful citizens (Hackett 2011).

A particular form of negative externalities can be found with open-access or common pool resources characterised by non-excludability as well as subtractability (Hackett 2011). This form of negative externality is called appropriation externality when the use of harvesting such common-pool resources leaves other appropriators of the resource with fewer or damaged resources (Hackett 2011). The process of excessive appropriation from common-pool resources has been called *the tragedy of the commons* (Hardin 1968). The solution for these known problems with common pool resources is seen in private ownership rather than in communitarian solutions (Spash 1999) or, in other words, where resources are limited, shared access will lead people to deplete the resource, which is why private ownership of natural resources and market-based solutions may lead to better management and conservation (Blowfield 2013: 28). However, Nemetz (2013) highlights that single ownership (i.e. private ownership) is not a guarantee against degradation or exhaustion of such resources and points to institutional mechanisms for successful management of commons, developed e.g. by the late Nobel Laureate Elinor Ostrom.

Daly and Cobb (1989) distinguish between localised and pervasive externalities and provide the coal industry as an example – black lung disease is a localised externality, whereas CO_2 build-up is a pervasive externality as it cannot be traced to one particular coal mine, unlike the miner suffering from the disease.

Daly and Cobb (1989: 37) call externalities *ad hoc corrections* that recognise neglected aspects of the real world in basic economic theory, which, according to them, is a "reasonable procedure" as long as it involves minor details. However, when such vital issues as the earth's capacity to support life are classed as externalities and are not included as an integral part of economic abstractions, a re-think is necessary, according to the authors (ibid.).

Likewise, a major criticism of the concept of externalities as a market failure underlines that it illustrates a divide between humankind and the environment – that an externality is external to human activities rather than incorporated into them. One major critique of the concept was by the little known economist K. William Kapp, who introduced his concept of social cost (1950) as a non-utilitarian response to neoclassical economists such as Coase and Pigou with regard to their response to externalities (Berger 2008). Kapp (1978: 13) defines social costs as "all direct and indirect losses sustained by third persons or the general public as a result of unrestrained economic activities" and argues that both social costs and environmental disruption have been neglected by economic theory (Kapp 1970). The complex interdependencies of society and economy which cumulatively lead to environmental problems and social costs have no solution in conventional economic theory. In order to meet this challenge, Kapp (1970) argues that the physical flows that stem from using natural assets and that take place outside the market and are therefore not registered in prices, need to be taken into account and be evaluated in real terms. In his criticism of mainstream (or orthodox) economic theory treating the economic sphere as closed and autonomous, as illustrated in Figure 2.3, Kapp (e.g. 1970) can be seen as a precursor to ecological economics (Berger 2008). Similarly, Georgescu-Roegen (1971) is in many ways considered to be the founder of ecological economics in using the laws of thermodynamics to provide a critique of conventional (neo-classical) economics in order to develop what he called "bioeconomics" (Mayumi 2001). It is this non-correspondence to natural laws that govern the physical world that is a fundamental flaw in the economic growth model (Robertson 2017).

As an alternative to conventional, more orthodox economics, approaches from ecological economics are part of a wider tradition of heterodox economics, which is outside the mainstream and therefore its proponents are less commonly known (see e.g. Neves 2016; Spash 1999; Røpke 2004, 2005 for a fuller exploration of the history of ecological economics, whilst Stratford 2020 looks at the field's contribution to transform economic institution to enable society to flourish within planetary boundaries).

Although environmental economics is considered to be a subset of neoclassical economics and thus substantially different from ecological economics (Nemetz 2013), what both environmental and ecological economists agree on is that prices should reflect environmental costs and thus tell the "environmental truth" (Cavagnaro and Curiel 2012: 46).

What then is the environmental truth? And who or what can tell us the environmental truth? This is where we can look to the concept of carrying capacity and ecological footprints to find out the truth (or a measure thereof) of humanity's impact on the environment and the environmental costs created that might allow us to account for externalities – despite the problems with this term we are using it here as it is more widely known.

CARRYING CAPACITY AND ECOLOGICAL FOOTPRINTS

Rachel Carson's book *Silent Spring* (1962) was mentioned in the previous chapter – her work on DDT illustrated how human-synthesised chemicals that had not existed before meant that no organisms had evolved in nature to break them down, thus leading to a build-up (accumulation) of the chemical, threatening living creatures (Cavagnaro and Curiel 2012). This raises the question of limits and carrying capacity – how can we measure it? And what might be the unintended consequences of the growth of business and technology, the economy and the human population?

What is the uppermost limit an ecosystem or habitat can sustain, provided nutrients are available and supplied? Hawken (1993: 139), having asked the question, states that sustainability can be defined in terms of the ecosystems' carrying capacity, using input-output models of energy and resource consumption:

> Sustainability is an economic state where the demands placed upon the environment by people and commerce can be met without reducing the capacity of the environment to provide for future generations.

This concept of the earth's carrying capacity is not a new one – it echoes Thomas Malthus's concerns about food availability for a growing population in his book *An Essay on the Principle of Population* (1798). Malthus was particularly concerned with the differences in growth – he expected population to grow exponentially (or geometrically) and natural resources to grow arithmetically, a situation which would eventually lead society to collapse, having exceeded the carrying capacity.[9] Advances in technology and productivity rendered his expectations void. Nevertheless, "Neo-Malthusians" have been influenced by his argument (Hackett 2011) e.g. the authors of *Limits to Growth* (Meadows et al. 1972, 2004) and of *The Population Bomb* (Ehrlich 1971). Meadows and colleagues (1972) highlight how natural absorptive mechanisms are being saturated and warn of ignoring the limits of earth's ability to absorb pollutants – theirs is also a systems approach to analysing the potential limits of humanity's impact on the system.

Everard (2011b) uses the term *biocapacity* and biological capacity in the context of humanity's footprint exceeding the planet's capacity to replenish its natural capital, demonstrating the links between the concepts already covered. The five capitals model illustrates that the economy is a subsystem of the ecosystem (Daly and Cobb 1989: 143) and that therefore the economy also has a proper scale, by which Daly and Cobb mean physical size: although the world is not static, the ecosystem does not grow as earth is finite. Conventional economic growth models therefore logically meet limits (see also Meadows et al. 1972). The growth model does not take into account real costs and does not put a price on ecosystem services and, if continued, will result in an ecological footprint that exceeds the carrying capacity of the earth (Robertson 2017). Likewise, Schumacher (1973: 26) argues that there can be growth towards a "limited objective" but unlimited, generalised growth is not possible, echoing the findings of Meadows and colleagues (1972).

The notion of a finite earth and limits to (economic) growth is contrary to one of the tenets of capitalism – that business is an open, linear system where growth is always possible through resource extraction and technology, provided capital is sufficient (Hawken 1993). Or as Everard (2011b: 35) puts it, "our contemporary model of economic growth has lost its connection to the 'real world' of natural and human productivity" where the market fails to tell both the ecological and human truth (e.g. through subsidising the externalities of artificially cheap fossil fuels).

What then is the carrying capacity and how is it linked to ecological footprints? Carrying capacity is defined by Nemetz (2013: 25) as the number of organisms within a particular species (human, animal or plant) a geographic area can support without irrevocably reducing its capacity to do so in the future, thus firmly echoing the definition of sustainable development provided in Chapter 1. Carrying capacity concerns thus earth as a receiving entity, whereas ecological footprint concerns humanity and its impact on earth.

Ecological footprints reflect a number of interlinked factors (Elkington 1997) – population, consumption, technology. It is the total amount of land, food, water and other resources that are used by a person or an entity's subsistence (Green 2013) – in other words, the use of natural capital is measured through its footprint which is usually measured in acres or hectares of productive land. The concept was developed by Mathis Wackernagel and William Rees in the early 1990s (Cavagnaro and Curiel 2012).

Wackernagel and Rees (1996) explain ecological footprint analysis as an accounting tool that allows an estimation of resource consumption and waste absorption requirements of a defined economy or population in terms of a corresponding productive land area. They argue that environmental footprint analysis in productive land area makes it consistent with the basic laws of physics, in particular the laws of mass balance and thermodynamics – linking economics and science (thus providing justification for this chapter to cover both fields, emphasizing that sustainability is inherently interdisciplinary). The concept shows the continuing material dependence of humans on nature, answering the question "how much land would be necessary to support a defined economy sustainably at its current material standard of living" (Wackernagel and Rees 1996: 12). They argue that the ecological footprint is an inversion from traditional carrying capacity by measuring the land area required per person (or population), rather than calculating the population per unit area. As such, the ecological footprint assumes that energy and material consumption as well as waste discharge require the productive or absorptive capacity of a finite area of land or water. This area may or may not coincide with the given population's home region (ibid.), thus making the measurement universally applicable, rather than locally confined.

Using publicly available data such as global and national statistics, Wackernagel and Rees (1996) calculate ecological footprint according to consumption categories (food, housing, transportation, consumer goods and services) and according to their conversion into areas of land use (energy land from energy consumption, consumed land from the built environment, farm land from food, and forest land from forest products). The Global Footprint Network that Wackernagel helped to found provides data for up-to-date footprints.

The difference between ecological footprints and carrying capacity is also explored by Nemetz (2013: 241) who calls a footprint also the "appropriated carrying capacity", a measure of the actual physical stocks of natural capital that are necessary to sustain a given population within a given geographic area. Furthermore, the carrying capacity is a limit beyond which any population will not be able to sustain itself in the long term (Meadows et al. 2004). Whilst the ecological footprint measures the human impact on earth, it is in itself not a limit, though it can indicate whether society is living beyond the earth's carrying capacity (ibid.) or as Robèrt (2002) points out, all the accumulated footprints from various activities are related to total carrying capacity. Calculating ecological footprints therefore allows the comparison to the biosphere's carrying capacity and helps to assess whether this capacity is overshot and natural capital is depleted (Robertson 2017) or whether there is still room to grow. This could be the case, particularly as there are vast differences in countries' ecological footprint and, in calculating the measure, it is clear that wealthier nations appropriate more than what could be argued to be their fair share of the planet's carrying capacity (Wackernagel and Rees 1996). Another way to understand the difference between ecological footprints and biocapacity is that the footprint looks at the demand side (what is required for consumption?) whereas biocapacity looks at the supply side (availability and productivity of natural resources) – or in other words, ecological footprint measures the "quantity of nature" we use and compares it to how much "nature" we have (carrying capacity) (Footprint Network 2017). In their original calculations, Wackernagel and Rees (1996) calculated a net figure of 1.7 hectares per capita of ecologically productive land, which compares with an ecological footprint per capita of 2.3 hectares, implying that human consumption exceeds nature's provision by 35% (Hackett 2011), as illustrated in the following graphs in Figure 2.4. By 2019, the so-called ecological

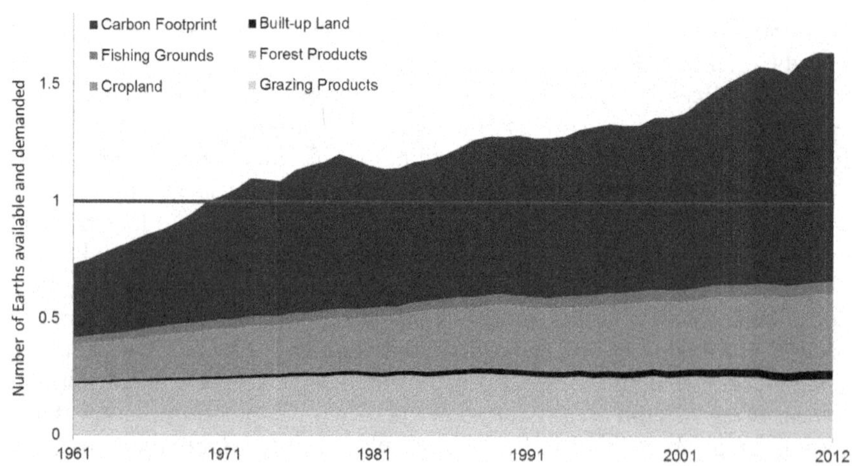

FIGURE 2.4A
World ecological footprint

Source: Footprint Network, used with permission.

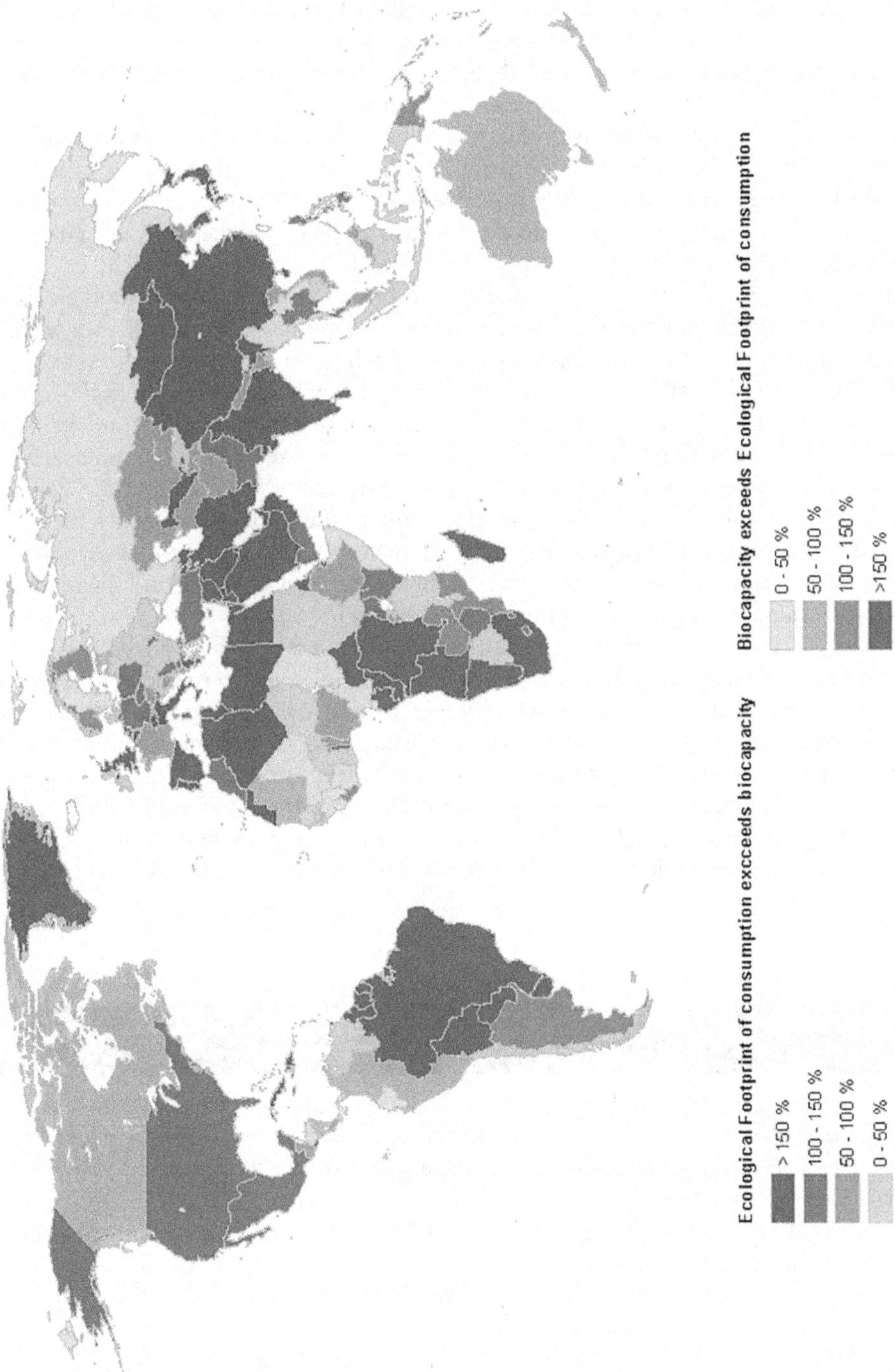

FIGURE 2.4B
(Continued)

overshoot of humanity using nature faster than it renews had reached 75%, meaning that nature's biological budget (or biocapacity) is overused by 75% (Wackernagel and Beyers 2019).

Despite ambiguities in the concept of carrying capacity, sustainable rates of natural resource use and environmental carrying capacity can be defined and therefore calculated in physical terms (Daly and Cobb 1989: 143). Daly and Cobb (1989) encourage the concept to be admitted into the "citadel" of economic theory. They also use it to estimate the establishment of optimal scale, which echoes the ability of the ecological footprint to be compared with carrying capacity to identify optimal allocation and distribution of resources.

As the concept of carrying capacity is closely linked to limits, Rockström and colleagues (2009a, 2009b) explored and attempted to quantify planetary boundaries, which, if crossed, would constitute a tipping point. In drawing on different branches of academia (ecological economics, earth system science as well as work on resilience) and studies on limits (Meadows et al. 1972, 2004), they identify nine planetary boundaries, of which three have already been exceeded (climate change, rate of biodiversity loss and human interference with the nitrogen cycle) and the others are (still) within a safe operating space (ocean acidification, stratospheric ozone depletion, phosphorous cycle – overlap with nitrogen cycle, global freshwater use, change in land use, atmospheric aerosol loading and chemical pollution) (Rockström et al. 2009a, 2009b). Kate Raworth (2017) also uses planetary boundaries in her work on doughnut economics, which is a criticism of mainstream economics and a sustainable alternative comprising social and environmental aspects, where the doughnut's outer boundaries represent the ecological ceiling and the inner ring represents the social foundation providing life's basics for people.

As equilibrium was covered for both science and economics, it is worth stating in this section that for renewable resources, the carrying capacity is a stable equilibrium if outside influences are absent, for example the biocapacity of fish stock is stable if there is no fishing (Anderson 2014).

Ecological footprint can be further broken down into e.g. water footprint, carbon footprint, and so on (material footprint), which is reflected in the challenges covered in part two of the book. Varying degrees of complexity are involved in calculating these (Whitfield and McNett 2014), which can be done on a global level, but also for regions, countries, industries, even down to the individual level. The most relevant here is of course industries and companies for which the footprint of products or processes also can be calculated.

BOX 2.3 WATER FOOTPRINT AND OTHER FOOTPRINT EXAMPLES

Water footprint is defined as the total volume of water that is used to produce a good across the full supply chain, including both direct and indirect water use. It is a multidimensional indicator as it shows water by source and type – e.g. blue, green and grey water which, respectively, represent freshwater, water from rainfall and water polluted from production processes and other uses. Water footprint is also known as virtual water content.

Examples of global average water footprints for selected products:

- 130 litres for 1 cup (125 ml) of coffee versus 27 litres for a cup (250ml) of tea
- 15,400 litres for 1kg of beef versus 4,330 litres for 1 kg of chicken meat
- 125 litres for 1 apple (150g) versus 230 litres for 1 glass (200ml) of apple juice

BOX 2.3 (CONTINUED)

- 2,500 litres for 1 kg of rice versus 290 litres for 1 kg of potatoes (but 1,040 litres for 1kg of potato chips!)
- 1,850 litres for 1 kg of dry pasta versus 3,020 litres for 1kg of olives and 214 litres for 1kg of tomatoes (but 530 litres for 1kg of tomato ketchup)

For product footprints, tomatoes offer an intriguing example: whilst tomatoes grown in greenhouses economically outperform tomatoes grown in open fields, they require much greater inputs and therefore leave a greater ecological footprint, leading to resource underpricing and unaccounted ecological costs.

For another footprint, that of plastic waste, see www.bbc.co.uk/news/science-environment-39279392.

Sources: Wackernagel and Rees 1996; Whitfield and McNett, 2014; Robertson, 2017, www.waterfootprint.org.

Related to ecological footprints is life cycle analysis (LCA) which companies can use to identify what goes in and out with regard to the production of a particular product or service. The significance of this is underlined by Blowfield (2013) – one major reason why the world is living beyond the ecological carrying capacity is business's inability to limit the throughput of materials used, irrespective of how this affects or depletes natural capital stocks, not to mention what the true costs are. Footprints can help industries understand their environmental impact and support decision-making. Together with life cycle analysis and other indicators of sustainability, such measurements can be used by decision-makers in corporations as well as in public life to assess what is or is not working or where different or additional efforts might be needed to achieve (environmental) sustainability (Robertson 2017). Life cycle analysis is discussed in greater detail in Chapters 3 and 9.

As ecological footprints and biocapacity are concerned with the supply and demand of natural capital, and thus resource availability, we return to the issue of capital and ask if biocapacity can be restored, if footprints can be reduced by replacing natural capital with created capital in order to keep overall capital stock constant. The theories of strong and weak sustainability can provide us with insights here.

STRONG VERSUS WEAK SUSTAINABILITY

If we remind ourselves of the five capitals model (see e.g. Figure 2.2), we recall that sustainability means keeping the overall capital stock constant, with all its components – natural and human-made capital. Both forms of capital are valuable and both can be degraded or improved over time – in the case of non-natural capital, repairs or replacements are possible, whereas for natural capital protection is possible but collective or government action is needed (Whitfield and McNett 2014). If the overall stock should be constant for sustainability to be achieved, the question then is: is natural capital substitutable by man-made capital? How we answer this question might give us an insight as to how strong sustainability is, as different responses are possible.

Daly and Cobb (1989) were among the first to use the terms *weak* and *strong* sustainability in discussing the substitutability of humanly created capital and natural capital, alongside Pearce, Markandya and Barbier (also 1989) (Wackernagel and Rees 1996). According to Daly and Cobb (1989) weak sustainability occurs when the depletion of natural capital is offset by accumulating an equivalent amount of humanly created capital ("high" substitutability), whereas strong sustainability requires the separate maintenance of the two types of capital, suggesting complementarity rather than substitutability. Conceptualising sustainability in this way combines the physical and natural dimension with the human dimension (Cavagnaro and Curiel 2012).

Central to sustainability is the proposition that the next generation should be left with a stock of capital no less than the current generation (Nemetz 2013), which we can interpret under the constant capital rule. The constant capital rule under weak sustainability permits the passing on of "less environment" to future generations as long as this is offset by increasing other stock of capital, whereas under strong sustainability perfect substitution is not a valid assumption (Nemetz 2013: 8). This constancy of the natural capital stock can take on different meanings, according to Pearce and colleagues (1989) – it is commonly interpreted in terms of constant *physical* capital stock but could alternatively also be seen in terms of a constant *economic* stock value.

Drawing on Daly and Cobb (1989), Cavagnaro and Curiel (2012) outline the following distinctions in how the sustainable use of natural resources is viewed in a matrix of strong and weak sustainability (Table 2.2).

The value in distinguishing strong and weak sustainability lies in operationalising sustainable development, for which Daly and Cobb (1989: 73) recommend strong sustainability, though acknowledge that "even weak sustainability" would mark an improvement over conventional practice. Pearce and Atkinson (1993) acknowledge the difficulties in operationalising sustainable development and proceed to propose a simple savings rule as a weak sustainability indicator where depletion of natural capital is acceptable within certain rules such as the Hartwick rule (or savings investment rule, Ang and van Passel 2012) which stipulates that for sustained levels of consumption, the gains from depleting natural capital must be reinvested in other forms of capital (Hackett 2011). Weak sustainability therefore implies that the loss of natural capital can be mitigated. Other indicators of weak sustainability include green national income (akin to GDP), genuine savings, genuine investment, as well as the GPI (genuine progress indicator) and the ISEW (Index of Sustainable Economic Welfare) developed by Daly and Cobb (1989; Pearce et al. 1996; Hackett 2011).[10]

The weak sustainability concept draws on neoclassical economics (e.g. Solow, Hartwick and others) – Ang and van Passel 2012) in its notion of offsetting the extraction of non-renewable resources with reproducible capital and is thus very much connected to the idea that technological progress can improve human well-being despite environmental damage. It is reminiscent of the empty world economics of Daly (2005), in contrast to the full world economics which is reflected in strong sustainability and its main proponent, the field of ecological economics (Ang and van Passel 2012). Here, natural capital, and in particular, critical natural capital as a subset, is viewed as non-substitutable, therefore requiring a separation in how humanly created and natural capital are measured in order to adhere to the constant capital rule (ibid.).

TABLE 2.2

Strong versus weak sustainability

	No decline of natural capital	No decline of overall capital
Natural capital not substitutable	Very strong sustainability	Moderate weak sustainability
Natural capital substitutable	Moderate strong sustainability	Very weak sustainability

Source: adapted from Cavagnaro and Curiel 2012: 89.

Why is natural capital considered non-substitutable under strong sustainability? Several factors come into play here. Firstly, many forms of natural assets do not have substitutes (e.g. the ozone layer or the climate-regulating functions of ocean phytoplankton) (Pearce et al. 1999). An extreme form of non-substitutability is irreversibility, for example in the case of species loss (ibid.). Secondly, uncertainty over the consequences of depleting natural capital makes it difficult to know how much humanly created capital might be necessary to compensate for declines in natural capital (Hackett 2011). Thirdly, the scale and discontinuities of natural systems are subject to threshold effects that, once exceeded, could lead to large-scale damage (Pearce et al. 1996; Hackett 2011). Instead of as substitutes, strong sustainability views the two types of capital as complements to each other[11] where natural capital is an essential ingredient in consumption, production and welfare (Anderson 2014). Related to this last point of welfare, Özkaynak and colleagues (2004) mention how inter- and intra-generational equity are added to ecosystem resilience as requirements to strong sustainability.

Indicators of strong sustainability include the ecological footprint discussed earlier and net primary product (NPP) which serve as a foundation for carrying capacity (Hackett 2011) as well as resilience which is connected to an ecosystem's ecological productivity (Pearce et al. 1996).

Operationalising sustainable development through indicators of weak and strong sustainability is also explored by Costanza and Daly (1992), arguing for the separation of human-made and natural capital in efforts to maintain overall capital stock rather than linking these and attaching substitutability. They also touch upon the differentiation of growth and development mentioned previously – growth occurs at the cost of natural capital and thus faces (economic) limits, whereas development does not come at the expense of natural capital. In doing so, the authors reverse the question of *how far can human-made capital substitute for natural capital?* with *how far can development substitute for growth?*, since the answer to the first question for them is "hardly at all" (Costanza and Daly 1992: 43).

In a similar line, a persuasive argument for strong sustainability is provided by Hawken and colleagues (1999) who ask their readers to imagine the technologies and manufactured capital that could replace natural capital and ecosystem services such as the production of oxygen, climate regulation, freshwater storage and distribution, water and air purification, and point to the experience and cost of Biosphere 2 ($200 million), an ambitious project to study life within a closed system (see also Juniper 2013).

Despite the difficulties and complexities in measuring sustainability (and we have not ventured into more economic territory such as discount rates, time preferences, net present values and the problems associated with using only monetary values in calculating natural capital (see e.g. Pearce et al. 1999 and box 3.5), or the problem of overshoot, Wackernagel and Rees 1996), in practice, so Hackett (2011) argues, a mixture might be possible as a guide to policy. In any complex society, policies consistent with either paradigm can be observed: some natural capital is sacrificed when creating constructed capital in the form of transportation infrastructure, factories and schools, but at the same time, sheltered wilderness areas are designated and regulatory systems are put in place to protect and conserve ecosystem services and wildlife.

SUMMARY

Businesses have an important role in society, and neither the book in general nor this chapter in particular should be read as anti-capitalist but as a recognition that business can play an even more important part if it recognises and accounts for its impacts on and use of nature (see also Blowfield 2013 for different perspectives of business in a sustainability context). Wackernagel and

Rees (1996: 147) put it very simply in the context of the bottom line – no ecosphere, no economy, no society, or even simpler: no planet, no profit. A reminder that the whole is more complex than what we are covering here – focusing on the environmental aspects of business, we should keep in mind that sustainability encompasses also equity and justice within the social and societal realm.

What should give both students, practitioners and companies hope is that underlying all ecological science is the fact that earth – given a chance – will eventually restore itself (Hawken 1993).

This chapter has chartered through both science and economics territory to explore why business should be concerned about environmental sustainability. As companies not only draw on created or man-made capital, human or financial capital, but also on natural capital, they make use of ecosystem services, mostly for free, and are responsible for negative externalities that are not reflected in the price of their products or services. How then should they act on this? The next two chapters provide more insight into this – the next chapter looking at what principles arise from this chapter (e.g. for the regulatory framework in which businesses operate), whilst the following chapter translates these into practical and applied models and strategies for sustainability.

SUGGESTED QUESTIONS AND ACTIVITIES

- Calculate your own and your fellow students' ecological footprints, using www.footprintcalculator.org/ or for your water footprint use https://waterfootprint.org/en/resources/interactive-tools/personal-water-footprint-calculator/.
- How would a global company start to consider measuring its footprints? What might be the challenges it faces?
- Should ecosystem services be monetised? In preparation for debating this, look at both proponents and opponents of valuing and paying for ecosystem services and goods.
- Weak and strong sustainability – which framework is more convincing? Which one might be more suitable for businesses to use?
- Are market-based systems better than government regulations at achieving sustainability?

NOTES

1 See Robertson 2017, Chapter 3, for a concise but more detailed description of these cycles and of the biosphere than is possible here.

2 Neither system though might be able to cope with a shock that passes a threshold or tipping point, contributed to by nonlinear system dynamics and feedback loops with positive feedback accelerating a system's move away from equilibrium in contrast to negative feedback which allows a system to balance back to equilibrium (Nemetz 2013). See also Meadows et al. (1972: 13) for a discussion of positive feedback loops, also called "vicious circles" (in some instances also "virtuous circles", Meadows et al. 2004: 25).

3 For a more general discussion on the (moral) limits of markets, the interested reader is pointed to Michael Sandel's excellent book *What money can't buy* (2012).

4 The terms mankind and man-made capital reflect a traditionally male perspective also in (economic) literature and are therefore replicated, though efforts are made to use more inclusive terms, such as humankind and human-made capital.

5 The other types of economics Chang (2014) includes beyond classical and neoclassical are Marxist, Developmentalist, Austrian, Schumpeterian, Keynesian, Institutionalist and Behaviourist schools of economics.

6 Daly and Farley (2011: 29) demonstrate the need for a different approach with an analogy, underlining that economists have ignored "metabolic throughput": "The circular flow vision is analogous to a biologist describing an animal only in terms of its circulatory system, without ever mentioning its digestive tract. Surely the circulatory system is important, but unless the

animal also has a digestive tract that connects it to its environment at both ends, it will soon die of starvation or constipation".

7 Economics also provides tools through game theory and the prisoner's dilemma to solve the problem of common-pool resources (tragedy of the commons) through cooperation (Hackett 2011).

8 Marginal analysis is central to economic analysis for identifying optima (Hackett 2011). The interested reader is pointed towards introductory economics textbooks.

9 Wackernagel and Rees (1996, Box 2.5) go even further back to illustrate how the relationship between people and land has occupied human thought, pointing back to Plato, Chinese and early Christian scholars.

10 It should be pointed out that some of these measures are controversial among economists as they are not as concrete and objective as traditional GDP measures seem to be (Hackett 2011).

11 Costanza, Cumberland and colleagues (1997: 85) illustrate this by asking the following questions: "what good is a sawmill without a forest? A refinery without petroleum deposits? A fishing boat without populations of fish?"

SOURCES AND FURTHER READING

Anderson, D. A. (2014) *Environmental Economics and Natural Resource Management*, London: Routledge, 4th ed.

Ang, F. and van Passel, S. (2012) Beyond the Environmentalist's Paradox and the Debate on Weak Versus Strong Sustainability, in: *BioScience*, 62(3): 251–259.

Berger, S. (2008) K. William Kapp's Theory of Social Costs and Environmental Policy: Towards Political Ecological Economics, in: *Ecological Economics*, 67(2): 244–252.

Blowfield, M. (2013) *Business and Sustainability*, Oxford: Oxford University Press.

Callicott, J. B. (2012) A Sand County Almanac: An Evolutionary-Ecological Worldview, in: Vaz, S. G. (ed.) *Environment – Why Read the Classics?* Sheffield: Greenleaf Publishing, pp. 36–71.

Carson, R. (1962) *Silent Spring*, Boston: Houghton Mifflin.

Cavagnaro, E. and Curiel, G. (2012) *The Three Levels of Sustainability*, Sheffield: Greenleaf Publishing.

Chang, H-J. (2014) *Economics: The User's Guide*, London: Penguin Books.

Costanza, R. and Daly, H. E. (1992) Natural Capital and Sustainable Development, in: *Conservation Biology*, 6(1): 37–46.

Costanza, R., d'Arge, R., de Groot, R., Faber, S., Grasso, M., Hannon, B., Limburg, K., Naeem, S. O'Neill, R. V., Paruelo, J., Raskin, R. G., Sutton, P. and van den Belt, M. (1997) The Value of the World's Ecosystem Services and Natural Capital, in: *Nature*, 387(15): 253–260.

Costanza, R., Cumberland, J., Daly, H., Goodland, R. and Norgaard, R. (1997) *An Introduction to Ecological Economics*, Boca Raton, FL: St Lucie Press.

Curry, P. (2011) *Ecological Ethics: An Introduction*, Cambridge, UK: Polity Press, 2nd ed.

Daly, H. E. (2005) Economics in a Full World, *Scientific American*, 293(3): 100–107.

Daly, H. E. and Cobb, J. B. (1989) *For the Common Good: Redirecting the Economy Towards Community, the Environment, and a Sustainable Future*, Boston: Beacon Press.

Daly, H. E. and Farley, J. (2011) *Ecological Economics: Principles and Applications*, Washington, DC: Island Press, 2nd ed.

Ehrlich, P. (1971) *The Population Bomb*, New York: Ballantine.

Elkington, J. (1997) *Cannibals with Forks: The Triple Bottom Line of 21st-Century Business*, Oxford: Capstone.

Elkington, J. (2012) *The Zeronauts: Breaking the Sustainability Barrier*, London: Routledge.

Everard, M. (2011a) The Earth's Natural Systems, in: Brady, J., Ebbage, A. and Lunn, R. (eds.) *Environmental Management in Organizations, The IEMA Handbook*, London: Earthscan/Routledge, 2nd ed., pp. 3–12.

Everard, M. (2011b) The State of the Global Environment, in: Brady, J., Ebbage, A. and Lunn, R. (eds.) *Environmental Management in Organizations, The IEMA Handbook*, London: Earthscan/Routledge, 2nd ed., pp. 13–38.

Everard, M. (2011c) Valuing Ecosystems and their Services, in: Brady, J., Ebbage, A. and Lunn, R. (eds.) *Environmental Management in*

Organizations, The IEMA Handbook, London: Earthscan/ Routledge, 2nd ed., pp. 63–70.

Footprint Network (2017) Global Footprint Network webpages, available at www.footprintnetwork.org (Retrieved 27.03.2017).

Georgescu-Roegen, N. (1971) *The Entropy Law and the Economic Process*, Cambridge, MA: Harvard University Press.

Green, T. (2013) *The Dictionary of Global Sustainability. With Case Studies on the Environmental, Economic, Technological, and Social Aspects of Sustainability*, New York: McGraw-Hill.

Hackett, C. (2011) *Environmental and Natural Resource Economics*, London: Sharpe, 4th ed.

Hardin, G. (1968) The Tragedy of the Commons, in: *Science*, 162(3859): 1243–1248.

Hawken, P. (1993) *Ecology of Commerce: A Declaration of Sustainability*, New York: Harper Business.

Hawken, P., Lovins, A. B. and Lovins, L. H. (1999) *Natural Capitalism: The Next Industrial Revolution*, London: Earthscan.

Holling, C. S. (1973) Resilience and Stability of Ecological Systems, in: *Annual Review of Ecology and Systematics*, 4: 1–24.

Juniper, T. (2013) *What Has Nature Ever Done for Us? How Money Really Does Grow on Trees*, London: Profile Books.

Kapp, K. W. (1950) *The Social Costs of Private Enterprise*, New York: Schocken.

Kapp, K. W. (1970) Environmental Disruptions and Social Costs: A Challenge to Economists, *Kyklos*, 23(4): 833–847.

Kapp, K. W. (1978) [1963] *The Social Costs of Business Enterprise*, second edition of The Social Costs of Private Enterprise, extensively revised and rewritten. Nottingham: Spokesman.

Kumar, S. (2012) Small Is Still Beautiful, in: Vaz, S. G. (ed.) *Environment – Why Read the Classics?* Sheffield: Greenleaf Publishing, pp. 114–133.

Lovelock, J. E. (1979/1987) *Gaia: A New Look at Life on Earth*, Oxford: Oxford University Press.

Malthus, T. (1798) *An Essay on the Principle of Population*, London: J. Johnson in St Paul's Church-yard.

Mayumi, K. (2001) *The Origins of Ecological Economics: The Bio-economics of Georgescu-Roegen*, London: Routledge.

McDonough, W. and Braungart, M. (2002) *Cradle to Cradle: Remaking the Way We Make Things*, New York: North Point Press.

Meadows, D. H., Meadows, D. L., Randers, J. and Behrens, W. W. III (1972) *The Limits to Growth, A Report for The Club of Rome's Project on the Predicament of Mankind*, London: Earth Island.

Meadows, D. H., Randers, J. and Meadows, D. L. (2004) *Limits to Growth: The 30-Year Update*, White River Junction, VT: Chelsea Green.

Millennium Ecosystem Assessment (MEA) (2005a) *Ecosystems & Human Well-being: Synthesis*, Washington, DC: Island Press.

Millennium Ecosystem Assessment (MEA) (2005b) *Ecosystems & Human Well-being: Opportunities & Challenges for Business & Industry*, Washington, DC: World Resources Institute.

Nemetz, P. N. (2013) *Business and the Sustainability Challenge: An Integrated Perspective*, London: Routledge.

Neves, V. (2016) What Happened to Kapp's Theory of Social Costs? A Case of Metatheoretical Dispute and Dissent in Economics, in: *Review of Political Economy*, 28(4): 488–503.

Özkaynak, B., Devine, P. and Rigby, D. (2004) Operationalising Strong Sustainability: Definitions, Methodologies and Outcomes, in: *Environmental Values*, 13(3): 279–303.

Pearce, D. W. and Atkinson, G. D. (1993) Capital Theory and the Measurement of Sustainable Development: An Indicator of "Weak" Sustainability, in: *Ecological Economics*, 8(2): 103–108.

Pearce, D. W., Hamilton, K. and Atkinson, G. D. (1996) Measuring Sustainable Development: Progress on Indicators, in: *Environment and Development Economics*, 1(1): 85–101.

Pearce, D. W., Markandya, A. and Barbier, E. (1999) *Blueprint for a Green Economy*, London: Earthscan (first published 1989).

Raworth, K. (2017) *Doughnut Economics: Seven Ways to Think Like a 21st-Century Economist*, London: Random House Business Books.

Repetto, R. (1989) *Wasting Assets: Natural Resources in the National Income Account*, Washington, DC: World Resource Institute.

Robèrt, K.-H. (2002) *The Natural Step Story: Seeding a Quiet Revolution*, Gabriola Island: New Society Publishers.

Robertson, M. (2017) *Sustainability Principles and Practice*, London: Routledge, 2nd ed.

Rockström, J., Steffen, W., Noone, K., Persson, Å., Chapin, F. S. III, Lambin, E., Lenton, T. M., Scheffer, M., Folke, C., Schellnhuber, H., Nykvist, B., De Wit, C. A., Hughes, T., van der Leeuw, S.,

Rodhe, H., Sörlin, S., Snyder, P. K., Costanza, R., Svedin, U., Falkenmark, M., Karlberg, L., Corell, R. W., Fabry, V. J., Hansen, J., Walker, B., Liverman, D., Richardson, K., Crutzen, P. and Foley, J. (2009a) Planetary Boundaries: Exploring the Safe Operating Space for Humanity, in: *Ecology and Society,* 14(2): 32, available at www.ecologyandsociety.org/vol14/iss2/art32/ (Retrieved 20.03.2017).

Rockström, J., Steffen, W., Noone, K., Persson, Å., Chapin, F. S. III, Lambin, E., Lenton, T. M., Scheffer, M., Folke, C., Schellnhuber, H., Nykvist, B., De Wit, C. A., Hughes, T., van der Leeuw, S., Rodhe, H., Sörlin, S., Snyder, P. K., Costanza, R., Svedin, U., Falkenmark, M., Karlberg, L., Corell, R. W., Fabry, V. J., Hansen, J., Walker, B., Liverman, D., Richardson, K., Crutzen, P. and Foley, J. (2009b) A Safe Operating Space of Humanity, in: *Nature*, 461(7263): 472–475.

Røpke, I. (2004) The Early History of Modern Ecological Economics, in: *Ecological Economics*, 50(3–4): 293–314.

Røpke, I. (2005) Trends in the Development of Ecological Economics from the Late 1980s to the Early 2000s, in: *Ecological Economics*, 55(2): 262–290.

Schumacher, E. F. (1973) *Small Is Beautiful: A Study of Economics as if People Mattered*, London: Abacus.

Spash, C. L. (1999) The Development of Environmental Thought in Economics, in: *Environmental Values*, 8(4): 413–435.

Stratford, B. (2020) The Threat of Rent Extraction in a Resource-constrained Future, in: *Ecological Economics*, 169, https://doi.org/10.1016/j.ecolecon.2019.106524.

Tansley, A. (1935) The Use and Abuse of Vegetational Concepts and Terms, in: *Ecology*, 16(3): 284–307.

Tyrrell, T (2013) My Verdict on Gaia Hypothesis: Beautiful but Flawed, in: *New Scientist*, available at www.newscientist.com/article/mg22029400-400-my-verdict-on-gaia-hypothesis-beautiful-but-flawed/ (Retrieved 17.03.2017).

Viederman, S. (1996) Sustainability's Five Capitals and Three Pillars, in: Pirages, D (ed.) *Building Sustainable Societies: A Blueprint for a Post-Industrial World*, Armonk, NY: M.E. Sharpe.

Visser, W. (2012) *The Quest for Sustainable Business: An Epic Journey in Search of Corporate Responsibility*, Sheffield: Greenleaf Publishing.

von Weizsäcker, E., Lovins, A. B. and Lovins, L. H. (1998) *Factor Four: Doubling Wealth, Halving Resource Use*, London: Earthscan.

Wackernagel, M. and Beyers, B. (2019) *Ecological Footprint: Managing our Biocapacity Budget*, Gabriola Island: New Society Publishers.

Wackernagel, M. and Rees, W. (1996) *Our Ecological Footprint: Reducing Human Impact on the Earth*, Gabriola Island: New Society Publishers.

Whitfield, R. and McNett, J. (2014) *A Primer on Sustainability: In the Business Environment*, New York: Business Expert Press.

Zolli, A. (2012) Learning to Bounce Back in: *New York Times*, available at www.nytimes.com/2012/11/03/opinion/forget-sustainability-its-about-resilience.html (Retrieved 20.03.2017).

CHAPTER 3

Principles, philosophies and policies for corporate environmental sustainability

From tensions to tools for regulators and companies

INTRODUCTION

This chapter builds on the previous one, asking how the foundations laid in Chapter 2 can be translated into principles, philosophies and policies for environmental sustainability, whilst the following chapter then asks how both can be translated and transformed into practical, applied models and strategies for business. What became evident in the previous chapter is that there are tensions in the field of sustainability e.g. in the use of free ecosystem services that are not paid for, externalities and environmental damage that are not accounted for in the price of goods and services, trade-offs between different forms of capital, and so on. We start by looking at these tensions across a number of sustainability aspects before we discuss how the regulatory framework addresses these tensions that might not be solved by business alone, as the following quote illustrates:

> The role of government is to assume those functions that cannot or will not be undertaken by citizens or private institutions.
>
> (Hawken 1993: 166)

As sustainability is a complex subject, characterised by interdependencies and connections, we also look at the relationship between the regulators and the regulated, i.e. business-government relations, before we explore frameworks that can guide both legislators and industry in the search for environmental sustainability.

"NATURAL" TENSIONS AND TRADE-OFFS IN BUSINESS AND ENVIRONMENTAL SUSTAINABILITY – TWO SIDES OF THE SAME COIN?

The relationship between business and environmental sustainability has in the past not been an easy one – the goals of profit maximisation and market share growth have seemingly been at odds with environmental protection and regulations. These are seen as costs and as market interferences to be avoided by industry. In turn, industry is seen as the problem by sustainability advocates. In the first part of this chapter, we explore the nature of these tensions before discussing responses to them.

Capital – natural versus human made

We concluded the previous chapter by comparing theories of weak and strong sustainability which differ in how they view the substitutability of natural capital versus human-created capital under the premise of keeping the overall capital stock constant. Particularly with regard to natural capital, there is a tension between mainstream economics approaches in viewing it as infinite and free, and more recent ecological economics in acknowledging the limits of natural capital and ecosystem services. Whichever side of the strong versus weak sustainability debate we fall, and however we end up viewing the substitutability between natural and other capital, it is clear that no constructed capital can ultimately exist without natural capital as without natural capital neither business nor life is possible. The tension here, we could argue, is therefore an artificial one.

Inside out versus outside in

It is thus clear, also from the five capitals model and the discussion of ecosystem services in the previous chapter, that the economy is not separate from the environment. There is a fundamental interdependence and necessary interaction between the two, because, as Pearce and colleagues state (1999: 4)

> Economies affect environments. Environments affect economies.

The questions emerging from this interdependence include *how do businesses affect the environment*? And *how do businesses react to environmental impacts*? Questions such as these have been conceptualised through *inside-out* and *outside-in* perspectives (Porter and Reinhardt 2007). Inside-out looks at understanding the impact of corporate activities on a company's environment, whereas outside-in reveals how changes in the environment and the regulatory framework can present opportunities and threats for firms (ibid.).

This approach was developed by Porter and Kramer (2006) to demonstrate and encourage a move towards interdependence between business and society – a relationship which has traditionally been seen as a tense one in the literature. The authors argue that corporate success and social welfare is not a zero-sum game and thus inside-out and outside-in practices should be integrated (ibid.). The notion of the relationship between corporate success and society as one of tension and as a zero-sum game is explored and contested by Porter and another co-author, van der Linde, in the particular context of environmental regulations.

Competitiveness and economic growth versus environmental protection and regulations

Conventional wisdom claims that economic growth and rigorous pollution abatement measures are mutually exclusive, as environmental measures will increase costs and lower profits (Wurzel 2002: 14; Simpson and Bradford 1996). Society's concern over negative environmental impacts of business activities has led to an increase in environmental regulations which has been met with opposition from industrial groups (Rugman and Verbeke 1998; Farzin 2003). Traditionally, the relationship between businesses' competitiveness and environmental objectives has therefore been conceived as a trade-off between private costs and social benefits, and thus as a kind of "arm-wrestling match" (Porter and van der Linde 1995a: 97). Researchers have argued that increased environmental regulations could possibly have unproductive investments, higher costs and possible losses of competitive advantage as a consequence (Walley and Whitehead 1994; Aragón-Correa and Sharma 2003). This assertion is contested by Porter and van der Linde (1995a, 1995b) through the so-called Porter Hypothesis, first proposed by Porter (1991), which states that tougher environmental standards – if properly designed – can lead to more innovation and thus to (more) economic benefits for corporate environmental leaders ("first-mover advantages"). The argument, that companies that proactively approach environmental issues can also be very profitable, has been an influential and seminal one (van der Tulder and van der Zwart 2006; Nemetz 2013).

According to the authors, pollution is fundamentally a "manifestation of economic waste and involves unnecessary, inefficient or incomplete utilisation of resources, or resources not used to generate their highest value" (Porter and van der Linde 1995a: 105). They argue that the framing of the environment-competitiveness debate as a struggle between ecology and the economy is incorrect and based on a static view of environmental regulations (ibid.). Instead, environmental improvement needs to be framed in terms of resource productivity, i.e. how efficiently and effectively companies use resources. This view allows environmental improvement and competitiveness to come together, particularly if environmental standards are "properly designed" (1995b: 98).

BOX 3.1 BENEFITS AND PRINCIPLES OF ENVIRONMENTAL REGULATIONS UNDER WIN-WIN SCENARIO BETWEEN COMPETITIVENESS AND ENVIRONMENTAL PROTECTION

The benefits of environmental regulations are that they

- represent a signal where companies might encounter resource inefficiencies and technological improvements
- raise corporate awareness
- reduce uncertainty
- are motivators for innovations and progress
- level the playing field
- improve environmental quality.

Principles for designing regulations that encourage innovation, resource productivity and competitiveness include

- Outcomes, not technology, should be the focus
- Regulations should be strict rather than lax
- Regulations should be as close to the end user as possible and encourage upstream solutions
- Phase-in periods should be employed
- Market incentives should be used
- Regulations in related fields should be harmonised or converged

BOX 3.1 (CONTINUED)

- Synchronise regulations with other countries or develop them ahead of them
- Regulatory process should be stable and predictable
- Industry should participate in standard-setting from the start
- Regulators should develop strong technical capabilities
- Regulatory process should use least amount of time and resources.

Source: Porter and van der Linde 1995a, 1995b.

The Porter Hypothesis has not been without critics (see e.g. Palmer et al. 1995) for one of the first and Ambec et al. 2011 for an overview), and empirical evidence has been mixed, although the theoretical argument has been strengthened (Ambec et al. 2011). Certainly, Porter and van der Linde's work – through their critique of the then mainstream position (Hilliard 2004) in suggesting that the environment-competitiveness nexus could be seen under a win-win light – has influenced discussions and decisions by policymakers and companies over the design and implementation of environmental regulations (Ambec et al. 2011; van der Tulder and van der Zwart 2006). The influence of the hypothesis has been particularly due to Porter's standing as a leading corporate strategy thinker (Dunphy et al. 2007).

According to Elkington (1997), the work encourages business executives to rethink the nature of their operations in order to be prepared for the so-called sustainability transition. Such a re-examination of production processes as a result of environmental regulations favours processes that are more productive, thus making firms more competitive (Hackett 2011). In this sense, Porter and van der Linde (1995a, 1995b) encourage a rethink in businesses – recognising the environment and how it is regulated not as a cost or a threat but as an opportunity for innovation. How industry responds to environmental problems is then a leading indicator of its overall competitiveness as innovation delivers competitive benefits, such as new technologies that reduce the cost of environmental compliance, improved resource productivity and price premiums for more environmentally friendly products and services (Porter and van der Linde 1995b).

The dichotomy and perceived trade-off between economic growth and the environment can also be applied to the broader, macro-economic context through the Environmental Kuznets Curve (EKC). The original hypothesis, named after economist Simon Kuznets (1995), posits that inequality and poverty initially rise during phases of economic growth and then fall when a sufficient level of welfare has been reached. This results in an inverted U-shaped curve between income per capita and inequality, the so-called Kuznets curve (Chang 2014). This concept has been translated to economic growth and environmental pollution, which – in very simple terms – gets worse before it gets better. The EKC (see Figure 3.1) explains why the worst pollution is often found in emerging markets as they industrialise, and as countries become richer, cleaner environments are "purchased" through more stringent environmental regulations (Hackett 2011: 317). It corresponds to the view that economies can "grow their way out of pollution" (Nemetz 2013: 45) which has led to perceptions of "pollution havens" and "race to the bottom" scenarios where pollution is outsourced from developed to developing countries – although there is no evidence for this (Stern 2004). Likewise, Arrow and colleagues (1995) point out that economic growth is not a panacea for environmental quality and that both rich and poor countries rely on ecosystem services.

Whilst the original Kuznets hypothesis has not held up to evidence, mainly because of the importance of economic policy in how inequality is addressed (Chang 2014), its environmental

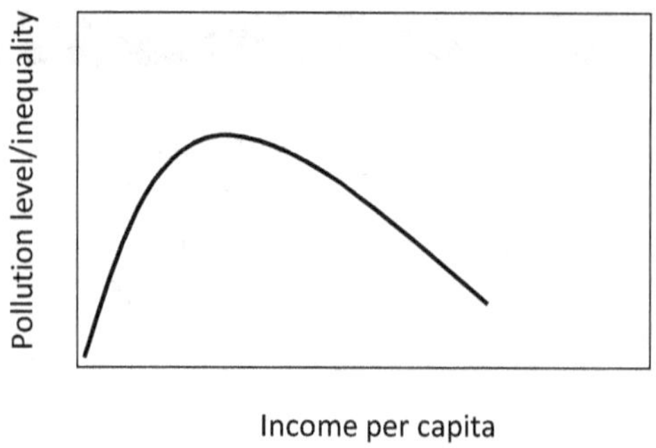

FIGURE 3.1
(Environmental) Kuznets curve

variant has found support. For pollution, for example, Grossman and Krueger (1995) do find evidence that for four pollutants examined, the levels initially increase but then decrease once a critical income level has been reached. Other studies have confirmed their findings (e.g. Dasgupta et al. 2002; Dinda 2004), though this does not hold true for all forms of environmental degradation where persistence, long-range transport and bioaccumulation are characteristics of certain pollutants (Nemetz 2013). For example in the case of climate change and CO_2 emissions the jury is still out (Stern 2006), and for waste, the direct relationship does not appear to decline with higher levels of income (Hackett 2011). The relationship is thus not a simple or a general one (Hackett 2011; Stern 2004) and differs between for example global emissions and local pollutants (Stern 2006, see Kedia et al. 2011 for more detailed coverage). Nemetz (2013: 48) goes further and calls the EKC "oversimplified" as it does not differentiate between local or global, immediate or delayed, modern or traditional risks and environmental burdens, and therefore needs to be revised to take into account environmental risk transition.

Technocratic versus ecocentric views of the environment

The tension between economic growth and environmental protection can be traced further, with a possible philosophical root in how humankind sees itself and its natural environment. Broadly, two modes of thought can be distinguished – anthropocentric and ecocentric. *Anthropocentrism* puts humans at the centre of how reality is viewed, whereas nature is seen through the lens of its usefulness to humans (Robertson 2017). In contrast, *ecocentrism* puts the earth at its centre in recognition that nature has value in and of itself and is thus worth preserving (Anderson 2014). This stance is therefore also called preservationist (Shrivastava et al. 2008). The corresponding conservationist position advocates the protection of nature as well but in light of its use and enjoyment by humans (ibid.).

This dichotomy also extends to how environmental issues should be addressed: the conservationist (anthropocentric) stance emphasises that nature can be controlled or managed and environmental problems can be solved with technology and institutions, whereas the preservationist (ecocentric) one argues that such problems need fundamental changes in the values and perspectives of modern culture and society (Shrivastava et al. 2008; Elliott 2013). In view of the importance of technology, anthropocentrism can also be called technocentrism, a stance further explored by O'Riordan (1981).

BOX 3.2 ANTHROPOCENE – HUMANITY'S IMPACT

The recognition of humanity's impact on nature is perhaps most strongly reflected in the naming of the Anthropocene (from the Greek anthropos for human and kainos, new) as the latest geological epoch after the Holocene, as suggested by Nobel laureate Paul Crutzen and his colleague Eugene Stoermer, and which has been recommended to the International Geological Congress for official declaration, though despite some progress and much debate, as of 2019 this had not been ratified.

Sources: Cavagnaro and Curiel 2012; Carrington 2016; Davison 2019.

These categories represent extreme positions across a simplistic dichotomy.[1] In practice, the modes are not fixed or exclusive as changing economic status and contextual factors play a role, too (Elliott 2013, drawing on O'Riordan 1981). Landrum (2018) equates technocentrism with weak and very weak sustainability, and ecocentrism with strong and very strong sustainability, as explored in the previous chapter.

The practical relevance of this dichotomy for businesses is explored by Shrivastava (1995: 126) who argues that anthropocentrism "asserts the separateness, uniqueness, primacy, and superiority of the human species". In his view, it reflects neo-classical economics (see previous chapter) and is part of the traditional management paradigm where the exploiting of natural resources by organisations is not only legitimate but also desirable. He proposes an ecocentric management paradigm, which recognises the ecological interdependence between society and ecosystems. This encourages a move away from the zero-sum game between competitiveness and environmental protection as encouraged by the Porter Hypothesis (see earlier discussion) in its emphasis on innovation.

Innovation is most commonly associated with technology, though it is not synonymous as there are other kinds of innovations (e.g. financial, economic and behavioural) (Blowfield 2013). In addressing sustainability, technology has an important role to play e.g. with regard to the resource productivity advocated by Porter and van der Linde (1995a, 1995b). However, its role is a double-edged sword as it can have unintended side-effects (Meadows et al. 1972). Whether technology contributes to environmental degradation or can help address it, represents therefore another kind of tension which we explore in the following section.

BOX 3.3 VALUES

It is not within the scope of the book to cover ethics and philosophy; it is, however, important to acknowledge that there is no value-neutral way of doing business. Doing business is often a negotiation between established values (or objectives), such as for example profit, and new or different values (or ethics), such as environmental protection. Reflecting on values might also lead to questioning and understanding the philosophical assumptions that underpin the way business operates; for example is it based on economic rationality or life-coherent rationality? The anthropocentric and ecocentric views concerning humanity and nature also give rise to different environmental philosophies. Whilst science provides us with understanding of the issues at hand, moral philosophy can guide us in considering our duty to inter-generational equity where future generations

BOX 3.3 (CONTINUED)

are treated as well as our own generation. Similarly, economics provides us with guidance that activities should continue until the marginal cost exceeds the marginal benefit, whilst ethical theories provide assistance with decisions on for example what kind of costs and benefits to consider.

A starting point to learn and read more about ethics in the context of sustainability and business is Crane and colleagues' *Business Ethics* (2019).

Sources: Helm 2003; Elliott 2013; Anderson 2014; Molthan-Hill 2017.

Role of technology as mitigator versus aggravator of environmental sustainability

The relationship between economic growth and the environment is closely linked to the role of technology (Anderson 2014). The technocentric mind-set introduced earlier sees technology as a solution to many of today's problems (Huesemann and Huesemann 2011). How technology is used can have a major impact on the environment, both positive and negative – whilst it offers opportunities to save resources and energy, it is not a "free gift" as it comes with its own set of environmental problems (Pearce et al. 1999: 22, see also Chang 2014). The relationship between environmental impact and the growth of the economy and the human population was first conceptualised by Ehrlich and Holdren (1971; Holdren and Ehrlich 1974) with the so-called *IPAT* equation:

$$I = P \times A \times T$$

where I equals the environmental impact, P population, A affluence and T technology, summarising the drivers of environmental degradation (Robertson 2017). Although the formula has been criticised for its simplicity (see e.g. Dietz and Rosa 1994; York et al. 2002), it illustrates how greater affluence in the form of material throughput or per capita consumption and population growth interact with technology changes to increase the human impact on the environment (Cavagnaro and Curiel 2012).[2]

Technology in the IPAT equation augments the environmental impact of population growth and increased consumption (Cavagnaro and Curiel 2012). This contrasts with the technological optimists' hope on the ability of technology to go beyond the limits to growth of population and capital (Meadows et al. 1972) and to find a technological solution to whatever problem arises (Pearce et al. 1999). Anderson (1998, 2009) calls such optimists *technophiles* who see technology as the saviour and compares them with the *technophobes*' (or technological pessimists') view of technology as the enemy. For Anderson, the IPAT equation is based on an old industrial revolution paradigm where technology is part of the problem (T1) as it is extractive, linear and wasteful – in other words, unsustainable. He restates the IPAT equation to make technology part of the solution (T2) in a new industrial paradigm:

$$I = \frac{P \times A}{T}$$

where technology moves from the numerator to the denominator and thus is harnessed to reduce the environmental impact of both population and affluence (Anderson 2009). In this scenario, technology is renewable rather than extractive, cyclical rather than linear, and focuses on resource productivity rather than labour productivity (Anderson 1998). In the revised IPAT equation, technological innovation offsets the increase in incomes (affluence) and population, and therefore the environmental impact (Laszlo and Zhexembayeva 2011).

Anderson (2009) asks how technology can shift from the numerator to the denominator, wondering whether market forces can phase out T1 and replace it with T2. Whilst acknowledging the power of the market, in line with Hawken and colleagues (1999), a redesign of commerce is needed, and Anderson advocates a change of laws and therefore incentives to industry. As there are limits to the law regarding the protection from negative and unintended effects of technologies since regulations often lag behind technological innovations (Huesemann and Huesemann 2011), minds and attitudes need to change as well (Anderson and White 2009).

Other limits to government regulations in the area of technologies are explored by Stern (2006) in the context of climate change: if regulations are based on existing technologies, there is a danger of *picking winners* and a *lock-in* of established technologies which could lead to path dependencies and prevent the development of alternative, potentially low-cost, technologies. As innovation is inherently unpredictable, policies should be technology neutral (Stern 2006). The options available for policymakers in terms of technology-based or technology neutral standards are discussed later in the chapter.

BOX 3.4 UNINTENDED CONSEQUENCES OF TECHNOLOGY – EXAMPLES

> Every new technology has side-effects.
> (Meadows et al. 1972: 145)

New technologies, however successful they may be, can have unintended consequences or side-effects. Meadows and colleagues (1972) give the example of social side-effects from the successful new technology represented in the Green Revolution – utilising new seed varieties in combination with fertilisers and pesticides was intended to be a technological solution to the problem of how to increase food production. The unintended consequence of the Green Revolution was an increase in economic inequality for example in India in the form of unemployment as the new seed varieties provided an incentive for greater mechanisation for harvesting and replanting.

Rachel Carson in her book *Silent Spring* (1962) illustrated the unintended consequences of the use of DDT (dichloro-diphenyl-trichloroethane) in agriculture – not just the death of pests and insects but also the death of wildlife, and in particular birds, were the side-effects of DDT use. In tropical medicine, DDT was used against malaria mosquitoes that quickly developed a resistance, thus rendering the chemical useless and instead of reducing instances of malaria contributing to its spread.

Antibiotic resistance are among the examples provided by Huesemann and Huesemann (2011), who distinguish between environmental consequences across different areas, such as agriculture, genetic engineering and medicine. In the case of the automotive industry, the technology of the internal combustion engine initially promised mobility and freedom – but has in many cases delivered the opposite, traffic congestion, which results in economic losses, time losses, noise and air pollution, not to mention its contribution to global warming. The automobile has not only had environmental but also social side-effects, according to the authors, in its impact on social interaction, urban sprawl, and resultant isolation and placelessness – the initially promised freedom has been replaced by an almost total dependence on the car (Huesemann and Huesemann 2011).

Despite the unintended negative environmental consequences of new technologies (Huesemann and Huesemann 2011), technological innovation plays a pervasive and significant role in addressing the sustainability challenge. They can do so in a number of ways and one helpful categorisation is provided by Blowfield (2013):

- Efficiency – technologies that use existing resources more efficiently, such as insulating homes, anaerobic waste digestion (see also eco-efficiency discussed later).
- Substitution – technologies that develop alternatives for harmful or exhausted resources, such as hydrogen fuel cells in transport, solar power for energy.
- Effectiveness – technologies that can add value in sustainability terms, such as for example green roofs (see also eco-effectiveness discussed later).
- Mitigation – technologies that counteract sustainability risk, for example geoengineering in the case of climate change.

Whilst innovation is in many cases technology-based, economic innovations play an important part in addressing sustainability through the use of market-based instruments and incentives by regulators (Blowfield 2013). The options available to policymakers in addressing environmental sustainability in terms of principles and types of regulations are the subject of the next section.

Summary of tensions and trade-offs

In this section, we continued where we left off in the previous chapter, reminding ourselves of the tension and seeming contradiction between natural and human-created capital and the differences between weak and strong sustainability. The interactions between business and the environment are necessarily two-sided, which the two perspectives of outside-in and inside-out can shed light on. By understanding the impact of business on the natural environment and exploring how business reacts to its impact (or not), the tension between competitiveness and environmental protection becomes less about black and white, less about threat or cost, and more about the opportunities inherent in environmental regulations. It is not just industry that has traditionally perceived the relationship between economy and environment as a zero-sum game with trade-offs between the two. The wider societal concern over sustainable development is illustrated by the Kuznets curve where environmental degradation is a seemingly necessary condition for economic growth. The background to this lies even deeper – how does humankind view its relationship with nature, ecocentrically or anthropocentrically and technocentrically? This philosophical question has practical business and management consequences – changing our view of nature can fundamentally transform how business approaches the natural environment – a zero-sum game with win-lose scenarios changes to a positive-sum game with win-win scenarios. The same is true for technology, which can be used to magnify the environmental impact of industrial activities, or it can become part of the solution when it is moved from numerator to denominator in the IPAT equation.

Although market forces are part of this shift – as Ray Anderson (1998: 20) writes, "There are new fortunes to be made in the next industrial revolution" after his carpet company Interface grew, having shifted its approach to the environment and technology – often, it has been environmental regulations alongside other stakeholder pressures that have been a driving force towards sustainability. We therefore turn our attention to the regulatory framework.

BOX 3.5 TENSE TIMES

One tension that is of particular significance to sustainability is the tension of timeframes between past, present and future. The Brundtland Commission defined sustainable development as meeting "the needs of the present without compromising the ability of future generations to meet their own needs" (WCED 1987: 43). This is reflected in two definitions of sustainability:

> that present and future generations can attain a high degree of economic security and achieve democracy while maintaining the integrity of the ecological systems.
>
> (Viederman 1996: 46)

> an economic state where the demands placed upon the environment by people and commerce can be met without reducing the capacity of the environment to provide for future generations.
>
> (Hawken 1993: 139)

Sustainability is intrinsically a long game which requires the understanding of interactions between sustainability pillars over a long-time horizon (Hackett 2011). Making this long-term view explicit and bringing the future into today's decision-making presents challenges for policymakers and corporations. Acting on environmental issues has not always been a priority since postponing action is always a possibility, because "future costs are less burdensome than current costs" (Pearce et al. 1999: 7). This stance is conceptualised in economics through *time preferences* and *discounting*. The trade-offs between time periods can be calculated through present-value calculations and the discounting of future costs and benefits (Anderson 2014).

Discounting can best be understood by comparing it to compounding – whilst compounding calculates how much today's investment or savings will be worth in the future, discounting tells us how much future benefits are worth at present (Brennan 1995). Discounting future benefits translates into preferring to receive them now, and placing a lower value on future benefits, whilst discounting future costs translates into less attention paid to future costs or preferring later costs rather than sooner costs (Anderson 2014). In other words, "the later a cost or benefit occurs, the less it matters" (Pearce et al. 1999: 132). The discount rate is in some ways an *intertemporal price* – the price of present against future consumption. The rate is based on time preference rates which are influenced by people's preference for the present ("pure impatience", Pearce et al. 1999: 134), as well as by risk and uncertainty. This natural desire for receiving benefits sooner rather than later is reflected in the time value of money. To calculate the present value economists use the discount rate and dynamic efficiency analysis – both Anderson (2014) and Hackett (2011) illustrate this in detail.

A crucial consideration is the choice of the discount rate and whether for example environmental benefits should receive particular treatment – differences in the rates used have major implications for decision-making (Anderson 2014). In the case of climate change as a long-term challenge, the discount rate is subject to complications (Stern 2006). Based on how future costs and benefits are viewed, we can distinguish between reactive and anticipatory approaches to environmental policies in policymaking (Pearce et al. 1999). In researching the automotive industry and environmental regulations, Wagner (2011) for example found that the perceived tension between competitiveness and environmental objectives could be solved by applying a temporal dimension to it – what is seen as a cost by companies in the short term can become an opportunity in the long term for environmental leadership (or sooner, as in the case of first-mover advantages).

Discounting is not without its critiques. For example, Daly and Cobb (1989: 155) reject

BOX 3.5 (CONTINUED)

the practice as it is based on "questionable judgments of both value and fact" and propose that any future reduction in the economic welfare enjoyed by the present generation should be counted as if that cost occurred in the present – this is known as the *equal standing* argument (Brennan 1995). Pearce et al. (1999) criticise the practice for shifting the burden of costs to future generations and precluding them from inheriting the natural wealth created. This reflects a question of intergenerational equity – equality between generations (Crane and Matten 2016) – demonstrating that environmental sustainability is closely interlinked with social sustainability.

Likewise, Brennan (1995) illustrates how discounting the future has not only economic implications but is subject to ethical considerations and normative questions (see e.g. Anderson 2014), such as how environmental benefits are viewed, how future (yet unborn) generations are treated in calculations of costs and benefits, and how uncertainty and risk are incorporated. Blowfield (2013) argues that it is only government that can manage the fundamental sustainability contradiction between the long-term interests of a society whose members we will never meet and the short-term biases of people today who prefer immediate benefits over future rewards.

REGULATORY FRAMEWORK

Addressing the trade-offs and tensions presented here requires both corporate and regulatory responses. Whilst regulations are often a response to market failures (e.g. in the case of externalities where pricing and costs are divorced in the market place, as the market has not incorporated the full environmental and social costs into prices), we need to acknowledge that there are instances of government failure where government intervention fails to address market inefficiencies (Hawken 1993; Hackett 2011). In practice, this means that environmental sustainability needs both hard and soft law, both voluntary and regulatory efforts, in other words, governments and businesses are equally tasked, at both global and local levels and in between.

As regulations provide the legal context and frame within which businesses operate, we will now look at these in detail – what principles underpin them, what kind of regulations and standard-setting philosophies are used. This helps us understand what guides environmental sustainability fundamentally before we ask what guides environmental sustainability in businesses practically and strategically in the following chapter. In the second and third part of the book, these foundations are then applied to different environmental challenges and different corporate functions.

Principles

The two most important principles discussed here are the *polluter pays principle* and the *precautionary principle*. Whilst other terms and principles are "out there" as well, these have been the most influential ones. Principles are often broad and general, and need to be translated into rules and regulations to be effective.

The polluter pays principle

One of the market failures tackled through government intervention comes in the form of externalities discussed in Chapter 2, where the market price for goods and services does not reflect the true value of all inputs and resources used to produce them (Pearce et al. 1999). The internalising of externalities gives rise to the polluter pays principle. It translates the economic concept of externalities into regulatory guidance.

The basic tenet of this policy principle is that the price of goods or services produced should fully reflect the total social cost of production, including the cost of all resources used, thus making polluters internalise externalities (Nemetz 2013). In other words, it equates the price charged for using environmental resources with the cost of damage to society as result of their use (Green 2013). Although the economist Pigou suggested solving the problem of negative externalities through taxes already in the 1920s (Hackett 2011), it was only in the 1970s that the polluter pays principle was widely adopted for environmental policymaking, particularly in member countries of the Organisation for Economic Co-operation and Development (OECD) (Nemetz 2013, see also OECD 1975). The use of the polluter pays principle marked a shift in regulatory instruments from command and control type regulations to using market-based incentives in environmental policy (Pearce et al. 1999). Whilst the principle only suggests who should pay for environmental protection but not how much should be paid (OECD 1995), market-based instruments attempt to put a monetary value on externalities. Among the basic mechanisms for making the polluter pay are standards, charges or taxes, and pollution permits according to a set standard which can then be traded (Pearce et al. 1999). We return to these in the next section where we look at how principles are enacted through regulations.

As the name suggests, the principle supports internalisation of costs in the specific case of pollution (Cavagnaro and Curiel 2012), but we should note that it can be applied more broadly generally as well in that prices of goods and services should fully reflect all costs associated with the process of their production and distribution.

Although the principle speaks of the polluter paying, in reality, costs are also passed on to the user or consumer by the producer, if possible (Pearce et al. 1999). Whilst making the consumer pay might seem at odds with the principle, Pearce and colleagues argue that this is what should happen as the revised price mechanisms signal the true cost of production. Such a *user pays principle* as part of the overall internalisation of costs could be used to increase environmental awareness (OECD 1995). In practice, depending on the curve and elasticity of demand, the distribution of the internalised cost will be split between producers and consumers (Nemetz 2013).

The precautionary principle

The second major principle of environmental regulations is the precautionary principle. It is based on the premise that we live in a world of uncertainty where society (and therefore industry) cannot wait for scientific certainty regarding environmental damage and therefore precaution should be the approach to this uncertainty (Nemetz 2013; Costanza et al. 1997). The concept of strong sustainability, as discussed in the previous chapter, gives rise to the principle of precaution due to uncertainty of the consequences of the use of natural resources and their non-substitutability.

The precautionary principle applies in cases where information about potential risks is incomplete (Green 2013). This pertains in particular to cause and effect, as the following definition of the principle in the Wingspread Statement underlines:

> When an activity raises threats of harm to human health or the environment, precautionary measures should be taken even if some cause and effect relationships are not fully established scientifically.
>
> (SEHN 1998)

The principle thus encourages anticipatory action rather than reactive action, although it offers no guidance on what kind of precautionary measures should be taken (Costanza et al. 1997). The costs of preventive action are normally tangible, can be clearly allocated and are mostly short term, whereas the costs of failing to act are problematic in terms of governance as they are less tangible, less clearly distributed and are normally longer term (EEA 2001).

Whilst policy decisions should be based on scientific knowledge or proof, the precautionary principle acknowledges that such certainty often comes too late for designing effective environmental policies (OECD 1995). With the precautionary principle, the burden of proof is shifted from the victim to the developer – before proceeding, the resource user has to demonstrate that activities are not harmful to the environment (O'Riordan 2000; OECD 1995). This marks a change from the past where risk-taking was accepted as necessary for progress (O'Riordan 2000).

BOX 3.6 RISK

Risk analysis is a tool widely used by regulators, businesses and even individuals, whether consciously or not. Risks are ubiquitous and the wish to reduce them and promote safety is part of modern society (which has also been described as the *risk society* e.g. by Beck 1992).

Risk can be defined as R = C x P, the product of accident consequences and accident probability. As both consequences and probability can exhibit different characteristics, we can differentiate the following categories of risk:

- *Certainty* where the consequences are known and the probability is 1.
- *Risk* where the consequences are known, but perhaps not precisely, and a probability distribution is used to determine the outcomes.
- *Uncertainty* where the consequences are known but the probability is unknown.
- *"Wicked" problems* where both consequences and probability are unknown.

Examples for these categories include asbestos as a risk, antibiotics in animal feed and bacterial resistance as uncertainty, and the revelation of CFCs (chlorofluorocarbons) destroying the ozone layer as a "wicked" problem.

Risk analysis makes use of various methodologies, whether it be probabilistic risk assessment or risk-benefit analysis (which combines probabilistic risk assessment and cost-benefit analysis), comparative risk assessment, revealed or expressed preferences. Both governments and corporations use the concept of risk and its analysis to inform decision-making and to manage risks. For example, in the US, the Environmental Protection Agency has developed risk assessments for different pollutants alongside guidelines for health and ecological risks. Risk analysis is subject to complexities and difficulties where the precautionary principle offers a decision rule in the context of ecosystem protection.

Sources: Nemetz (2013), Hackett (2011), Gerrard (2000).

Although the precautionary principle has been criticised e.g. on economic grounds for departing from dynamic efficiency, Anderson (2014) argues that in taking into account uncertainty and risk aversion, it is a rational criterion for policymaking. Other criticisms include that it is ill-defined and holds back progress in its perceived risk aversion.[3] Its acceptance is nevertheless evident in its use in a plethora of international environmental agreements and across other levels of environmental policymaking (see e.g. Nemetz 2013; Anderson 2014) for examples). For example, in Germany, an early acceptance of the principle legitimised the adoption of pollution abatement measures in the absence of scientific proof if there is (significant) risk that (irreversible) environmental damage would otherwise occur (Wurzel 2002).

Other principles

The philosophy of *prevention*, which arguably is the most sustainable option in the first place, is embedded in the precautionary principle (Robertson 2017). With regard to the (scientific) uncertainty referred to in the previous section, prevention of pollution and environmental damage is then an appropriate policy (Costanza et al. 1997). However, O'Riordan (2000) distinguishes prevention from precaution in that he calls prevention a regulatory measure aimed at eliminating an established threat, whereas precaution means the duty of care on all action. Everard (2011) also differentiates prevention from precaution, as the prevention principle does not go as far as the precautionary principle. Nevertheless, it allows early protective action to prevent damage rather than repair damage that has already occurred.

The *precautionary polluter pays principle* puts the two principles of precaution and polluter pays together. Costanza and colleagues (1997) argue that linking the two helps to deal with uncertainty in a way that is both economically efficient and ecologically sustainable. In practice this would resemble a deposit-refund scheme and involve an environmental assurance bond system that incorporates known and uncertain environmental costs (ibid.). Polluters or risk-takers with natural resources would post a bond that would cover environmental damages in a worst-case scenario (Anderson 2014), forcing the internalisation of eventual externalities. The bond covers the damage but is returned to the resource user should no damage occur (ibid.). The combination of the two principles means that the polluter pays not as part of a damage done but as a precaution.

The *proximity principle* stipulates that for example waste should be treated and disposed of as near to the point of production as possible, under technical and environmental considerations (Everard 2011). The application of this principle allows the management of the risks and costs involved in the transport of waste (ibid.).

Other principles that also involve a spatial dimension include *rectification of damage at source* and *end of pipe* solutions. The boundaries to standard-setting philosophies (discussed later) are somewhat blurry so that end of pipe solutions are not only a principle but also technologies such as for example catalytic convertors in cars or scrubbers on smokestacks that remove or reduce pollutant emissions after they have formed (Green 2013). Such end of pipe solutions could therefore be seen as "fixes" that are applied after the environmental damage has occurred (Robertson 2017: 305). The rectification of damage at source principle is related to the prevention principle – in the case of prevention failing, it steps in to combat damage at an early stage, i.e. at the source, as the name suggests (van Calster and Reins 2017). As such, we could argue, it is also related to the proximity principle.

Summary of principles

Exploring the principles that underpin regulations gives us a better understanding of the regulatory framework in which businesses operate. Whilst we can distinguish principles such as precaution and polluter pays as separate, in practice, regulations often apply these principles comprehensively. Stookes (2011) gives the example of EU environmental policy that is based on the precautionary principle, encourages preventive action and rectification of damage at the source and makes the polluter pay. In the broader context, Everard (2011: 48) argues that the reality overarching all of these principles is that "we are all part of the biosphere of planet Earth" and that neither individuals nor corporations are independent of the biospheric system – a useful reminder as we look at how these principles are used in the regulatory framework.

Environmental regulations

We can distinguish broadly between three types of regulations – command and control regulations, market-based regulations and voluntary regulations by industry. In today's practice, a combination of market mechanisms, policies and legislation as well as self-regulation governs issues of environmental sustainability (Blowfield 2013). Areas generally covered under environmental legislation include waste, noise, air, soil, nature, water and chemicals, although classifications may vary. Due to the transnational nature of environmental problems, environmental law often starts at a global level with broad principles and conventions and is then cascaded down to be delivered and enacted at the national or local level through regulations (Stookes 2011).

Command and control regulations: Standards

This kind of regulation sees the government "command" businesses to meet certain standards and "control" the methods (e.g. the technology) needed to achieve these standards (Green 2013).

The initial regulatory response to environmental degradation in the late 1960s was one of *command and control* (Nemetz 2013), which we could also call a 'thou shalt not' approach. It is mostly a *direct* regulatory system, requiring e.g. uniform reductions or specific technologies (Anderson 2014). This means that there is often greater certainty that environmental goals will be achieved (Hackett 2011). For policymakers they are also easier to implement.

In this conventional approach to regulation, standards are a frequently used instrument where we can distinguish between different kinds of standards (Anderson 2014; Hackett 2011):

- Ambient standards concern specified locations of measurements where targets for pollution concentrations are set. In order to achieve ambient standards, emissions and technology standards may be set.
- Technology standards require the use of particular methods or equipment to achieve targets.
- Emission standards come either as a set target or amount, or as a designated percentage of reductions.
- Performance standards set uniform standards for all regulated firms, but unlike technology standards, firms have some choice over how they can meet the target.

A problem with standards is that they are static while technology is dynamic (von Weizsäcker et al. 1998), leading them to become obsolete before long or to be misinterpreted as a ceiling rather than as a minimum standard (Hawken et al. 1999) with no incentive for companies to go beyond the standard (Nemetz 2013). The inflexibilities of command and control policies in setting both targets and methods can lead to inefficiencies (Anderson 2014), such as higher costs and the lock-in of technologies preventing the development of new and improved methods (Hackett 2011). In order to address such inefficiencies, incentive-based policies can be used.

BOX 3.7 STANDARD PHILOSOPHIES

All three types of regulations involve standard setting. Just as different principles underpin regulations, standard setting can be approached with different *philosophies*, which mirror the considerations and principles for regulations more generally. Wurzel (2002) contrasts the standard-setting philosophies of Germany and the United Kingdom in their approach to environmental regulations at the national and European Union level. Uniform emission limits (UELs) – as favoured by Germany – focus on the sources of emissions, whilst environmental quality objectives (EQOs) – as advocated by the UK – focus on the receiving environment.

Germany, as an early adopter of the precautionary principle (*Vorsorgeprinzip*), legitimised pollution abatement measures in the absence of scientific proof on whether there is (significant) risk that (irreversible) environmental damage would occur otherwise. Setting pollution limits at the source is based on the principle of *best available technology* (BAT) though we can also see the proximity and the rectification of damage at source principles reflected in this approach.

This is in contrast to the principles of *best practicable means* (BPM) and the *best practicable environmental option* (BPEO) which illustrate the UK's preference for cost-effective environmental measures that are based on scientific proof. The approach to environmental policy here is science-driven and makes use of the carrying capacity of the environment, for example through the use of 'high chimney' policies (seemingly an end of pipe solution).

Whilst UELs are easier to monitor than EQOs, EQOs are better for determining the most cost-effective policy measures. The two approaches could easily be perceived as mutually exclusive since the UEL approach is prescriptive and the EQO pragmatic (even "light touch"). They can, however, be combined in parallel and complementary ways by first regulating emissions at the source and by secondly monitoring the quality of the receiving environment, whether it be air, water or soil. At the EU level, environmental policymaking has been approached through the compromise of the BATNEEC principle, which can stand for both *best available technology* as well as *techniques not entailing excessive costs.*

Sources: Wurzel (2002), Lees (2008).

Market-based regulation: incentive-based policies

In contrast to command and control regulation, market-based (or incentive-based) approaches have general goals established by the government but allow businesses to choose the methods to achieve them based on cost-effectiveness (Green 2013). Incentive regulation is mostly an *indirect* regulatory system that makes use of economic instruments such as taxes, prices, subsidies or markets (Hackett 2011). Another term to describe such approaches is "carrot and stick".

The classic incentive-based instrument is a Pigouvian tax – a tax, fee or charge per unit of pollution emitted (Costanza et al. 1997). The tax is named after the economist A. C. Pigou who established that negative externalities can be internalised through the use of pollution taxes[4] (Hackett 2011). The incentive here is that reducing emissions or pollution results in a tax reduction and therefore motivates firms to invest in research and development to find cost-effective ways of producing in a less polluting way e.g. through developing new pollution abatement technologies (see e.g. Nemetz 2013). Pigouvian taxes may form part of ecological tax reform where public finance is shifted from productive activities such as income and employment to environmentally destructive activities (von Weizsäcker et al. 1998; Hackett 2011). Another form of tax is a natural capital depletion tax (Costanza et al. 1997) which is consistent with strong sustainability.

The other major economic incentive is in the form of marketable or tradable pollution permits (Nemetz 2013). Two types of allowance markets can be distinguished – *tradable pollution credits* for *individual* pollution sources and *cap-and-trade* programmes where an *aggregate* emissions limit (cap) is set (Hackett 2011). Both types cap emissions which are therefore made scarce and valuable. Cap and trade has been a focus in recent years because of its ability to create a win-win situation of generating environmental improvements and reducing compliance costs (Hackett 2011). It involves the setting of an acceptable pollution level for a particular geographic area for which permits are then issued. Permit holders (as sellers) and other polluters (as buyers) can form a market within which to trade permits (Nemetz 2013). Like taxes, trading programmes provide incentives for a reduction of pollution levels – when the price for cutting emissions is less than the permit price, businesses will cut emissions and sell the permit not needed at a profit (Anderson 2014).

Permits can either be allocated through auctions or based on the status quo or historic emissions, so-called grandfathering (Hackett 2011). Other mechanisms used in trading schemes include banking, bubble provisions, offsets and netting (Anderson 2014).

Beyond Pigouvian taxes and marketable permits, economic incentive-based instruments for environmental sustainability include subsidies, grants or loans, property rights, liabilities, fees, and bonds (see e.g. Nemetz 2013; Hackett 2011; Costanza et al. 1997 for details).

Despite their advantages over command and control instruments in terms of economic efficiency, market-based incentives come with their own problems, such as perceptions of "licenses to pollute" and how permits are allocated (e.g. "grandfathering" benefits current heavy emitters) (Costanza et al. 1997: 202). Sandel (2012) also points to tradeable permits being perceived as letting companies buy the right to pollute and as constituting outsourcing of an obligation rather than making meaningful changes to wasteful habits. In the previous chapter we touched upon the difficulties of putting a price on nature's services and criticisms of the practice (e.g. McDonough and Braungart 2002), which provide challenges for policymakers. There are thus difficulties in determining the optimal level of Pigouvian taxes and charges, and a marketable permit system does not work in all circumstances and depends on the characteristics of pollutants, for example whether they have local, regional or global effects, or to what extent they accumulate (Nemetz 2013).

Both taxes and a cap-and-trade system are subject to a whole range of design considerations, both technical and political (e.g. what price to set, what limit to set, political acceptance). At an abstract level of analysis, the two instruments are equivalent in terms of cost-effectiveness, but in reality there are additional complexities and complications (Hackett 2011). While taxes provide certainty over the price of emissions, there is greater uncertainty over their level. Equally, although a cap-and-trade system gives greater certainty over the quantity of emissions, there is less certainty over the price of pollution (Hackett 2011).

Nevertheless, market-based incentives allow more flexibility and creativity for firms to achieve environmental targets as differing circumstances will encourage different ways to address environmental problems (Anderson 2014). An alternative can be found in voluntary regulation.

Voluntary regulations

Although we conventionally associate regulations with governments, there is regulation that is issued by companies and industry associations, also referred to as private regulation (Crane and Matten 2016). Self-regulation has advantages as it is less costly than law enforcement from the side of government and offers businesses flexibility (Cavagnaro and Curiel 2012). Although self-regulation by businesses could be seen as avoiding, forestalling or softening potential laws

(Crane and Matten 2016), it can secure early buy-in by industry ahead of formal regulations and could therefore achieve higher commitment (Bigg 2011). Such commitment is reflected in the term *covenant*, what non-regulatory initiatives are also called (Cavagnaro and Curiel 2012). Voluntary regulation allows for greater acceptance by industry compared to legislation, but questions remain over compliance and enforcement, particularly in the case of serious risks (Bigg 2011). One type of voluntary effort underpinned by economics is Coasian contracting or bargaining which allows private solutions to environmental problems, though only in the case of a limited number of parties (Anderson 2014).

Private regulation comes in various forms – there is pure self-regulation by business and industry and collaborative activities where businesses work together with governments and/or civil society organisations to develop programmes, initiatives and codes of conducts (Crane and Matten 2016). This reflects therefore a move away from command and control approaches ("thou shalt not") to cooperative approaches ("let's work together" and "carrot and stick").

As regulation and environmental sustainability is not a one-sided affair, it is worth looking into the relationship between business and government more closely.

BOX 3.8 EXAMPLES OF PRINCIPLES AND REGULATIONS IN ACTION

As the effects of environmental degradation vary across local, regional, national or global levels, they are addressed at these different levels, depending on effectiveness and suitability. We thus find environmental regulations across all levels of policymaking. The following provides selected examples of how the principles and regulations covered in this chapter have been put into action.

Polluter pays principle

The principle is used in international environmental law, such as for example the Oil Pollution Response Convention 1990. At the national level it can be found underpinning UK legislation in the area of environmental crimes where strict liability is applied. One example of how externalities can be dealt with is the case of forcing users of river water to include the cost of pollution (or depletion) in the pricing of manufacturing by requiring the location of the intake pipe to be downstream from the effluent pipe. As the principle is based on internalising the environmental costs of negative externalities, it is put into action through incentive-based or market-based regulations, more examples of which can be found later.

Precautionary principle

Variations of the principle have been used in international environmental treaties and agreements. Examples include the 1987 Montreal Protocol on Substances that Deplete the Ozone Layer and the 2001 Stockholm Convention on Persistent Organic Pollutants at the global level, whilst at the regional level the 1987 London Declaration (Protection of the North Sea) advocates the principle and the European Union bases environmental policies on the principle where there is sufficient scientific uncertainty. At a national level, the United States applied the precautionary principle in its early ban on CFCs in aerosols.

Command and control regulations

Technology-based standards and uniform emission limits are among conventional instruments that are for example used to regulate exhaust emissions from motor vehicles – in the case of the European Union through the so-called Euro norms which set mandatory limits for carbon monoxide (CO), hydrocarbons (HCs), nitrous oxide (NOx) and particulate matter (PM). The original Clean Air Act of 1970 in the United States

BOX 3.8 (CONTINUED)

required all coal-burning electricity-generating facilities to install costly scrubbers. Other tools include prohibitions, production or process, and product standards. At a local level, for example, San Francisco was the first US city to ban plastic bags in supermarkets in 2007, and cities such as Athens, Paris, Madrid and Mexico City have announced plans to ban diesel cars from their roads by 2025.

Incentive-based regulations

Most OECD countries have used the polluter pays principle in designing market-based regulations that use economic incentives. A prime use is found in the case of cars where there is differentiated taxation of motor fuels such as petrol and diesel, and the taxation of vehicles according to fuel consumption and thus carbon dioxide (CO_2) emissions. Eco-taxes and wider ecological tax reform come under this type of regulations as well. Deposit-refund systems are widely applied in the case of beverage containers (cans or bottles), such as for example in Norway, Germany and South Africa, where consumers pay a deposit when purchasing a drink, which is refunded to them when they return the container.

The United States has successfully used a cap-and-trade system for sulphur dioxide emissions as part of their acid rain programme at the national (federal) level; whilst at the regional level, several states have used pollution taxes or fees such as California's E-Waste recycling fee and Iowa's pesticide and fertiliser tax. On a regional, supranational level, the EU emissions trading system provides another example of the use of marketable permits.

Other examples for market-based regulations include emission charges, charges for noise or pollution, user charges and subsidies. Local authorities in Germany, for example, charge residence a waste collection fee, which some councils use to differentiate for reduced frequency of collection, thus creating incentives to save costs by reducing waste.

Voluntary regulations

At the global level, we find multipartite agreements between government, business and civil society, such as the UN Global Compact, or the OECD guidelines for multinational enterprises, or initiatives that result from civil society and industry collaborating, such as the Marine Stewardship Council, the Ethical trading initiative or the Ceres (Coalition for Environmentally Responsible Economies) Principles. Sector-wide business-driven self-regulation include for example the Responsible Care programme initiated by the International Council of Chemical Association, or the Extractive Industry Transparency Initiative or the Global Sustainable Tourism Criteria. Across industries, the ISO 14000 series of the International Standard Organisation represents global standards for environmental management systems. Other voluntary regulations come in the form of codes, such as, for example, the Global Sullivan Principles of Corporate Social Responsibility or the initiatives developed by the World Business Council for Sustainable Development (WBCSD).

Sources: Crane and Matten (2016), Harvey (2016), Scientific American (2015), Nemetz (2013), Cavagnaro and Curiel (2012), Hackett (2011), Stookes (2011), Wagner (2011), OECD (1994, 2010), Meadows et al. (1972).

Relationship between legislators and companies

On the one hand, we could see the relationship between regulators and companies as one more natural tension not covered in the first section of the chapter – after all, government intervention often only takes place in the case of market failure, and regulations are seen as interfering with companies' quest for competitiveness and profit maximisation. On the other hand, achieving environmental sustainability is such a complex and wide-ranging challenge that we could argue that only in collaboration between businesses and regulators (and civil society) is success possible. Going back to the Porter Hypothesis, environmental regulations – if designed properly – provide opportunities for environmental leadership rather than simply another cost. Paul Hawken (1993: 165) puts it like this: "in order for business to function both effectively and ecologically, the contradictions between guardian and commercial interests must be reconciled" and encourages a consensus-building, collaborative approach that both sides can support.

Corporate attitudes to environmental sustainability are reflected in their relationship and interactions with policymakers and their response to regulations. New laws and regulations add to the pressures shaping the business environment, influencing factors such as prices, input costs, demand, competitiveness, R&D progress, and the commercialisation of new technologies (Marcus et al. 2011).

For example, Wagner (2011) found in researching industry's attitudes to environmental regulations that companies do indeed view these as a cost, but that a distinction can be made where this is particularly the case in the short term, whereas in the long term such policies may be viewed as an opportunity by companies.

Broadly speaking, there are two corporate responses to environmental policy (cf. Hart 1995): firstly, a compliance strategy, and secondly, going beyond compliance with an emphasis on prevention. Aragón-Correa and Sharma (2003) call these two modes reactive (compliance) and proactive (prevention), and add environmental leadership to this typology of corporate environmental strategies, which goes further than the prevention aspect of proactive strategies (Aragón-Correa et al. 2008).

With a compliance strategy, there is a tendency towards legislative and political lobbying aimed at fighting regulations or slowing down the pace of environmental legislation. Contrastingly, within the prevention policy, firms develop skills helping them to adopt external technologies in order to meet society's demands and possibly to "raise the bar", whereas compliance-oriented firms develop political acumen to fight against these imperatives. Such a reactive focus could however be seen as inferior to one of prevention (Russo and Fouts 1997). Demonstrating the choice for business, Laszlo and Zhexembayeva (2011: 24) ask whether a company will "ride the wave of legislative changes to get ahead of the curve, or wait until the waters drag it under?" Compliance might no longer be sufficient as expectations on companies rise.

What companies can do is try to proactively shape government regulations (or private industry standards). Making strategic use of regulation may for example raise competitors' costs and thus provide opportunities for both competitive and environmental leadership (see e.g. McWilliams et al. 2002; Laszlo and Zhexembayeva 2011). Such shaping of government policy comes in the form of lobbying, which is the intended influence of political decision-making processes by persons who are not involved in these decisions (Köppl 2005). Although the term lobbying is perceived negatively in some languages and countries, it is a legitimate activity that is both legal and beneficial as it allows institutions to receive inputs necessary for decision-making that is adequate and equitable (Suder 2011).

The relationship between businesses and governments is not a homogenous one, as there are corporate environmental leaders as well as corporate environmental laggards (Wurzel 2002), and there are differences in how different sectors are affected by environmental policies. For pollution-intensive industries in manufacturing and energy production (e.g. chemicals and petroleum), environmental legislation may be costly, and political involvement therefore happens for obvious reasons. There are also significant differences across regulatory systems, where for example in Europe a more consensus-based approach is taken to environmental policymaking, whereas a more adversarial tradition is prevalent in the US, based on different perceptions of risk, government and environmental protection (Dunphy et al. 2003; Mahoney 2007). Furthermore, individual companies differ in how they perceive the political decision-making process – as "illegitimate meddling in corporate affairs" or as legitimate, worthy of engagement and resources at the two ends of the spectrum or somewhere in the middle as illegitimate but necessary or legitimate if based on facts (Anastasiadis 2014: 275). This influences how companies approach policymakers in their lobbying (e.g. with condescension and a sense of entitlement or as a competent partnership with respect) (ibid.)

To conclude our brief look at the relationship between regulators and regulated we can summarise that whether firms are cost- and compliance-oriented, or prevention- and opportunity-oriented, or even prepared to take on environmental leadership is thus also dependent on how they perceive regulations and their makers.

Regulatory compliance

Regulations can only be meaningful if there is compliance and enforcement – if legislation is not worth the paper it is written on, then what's the point? we might ask. Effective enforcement involves having mechanisms in place for monitoring compliance and penalising those who breach the rules, which also acts as a deterrent. Here, the difference between types of regulations warrants a closer look so we can distinguish between so-called hard and soft laws.

In international legislation, hard law includes for example formal conventions such as protocols, treaties and statutes and other written agreements between countries, whereas soft law includes declarations, statements and resolutions. Although soft law is non-binding, it still has legal relevance, whereas hard law consists of legally binding obligations that are often precise and unambiguous (Skjærseth et al. 2006). Soft law has the advantage that it is more likely to be adopted as it is not legally binding, which also means, however, that there are no enforcement mechanisms as in the case of hard law, such as for example dispute settlement mechanisms (Stookes 2011). Environmental regulations are less influential when the mechanisms for enforcement are weak (Anderson 2014).

Voluntary regulations as a form of soft law are unlikely to be effective unless underperformance can be successfully penalised. The potential or threat of regulatory action in the case of self-regulation failing may not always be enough to motivate compliance (Bigg 2011). Van Tulder and van der Zwart (2006) confirm that many international codes of conduct, whether drawn up by industry itself or in collaboration with civil society or government organisations, are often weak when it comes to compliance, sanctions and therefore effectiveness.

Effective environmental governance depends on serious implementation, which is more likely to be the case for hard law, which also features improved compliance (Skjærseth et al. 2006). Compliance in most cases means meeting the requirements set by law, standards or rules, which will depend on the country and industry sector in which a company is operating (Dunphy et al. 2003). Demonstrating compliance offers a minimum legal *licence to operate* for business (Reinmoeller 2012) and could be called the "base platform" on which sustainability can be built (Dunphy et al. 2003: 252). Going beyond compliance (or "compliance plus", Dunphy et al. 2003) and exceeding minimum legal requirements offers proactive companies the chance for competitive advantage and environmental leadership[5] (Reinmoeller 2012).

Returning to the term *licence to operate*, meeting legal and regulatory standards could be said to be the basic conditions for continued operation (Friedman 2011). However, pollution taxes or fines for breaching standards are not simply to be paid in order to continue business as usual, as a form of a *licence to pollute*. Stookes (2011) emphasises that the polluter pays principle is not an opportunity for rich companies or individuals to pay to pollute, and needs to be seen as a deterrent and as a response to environmental pollution. In a similar line, Friedman (2011) points out that enforcement of environmental laws is also important for business people not just as a matter of public policy but also to prevent competitive advantage for those companies ignoring and thus breaching the law.

Non-compliance constitutes a significant business risk as companies may be faced with liability costs, lawsuits over environmental crime and damages to corporate reputation (Dunphy et al. 2003). The potential penalties for violating environmental legislation act as a deterrent (Anderson 2014), particularly if the expected penalty exceeds the gain from violating such law, i.e. when crime does not pay (Hackett 2011). As an aside, Anderson (2014) points out that deterrence can be excessive and lead to disproportionate precaution when for example regulation and liability are combined e.g. in the case of pollution taxes and the threat of litigation.

It is the failure of corporations to comply with regulations that is probably the most important aspect of breaches of environmental law (Stookes 2011). Sanctions for breaching the law include financial ones (fines, compensation, liability for court costs) and imprisonment but also fewer tangible sanctions where the reputation and the image of an organisation are affected (ibid.). The internet provides an opportunity to disseminate information at a low cost, which can facilitate reputation deterrence, for example through naming and shaming polluters (Hackett 2011; Jacquet 2015). Thus, market reputation acts as an incentive for firms to demonstrate how environmentally friendly they are with the avoidance of consumer boycotts and voluntary over-compliance as examples (Hackett 2011).

For companies, compliance is only a small part of good environmental management, and going beyond it offers the opportunity to devise new products and materials that create market demand while at the same time addressing environmental concerns (Friedman 2011, based on Piasecki and Fletcher 1996).

Summary of regulatory framework

In this section, we have looked in detail at what guides environmental regulations in terms of principles and then how these are put into practice in different kinds of regulations against the background of how regulators and companies interact with each other. For good environmental management, it is useful for managers, regardless of their background – business, engineering, law or other – to have a basic understanding of environmental law and what underpins it (Friedman 2011).

In summarising the section we want to remind ourselves about the advantages and disadvantages of the two major kinds of regulation used in environmental legislation (Hackett 2011; Costanza et al. 1997). Command and control regulation provides a familiar and accepted framework that is direct and seemingly simple, through specifying uniform standards for firms, often mandating how pollution is to be reduced, and offers greater certainty of reaching environmental goals. The drawbacks however are that it risks technology lock-in through its rigid nature and that it is very costly in terms of information requirements for regulators, monitoring, enforcement and compliance costs, and lacks incentives to go beyond the mandated levels. Market-based regulation provides a more flexible framework that uses market signals to encourage environmentally friendly behaviour through economic incentives. It is more cost-effective as regulators do not need the same technical information for efficient regulation, by being consistent with the polluter pays principle and as it encourages both consumers and producers to change behaviour. It is indirect, does not specify methods or equipment and thus provides dynamic incentives for companies to research and develop cheaper and cleaner technologies. Nevertheless, incentive regulation is subject to complex and in some cases contentious design considerations with which policymakers have to contend.

In practice, traditional, conventional standard-setting should be supplemented with more market-based approaches (Pearce et al. 1999). Governments need to consider variables beyond economic efficiency, such as administrative feasibility and costs, political acceptability, informational requirements, effect on different regions and income levels, as well as reliability and dynamic incentives for continuous improvements (Nemetz 2013). Whilst market-based regulations provide cost-effective solutions, in cases where urgent action is required or particularly serious risks or toxicity are at play, command and control have a place in the armoury of government policy (ibid.). Costanza and colleagues (1997) point out that it depends on the level of scientific certainty or uncertainty with regard to whether incentive or command and control systems are more appropriate and which principle is more significant – for example in the case of uncertainty, prevention is more suitable.

One example that combines both approaches in practice is the Euro norms regulating pollutant emissions from motor vehicles in the EU – whilst they set mandatory limits for cars to achieve within a certain time frame, they also allow for fiscal incentives at the national level to encourage early adoption of cars complying with the standard.

Environmental regulations do not operate in isolation, they interact and interfere with other kinds of policies across local, national and global levels. What market-based regulations based on incentives for example do not address is the trade-off between equity and efficiency (Costanza et al. 1997) or the argument of fairness[6] (Hackett 2011). As we discussed in the first chapter, sustainability in its full extent does not only cover environmental aspects but also societal aspects. Environmental policy therefore does not happen in isolation but needs to be integrated with other considerations, both in terms of sectors and issues (e.g. agriculture and transport), but also more broadly with regard to wider societal goals.

Such integration also extends to cooperation between regulated and regulators – only with input from industry can environmental policy be realistic and feasible, as well as acceptable. Friedman (2011) argues that, as the costs of environmental programmes are high, it is essential that companies and policymakers work together to ensure the money is well-spent and environmental goals achieved efficiently.

In order to be effective, environmental regulation needs to be politically acceptable (secure enough votes in the decision-making body), technologically realistic (ambitious and stretching but not beyond reach for companies), economically efficient (financially viable without astronomical costs) and monitoring and enforcement needs to be feasible and effective. In designing policy, regulators need to meet society's need for a natural environment that is clean and able to provide ecosystem services and goods, and they need to accommodate business needs for a fair level playing field where companies are able to compete fairly, have time to adapt and are motivated to go beyond simple compliance. The following section explores some of the tools that can guide regulators as well as companies in this endeavour for environmental sustainability.

FRAMEWORK OF FOUNDATIONS AND TOOLS TO GUIDE COMPANIES AND REGULATORS

As the question underpinning this chapter is what guides environmental sustainability at a fundamental level, we explore what kind of tools and concepts can be used in preparing and implementing environmental regulations. The following serves as a general and broad overview on how these guidelines can be used by both regulators and businesses. Some of the tools will be covered in greater detail in other chapters, as they can be used in a more focused and applied manner by companies in addition to their broad, principled use by policymakers and industry in general.

Environmental impact assessment

One of the most established and accepted tools used by companies and regulators is the environmental impact assessment (Fuller 2011a). The use of environmental impact assessments (EIAs) was first introduced in the United States in 1969 with the National Environmental Policy Act (O'Riordan 2000) and has been adopted widely since then, often as a prerequisite for major projects to assess the possible effects of policies or developments (Anderson 2014). A distinguishing feature of the tool is that it is anticipatory in trying to predict environmental consequences of major projects and thus possibly acting as an early warning for developers (Fuller 2011a). Other terms for EIA include environmental impact studies (Anderson 2014), environmental impact statements (Costanza et a. 1997) and environmental impact reports (Friedman 2011).

EIAs may draw on earth, life and social sciences, including all of their sub-branches, and therefore take a multidisciplinary approach (O'Riordan 2000) in line with the integrated and holistic nature of sustainability in general. EIAs not only cover the environment but also the general public and local communities affected by projects. Costanza and colleagues (1997) call the inclusion of the public an important policy innovation in a process previously tightly controlled by polluters and their seemingly captive regulators. Friedman (2011) encourages the early inclusion of all affected parties, including agencies and environmental and local groups in the preparation of environmental impact statements to avoid delays and litigation. Fuller (2011a) calls it a democratic tool as it involves a number of different stakeholders, serving public authorities, developers and local communities as an information-providing tool, as a design or risk management tool and as a means to respond to concerns of affected citizens.

The major steps involved in an EIA are the proposal identification, which includes the consideration of alternatives, screening, scoping, assessing impacts, mitigation, review, decision-making and follow-up (Fuller 2011a). In assessing the impacts, three elements are of significance – understanding the baseline conditions, predicting the magnitude of the impact and assessing the significance of those impacts (Fuller 2011a). Although the principle might seem simple, the completion of EIAs is a complex matter (Friedman 2011).

In the case of the EU, the assessment needs to relate to impacts on human beings, flora and fauna, soil, air, water, climate, landscape, material assets, cultural heritage as well as interaction between the factors mentioned (Fuller 2011a). There is, however, no requirement to assess these impacts in monetary forms (Pearce et al. 1999). The EU differentiates between projects always requiring EIAs (such as for example oil refineries, large waste and wastewater disposal and treatment facilities, large intensive agricultural projects, other industrial facilities) and projects that require an EIA in case of significant environmental impact (e.g. extraction projects above a certain size, wind farms, food industry, coastal defence, tourism and leisure projects). Criteria to determine the nature include the project's characteristics, its location and the characteristics of the potential effects (Fuller 2011a).

The EU also uses the related strategic environmental assessment (SEA), which is a higher level tool aimed at the plan, programme or policy level, thus responding to the limitations of EIAs at the project level (Fuller 2011b). Limitations of the EIAs include for example that they only address direct effects of construction and operation and not any sourcing issues in the process, that cumulative effects are not addressed and that environmental performance criteria are not set

(Fuller 2011a). Whilst EIAs are a useful tool to manage negative impacts of proposals, strategic environmental assessments (SEAs) support the proactive identification of the best environmental solution (Fuller 2011b).

In assessing the environmental impact and in considering alternative proposals, three other tools are available to policymakers and companies which may be incorporated into EIAs. We will thus briefly introduce life cycle assessment, cost-benefit analysis and risk management, as they are also covered elsewhere in the book, before turning to other guiding tools.

Risk analysis, assessment and management

Risk analysis was introduced in Box 3.6 in connection with the precautionary principle, which, in essence, advocates a risk averse approach to decision-making. At the early stages, EIAs might be used as a business risk management tool in considering alternatives (Fuller 2011a). What EIAs and SEAs do not cover, however, despite some similarities, is the distinction between environmental impacts and the probability of their actual occurrence (Pollard 2011). Environmental risk assessment offers a systematic process to manage risk from or to the environment and enables better decision-making based on robust evidence (ibid.). The tool plays a direct role in how environmental policy is formulated and assessed on economic grounds (Hackett 2011). In economics, expected-utility analysis is a standard tool for working with risk, which is not the same as uncertainty – whilst risk allows some assessment of probabilities, this is not possible for uncertainty (Stern 2006).

Gerrard (2000) points out that there is no universally accepted definition of risk and hazard. There are a range of terms and stages in risk management, which include hazard and hazard identification, risk estimation, consequence analysis, risk assessment, risk characterisation, risk migration, risk monitoring, risk policy evaluation, alongside harm, probability, source, receptor and pathways (Gerrard 2000; Pollard 2011). As with environmental impact assessment, a variety of disciplines are involved and aspects covered range from scientific and technical ones to cultural and social ones with different stakeholders attaching significance to the various factors based on both facts and value judgements (Gerrard 2000). Whichever terms or stages are used, risk analysis can be used effectively by governmental and corporate decision-makers alike (Nemetz 2013), also for communication to stakeholders about risk (Green 2013; Gerrard 2000).

A formalised process of risk assessment is valuable when uncertainty increases and the probability of serious consequences is not very clear. The value lies in a structured approach which establishes a logical basis for the management of risk where a distinction is made between impact and likelihood, decisions are recorded and the decision-making process is transparent (Pollard 2011). The latter is particularly important as public involvement and stakeholder engagement are increasingly expected with regard to decisions on risk (ibid.). The process of risk assessment is therefore as important as the result and the procedure used (Gerrard 2000). Various techniques used in risk assessment include for example event trees, source-pathway-receptor models, conceptual models and risk matrices, though a good starting point is qualitative risk assessment (Pollard 2011).

Horstman and colleagues (2008) point out that innovations related to sustainability need significant investments and can be a considerable risk for companies. How a company deals with risk depends on its attitude to it – whether it is risk-neutral, risk-averse or risk-loving and what it is willing to pay to avoid uncertainty (risk burden versus risk premium, Anderson 2014). This is very much connected to what the expected costs and benefits are for a firm for which there is another, related tool available – cost-benefit analysis, often used in risk management as this involves a balancing act between the cost of risk reduction and the benefit from the amount of risk reduction[7] (Gerrard 2000).

Cost-benefit analysis

Very simply put, cost-benefit analysis (CBA) operationalises the idea that decisions should be taken based on weighing up the (monetary) advantages and disadvantages of a particular course of action (Pearce et al. 1999). It is a tool used by governments as a conceptual framework for assessing competing projects and policies with an environmental impact but may also be used by other stakeholders and decision-makers for information on the consequences of for example regulations or administrative actions that change the current status (Nemetz 2013; Hackett 2011). For each (policy) alternative, the (net) present value of benefits and costs are calculated in CBA (or benefit/cost analysis, as Hackett 2011 calls it). A common decision rule of CBA is that the project with the highest net present value is chosen (Nemetz 2013). Central to CBA is that all costs and benefits of relevance are included and given a monetary value, and that all future costs and benefits are discounted appropriately (Nemetz 2013). The concept of discount rates was introduced in Box 3.5. Its choice is particularly critical regarding policies that address multigenerational sustainability (Hackett 2011). Nemetz (2013) highlights the different types of discount rates that all lead to different results in CBA: personal, corporate and social discount rates of which the latter most take into account future generations whereas the first two are more focused on short-term benefits. Sustainability can be included in CBA by introducing constraints on how natural capital stocks are depleted and degraded, thus integrating environmental values (Pearce et al. 1999). We discussed the difficulties of assigning monetary values to environmental benefits when we covered ecosystem services in the previous chapter. Estimating the environmental benefits in CBA remains an intractable problem (Nemetz 2013) alongside other limitations in CBA, such as the questions of commensurability (not everything is measurable), complexity (making CBA expensive), and intergenerational equity (discount rates), amongst others (Hackett 2011).

Despite the challenges in monetising costs and benefits, a number of methodologies have been developed to estimate benefits. Conventionally, benefits are defined as the value of the environmental damages avoided whereas costs are those paid to control pollution, which are generally easier to estimate (Nemetz 2013). The methodologies for estimating benefits fall broadly under two categories: market-based and non-market based (or demand and non-demand curve based).[8]

Non-market based approaches include dose-response methods, replacement costs, mitigation costs and opportunity costs (Nemetz 2013). Under the marked-based umbrella, monetary valuation methods and techniques can be divided into indirect approaches based on revealed preferences and direct approaches based on stated preferences (Adger 2000). Revealed preference methods include the travel cost method and hedonic pricing method, whereas stated (or expressed) preference methods include primarily contingent valuation as well as conjoint analysis, production functions and benefits transfer (Nemetz 2013). An overview of steps taken in CBA is provided in Table 3.1.

While it is a useful tool that makes explicit what is involved in decisions, it is limited to economic factors only (Adger 2000) and is inappropriate as the single deciding factor as social and environmental impacts need to be integrated as well (Hackett 2011). In a similar line, Nemetz (2013) points out that CBA is biased towards projects with easily quantifiable benefits over projects where values are not easily monetised. To compensate, he suggests supplementing CBA with other decision methodologies, such as for example multi-attribute decision-making, and using biophysical approaches that focus on ecological values to complement preference-based

TABLE 3.1

Steps in cost-benefit analysis

Step 1	Define proposed reallocation or change in resource use
Step 2	Delineate relevant costs and benefits
Step 3	Quantify environmental and physical impacts of proposed or observed change
Step 4	Value relevant effects
Step 5	Discount temporal cost and benefit flows
Step 6	Decide desirability of proposed change and policy recommendations
Step 7	Sensitivity analysis and scaling activities

Source: Adger 2000.

approaches that largely take an anthropocentric perspective (ibid.). Such biophysical approaches take a number of forms, for example ecological footprint (introduced in the previous chapter) and industrial ecology (covered in Chapters 8 and 9), and related to both, life cycle analysis (LCA).

Life cycle analysis

Life cycle analysis or assessment (LCA) was developed in the same time period as ecological footprint tools, offering a method of measuring the impact of a product on the environment through its entire life cycle, from "cradle to grave", comprising energy and material flows and the impacts of extraction, acquisition, production, use and disposal (Robertson 2017). LCA can be used by companies, consumers, activists and policymakers; for example, it allows a firm or industry to identify areas where improvements and cost savings may be achieved or to provide a basis for claims about how environmentally friendly or superior a product is (Hackett 2011). LCA normally consists of two stages, firstly, an inventory and secondly, the impact assessment, although Elkington (1997) includes four stages: initiation, inventory, impact analysis and improvement.

While LCA covers both inputs and outputs, such as water, air, energy and raw materials and then emissions and wastes, it is not the same as input-output analysis, which takes a broader perspective (Nemetz 2013). Input-output analysis addresses one of the drawbacks of LCA, that of where boundaries are drawn within the system analysed. It does so by addressing sector- and economy-wide emissions, pollutants or waste throughout the whole chain of production, not just of one product or company (Nemetz 2013). This drawback of LCA is a challenge to companies, as much of the impact during the life of a product lies outside company borders, which is often recognised through LCA (Laszlo and Zhexembayeva 2011; Robertson 2017) – we touch upon this challenge in the chapters on the supply chain, production and on marketing where we briefly look at consumption, as often the use of a product is responsible for most of its environmental impact. Other disadvantages include its requirement of management time and resources for a complex and challenging accounting process (Laszlo and Zhexembayeva 2011; Robertson 2017).

We will return to LCA in Chapter 9 on production to see how companies actively use it in shifting production from cradle-to-grave to cradle-to-cradle perspectives. In the wider context of this chapter, LCA enables companies to manage sustainability risks (Laszlo and Zhexembayeva 2011) and forms an essential part in corporate strategy and decision-making with regard to sustainability (Nemetz 2013). The cyclical nature of LCA is also reflected in the moves towards a circular economy, which has gained increasing interest within policymaking, industry and research with the help of the Ellen MacArthur Foundation as an alternative to the traditional take-make-waste linear economy (Geissdoerfer et al. 2017; Bocken et al. 2017). We return to the concept in Chapters 7, 8 and 9.

Apart from understanding the whole life cycle of a product, another goal of conducting LCAs is to cut the environmental impact of a product, which can be achieved by reducing for example energy and water use as well as reducing, reusing and recycling materials. The overarching goal is therefore eco-efficiency over the whole life cycle using the so-called eco-efficiency triad of reduce, reuse, recycle (Cavagnaro and Curiel 2012). LCAs help identify priorities (Elkington 1997) and can highlight eco-efficiency gains and support companies in improving their environmental performance whilst also cutting costs (Whitfield and McNett 2014).

Eco-efficiency

Eco-efficiency is generally known as "doing more with less". The concept was introduced by the World Business Council for Sustainable Development (WBCSD 2006) which defines eco-efficiency as

> the delivery of competitively priced goods and services that satisfy human needs and bring quality of life, while progressively reducing ecological impacts and resource intensity throughout the life-cycle to a level at least in line with the Earth's estimated carrying capacity.

This echoes the ecological footprint[9] and carrying capacity concepts covered in the previous chapter and LCA introduced earlier. A principal concern is to create more value with less impact (WBCSD 2006), which can be done by measuring the ratio between a product and its environmental impact, whether it be resource consumption, waste generation or pollutant emissions (Robertson 2017; Fiksel 2012).

Eco-efficiency as a concept can be used not just by companies, but also by individuals, governments and other organisations in order to become more sustainable (WBCSD 2000). It simultaneously addresses economic efficiency and environmental responsibility (Green 2013) and, according to Elkington (1997), is easy for businesses to grasp and manage.

Related to eco-efficiency is improved resource productivity which von Weizsäcker and colleagues (1998) explore in their book *Factor Four* where wealth is doubled and resource use halved (the "efficiency cure" of doing more with less). Similarly, Hawken and colleagues (1999) concentrate on efficiency and resource gains. One particular form of eco-efficiency is industrial ecology, where there is an emphasis on waste reduction and output recycling in order to facilitate closed-loop manufacturing (Dunphy et al. 2003), a theme we return to in Chapters 8 and 9.

Eco-efficiency has been criticised for being perceived by businesses to be enough for moving towards environmental sustainability (Elkington 1997); in particular, the WBCSD is seen to promote eco-efficiency as its "own brand of sustainability" based on its perceived profitability (Anderson 2014: 203). The WBCSD (2006) identifies seven approaches to eco-efficiency (e.g. reducing the material and energy intensity as well as toxic dispersion, and increasing recyclability, durability and service intensity of products and services) which can be achieved across four areas (re-engineering processes, revalorising by-products, redesigning products, and rethinking markets).

Despite the obvious importance of these approaches and the value of eco-efficiency as a necessary condition for sustainable development, it is not sufficient (Elkington 1997). In particular, McDonough and Braungart (2002) argue that it is not a strategy for the long term as it does not reach far or deep enough, as "less bad" is still bad and based on a destructive system, which means that "efficient is not sufficient" (Nemetz 2013: 298). Everard (2011: 51) uses the term "eco-efficient unsustainability" to describe this more efficient linear use of resources that does not take the whole life cycle into account. Blowfield (2013) identifies eco-efficiency as a driver of, as well as a constraint of, corporate engagement with sustainability – when good environmental management means cost savings, companies are motivated to engage, but when costs are greater and no regulatory pressure exists, eco-efficiency is more challenging.

Piasecki and colleagues (1999) also observe – and lament – that corporate environmental efforts have exclusively focused on efficiency at the cost of effectiveness. As an alternative to eco-efficiency, McDonough and Braungart (2002) propose eco-effectiveness which goes beyond eco-efficiency's *doing less harm but not yet doing good* and which moves towards a cradle-to-cradle approach based on nature, discussed in more depth in Chapters 4 and 9.

Whilst eco-efficiency addresses both the economic and environmental part of sustainability, it does not incorporate the social aspect – an omission also recognised by the WBCSD (2006) and one that the so-called triple bottom line addresses.

The triple bottom line

In response to companies identifying sustainability as eco-efficiency and thus comparing sustainability with economising on natural resources, John Elkington (1997) created the concept of the triple bottom line (TBL) in order to capture the full business agenda (Cavagnaro and Curiel 2012; Robertson 2017). Whilst eco-efficiency is an important stepping stone towards sustainability, it leaves other areas unaddressed. The TBL encapsulates the three pillars of sustainability – environment, economics and equity (or ecology, economics, equality) at a societal level – which at a corporate level equate to the 3Ps of people, planet and profit. Using the business metaphor of the bottom line (economic profit) and extending it to social and environmental bottom lines helps companies to measure and compare sustainability performance (Whitfield and McNett 2014). Although for example the WCED incorporated all three aspects into its work on sustainable development, in the context of corporations, Porter and van der Linde (1995b) are considered to be early explorers of the idea that companies gain a sustainable competitive advantage by including financial but also social and environmental considerations and goals into their strategic planning (Nemetz 2013).

The TBL reflects some of the thinking underpinning the five capitals model in that a business needs to build up and manage different kinds of capital in order to be sustainable (Blowfield 2013). Although this could be seen as a balancing act that involves trade-offs, Elkington (1997) emphasises that the three dimensions are all equally important, and sustainable organisations pursue activities that have a positive impact on all 3Ps (Cavagnaro and Curiel 2012). There is thus a recognition of interconnectedness, a theme running through sustainability. Adopting the TBL often involves multiple stakeholder engagement in identifying what environmental and social performance criteria are (Dunphy et al. 2003).

In the context of this chapter where we have covered the regulatory framework in detail, Elkington (1997) emphasises the importance of regulation in ensuring effective markets that allow the pursuit of TBL objectives. The TBL approach enables companies to internalise external TBL costs on a voluntary basis, but regulation is needed for all companies to follow suit.

As we identified in the previous chapter, valuing environmental services is challenging, and this is one of the criticisms connected with the TBL approach, too, as there are no standard measurements for the "non-traditional" social and environmental elements of the TBL (Molthan-Hill

2017: 89). Although TBL accounting and reporting has been adopted in one form or another (Nemetz 2013), Blowfield (2013: 229) argues that it is best seen as a "framing device" to understand business relationships rather than as a foundation for a new kind of accounting. Whilst we will discuss accounting for sustainability in Chapter 12, we will return to TBL in the next chapter when we look at the seven revolutions Elkington (1997) identifies for sustainable corporations. As the next chapter covers sustainable business models and strategies, it is worth noting that McDonough and Braungart (2002) suggest a triple *top* line approach as a strategic tool to address all 3Ps up front rather than after the fact, which often ends up focusing on economic consideration with social or environmental benefits considered as an afterthought rather than in equal measures.

The Natural Step and systems conditions

Similarly to the TBL, which is based on many of the same tenets as the Brundtland report, The Natural Step was launched two years after the report, setting out four systems conditions or principles for sustainability (Elkington 1997; Blowfield 2013). It is based on the work of Swedish cancer scientist Karl-Henrik Robèrt, who used the agreement on the basics of the smallest unit of life, the cell, as a platform to understand what makes life possible and what is the root cause of environmental damage (Elkington 2012; Hawken 1993). In asking systemic questions, The Natural Step allows consensual agreement from science, industry, politics, environmental organisations and even religion (Hawken 1993). Drawing on the laws of thermodynamics (see previous chapter), four systems conditions help make sense of the complexity of sustainability by translating ecological principles into action at the individual, organisational and community levels (Nemetz 2013; Cavagnaro and Curiel 2012).

We will cover the precise steps of the framework in the next chapter in the context of sustainable business models, but we outline the systems conditions here as they can be used as general guidance not only by companies but also by regulators (as happened with The Natural Step when it was first used in Sweden). The systems conditions cover fossil fuels and minerals, toxins, resource extraction as well as equity and economics (Robertson 2017), and require the elimination of humanity's contribution to the progressive

- build-up of substances extracted from earth (e.g. metals, fossil fuels)
- build-up of chemicals and compounds produced by society (e.g. PCBs, dioxins)
- physical degradation and destruction of nature and natural processes (e.g. deforestation, habitat loss)
- conditions that undermine people's capacity to meet basic human needs (e.g. unsafe working conditions) (Blowfield 2013; Robèrt 2002).

The result is a sustainable society where flows of matter are balanced and natural cycles define the limits within which society operates (Robèrt 2002), allowing organisations to manage their social and environmental impact.

Mission zero

In some way or another, the culmination of the principles, regulations, foundations and guidelines is found in the concept of zero – sustainability means moving towards zero waste, zero emissions, zero pollution, in other words towards a footprint of zero. While these targets could – rightly – be considered physically impossible goals (von Weizsäcker et al. 1998), which few companies have hope of reaching, they set "stretch" goals without which sustainability might not be possible (Elkington 1997). The term "mission zero" stems from the carpet company Interface which, led by Ray Anderson after having read Paul Hawken's *Ecology of Commerce* (1993), applied The Natural Step framework in its quest for sustainability, equating the achievement of mission zero with 100% sustainability (Anderson 2009, see also Chapter 4).

Mission zero involves dematerialisation which can be achieved by applying cradle-to-cradle principles (Cavagnaro and Curiel 2012, see Chapter 9 later in the book for such life cycle approaches). This underpins for example the zero waste concept where waste of one organisation is considered as a potential resource or input for another one, using the cyclical model of natural systems (Robertson 2017). For companies using total quality management, zero defects, zero inventory and zero accidents are already familiar concepts, from which it is then another step to zero waste, zero emissions and zero pollution. Elkington (2012) explores these connections further in his book *The Zeronauts,* which details how to achieve a zero-impact growth economy.

The role of the regulatory framework with regard to mission zero can for example be seen in California's Zero Emission Vehicle (ZEV) requirements which require a certain percentage of cars produced to be ZEVs (Anderson 2014), though carmakers can offset a lack of ZEVs in their portfolio through credit balances. Other applications of the zero concept include e.g. zero carbon buildings or zero carbon retail (Blowfield 2013).

We return to the concepts and mission of zero carbon, zero pollution and zero waste in the second part of the book when we look at the environmental challenges of climate change, pollution and waste in chapters 5, 6 and 7.

SUMMARY

In this chapter, a great deal of material was necessarily covered. The complexities and interdependencies inherent in (environmental) sustainability require knowledge of a broad range of areas for budding sustainability professionals. As they are faced with difficult decisions and uncertain outcomes, it was important to cover the "natural" tensions integral in the field before discussing how the regulatory framework handles the perceived tension between competitiveness and environmental protection, which some companies use to their advantages by becoming environmental leaders. As all companies have to operate within the law, a basic understanding of the principles and types of environmental regulations is important whilst also managing relationships with legislators who may value corporate perspectives in designing feasible and ambitious policies.

Both companies and policymakers can draw on common concepts and tools that were covered in the last part of the chapter. With the tools and guidelines introduced there, it is important to keep in mind that no single method can provide a comprehensive analysis or complete overview, and it is therefore important to triangulate tools used to work towards environmental sustainability (Nemetz 2013). Beyond tools and techniques, what is also required, is a shift in thinking, as e.g. explored regarding the role of technology – shifting it from part of the problem to part of the solution and e.g. with regard to environmental regulations – viewing them as opportunities for competitiveness and environmental leadership instead of as costs, thus solving those tensions through re-thinking them. We will come across other shifts in thinking throughout the book, also in the next chapter which provides more applied, in-depth discussions of how companies can move towards environmental sustainability.

SUGGESTED QUESTIONS AND ACTIVITIES

- Do you agree with the Porter hypothesis? Why/why not?
- Should ecosystem services be monetised through the state or regulated differently?
- Are market-based systems better than government regulations at achieving sustainability?
- Is the precautionary principle needed? Is the precautionary principle a sound approach to risk analysis?
- What philosophical approach do you take in your business? A technocentric or ecocentric one or a combination?
- Value reflection: which values and philosophical assumptions guide your decisions?
- Watch this video that explains sustainability with simple natural science: https://youtu.be/eec0UYGIeo4 – which framework covered in the chapter does it relate to?

NOTES

1 It is worth noting here that other perspectives and terms exist in this spectrum. For example, O'Riordan (1981) puts deep ecology under ecocentrism and cornucopian views under technocentrism – the former a moral philosophy developed in response to shallow ecology by Arne Naess (Anderson 2014), the latter an essentially optimistic view of technology and largely business as usual (O'Riordan 1981). Callicott (2012) contrasts an Abrahamic/biblical worldview with Leopold's evolutionary-ecological worldview and resulting land ethic. Daly and Cobb (1989: 379, 381) call Leopold's view "biocentric" and Lovelock's earth system thinking "geocentric". The authors acknowledge Christianity's and in particular Protestantism's role in anthropocentrism and develop a *theocentrist* underpinning of biospheric perspectives as an alternative. Dunphy and colleagues (2007) use the terms cosmocentric and egocentric, which parallel ecocentric and anthropocentric. Another perspective is provided by ecofeminism (coined by d'Eaubonne) which equates environmentalism with feminism, associating anthropocentrism with androcentrism (man rather than humankind at the centre) and thus links the domination of women in patriarchy with the exploitation of nature (Anderson 2014; Green 2013).

2 Meadows et al. (2004) draw the link between the IPAT equation and the ecological footprint discussed in the previous chapter – each term in the equation reveals ways in and extents to which the ecological footprint can be reduced: affluence is determined by consumption and thus the material, energy and emissions associated with it and technology can be seen as the energy and material needed to enable such consumption.

3 See Easton (2014, unit 1.1) for an in-depth discussion of both perspectives of the precautionary principle.

4 For details on how pollution taxes are calculated (based on marginal external costs), Hackett (2011) as well as Anderson (2014) or other environmental economics textbooks provide more insight. Daly and Cobb (1989) point out that in the case of positive externalities, a Pigouvian tax would be negative and therefore constitute a subsidy.

5 Friedman (2011) highlights that going beyond compliance is challenging when complex laws and regulations make it difficult to define what constitutes compliance. With the development of sophisticated instruments for detecting pollution, the difference between *detectable* quantities and *harmful* quantities might not be recognised or understood by the public, thus creating challenges for regulators trying to apply the precautionary principle and difficulties for companies trying to meet regulations (ibid.). Other terms used in this context are (socially) *optimal* quantities of pollution and *zero* pollution, the latter of which Anderson (2014) points out is neither optimal nor efficient.

6 It could be argued that the Pareto principle introduces fairness considerations into economic policy analysis (Costanza et al. 1997; Hackett 2011).

7 Nemetz (2013) points out that risk-benefit analysis is a modification of cost-benefit analysis that more explicitly accounts for risk.

8 This only serves as an overview of the approaches; more detail can be found in any of the sources used here.

9 In the context of companies, von Weizsäcker and colleagues (1998) explore the concept of "ecological rucksack", akin to ecological footprint, but more focused on the material intensity per service.

SOURCES AND FURTHER READING

Adger, W. N. (2000) Environmental and Ecological Economics, in: O'Riordan, T. (ed.) *Environmental Science for Environmental Management*, Harlow: Pearson, 2nd ed., pp. 93–118.

Ambec, S., Cohen, M. A., Elgie, S. and Lanoie, P. (2011) The Porter Hypothesis at 20: Can Environmental Regulation Enhance Innovation and Competitiveness? Resources for the Future, Discussion Paper, RFF DP 11–01.

Anastasiadis, S. (2014) Toward a View of Citizenship and Lobbying: Corporate Engagement in the Political Process, in: *Business and Society*, 53(2): 260–299.

Anderson, D. A. (2014) *Environmental Economics and Natural Resource Management*, London: Routledge, 4th ed.

Anderson, R. C. (1998) *Mid-Course Correction. Toward a Sustainable Enterprise: The Interface Model*, White River Junction, VT: Chelsea Green.

Anderson, R. C. and White, R. (2009) *Confessions of a Radical Industrialist: How Interface Proved that You Can Build a Successful Business Without Destroying the Planet*, London: Random House.

Aragón-Correa, J. A., Hurtado-Torres, N., Sharma, S. and García-Morales, V. J. (2008) Environmental Strategy and Performance in Small Firms: A Resource-Based Perspective, in: *Journal of Environmental Management*, 86(1): 88–103.

Aragón-Correa, J. A. and Sharma, S. (2003) A Contingent Resource-Based View of Proactive Corporate Environmental Strategy, in: *Academy of Management Review*, 28(1): 71–88.

Arrow, K., Bolin, B., Costanza, R., Dasgupta, P., Folke, C., Holling, C. S., Jansson, B.-O., Levin, S., Mäler, K.-G., Perrings, C. and Pimentel, D. (1995) Economic Growth, Carrying Capacity, and the Environment, in: *Science*, 268(5210): 520–521.

Beck, U. (1992) *Risk Society: Towards a New Modernity*. London: Sage.

Bigg, M. (2011) Pollution Prevention and Control, in: Brady, J., Ebbage, A. and Lunn, R. (eds.) *Environmental Management in Organizations, The IEMA Handbook*, London: Earthscan/Routledge, 2nd ed., pp. 212–238.

Blowfield, M. (2013) *Business and Sustainability*, Oxford: Oxford University Press.

Bocken, N. M. P., Olivetti, E. A., Cullen, J. M., Potting, J. and Lifset, R. (2017) Taking the Circularity to the Next Level: A Special Issue on the Circular Economy, in: *Journal of Industrial Ecology*, 21(3): 476–482.

Brennan, T. J. (1995) Discounting the Future: Economics and Ethics, in: *Resources for the Future*, Summer, 120: 3–6.

Callicott, J. B. (2012) A Sand County Almanac: An Evolutionary-Ecological Worldview, in: Vaz, S. G. (ed.) *Environment – Why Read the Classics?* Sheffield: Greenleaf Publishing, pp. 36–71.

Carrington, D. (2016) The Anthropocene Epoch: Scientists Declare Dawn of Human-Influenced Age, *The Guardian*, 29.08.2016, available at www.theguardian.com/environment/2016/aug/29/declare-anthropocene-epoch-experts-urge-geological-congress-human-impact-earth (Retrieved, 31.03.2017).

Cavagnaro, E. and Curiel, G. (2012) *The Three Levels of Sustainability*, Sheffield: Greenleaf Publishing.

Chang, H-J. (2014) *Economics: The User's Guide*, London: Penguin Books.

Costanza, R., Cumberland, J., Daly, H., Goodland, R. and Norgaard, R. (1997) *An Introduction to Ecological Economics*, Boca Raton, FL: St Lucie Press.

Crane, A. and Matten, D. (2016) *Business Ethics*, Oxford: Oxford University Press, 4th ed.

Crane, A., Matten, D., Glozer, S. and Spence, L. (2019) *Business Ethics*, Oxford: Oxford University Press, 5th ed.

Daly, H. E. and Cobb, J. B. (1989) *For the Common Good: Redirecting the Economy towards Community, the Environment, and a Sustainable Future*, Boston: Beacon Press.

Dasgupta, S., Laplante, B., Wang, H. and Wheeler, D. (2002) Confronting the Environmental Kuznets Curve, in: *The Journal of Economic Perspectives*, 16(1): 147–168.

Davison, N. (2019) The Anthropocene Epoch: Have We Entered a New Phase of Planetary History? *The Guardian*, 30.05.2019, available at www.theguardian.com/environment/2019/may/30/anthropocene-epoch-have-we-entered-a-new-phase-of-planetary-history (Retrieved, 02.11.2019).

Dietz, T. and Rosa, E. A. (1994) Rethinking the Environmental Impacts of Population, Affluence and Technology, in: *Human Ecology Review*, 1(2): 277–300.

Dinda, S. (2004) Environmental Kuznets Curve Hypothesis: A Survey, in: *Ecological Economics*, 49(4): 431–455.

Dunphy, D., Griffiths, A. and Benn, S. (2003) *Organizational Change for Corporate Sustainability: A Guide for Leaders and Change Agents of the Future*, London: Routledge.

Dunphy, D., Griffiths, A. and Benn, S. (2007) *Organizational Change for Corporate Sustainability: A Guide for Leaders and Change Agents of the Future*, London: Routledge, 2nd ed.

Easton, T. (ed.) (2014) *Taking Sides: Clashing Views on Environmental Issues*, Dubuque, Iowa: McGraw-Hill, 16th ed.

EEA (2001) Late Lessons from Early Warnings: The Precautionary Principle 1896–2000, Environmental issue report No 22/2001, European Environment Agency, available at www.eea.europa.eu/publications/environmental_issue_report_2001_22 (Retrieved 09.04.2017).

Ehrlich, P. R. and Holdren, J. P. (1971) Impact of Population Growth, in: *Science*, 171(3977): 1212–1217.

Elkington, J. (1997) *Cannibals with Forks: The Triple Bottom Line of 21st-Century Business*, Oxford: Capstone.

Elkington, J. (2012) *The Zeronauts: Breaking the Sustainability Barrier*, London: Routledge.

Elliott, J. A. (2013) *An Introduction to Sustainable Development*, London: Routledge, 4th ed.

Everard, M. (2011) Sustainability and Sustainable Development, in: Brady, J., Ebbage, A. and Lunn, R. (eds.) *Environmental Management in Organizations, The IEMA Handbook*, London: Earthscan/ Routledge, 2nd ed., pp. 39–61.

Farzin, Y. H. (2003) The Effects of Emissions Standards on Industry, in: *Journal of Regulatory Economics*, 24(3): 315–327.

Fiksel, J. (2012) *Design for Environment*, New York: McGraw-Hill, 2nd ed.

Friedman, F. B. (2011) *Practical Guide to Environmental Management*, Washington, DC: Environmental Law Institute, 11th ed.

Fuller, K. (2011a) Environmental Impact Assessment, in: Brady, J., Ebbage, A. and Lunn, R. (eds.) *Environmental Management in Organizations, The IEMA Handbook*, London: Earthscan/ Routledge, 2nd ed., pp. 469–481.

Fuller, K. (2011b) Strategic Environmental Assessment, in: Brady, J., Ebbage, A. and Lunn, R. (eds.) *Environmental Management in Organizations, The IEMA Handbook*, London: Earthscan/ Routledge, 2nd ed., pp. 483–493.

Geissdoerfer, M., Savaget, P., Bocken, N. M. P. and Hultink, E. J. (2017) The Circular Economy – A New Sustainability Paradigm? in: *Journal of Cleaner Production*, 143(1): 757–768.

Gerrard, S. (2000) Environmental Risk Management, in: O'Riordan, T. (ed.) *Environmental Science for Environmental Management*, Harlow: Pearson, 2nd ed., pp. 435–468.

Green, T. (2013) *The Dictionary of Global Sustainability. With Case Studies on the Environmental, Economic, Technological, and Social Aspects of Sustainability*, New York: McGraw-Hill.

Grossman, G. M. and Krueger, A. B. (1995) Economic Growth and the Environment, in: *The Quarterly Journal of Economics*, 110(2): 353–377.

Hackett, C. (2011) *Environmental and Natural Resource Economics*, London: Sharpe, 4th ed.

Hart, S. L. (1995) A Natural-Resource-Based View of the Firm, in: *Academy of Management Review*, 20(4): 986–1014.

Harvey, F. (2016) Four of World's Biggest Cities to Ban Diesel Cars from Their Centres, *The Guardian*, 02.12.2016, available at www.theguardian.com/environment/2016/dec/02/four-of-worlds-biggest-cities-to-ban-diesel-cars-from-their-centres (Retrieved, 11.04.2017).

Hawken, P. (1993) *Ecology of Commerce: A Declaration of Sustainability*, New York: Harper Business.

Hawken, P., Lovins, A. B. and Lovins, L. H. (1999) *Natural Capitalism: The Next Industrial Revolution*, London: Earthscan.

Helm, D. (2003) The Assessment: Climate-Change Policy, in: *Oxford Review of Economic Policy*, 19(3): 349–361.

Hilliard, R. (2004) Conflicting Views: Neoclassical, Porterian, and Evolutionary Approaches to the Analysis of the Environmental Regulation of Industrial Activity, in: *Journal of Economic Issues*, 38(2): 509–517.

Holdren, J. P. and Ehrlich, P. R. (1974) Human Population and the Global Environment: Population Growth, Rising Per Capita Material Consumption, and Disruptive Technologies Have Made Civilization a Global Ecological Force, in: *American Scientist*, 62(3): 282–229.

Horstman, N. A. M., de Bakker, F. G. A., Masurel, E. and van Hemert, P. P. (2008) Monkey See, Monkey Do? Some Observations on Sustainable Innovations in Zoos, in: Wankel, C. and Stoner, J. A. F. (eds.) *Innovative Approaches to Global Sustainability*, New York: Palgrave Macmillan, pp. 105–122.

Huesemann, M. H. and Huesemann, J. A. (2011) *Techno-Fix: Why Technology Won't Save us or the Environment*, Gabriola Island: New Society Publishers.

Jacquet, J. (2015) *Is Shame Necessary? New Uses for an Old Tool*, London: Allen Lane.

Kedia, B. L., Clampit, J. and Gaffney, N. (2011) Institutions, MNEs, and Sustainable Development, in: Jain, S. C. and Kedia, B. L. (eds.) *Enhancing Global Competitiveness Through Sustainable Environmental Stewardship*, Cheltenham: Edward Elgar, pp. 70–108.

Köppl, P. (2005) Lobbying, in: Althaus, M., Geffken, M. and Rave, S. (eds.) *Handlexikon Public Affairs*, Münster: LIT Verlag, pp. 191–195.

Kuznets, S. (1955) Economic Growth and Income Inequality, in: *The American Economic Review*, 45(1): 1–28.

Landrum, N. E. (2018) Stages of Corporate Sustainability: Integrating the Strong Sustainability Worldview, in: *Organization & Environment*, 31(4): 287–313.

Laszlo, C. and Zhexembayeva, N. (2011) *Embedded Sustainability: The Next Big Competitive Advantage*, Sheffield: Greenleaf Publishing.

Lees, C. (2008) Environmental Policy in the United Kingdom and Germany, in: Green, S. and Turner, E. (eds.) *Policy Convergence in the UK and Germany: Beyond the Third Way?* pp. 164–183.

Mahoney, C. (2007) Lobbying Success in the United States and in the European Union, in: *Journal of Public Policy*, 27(1): 35–56.

Marcus, A., Aragón-Correa, J. A. and Pinkse, J. (2011) Firms, Regulatory Uncertainty, and the Natural Environment, in: *California Management Review*, 54(1): 5–16.

McDonough, W. and Braungart, M. (2002) *Cradle to Cradle: Remaking the Way We Make Things*, New York: North Point Press.

McWilliams, A., Van Fleet, D. D. and Cory, K. D. (2002) Raising Rivals' Costs through Political Strategy: An Extension of Resource-Based Theory, in: *Journal of Management Studies*, 39(5): 707–723.

Meadows, D. H., Meadows, D. L., Randers, J., Behrens, W. W. III (1972) *The Limits to Growth, A Report for The Club of Rome's Project on the Predicament of Mankind*, London: Earth Island.

Meadows, D. H., Randers, J. and Meadows, D. L. (2004) *Limits to Growth: The 30-Year Update*, White River Junction, VT: Chelsea Green.

Molthan-Hill, P. (ed.) (2017) *The Business Student's Guide to Sustainable Management*, Sheffield: Greenleaf Publishing, 2nd ed.

Nemetz, P. N. (2013) *Business and the Sustainability Challenge: An Integrated Perspective*, London: Routledge.

OECD (1975) *The Polluter Pays Principle: Definition, Analysis, Implementation*, Organisation for Economic Cooperation and Development, Paris: OECD.

OECD (1994) Economic Instruments in Environmental Policy: Lessons from the OECD Experience and their Relevance to Developing Economies, Organisation for Economic Cooperation and Development, Working Paper No. 92 by J.-P. Barbe, OCDE/GD(93)193, Paris: OECD.

OECD (1995) *Environmental Principles and Concepts*, Organisation for Economic Cooperation and Development, OCDE/GD(95)124, Paris: OECD.

OECD (2010) *Taxation, Innovation and the Environment*, Organisation for Economic Cooperation and Development, Paris: OECD.

O'Riordan, T. (1981) Environmentalism and Education, in: *Journal of Geography in Higher Education*, 5(1): 3–17.

O'Riordan, T. (2000) Environmental Science on the Move, in: O'Riordan, T. (ed.) *Environmental Science for Environmental Management*, Harlow: Pearson, 2nd ed., pp. 1–27.

Palmer, K., Oates, W. E. and Portney, P. R. (1995) Tightening Environmental Standards: The Benefit-Cost or the No-Cost Paradigm? in: *Journal of Economic Perspectives*, 9(4): 119–132.

Pearce, D., Markandya, A. and Barbier, E. (1999) *Blueprint for a Green Economy*, London: Earthscan (first published 1989).

Piasecki, B. W. and Fletcher, K. A. (1996) The Changing Face of the Environmental Strategist, in: *Corporate Environmental Strategy*, 3(4): 25–30.

Piasecki, B. W., Fletcher, K. A. and Mendelson, F. J. (1999) *Environmental Management and Business Strategy: Leadership Skills for the 21st Century*, New York: John Wiley & Sons.

Pollard, S. (2011) Environmental Risk Governance, in: Brady, J., Ebbage, A. and Lunn, R. (eds.) *Environmental Management in Organizations, The IEMA Handbook*, London: Earthscan/Routledge, 2nd ed., pp. 495–509.

Porter, M. E. (1991) America's Green Strategy, in: *Scientific American*, 264(4): 168.

Porter, M. E. and Kramer, M. (2006) Strategy and Society: The Link Between Competitive Advantage and Corporate Social Responsibility, in: *Harvard Business Review*, 84(12): 78–92.

Porter, M. E. and Reinhardt, F. L. (2007) A Strategic Approach to Climate, in: *Harvard Business Review*, 85(10): 22–26.

Porter, M. E. and van der Linde, C. (1995a) Toward a New Conception of the Environment-Competitiveness Relationship, in: *Journal of Economic Perspectives*, 9(4): 97–118.

Porter, M. E. and van der Linde, C. (1995b) Green and Competitive: Ending the Stalemate, in: *Harvard Business Review*, 73(5): 120–134.

Reinmoeller, P. (2012) Embedding Corporate Sustainability as a Knowledge-Creation Journey, in: Grayson, D. and Exter, N. (eds.) *Cranfield on Corporate Sustainability*, Sheffield: Greenleaf Publishing, pp. 27–52.

Robèrt, K.-H. (2002) *The Natural Step Story: Seeding a Quiet Revolution*, Gabriola Island: New Society Publishers.

Robertson, M. (2017) *Sustainability Principles and Practice*, London: Routledge, 2nd ed.

Rugman, A. M. and Verbeke, A. (1998) Corporate Strategies and Environmental Regulations: An Organizing Framework, in: *Strategic Management Journal*, 19(4): 363–375.

Russo, M. V. and Fouts, P. A. (1997) A Resource-Based Perspective on Corporate Environmental Performance and Profitability, *Academy of Management Journal*, 40(3): 534–559.

Sandel, M. (2012) *What Money Can't Buy: The Moral Limits of Markets*, London: Penguin Books.

Scientific American (2015) Do Plastic Bag Bans Work? available at www.scientificamerican.com/article/do-plastic-bag-bans-work/ (Retrieved 12.04.2017).

SEHN (1998) Wingspread Conference on the Precautionary Principle, Science and Environmental Health Network (SEHN), available at http://sehn.org/wingspread-conference-on-the-precautionary-principle/ (Retrieved 09.04.2017).

Shrivastava, P. (1995) Ecocentric Management for a Risk Society, in: *The Academy of Management Review*, 20(1): 118–137.

Shrivastava, P., Allan, D. E. and Hiller, T. B. (2008) Designing Undergraduate Education on "Managing for Sustainability", in: Wankel, C. and Stoner, J. A. F. (eds.) *Innovative Approaches to Global Sustainability*, New York: Palgrave Macmillan, pp. 159–173.

Simpson, R. D. and Bradford, R. L. I. (1996) Taxing Variable Cost: Environmental Regulation as Industrial Policy, in: *Journal of Environmental Economics and Management*, 30(3): 282–300.

Skjærseth, J. B., Stokke, O. S. and Wettestad, J. (2006) Soft Law, Hard Law, and Effective Implementation of International Environmental Norms, in: *Global Environmental Politics*, 6(3): 104–120.

Stern, D. I. (2004) The Rise and Fall of the Environmental Kuznets Curve, in: *World Development*, 32(8): 1419–1439.

Stern, N. (2006) *The Economics of Climate Change: The Stern Review*, Cambridge, UK: Cambridge University Press.

Stookes, P. (2011) Overview of the Law and International Legislation, in: Brady, J., Ebbage, A. and Lunn, R. (eds.) *Environmental Management in Organizations, The IEMA Handbook*, London: Earthscan/Routledge, 2nd ed., pp. 71–81.

Suder, G. (2011) *Doing Business in Europe*, London: Sage, 2nd ed.

van Calster, G. and Reins, L. (2017) *EU Environmental Law*, Cheltenham: Edward Elgar.

van Tulder, R. and van der Zwart, A. (2006) *International Business-Society Management: Linking Corporate Responsibility and Globalization*, London: Routledge.

Viederman, S. (1996) Sustainability's Five Capitals and Three Pillars, in: Pirages, D. (ed.) *Building Sustainable Societies: A Blueprint for a Post-Industrial World*, Armonk, NY: M.E. Sharpe.

von Weizsäcker, E., Lovins, A. B. and Lovins, L. H. (1998) *Factor Four: Doubling Wealth, Halving Resource Use*, London: Earthscan.

Wagner, S. M. (2011) The Corporate Political Activities of Multinational Enterprises: The Automotive Industry and Environmental Regulations in the European Union, PhD Thesis, Loughborough University.

Walley, N. and Whitehead, B. (1994) It's Not Easy Being Green, in: *Harvard Business Review*, 72(3): 46–52.

WBCSD (2000) Measuring Eco-efficiency: A Guide to Reporting Company Performance, World Business Council for Sustainable Development, available at www.bcsd.org.tw/sites/default/files/node/domain_tool/110.file.128.pdf (Retrieved 25.04.2017).

WBCSD (2006) Eco-efficiency Learning Module, World Business Council for Sustainable Development, available at www.wbcsd.org/Projects/

Education/Resources/Eco-efficiency-Learning-Module (Retrieved 02.11.2019).

WCED (1987) *Our Common Future: World Commission on Environment and Development*, Oxford: Oxford University Press.

Whitfield, R. and McNett, J. (2014) *A Primer on Sustainability: In the Business Environment*, New York: Business Expert Press.

Wurzel, R. K. W. (2002) *Environmental Policy-Making in Britain, Germany and the European Union*, Manchester: Manchester University Press.

York, R., Rosa, E. A. and Dietz, T. (2002) Bridging Environmental Science with Environmental Policy: Plasticity of Population, Affluence, and Technology, in: *Social Science Quarterly*, 83(1): 18–34

CHAPTER 4

Strategies and business models for sustainability

INTRODUCTION

Strategy is originally a military term but has been adopted by business, which has broadened its meaning, making it a difficult concept to pin down. Whilst there are plenty of definitions of what strategy is or should be, whether it is a plan, ploy, pattern, position or perspective (Blowfield 2013), in the context of sustainability, Laszlo and Zhexembayeva (2011) emphasise strategy as a learning process above all else. This fits well with the notion that there is not a right or wrong framework for sustainability as it is always a work in progress through which a company learns its way and develops its strategy towards mission zero. There are also different understandings of what a sustainability strategy or strategy for sustainability means – Blowfield (2013) for example offers three interpretations:

- Sustainability strategies where the main focus is the sustainability of the company (which responds to external threats) with sustainability consisting of a number of ad hoc issues for the company to address. Sustainability is an add-on to existing strategy.
- Sustainability integrated into the overall corporate strategy where the external environment presents a coherent entity of interconnected environmental and social factors. The company looks for the sustainability sweetspot where business objectives overlap with sustainability benefits.
- Sustainability as the determinant of change, central to the vision and direction of the company. The focus shifts beyond minimising environmental impact to making a positive contribution for example by maintaining and expanding natural capital stocks.

We will see these interpretations reflected in the models and strategies we cover in this chapter. Whichever way a company approaches sustainability in its strategy and chooses to address the tensions inherent in sustainability discussed in the previous chapter – which we will come across again – if a business builds its strategy without taking into account the changing realities of its environment, it risks failure.

In the following we do not provide all the answers (sorry!) but give an overview of tools and techniques to devise strategies for an environmentally sustainable business. Both conventional and sustainability-focused frameworks are introduced before we look at how change can be achieved in a company. The inherent tensions of sustainability are acknowledged, and emphasis is placed on the changes and shifts needed under good leadership.

SUSTAINABILITY AT WORK

The shift to sustainable strategic management begins with a shift in thinking about the basic relationship between the economy, society and the natural environment (Stead and Stead 2014). Whilst recognising the holistic and systemic nature of sustainability, here we focus on the relationship between business and the natural environment. What a systems perspective enables us to do is to resolve the contradictions and tensions between economy and ecology or profit and planet into mutually reinforcing pressures moving business forward towards sustainability.

This shift in thinking includes changing the question from "does it pay to be green" to "when and how does it pay to be green"[1] (Russo and Minto 2012, based on Reinhardt 1998), moving from reactive compliance to proactive prevention, from seeing sustainability as a cost to seeing it as an opportunity. Other tensions that might not necessarily need a clear answer but that can guide the move towards sustainability are formulated by Laszlo and Zhexembayeva (2011) as follows:

- Growth or no growth? Is growth in itself sustainable or is there a limit to growing ourselves out of problems?
- Stopping the bad or creating the good? Is business part of the problem that needs to be regulated or is the market part of the solution that needs to be harnessed?
- Evolution or revolution? What kind of change is needed?
- Fear or enlightened self-interest? What motivates business to change – fear of disaster or a positive vision of the future?

Laszlo and Zhexembayeva (2011) advocate a both/and perspective of whole systems, where there are no wrong answers to these questions, which means that each organisation has to find its own way to sustainability, based on available frameworks and tools.

The questions asked here point to some fundamental assumptions underlying sustainability and business: systems thinking, values reflection and philosophical assumptions, which Molthan-Hill and colleagues (2017) call key sustainability literacy skills as applied to business and management. We have already touched upon systems thinking as key to sustainability – it encourages acceptance of complexity and helps to deal with the interconnectedness of sustainability issues in the context of multiplicity, ambiguity, interdependence and flux (Whitfield and McNett 2014). Although managers are reluctant to talk about the moral dimensions of their work, business is not value-neutral but often involves negotiating between established values (or objectives, such as profit) and new or different values (e.g. ethics or environmental protection). The Giving Voice to Values initiative helps to develop this sustainability literacy skill (Molthan-Hill et al. 2017). Values reflection forms part of understanding philosophical assumptions, which we covered briefly in the previous chapter. These may be explicit or implicit, and discussing these could be starting points to reflect on existing tools and how these may be used for sustainability.

In light of the questions and tensions mentioned earlier and the acknowledgement that for some of these questions there is no clear answer, we are reminded that the sustainability profession is an evolving one, with changing job descriptions and positions being created by practitioners as they go along (Robertson 2017). It is a profession that requires both broad, systems-level knowledge as well as specific technical skills, underpinned by communication skills that are essential in work that involves collaboration across the whole organisation. Putting sustainability initiatives into practice involves goal definition, measurement and reporting, and the implementation of change (Robertson 2017), all of which involve decisions regarding for example scale (low-hanging fruit versus large scale), monitoring (what's working, what isn't, what are the priorities), indicators, what type of change and so on. As strategies and tactics are developed, frameworks can provide common language and understanding. In this chapter we introduce a number of frameworks that may be used to guide decision-making on strategy and models whilst other frameworks are more concerned with standards and principles (see e.g. the previous chapter) or with assessments and measurements for example that will be covered in part two and three of the book.

We will start with the general level of strategic management. Whilst moves towards environmental sustainability often start at a strategic and high level, sustainable business can only become a reality if enacted across all levels and functions in an organisation, which will be covered in subsequent chapters in part three.

SUSTAINABLE STRATEGIC MANAGEMENT

Traditionally, there have been two schools of thought in strategic management, outside-in and inside-out perspectives, which we briefly referred to in the previous chapter in terms of how companies interact with their environment. For strategic management, the question is whether competitive advantage is based on internal resources and strengths, as advocated by the resource-based view (e.g. Barney 1991), or whether it is based on positioning vis-à-vis external threats and opportunities, as maintained by the industrial organisations model (e.g. Porter 1980).[2]

Sustainable strategic management goes beyond conventional or traditional strategy management as it is integrative (Stead and Stead 2014), combining outside-in and inside-out approaches. It therefore needs to take into account both internal capabilities and competencies as well as the external environment, which includes not just the natural environment, but also the economic, competitive, political, legal, technological, social and cultural environment.

Whilst we may draw on or refer to established frameworks of strategy, the emphasis here is on models and frameworks that move a company towards environmental sustainability. This involves the use of conventional frameworks but is not limited to them.

The strategy process often begins with strategic analysis regarding the external and internal environments of a firm before formulating the vision, mission and goals. This is followed by strategy formulation and then by strategy implementation, which often requires organisational change under strategic leadership (Stead and Stead 2014).

Although the strategy process is often hierarchical, all levels of an organisation are part of strategy – enterprise level, corporate level and down to business unit and functional level. "Level" thinking is crucial to sustainability – across the formal organisational levels as in the previous sentence but also in terms of the spatial level – at the local, factory level, regional level, national, subsidiary level or global, headquarter level. This echoes the previous chapter in which we explored how environmental regulations are designed and implemented across all of these levels, and reminds us of the first chapter where the societal, organisational and individual levels illustrated the interdependencies and interlinkages of sustainability (Cavagnaro and Curiel 2012). We will come across this level thinking again in the next part on challenges and in the corporate functions as sustainability cuts across not only levels but also issues and functions.

Strategic responses to sustainability – cost versus value

In the previous chapter we explored the tension of competitiveness and economic growth versus environmental protection, which have traditionally been perceived as a trade-off, in an either/or fashion. The conventional strategic response to sustainability has therefore been to see it as a cost rather than as an opportunity. Viewing sustainability as an (added) cost effectively sees it as value destruction and nothing else (Laszlo and Zhexembayeva 2011). In contrast, viewing sustainability as value creation provides more than one opportunity for competitive advantage – Laszlo and Zhexembayeva (2011) list seven ways how sustainability-driven business value can be created by adopting both/and perspectives:

- *#1 risk mitigation*: As a starting point, value is not necessarily created here, but protected from destruction by managing potentially costly liabilities in the form of e.g. environmental risks such as oil spills and subsequent business risks such as reputational damage and fines.
- *#2 efficiency opportunity*: Here, eco-efficiency measures create opportunities for cost reduction by doing more with less, as discussed in the previous chapter. Efficiency gains could be achieved as one-off in the short-term or lead to longer-term savings.
- *#3 a factor of differentiation*: For companies pursuing a differentiation strategy rather than a low-cost strategy (see also the later discussion), environmental sustainability attributes add another dimension to quality or performance on which basis they can charge a price premium.
- *#4 a pathway to new markets*: New markets consist of providing solutions for environmental problems for example through pollution control equipment or clean water options. New markets here come in the form of new products or in the form of new customers e.g. at the so-called base of the pyramid, that might have previously been excluded from consumption.
- *#5 a way to protect and enhance the brand*: As corporate value is more and more driven by intangible assets such as reputation and stakeholder trust, environmental performance needs to be complemented by how well a company is known and perceived for such performance amongst stakeholders.
- *#6 influencing industry standards*: Environmental leadership enables companies to go beyond compliance and raise the bar for competitors by advocating stricter environmental regulations or influencing voluntary industry standards, both of which could create barriers to entry.
- *#7 a driver of radical innovation*: This is arguably the most important value-creator, as it encompasses the other six of risk, process, product, market, brand and business context. It is therefore a complex composite and has the potential for remarkable creativity and deep innovation that can help to fundamentally rethink the nature of a company's business, particularly if sustainability is embedded into its strategy rather than a bolt-on to existing strategies (Laszlo and Zhexembayeva 2011).

Shifting the strategic lens on sustainability from seeing it as a cost to seeing it as an opportunity for value creation makes the business case for sustainability ("does it pay to be green?") much more tangible. Establishing the business case is a key part of strategic management and involves both tangible and measurable aspects such as for example waste reduction and energy efficiency, as well as challenging ones (new markets) and less tangible ones (e.g. brand development and reputation) (Bourne et al. 2012).

Rethinking sustainability from cost to opportunity requires a number of shifts in how ecological factors are viewed, as Laszlo and Zhexembayeva (2011) summarised in the following Table 4.1:

TABLE 4.1

Shifts in thinking – how ecological factors are treated

From sustainability as cost	→	To sustainability as opportunity
Defensively		Offensively
In isolation		Systemically
In terms of harm reduction		In terms of zero harm and positive benefits
As a niche		As mainstream
Through trade-offs (win-lose)		Through innovation (win-win)
As if they require green price premium		As if they must be competitively priced
In light of incremental change		In light of disruptive change
As a subset of existing customer needs		As an entirely new class of buyer needs

Source: adapted from Laszlo and Zhexembayeva 2011: 71–73.

If changing the perception of sustainability necessitates such shifts in thinking, are conventional strategy frameworks adequate or is a shift towards new frameworks needed? In the following we briefly review some established strategic management frameworks through the lens of sustainability before turning our attention to frameworks that have been specifically developed to incorporate sustainability. We re-iterate here again that there is no right or wrong framework for companies – strategy is a learning process, as Laszlo and Zhexembayeva (2011) emphasise, which is often nonlinear and iterative.

Conventional strategy frameworks for sustainability purposes

For Laszlo and Zhexembayeva (2011) outside-in approaches to strategy are more appropriate in the context of sustainability as the interconnected outside trends of declining resources, radical transparency and increasing expectations pressurise businesses to redefine how they create value. We briefly review three of these approaches: generic strategies, blue ocean strategy and disruptive innovation.

Generic Strategies

Most students of business and management will be familiar with Michael Porter's three generic strategies which form part of a larger framework including industry analysis (the Five Forces framework) and value chain analysis as well as national competitive advantage (the Diamond model). The three strategies of cost leadership, differentiation and focus are devised to outperform competitors, positioning a company within an industry to face the competitive forces characterising the industry (threat of new entrants, bargaining power of suppliers and buyers, threat of substitutes and rivalry among existing firms) (Porter 1980). Laszlo and Zhexembayeva (2011) re-examine the three strategies through a sustainability lens.[3] For cost leadership, sustainability can be a strong driver as it demands fewer materials, less energy, less (or zero) waste and redefines the business model with reduced negative (read: costly) externalities. For differentiation, companies are able to charge a price premium for green attributes in terms of performance, quality and other characteristics that reinforce product uniqueness. A focus strategy is based on using sustainability for market segmentation either built on cost or on differentiation.

BOX 4.1 EXAMPLES OF SUSTAINABILITY-DRIVEN GENERIC STRATEGIES

Efficiency savings (cost-driven):

Walmart is the world's largest retailer and a cost leader in its industry, with a broad scope. Its goals to make energy supplies 100% renewable and to aspire to zero waste have led to improvements in truck fleet fuel efficiency by 30%, thus saving costs and reducing the use of fossil fuels. The company has developed different packaging solutions such as for example lighter bottles and non-tray packaging for meats and seafood, which has helped to reduce the volume and weight of waste.

The UK's largest supermarket Tesco is combining cutting costs and reducing its environmental impact through its F plan, which stands for fuller pallets, fuller trucks, fewer miles and greater fuel economy. Through using double-deck trailers, more trains, relocating depots, training drivers and applying transport management systems, it has cut costs and its carbon footprint. In 2017 it was exceeding its glide path trajectory of halving its carbon intensity per area of stores and distribution centres by 2020 compared to a 2006/2007 baseline, having made 40% progress.

In manufacturing, 3M's Pollution Prevention Pays (3P) has been running for over 40 years and has saved the company $1.9 billion in costs and has prevented over 2 million tons of waste. The company has also set ambitious goals and targets to maintain its position as a leader in environmental stewardship in its industry.

Differentiation:

Within the luxury car market, Lexus hybrid cars provide a differentiator in terms of the quietness and cleanness of the electric engines whilst not compromising on engine performance and fuel efficiency.

The Clorox company's Green Works brand offers plant-based cleaning products that are biodegradable, making the company a mainstream sustainability leader as a broad-scope differentiator.

Focus:

In focus strategies, sustainability is utilised for market segmentation, appealing to growing segments of consumer groups known as for example LOHAS (Lifestyles of Health and Sustainability), or True Blue Greens.

One example is outdoor clothing and equipment manufacturer Patagonia, which incorporated environmental ethics from its start and is able to charge a price premium as a focused differentiator seen as 100% green attuned. A subset of sustainability customers is willing to pay more for specialised products of a company that lives its sustainability values.

An example of cost focus is Mexican producer of inexpensive hair gel Gomina, whose reduction of packaging has enhanced its competitive position in a localised market.

Sources: Laszlo and Zhexembayeva 2011; FT 2011; Laasch and Conway 2015, company websites (http://corporate.walmart.com/global-responsibility/sustainability/, www.tescoplc.com/tesco-and-society/sourcing-great-products/reducing-our-impact-on-the-environment/, www.3m.com/3M/en_US/sustainability-us/goals-progress/, www.greenworkscleaners.com/about-us/, http://eu.patagonia.com/enPL/footprint, www.gomina.com.mx/default.asp).

Blue Ocean Strategy

Whilst strategic positioning aims to strengthen a company's existing strategy, blue ocean strategy aims to change it. Kim and Mauborgne (2004: 77) conceptualise the business environment as two oceans, red and blue, where the red ocean represents the known market space – all currently existing industries competing for market share in an often-crowded market where competition "turns the water bloody". The blue ocean represents the unknown market space where there is little or no competition where demand is created rather than fought over. A cornerstone in blue ocean strategy is value innovation (Laszlo and Zhexembayeva 2011) where the value/cost trade-off is broken and competition becomes irrelevant by creating a leap in value for both customers and the company itself (Kim and Mauborgne 2004). The concept of value innovation fits strongly with a sustainability-driven business strategy as the ecological challenges society and industry are faced with today demand system-level change that the blue ocean strategy advocates in pursuit of both differentiation and low cost. It encourages companies to go beyond incremental changes and tweaks to existing products and to re-imagine the market space.

BOX 4.2 EXAMPLES OF SUSTAINABILITY-DRIVEN BLUE OCEAN STRATEGIES

GE's Ecomagination programme has seen the company migrate from being considered as one of the world's worst polluters to one of the world's top global green brands. The programme enabled the company to open new market spaces in its traditional areas of power, aviation and lighting. A total investment of $20 billion in cleaner technology has so far resulted in a return of over $250 billion in revenues alongside reductions in greenhouse gas emissions for example.

We include an example of a blue ocean strategy for social sustainability for its economic scale and success: Mexican cement producer Cemex's initiative "patrimonio hoy" saw the company offer micro-credits and technical assistance alongside its core product of building materials to the underserved market of low-income people, enabling them to complete housing projects in line with their ability to repay their debts. The repayment rate is 99%, and the programme has benefitted half a million families while providing Cemex with multi-million-dollar revenues per year at double-digit growth rates.

Sources: Laszlo and Zhexembayeva 2011, company websites (www.gesustainability.com/performance-data/ecomagination/, http://archive.cemex.com/MediaCenter/Story/Story20151005.aspx).

Disruptive Innovation

Similarly to blue ocean strategy, disruptive innovation focuses on radical rather than incremental business solutions and technologies, which are particularly relevant to sustainability (Laszlo and Zhexembayeva 2011). In the theory of disruptive innovation, Clayton Christensen (1997) differentiates sustaining technologies from those that are disruptive. Sustaining technologies are mostly incremental in nature (though some can be discontinuous) and offer improved performance of established products in line with the demands of mainstream buyers in major markets. In contrast, disruptive technologies enable the emergence of new markets but are often unattractive to the mainstream, which might therefore overlook value proposition in non-traditional markets. Christensen and Overdorf (2000) posit that an organisation's resources, processes and values determine what kind of change a company is able to embrace. Resources are both tangible and intangible (e.g. people and equipment versus brands and stakeholder relationships), whilst processes might be formal or informal, such as communication and budgeting versus ways of working and routines which provide employees with consistent ways of performing tasks. Values in the authors' view are not the conventional ethical corporate values but standards through which employees prioritise and assess the attractiveness of an order, the importance of a customer or the potential of a new idea or project. The dilemma of the innovator, as Christensen (1997) calls it, is that processes and values are not flexible and therefore the capabilities of an organisation can become its disabilities from one context to another. We will return to the question of capabilities and resources later.

In the context of sustainability, industry leaders have difficulties in dealing with disruptive technological change, which opens the door for new entrants, for which Laszlo and Zhexembayeva (2011) offer the example of the electric car, which was initially developed by GM but found little interest in the 1990s, whereas by 2010 more than two dozen electric car companies had been formed that were all new to the automotive industry (although many included frustrated engineers from incumbent firms). Established firms joined the trend only belatedly. For other sustainability-driven innovations, this raises questions for firms – should major sustainability challenges be approached from a perspective of disruptive or sustaining technological change?

Embedded Sustainability

Laszlo and Zhexembayeva (2011) bring these three approaches together into one framework of embedded sustainability to examine how sustainability contributes to business strategy. Embedding sustainability through incorporating environmental values into the core business with no trade-off in quality or price can improve the strategic positioning via the three generic strategies, can lead to entirely new market opportunities through blue ocean strategies and may happen on the basis of incremental or radical technological change – disruptive innovation. Embedded sustainability differs from bolt-on sustainability, where sustainability is a separate strategy in addition to shareholder value pursuit, whereas embedded sustainability transforms a company in pursuit of sustainable value based on principles of value creation, relationships and capabilities development (Laszlo and Zhexembayeva 2011). Whilst these frameworks are mainly outside-in approaches, we turn to inside-out approaches, as sustainability requires attention to both the external and internal corporate environment.

Resource-based view

Inside-out approaches focus on internal mechanisms to explain and guide strategic management and the pursuit of competitive advantage where resources are the strengths of a firm used to design and implement strategies (Barney 1991). Two assumptions underpin the resource-based view – that of heterogeneously distributed resources and that of immobile resources which cannot be transferred between firms without cost. From these assumptions follow two fundamental arguments – resources that are valuable and rare may yield competitive advantages, and if these resources are also not imitable, not substitutable and not transferable, then the competitive advantage may be sustainable[4] (Barney 1991). These arguments are also summarised as the "VRIN Framework". Other areas that contribute to modern resource-based theory beyond the resource-based view include dynamic capabilities, commitment and the knowledge-based view (Kor and Mahoney 2000).

BOX 4.3 SUMMARY OF THE RESOURCE-BASED FRAMEWORK

Rouse and Daellenbach (2002: 966 italics in original) put the different terms and concepts of the resource-based framework together as follows – their quote is replicated in full due to its useful comprehensiveness:

"The framework is essentially one that privileges:

a) *resources* (tangible and intangible) which are bundled, linked, incorporated, converted and *organized* into
b) sociotechnical *processes* (knowledge, routines, structures of relationships, cultures, and so on) some of which are *rare, inimitable* (or costly to duplicate), and *non-substitutable* that form
c) *capabilities* and *core competencies*. These then become sources of *competitive advantage* which when leveraged into products and services generate
d) value and *competitive advantage* which are indicated by their *performance* consequences".

Although often no distinction is made between resources and capabilities, as they all contribute to competitive advantage (Barney 2003), it is worth stating that capabilities are more complex and result from teams of resources working together. Capabilities, in turn, are building blocks that combine into competencies (Laszlo and Zhexembayeva 2011). Core competencies, in particular, are the collective learning in a company, the coordination of diverse (production) skills and the integration of technologies (Prahalad and Hamel 1990).

Sustainability capabilities reflect the complex nature of sustainability and feature scientific, technological, organisational and social dimensions. They require not only environmental capabilities, but also complementary ones (Laszlo and Zhexembayeva 2011). Competitive advantage is built through the combination of complementary capabilities into competencies that are hard to imitate for competitors. As the sustainability profession is still an evolving one, it is dynamic capabilities that are particularly significant for companies to acquire and develop. Building environmentally oriented capabilities requires the integration of environmental issues into strategic planning (Russo and Minto 2012). One approach in the strategic management field is an extension of the resource-based theory which explicitly includes the natural environment and ecological limits – the natural resource-based view.

BOX 4.4 RESOURCE-BASED VIEW EXAMPLE

Toyota was early out in its effort and successful development of the Prius as the first mass-produced hybrid vehicle, which provided the company with competitive advantage based on its technological learning, brand image and public relations through celebrities as early adopters of the technology. This competitive advantage is to some extent sustainable, though as other competitors are imitating the technology, Toyota is no longer the only player in the mass-produced hybrid field.

Source: Laasch and Conway 2015.

Natural resource-based view

In proposing a natural resource-based view, Hart (1995) addresses the limitations of the resource-based view in ignoring the natural environment and the constraints it poses on companies. He argues that the challenges of the biophysical environment are one of the most important drivers of developing new resources and capabilities for firms. This results in strategy and competitive advantages rooted in capabilities enabling environmentally sustainable economic activity. Hart's conceptual framework consists of three strategies: pollution prevention, product stewardship and sustainable development, which are all driven by different environmental forces such as minimising emissions, waste, lifecycle costs and environmental burden of firm growth. Pollution prevention is based on the key resource of continuous improvements and leads to lower costs as competitive advantage, whereas product stewardship is based on stakeholder integration and pre-empts competitors. Sustainable development concerns the future position of a company and is based on a shared vision. The three strategies are interconnected – not separate as Porter's three generic strategies. They are also characterised by path dependence and embeddedness. The path dependence is clearer in *Beyond Greening* where Hart (1997) further develops these strategies into stages, where pollution prevention and product stewardship are stage one and two in today's environment, and clean technology is stage three on the path to sustainability vision in tomorrow's world. A company needs to balance its sustainability portfolio across all stages to meet both current and future needs, whilst also balancing the internal aspects of pollution prevention and clean technology, and the external aspects of product stewardship and sustainability vision.

BOX 4.5 EXAMPLES OF COMPANIES GOING *BEYOND GREENING*

Early adopters of a product stewardship strategy include Xerox, manufacturer of photocopiers, scanners and printers. The company developed an infrastructure for its pioneering leasing and remanufacturing of office equipment, which allowed it to recondition, test and reassemble components and parts into "new" machines. Globally, its combined returns programme now prevents over 40,000 metric tons of waste from landfill.

Pollution prevention has been adopted by many companies, particularly in the chemical industry, such as for example DuPont, which first started to set environmental goals in 1989. The goals it set in 2006 for 2015 were met two years earlier, enabling the company to set tougher goals for reducing its environmental footprint. The company 3M's pollution prevention programme was introduced in a previous box (4.1).

A move towards clean technology can for example be seen in the chemical industry and agriculture, where increasingly using bioengineering and biotechnology enables a sustainable path to agricultural yields without chemical pesticides or fertilisers.

Sources: Hart 1997, company websites (www.xerox.com/en-us/about/ehs, www.dupont.com/corporate-functions/sustainability/sustainability-commitments/goals-progress/footprints-goals.html).

In more recent work, the strategy of sustainable development has been separated into the two distinct strategies of clean technology and serving the bottom of the pyramid (Hart and Dowell 2011, see also Prahalad and Hart 2002; Hart 2010), which acknowledges the environmental and social dimensions of sustainable development (see also Chapter 1). For both strategies, Hart and Dowell (2011) emphasise the importance of developing dynamic capabilities in light of the complex environment in which firms operate. These capabilities need to become native in such environments in order for companies to embed innovation and move towards true sustainability and inclusive commerce, a path more fully explored in Hart's book *Capitalism at the Crossroads* (2010), which covers all three aspects of sustainability.

Russo and Fouts (1997) differentiate between compliance and (pollution) prevention as prototypical strategies with regard to environmental performance and argue that a prevention strategy leverages capabilities that are difficult to imitate and therefore serve as a foundation for competitive advantage (see also Russo and Minto 2012). Aragón-Correa and Sharma (2003) call these two modes reactive (compliance) and proactive (prevention), and add environmental leadership to this typology of corporate environmental strategies, which goes further than the prevention aspect of proactive strategies (Aragón-Correa et al. 2008). They also emphasise how exogenous factors such as environmental regulations impact on the development of internal environmental capabilities, thus again emphasising how sustainability needs both outside-in and inside-out approaches.

Hart and Dowell (2011) observe that most companies engage in incremental strategies such as pollution prevention and product stewardship but argue that these might not be sufficient in meeting the environmental challenges industry is facing. They encourage the development of breakthrough strategies to provide actual solutions to environmental problems rather than simply lessening the negative impacts from their operations. This is similar to the disruptive innovation advocated by Christensen (see earlier discussion) and the creative destruction Joseph Schumpeter described (Hart 2010). The choice Hart and Dowell present reflects the question Laszlo and Zhexembayeva (2011) ask – is it about creating the good or stopping the bad? The authors call the advocates of these choices market optimists and market pessimists, or bright greens and dark greens, respectively.

Companies are thus faced with a range of choices and tensions in how they plan their journeys and strategies towards environmental sustainability.

Choices and tensions in deciding on strategies for sustainability

Whilst we explored some of the broader tensions inherent in sustainability in the previous chapter, here we have come across concrete choices and tensions companies face in their quest for sustainability. Van der Byl and Slawinski (2015) summarise the perspectives of how the tensions inherent in corporate sustainability are approached as win-win, trade-off, integrative or paradox. Hahn and colleagues (2018) argue for a paradox perspective that accepts and integrates tensions in sustainability, which the business case tries to eliminate by aligning the often-conflicting strands of financial, environmental and social performance (win-win). A paradox approach allows decision-makers to achieve competing objectives simultaneously and allows environmental concerns to have intrinsic value rather than being a means to a financial end (Hahn et al. 2018).

The business case question of whether it pays to be green poses sustainability as a choice between cost and opportunity for value creation. As demonstrated earlier, the options for value creation are multiple, so the next choice is then what kind of value creation from sustainability?

In the past, environmental strategy has often been in the form of piecemeal projects to control or prevent pollution, with few companies truly incorporating and integrating sustainability into their strategic thinking (Hart 1997). The choice for companies is thus between bolt-on sustainability or embedded sustainability. Whilst there is no right or wrong answer here, embedding sustainability into existing business priorities might prove a smarter business model (Laszlo and Zhexembayeva 2011), even though a bolt-on approach might be the first step for a company just exploring what it can do in this respect. Similarly, although Hart and Dowell (2011) encourage breakthrough strategies, incremental strategies offer companies a foot in the door in the form of low-hanging fruit, where being less unsustainable is a step in the right direction even though it is still not the same as being more sustainable. Other tensions include outside-in versus inside-out approaches as a focus for sustainable strategic management or a combination of the two.

Viewing sustainability as value destruction or creation is similar to what Blowfield (2013) terms defensive and offensive approaches to sustainability. Whereas a defensive sustainability strategy focuses on regulatory compliance, an offensive approach moves from risk mitigation to using corporate competencies to address sustainability. Blowfield argues for an integration of both approaches as they are complementary – on the one hand a defensive approach may protect the brand or reputation of a company, but it cannot enhance it, whilst on the other hand, an offensive approach can do just that, yet cannot protect it (Blowfield 2013, based on Kramer and Kania 2006).

In choosing a strategy or model for sustainability, companies can choose between familiar and established frameworks or entirely new frameworks. In the section here, we looked at sustainability through the lens of conventional strategy frameworks and extensions thereof. In the following we introduce frameworks that have as their core objective sustainability. Whether companies use traditional or new frameworks, they all require organisational change, which we discuss subsequently.

Dedicated sustainability strategy frameworks

The Natural Step

In the previous chapter we outlined The Natural Step's four sustainability principles of eliminating humanity's progressive contribution to extraction and production of substances, to physical degradation and to the undermining of basic needs being met. The principles derive from the four systems conditions covering fossil fuels and minerals, toxins and resource extraction as well as equity and economics. These systems conditions are displayed in Figure 4.1.

From this foundation of principles and systems conditions, for which there is scientific agreement, The Natural Step (TNS) framework proceeds with *backcasting* and the ABCD method. The scientific agreement to the principles is an important underpinning of the collaborative and consensus-building approach that TNS has at its core (Robèrt 2002). Such common ground enables people with diverse values and backgrounds to work together (Robertson 2017), where many brains working in teams function better than brains on their own.

Backcasting is strategic planning that starts from a point of success in the future and counters the drawbacks of traditional forecasting, which Robèrt (2002), the initiator of this framework, likens to planning by looking in the rear view mirror – in other words it extrapolates from the past and could thus lead to solutions that created the problem in the first place (Cavagnaro and Curiel 2012). Backcasting, in contrast, starts with a future goal and strategic visioning of a successful result in the future and then asks – what can we do today to reach that goal? Backcasting is deemed to be more effective and, in particular, more creative than forecasting (TNS websites); this might be because it provides a way of thinking about what a company can do rather than what it cannot (Blowfield 2013).

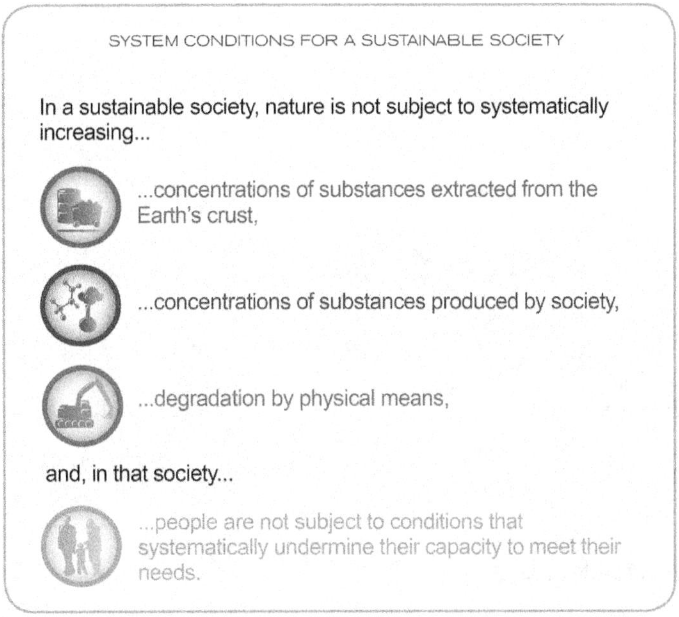

FIGURE 4.1
The Natural Step Systems Conditions

Source: www.naturalstep.ca/four-system-conditions, used with permission from The Natural Step Canada.

As displayed in Figure 4.2, the ABCD method then consists of the following steps which we will discuss in turn based on the TNS websites and Robèrt (2002):

- Awareness and vision
- Baseline mapping
- Creative solutions
- Decide on priorities

In the first step of *awareness and vision*, the framework is shared among all participants of the planning process. Before the four system conditions and principles are covered, the image of a funnel is used as a metaphor for the awareness of the problem of non-sustainability. It visualises the economic, social and environmental pressures that demonstrate that the narrower the funnel, the less room to manoeuvre. The shift towards sustainability opens the walls of the funnel through innovation and creativity. A whole-systems approach is taken here, using the tree as a metaphor for complex systems – this enables simplicity without reduction. Creating a common understanding of the problem and setting ambitious goals in this step encourages the emergence of a shared mental model that can be built on in the next steps. The use of graphs and metaphors in TNS makes the planning process practical and accessible.

The second step of *baseline mapping* asks how does the organisation look today? It is a gap analysis using the four sustainability principles to see how an organisation's flows and impacts

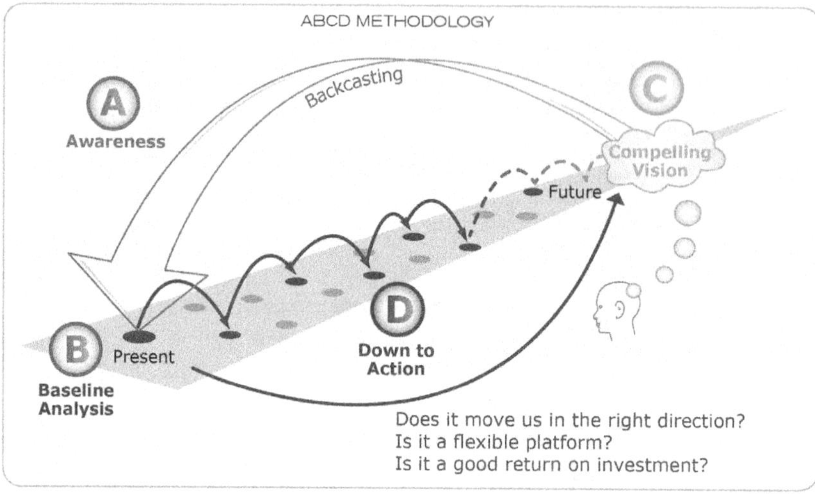

FIGURE 4.2
ABCD planning in The Natural Step

Source: www.naturalstep.ca/abcd, used with permission from The Natural Step Canada.

violate the systems conditions. Critical sustainability issues and their business implications and what options for change there are may be identified in this stage.

The third step of *creative solutions* asks how does the organisation look in a sustainable society? The B-list of the previous step triggers creativity for identifying solutions. Brainstorming leads to the so-called C-list of theoretically possible solutions with no restrictions other than the systems conditions. It is here that backcasting identifies opportunities and potential solutions by looking backwards from the vision. In this step, backcasting prevents the development of strategies that solve today's problems rather than look to tomorrow's solutions.

In the last step of *deciding priorities* the solutions from step C are prioritised into early actions for change based on the question what are the measures that move the organisation towards sustainability? Here, three strategic prioritisation questions help in terms of direction, platform and low-hanging fruits:

- Is this action moving you towards or away from your sustainability vision?
- Is this action a flexible platform towards your sustainability vision?
- Will this action offer an adequate return on investment?

Beyond these questions, companies will develop additional questions specific to their context to help them prioritise in designing a strategic programme for sustainability that works in both the short and long term. Backcasting is still used in this step to continue assessing decisions and actions to check that the organisation is moving in the right direction.

> What's considered realistic today never determines the direction of change, only its pace.
> (TNS websites)

The strengths of the TNS approach include its foundation on robust scientific principles, its comprehensiveness of all aspects of sustainability, its applicability across industries and geographical scales, and its ability to support backcasting and strategic planning within conventional business planning cycles (Everard 2011). The strongest endorsement for its achievements is in the form of companies that have successfully applied TNS (see the accompanying boxes), and has led to the framework being used across the world (TNS websites).

BOX 4.6 RAY ANDERSON AND INTERFACE – APPLYING THE NATURAL STEP IN PRACTICE

Ray Anderson was a pioneer in executing The Natural Step after having read about it in Paul Hawken's (1993) *Ecology of Commerce*, whose message Anderson calls his spear in the chest. He transformed his carpet company Interface, which specialises in carpet tiles, and told his story in *Mid-Course Correction* (1998) and *Confessions of a Radical Industrialist* (2009, with Robin White).

In reformulating Interface's business, the resulting vision was of a firm with no pipes and no chimneys, which was translated into Mission Zero. Ray Anderson compared the journey to become a fully sustainable company like summiting a mountain higher than Everest – difficult, but not impossible with careful planning and attention. This image resonated and became known as Mount Sustainability that needed to be climbed in order to achieve Mission Zero by 2020. Seven key areas were identified to achieve this. These became known as the "seven fronts" of mount sustainability, where progress was tracked and measured:

1. Waste elimination: eliminate all forms of waste in every area of business.
2. Benign emissions: eliminate toxic substances from products, vehicles and facilities.
3. Renewable energy: operate facilities with 100% renewable energy.
4. Resource recycling: redesign processes and products to close the technical loop using recovered and bio-based materials.
5. Resource-efficient transportation: transport people and products efficiently to eliminate waste and emissions.
6. Stakeholder sensitisation: create a culture that uses sustainability principles to improve the lives and livelihoods of all of our stakeholders – employees, partners, suppliers, customers, investors and communities.
7. Redesign of commerce: create a new business model that demonstrates and supports the value of sustainability-based commerce.

The pledge to eliminate negative impacts on the environment corresponds to the first principle of TNS, whilst the reduction of damaging emissions to the second TNS principle. In moving away from the old industrial revolution paradigm to a new industrial revolution paradigm, Interface was envisioned as a dematerialised, zero-waste, no-emissions company, achieved by applying cradle-to-cradle principles and closed loop systems. It moved technology from the numerator to the denominator in the so-called *IPAT* equation (see Chapter 3) and moved from being a linear take-make-waste model to a cyclical resource-efficient, solar-driven, equitable one. Going beyond Mission Zero, the aim is to become a restorative enterprise committed to a whole systems approach of embeddedness encompassing people, process, product, place and profits. Being restorative involves giving back more than they take which means going beyond Interface and helping others to achieve the same.

At its carpet factories, as of 2018, Interface had reduced greenhouse gas emissions by 96%, water use per unit of product by 89%, increased use of recycled and biobased raw material by over 50%, improved energy efficiency by 46% and had made 89% of energy used renewable since 1996 alongside significant reductions in waste.

Ray Anderson died in 2011 but left behind a crucial legacy for sustainability not just for his own company. Watch him for yourself tell his

BOX 4.6 (CONTINUED)

story, making the business case (or logic) for sustainability: www.ted.com/talks/ray_anderson_on_the_business_logic_of_sustainability.

Sources: Blowfield 2013; Cavagnaro and Curiel 2012; Anderson 1998; Anderson and White 2009, company websites (www.interface.com, www.interface.com/EU/en-GB/sustainability/our-journey-en_GB).

BOX 4.7 OTHER EXAMPLES OF THE NATURAL STEP

Whilst Ray Anderson is maybe the best-known CEO to use TNS for his carpet company Interface (see Box 4.6), other well-known companies have also used the framework, such as IKEA, Scandic Hotels and Electrolux from the home of TNS, Sweden, as well as Starbucks and Nike.

IKEA first engaged with TNS framework in 1990, initially offering one product, then rolling out the Gustavian line that was produced in line with the sustainability principles of TNS. Its first environmental action plan covered six implementation areas: management and personnel, products and materials, customers, suppliers, buildings equipment and consumable materials, and transport.

In waste reduction its "Trash is Cash" programme led initially to savings in the cost of management but then turned to a profit-making opportunity for the company. In 2016, IKEA offered desk protectors, door mats and spray mats made from recycled plastic film – used packaging material. In terms of the materials used, IKEA minimised its use of harmful materials such as formaldehyde or PVC and is increasingly using natural materials such as cork, rattan, and bamboo. These materials are not only renewable but have other features that add to a product's quality when used – e.g. bamboo has the ability to cleanse air from substances such as benzene, formaldehyde, chloroform, carbon monoxide and xylene, whilst rattan is durable in humidity, and cork is non-absorbing, as emphasised in its 2016 highlights.

The following pictures illustrate how TNS might work in practice – the author bought the so-called *Expedit* bookshelf from IKEA in 2009, and in 2014 when looking to buy another bookshelf following a house move, searching for *Expedit* on the company's webpage showed up the so-called *Kalläx* range. The book space in both shelves is the same, but the material required has been reduced, requiring fewer resources and enabling more efficient transport. On the left-hand side of the pictures you will see *Expedit*, on the right-hand side *Kalläx*.

Sources: Robèrt 2002; Nemetz 2013, http://thenaturalstep.nl/ikea/, https://highlights.ikea.com/2016/.

The framework situates itself within a 5-level framework for planning in complex systems where TNS is applied to levels 2 and 3 (TNS websites, Robèrt 2002). The five levels and its contents are:

1 Systems level – use of funnel metaphor, scientific underpinning
2 Success level – sustainability principles and systems conditions
3 Strategic level – backcasting and ABCD method
4 Action level – concrete activities towards sustainability
5 Tools level – metrics and concepts to manage path towards sustainability.

The TNS framework is applied to guide level 4 and 5, activities and metrics, which are not part of the TNS framework that provides overall guidelines. The tools used differ according to context but can create synergies when used within TNS. They include for example ISO 14001 and environmental management systems, life cycle assessment and ecological footprinting, zero emissions, factor 10, natural capitalism. We have come across LCA and footprinting in previous chapters and will return to them in Chapters 8 and 9, whilst we cover Environmental Management Systems and ISO 14001 in Chapter 11. For now, we turn our attention to natural capitalism as another strategic tool to guide a company towards environmental sustainability. Blowfield (2013) calls the TNS framework an interpretation of eco-capitalism, which is also called natural capitalism. Whilst business is often portrayed as the problem and therefore the "villain" in the sustainability story, not all sustainability advocates are anti-capitalist, but rather critical friends of the market. Eco-capitalism posits the beneficial nature of capitalist enterprises if they recognise and account for the true value of the assets they use and the impacts they create (Blowfield 2013: 59–60).

Natural Capitalism

In *Natural Capitalism*, Hawken et al. (1999) propose a road map for integrating ecological with economic goals, taking a systems view of how business and society relate to the environment. The conventional industrial system defines capital as accumulated wealth in the form of factories, investments and equipment. Similar to the five capitals model discussed in Chapter 2, the authors remind us that an economy needs four types of capital to function properly: human, financial, manufactured and natural capital, where the first three transform natural capital into what makes up our daily lives – cars, roads, cities, buildings, food, medicine. In using natural resources and living systems, capitalism, as practiced in the 20th century, does not assign value to the largest stock of capital it uses, thus violating its own accounting principles. The authors ask the question: what would the economy look like if all forms of capital were fully valued? Their answer is natural capitalism, which requires the next industrial revolution. This means a move from the conventional capitalism that originated in the industrial revolution, where labour was historically scarce and natural capital abundant, to natural capitalism where labour is abundant and natural capital scarce.

Four strategies (or principles) are central to natural capitalism which are akin to four major shifts in business practices. We will discuss these four in turn, based on Lovins et al. (1999) and Hawken et al. (1999):

1 radical resource productivity
2 biomimicry
3 service and flow economy
4 investing in natural capital.

Significantly increased *resource productivity* is the cornerstone on which natural capitalism rests as it produces three major benefits in slowing resource depletion, lowering pollution and providing increased employment. Whilst the first industrial revolution produced enormous increases in labour productivity (human capital), which were previously unimaginable, the next industrial revolution will similarly see the productivity of natural resources rise. The first such call came in 1994 with the "Carnoules Declaration" by a group of economists, scientists, government and business representatives who called themselves The Factor Ten Club for the ten-fold increase in efficiency that could be achieved in energy, natural and other resources. Such an increase in efficiency would result in a 90% reduction in energy and material intensity, whilst Factor Four would equate to a 75% reduction. The latter factor was extensively explored by two of the same authors of natural capitalism (Lovins and Lovins) alongside von Weizsäcker (1998) with examples of energy, materials and transport productivity. Lovins and colleagues (1999) even argue for a factor of 100, made possible by revolutionary leaps in design and technology, which are the two main ways companies can increase their resource productivity at a profit: implementing whole-system design and adopting innovative technologies. Other methods include controls in the form of information technologies and large-scale optimisation, changing corporate culture to one of a learning organisation, new processes and saving materials, where the authors introduce the concept of "born-again materials" and remanufacturing.

Biomimicry involves redesigning industrial systems to mimic nature, which leads to reducing the throughput of materials that are wasteful. Such biomimetic production uses closed-loop manufacturing that prevents waste, such as for example toxic materials that require disposal at a cost, thus demonstrating the synergies with resource efficiency. We will return to biomimicry as a model for sustainable design and the work of Benyus (1997) in Chapter 9 on production and operations. Box 4.8 outlines some general strategies and principles businesses can use by finding inspiration in nature.

BOX 4.8 GENERAL BIOMIMICRY PRINCIPLES

Biomimicry not only has lessons for how products should be designed, but also more generally for business, in how to green businesses and how to conduct business based on how organisms in a mature ecosystem operate, thus emulating nature's economics.

Benyus (1997) outlines the following ten principles:

1. Waste used as a resource – closing loops and extending producer responsibility beyond the factory gate.
2. Diversification and cooperation to make full use of the habitat – precompetitive cooperation and alliances to develop (e.g. common standards for materials and labelling).
3. Energy collected and used efficiently – use current sunlight in the form of renewable energy sources (solar, wind, tidal) rather than ancient sunlight (fossil fuels).
4. Optimisation over maximisation – shift focus to durability and maintenance (including remanufacturing and upgrading) from cheap, mass-produced, short-lived modus requiring new models annually.
5. Materials used sparingly – move away from planned obsolescence to dematerialisation and leasing as a way of life.
6. Not fouling the nest – emission of pollutants should not be greater than the planet's capacity of assimilation, thus encouraging moves to near zero emissions, chemicals-on-demand and just-in-time production to avoid any risks of leakage.
7. Resources not drawn down – non-renewable resource used in line with rate of development

BOX 4.8 (CONTINUED)

of substitutes, renewable resources used in line with rate of regeneration.

8 Industrial ecosystem in balance with biosphere – stock and flow of nutrient cycle as give-and-take to maintain conditions needed for life.

9 Run on information – feedback mechanisms between businesses and from environment to avoid overshooting and system breakdown.

10 Local shopping – nature operates in localised way in close connection to space and time; adapt economy to suit natural environment and local attributes.

The collaborative initiative *Biomimicry for Creative Innovation* distilled these principles further into a concise set of nature's business principles for the firm of the future which are: build resilience, optimise, adapt, integrate systems, navigate by values and support life. These are underpinned by existing business theories and form a framework for transforming companies towards a sustainable future (The Nature of Business 2013). Examples where biomimicry principles are applied by business include living buildings, machine learning and artificial intelligence, biological manufacture and biophilic design (Woolley-Barker 2016).

The *service and flow economy* fundamentally changes the relationship between production and consumption, between producers and consumers. It marks a shift from an economy of goods and purchases to one of service and flow. In the ten-year anniversary edition of their book, the authors call this principle *the solutions economy*. The logic of lean thinking and cradle-to-cradle production are combined here (which we will return to as well in Chapter 9), emphasising the elimination of waste. Instead of selling a product that a customer hopes will provide a required service, the service is provided directly. It shifts the view of products as an end to products as a means. For manufacturers, this change in the business model and relationship with consumers reduces the volatility that is inherent in the turnover of durable goods, as a service model results in a continuous flow of revenues.

Reinvestment in natural capital serves to reverse environmental destruction by restoring, sustaining and expanding natural capital. Here, Hawken and colleagues (1999) draw on the foundation of capitalism which requires the reinvestment of earnings in productive capital. Natural capital for them is the most important form of capital as it cannot be produced by human activity and as it underpins life. Blowfield (2013) calls such a strategy where companies maximise their contribution to the maintenance and expansion of natural capital stocks a genuine sustainability strategy, which is often cased around a cradle-to-cradle approach, taking sustainability into account across the whole lifetime of a product.

BOX 4.9 NATURAL CAPITALISM STRATEGIES IN ACTION

Radical resource productivity:

- Carpet maker Interface's engineer Jan Schilham made two simple design changes to pump and pipe sizes which reduced its energy requirements by 90%. Pipe sizes were increased (fat rather than thin, and short and straight rather than long and crooked) which enabled liquids to flow with less friction and less power.

BOX 4.9 (CONTINUED)

- Furniture maker Herman Miller remanufactures furniture and uses the rePurpose programme to find new life for used furniture and other assets, helping other companies to dispose of existing furniture responsibly.

Biomimicry:

- Velcro is one of the oldest examples where nature was used as a model, mimicking how burdock burrs attach to animals.
- A more recent example is the redesign of the nose of Japan's high-speed train Shinkansen modelled on the kingfisher which led to a reduction in noise and power use (15% less electricity) whilst enabling greater speeds (by 10%).

Service and flow:

- We already came across Xerox's leasing of office equipment in Box 4.5. Other companies that have similarly used leasing include for example Carrier, which shifted its mission from selling air conditioners to renting out comfort.
- Dow Chemicals is using Chemical Leasing to sell performance rather than product, which aligns incentives between supplier (now service provider) and buyer – in the traditional relationship the supplier wants to increase product supply, whilst the buyer wants to decrease costs, whilst in the new model both want to decrease the lifecycle costs of material, labour and waste management.

Investing in natural capital:

- Dow Chemicals is also an example of a company investing in natural capital. When faced with the decision on how to meet wastewater quality regulations, the company could have invested $40 million in a treatment plant but opted to create a wetland that filters water naturally at much lower costs of $1 million for construction and operations.
- Water is also a crucial natural capital for Coca-Cola, which has committed to return an amount of water to nature and communities equal to what it uses in its drinks. It does so through replenishment, treatment, increased efficiency and mitigating risk.
- Industries such as fishing, forestry and farming that harvest living systems directly provide other examples of evolving practices in natural capital reinvestment.

Sources: Lovins et al. (1999), corporate and other websites (www.hermanmiller.com, www.bloomberg.com/news/photo-essays/2015-02-23/14-smart-inventions-inspired-by-nature-biomimicry, https://biomimicry.org/biomimicry-examples/, www.carrier.com, www.dow.com, www.greenbiz.com/article/coca-cola-apple-dow-see-fertile-ground-investing-natural-capital, www.coca-colacompany.com).

The four strategies are interrelated and interdependent, mutually reinforcing the reduction of environmental impact, the creation of economic growth and of meaningful employment (Lovins et al. 1999).

Most of these strategies can be pursued by companies themselves, but as there are limits to what the market and its forces can do,[5] it also requires the involvement of the regulatory framework in the form of governmental policies, for example to provide an appropriate fiscal regime that taxes bad (pollution/resource depletion) rather than good (jobs/income) and that phases out perverse subsidies incentivising disinvestment in natural capital rather than encouraging positive outcomes (e.g. subsidies for fossil-fuel industries).

The recognition that natural capitalism requires more than individual organisations and needs collaboration is reflected in the multi-stakeholder initiative of the Natural Capital Coalition (http://naturalcapitalcoalition.org) which aims to harmonise natural capital approaches by bringing together international business, governments and civil society organisations. The coalition's protocol provides a framework that helps managers make informed decisions with regards to operations, legal and regulatory issues, finances, reputation and marketing as well as relationship with society.

Green to Gold

In their book *Green to Gold* Esty and Winston (2009) classify initiatives for strategic environmental management around four categories: costs, revenues, risks and intangibles (Valentine 2010). Esty and Winston identify eight strategies that companies pursue to manage the downside of cost (1-3) and of risk (4), and to build the upside of revenues (5-7) and of intangibles (8):

1 *Eco-efficiency*: improved resource productivity, such as for example of water, material and energy.
2 *Eco-expense reduction*: reduce environmental costs and regulatory burden, for example landfill fees by cutting waste.
3 *Value chain eco-efficiency*: eco-efficiency extended to lower costs upstream and downstream, with the help of life cycle assessment, for example.
4 *Eco-risk control*: managing business risk driven by the environment (e.g. anticipating regulatory changes or going beyond compliance for competitive advantage).
5 *Eco-design*: meet customers' environmental needs, redesign products and processes.
6 *Eco-sales and marketing*: green attributes as foundation to build product position and customer loyalty, find niches, use eco-labelling.
7 *Eco-defined new market space*: value innovation is promoted and breakthrough products are developed (e.g. through servicising).
8 *Intangible value*: corporate reputation is built alongside trusted brands.

These strategies, which are driven by environmental challenges ("natural forces") and stakeholders ("players"), build on a foundation of what the authors call an *eco-advantage mind-set*, which is supported by eco-tracking, culture and redesign. This mind-set provides a principled, broad and holistic picture, whilst eco-tracking systematically captures detailed data allowing better management. Corporate culture is a significant foundation for promoting and driving environmental thinking, for example by setting goals, encouraging ownership and engagement and providing a narrative for internal and external stakeholders. We return to the subject of culture in the context of changes needed later and in Chapter 13 on human resource management (HRM). Similarly, as several of the eight strategies cover different corporate functions, we will come across them again in one form or another in the third part of the book.

Usefully, Esty and Winston (2009) also cover the hurdles companies face in their strategic quest for eco-advantage and sustained value. Failures of planning (e.g. misunderstanding the market or customers, ignoring stakeholders, expecting a price premium, seeking perfection), of organisation (e.g. silo thinking, middle management squeeze, isolation of environmental professionals), and of substance (e.g. greenwashing, unexpected consequences, not communicating) all have the potential to upset a company's sustainability efforts. However, these would-be pitfalls can all be addressed through a variety of solutions and tools. For example, silo thinking can be addressed through life cycle assessment and design for the environment, which we cover in Chapter 9, whilst isolation of environmental managers can be addressed through broad-based top management commitment and ownership at the operational level.

Redesign as one of the foundations requires companies to look at environmental issues in a new way, moving from the known priorities of reduce, reuse, recycle to redesign and reimagine what they do and how they do it (Esty and Winston 2009). As Hawken (1993: 12) puts it, the three basic issues business faces are "what it takes, what it makes, and what it wastes", which are all inextricably linked in a linear fashion. In a circular economy, one company's waste becomes another one's input, and the economy is redesigned to "bend" the linear model into a circular one, resulting in a restorative and regenerative economy (EMF 2013). The concept has been influenced by industrial ecology and cradle-to-cradle, to which we now turn.[6]

Eco-effectiveness and cradle-to-cradle

This re-imagination of how companies manage business is well illustrated in *Cradle To Cradle*, McDonough and Braungart's (2002) approach to *remaking the way we make things* (subtitle of their book). The authors argue that eco-efficiency, advocated by many and covered earlier, is not good enough, as it may result in making the wrong things more efficiently, and as *less bad* is still not good.[7] Their alternative to eco-efficiency is eco-effectiveness: working on the right things instead of "making the wrong things less bad" (McDonough and Braungart 2002: 76). Their remaking of business involves a move from the conventional linear, cradle-to-grave way to a cyclical cradle-to-cradle model where the waste of (biological and technical) material flows equals food, to be used (again) through upcycling rather than downcycling. The cyclical model also is reflected in the circular economy, which draws on cradle-to-cradle principles.

We briefly cover the steps towards eco-effectiveness here but return to it in Chapter 9 on design and production, as design is central to the cradle-to-cradle approach, which represents a move from built-in obsolescence to mimicking nature's effectiveness (akin to biomimicry, see earlier discussion and Chapter 9). Like natural capitalism, eco-effectiveness is part of the next industrial revolution, redesigning the industrial system for the 21st century (McDonough and Braungart 1998).

To put eco-effectiveness into practice McDonough and Braungart (2002) suggest the following five steps:

1. *Get "free of" known culprits*: harmful substances are "filtered" out of the design of products and processes, ensuring that for example a "lead free" or "phosphate free" product does not contain other hazardous components that have replaced lead or phosphate.
2. *Follow informed personal preferences*: when limited data is available (e.g. from suppliers), decision-making may be based on best information available, aesthetics, ecological intelligence, respect, creativity and fun.
3. *Create a "passive positive" list*: when phasing out substances that are most problematic and urgent ("X list") and harmful but less urgent ("grey list"), a positive list of preferred substances that are healthy and safe for use is drawn up. Substitutions are made where possible.
4. *Activate the positive list*: redesign of processes, moving from less bad to good by choosing new materials that go through biological and technical material cycles safely as nutrients.
5. *Reinvent*: going beyond material ingredients and reimagining products that fulfil needs in a dynamic context of culture and technology.

The authors caution that this process takes effort, time, financial resources and creativity, but provide five principles that guide the transformation to an eco-effective vision:

- signal intention – beyond incremental improvement,
- restore – economic growth to good growth and design restorative products,
- be ready to innovate further – think beyond perfecting existing products,
- understand and prepare for the learning curve – innovative design takes time,
- exert intergenerational responsibility – leave behind a positive design legacy.

The principle of the learning curve reminds us of the definition of strategy presented earlier as a learning process. This is also reflected in the image of businesses learning to ride the green wave that Esty and Winston (2009) use. We discuss the developmental aspect of how companies move towards sustainability after covering business models for sustainability.

BUSINESS MODELS FOR SUSTAINABILITY

Business models are an emerging concept that sum up the key dimensions of successful market introductions (Boons and Lüdeke-Freund 2013). They describe the rationale of how an organisation creates, delivers and captures value (Osterwalder and Pigneur 2010: 14) and can be used to analyse, compare, assess, manage, communicate and innovate how a company does business and defines its competitive strategy (Bocken et al. 2014). Boons and Lüdeke-Freund (2013) distinguish between four elements that make up a generic business model concept: value proposition (product/service offering), supply chain (upstream relationships), customer interface (downstream relationships) and financial model (costs and benefits from first three elements). This draws on the four main business areas of customers, offer, infrastructure and financial viability that Osterwalder and Pigneur (2010) use as a blueprint for nine basic building blocks that underpin a company's logic.[8] Central to their business model "canvas" is the value proposition, which is flanked by key activities, key resources and key partners on the upstream side (infrastructure, focus on efficiency), and by customer relationship, channels and customer segments on the downstream side (focus on value). These seven blocks are supported by the cost structure and revenue streams. The canvas can be used to chart both new and existing business models.

The concept of business models is significant in the context of sustainability as the creation of new business models through innovation can encourage sustainability by embedding it and using it as a driver for competitive advantage (Bocken et al. 2014). Conventional business models often focus on customer needs, financial return and regulatory compliance and might only add sustainability to the model, but sustainability requires a more holistic approach that integrates social and environmental goals to create sustainable value (Pedersen 2017; Bocken et al. 2015). Sustainable business models and related concepts such as green, social, triple layers, shared value, circular economy or flourishing business models, to name but a few, therefore feature core characteristics of a broader understanding of value and stakeholders, long-term orientation and a systems perspective as well as circular thinking (Pedersen 2017). Stubbs and Cocklin (2008) were among the first to conceptualise sustainable business models which transform the neoclassical economic model of the firm as a profit-maximising entity for shareholders into an organisation with a redefined purpose in collaboration with key stakeholders based on structural and cultural capabilities, amongst others.

Bocken and her colleagues (2014) use the business model concept, under which they include three elements (value proposition, value creation and delivery, and value capture) to categorise the mechanisms that deliver sustainability. Based on technological, social and organisational innovations, seven archetypes of sustainable business models are developed, as shown in Figure 4.3.

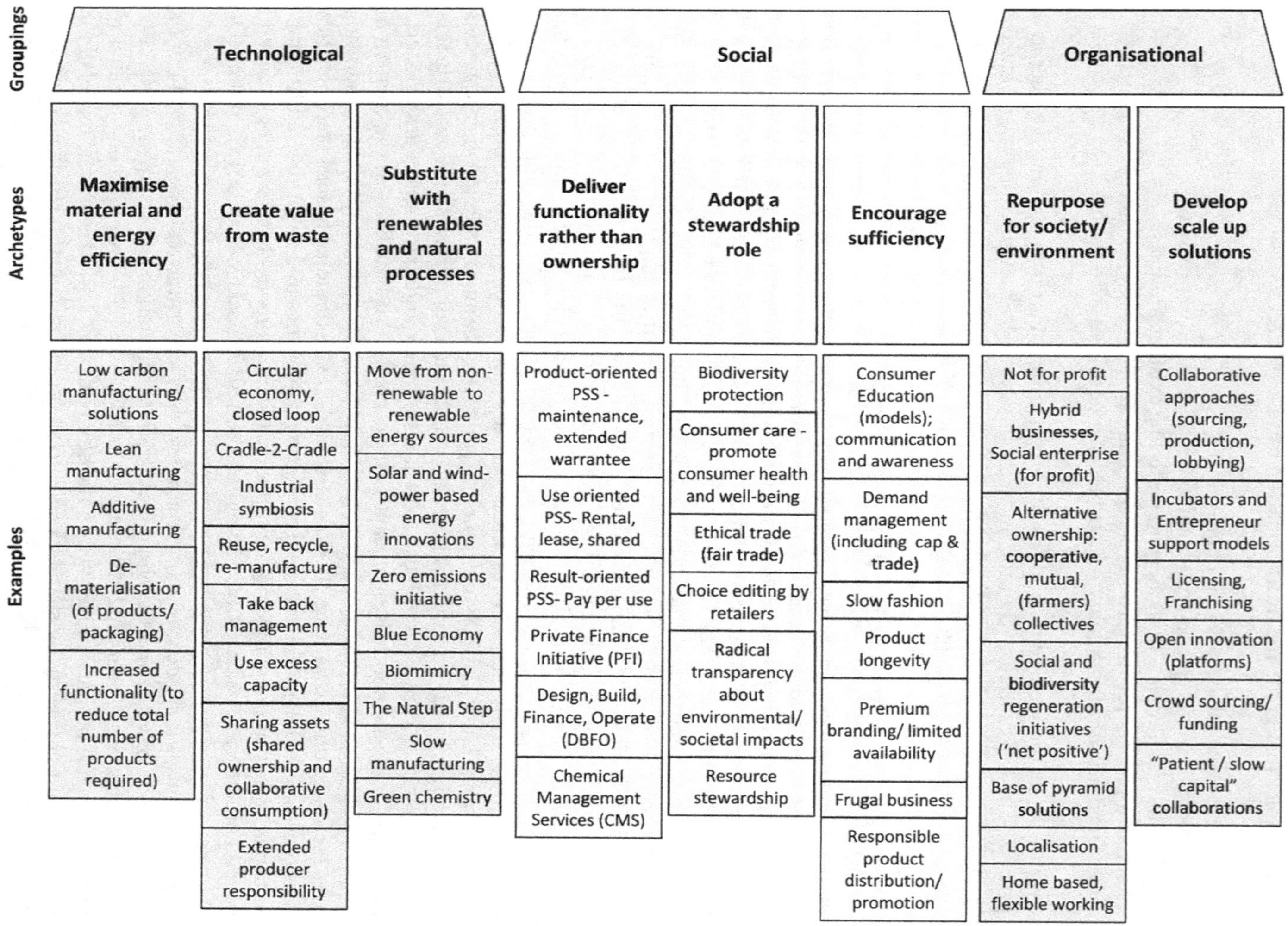

FIGURE 4.3
The sustainable business model archetypes

Source: Bocken et al. (2014), used with permission from Elsevier.

To help companies rethink their business model for sustainability, the tool of value mapping can be used which enables them to chart the value captured, missed and destroyed in the wider context of stakeholders with the help of four brainstorms (Bocken et al. 2015). The transformation of business models towards sustainability may be in the form of adjustment, innovation or redesign (Gauthier and Gilomen 2016).

Schaltegger and colleagues (2016: 6) define a business model for sustainability as follows, echoing Bocken et al. (2014) in their emphasis on value as well as reflecting *natural capitalism* with its focus on other forms of capital:

> A business model for sustainability helps describing, analyzing, managing, and communicating (i) a company's sustainable value proposition to its customers, and all other stakeholders, (ii) how it creates and delivers this value, (iii) and how it captures economic value while maintaining or regenerating natural, social, and economic capital beyond its organizational boundaries.

This definition also highlights that working towards sustainability is not confined to the firm, but goes outside, reflecting the importance of inside-out and outside-in perspectives emphasised earlier in the chapter. We will cover this boundary-spanning aspect in more detail in Chapter 8 on supply chains when we explore the importance of collaborative relationships with suppliers for sustainability. This aspect is also included in the categorisation of sustainability buzzwords and concepts of Hart (2010) shown in Table 4.2 who argues that businesses that want to create

TABLE 4.2

Corporate sustainability buzzwords

	Internal capabilities	External constituencies
Focus: tomorrow, build, opportunity	*Emphasis: Clean technology*	*Emphasis: Sustainability/BoP*
	Eco-effectiveness	Sustainable development
	Biomimicry	Base of the pyramid (BoP)
	Leapfrog technology	Urban reinvestment
	Sustainable technology	Brownfield redevelopment
	Knowledge and service intensity	Inclusive capitalism
	Cradle-to-cradle	Pro-poor business
	Closed loops	Social entrepreneurship
	Restorative technology	Radical transactiveness
	Systems thinking	B24B (business to four billion)
Focus: today, manage	*Emphasis: resource efficiency*	*Emphasis: product stewardship*
	Environmental management systems	Corporate social responsibility
	Greening	Industrial ecology
	Pollution prevention	Stakeholder management
	Eco-efficiency	Life cycle management
	Risk management	Design for environment (DfE)
	Environmental management	Green design
	ISO 14001	Corporate citizenship
	Waste reduction	Full-cost accounting
	Resource productivity	Take-back
		Transparency

Source: adapted from Hart (2010, exhibit 3.3 The Buzzword Sort, and Hart 1997).

sustained value need to address both internal and external capabilities, and stakeholders, as well as keep both the present and the future in mind.

Drawing together the overviews of both Bocken and colleagues (2014) and Hart (2010), it is clear that corporate (environmental) sustainability is a complex and multidimensional field with a seemingly confusing array of choices for companies to make. The classifications provided help make sense of these choices – for example, Bocken and colleagues (2014) emphasise that their archetypes of sustainable business models provide a mechanism to support the innovation process that embeds sustainability in business models. They further detail this mechanism through value mapping (Bocken et al. 2015). An alternative taxonomy of sustainable business models based on 45 value-creating patterns along ecological, social and economic sustainability dimensions is provided by Lüdeke-Freund and colleagues (2018).

It is beyond the scope of the book to cover all the depicted approaches, tools and techniques in detail – as emphasised in the introduction to the chapter and book, the intention here is not to provide all the answers but to equip readers with a broad understanding of corporate environmental sustainability and management. We nevertheless want to stress that in one way or another these concepts have been introduced or covered earlier (with the exception of those falling under social sustainability) even though not necessarily using the same terms – for example delivering functionality rather than ownership is called servicising by Esty and Winston (2009), and service and flow by Hawken and colleagues (1999). Other tools and examples beyond the ones presented in this chapter (e.g. The Natural Step, resource productivity, eco-effectiveness) will be explored further through the corporate functions to which they are most relevant (e.g. lean manufacturing and closed loop in Chapter 9 on production and operations).

Regardless of which tool or strategy companies decide to use, simply choosing an approach is not the endpoint but a significant starting point from which other choices become clearer. How companies move towards sustainability in stages is the subject of the next section, following which we explore what kind of changes and re-thinks are needed for such moves.

PHASES, WAVES AND STAGES IN MOVING TOWARDS SUSTAINABILITY

> If we are to move corporations towards full sustainability, we must be able to identify the stage where they are now so we can determine how to move forward.
>
> (Dunphy et al. 2007: 13)

Whichever overall strategy or model a company adopts, and whichever specific tools and methods it uses within its corporate functions, businesses are at different levels of maturity as environmental management differs in its evolvement and development within companies. The terminology and number of stages varies,[9] though classifications of how companies are placed in their journey towards sustainability are often grouped into passive or reactive, active or preventive, and proactive (Ormazabal and Sarriegi 2014). Grayson (2012) for example identifies the following maturity stages for corporate responsibility: denier, complier, risk mitigator, opportunity maximiser, champion or civil corporation. Hoffman (2008) uses Hunt and Auster's (1990) stages of corporate environmental strategy development that range from zero to maximum protection from environmental risk and from minimal to active resource and management commitment: beginner, firefighter, concerned citizen, pragmatist, proactivist.

Using such models of classifications allows companies and their stakeholders to place them on their journeys and map out what they need to address in order to move to the next stage. This journey is not necessarily uni-directional and can be non-linear, influenced by internal as well as external factors and their interaction (Ormazabal et al. 2017). Landrum (2018) differentiates between models of corporate social responsibility (CSR), corporate sustainability, environmental management and sustainable development. In the following we introduce three models of stages or waves – the phase model of Dunphy and his colleagues (2007) represents a path forward to corporate sustainability, whilst Ormazabal and her colleagues (2017) put forward a framework for how environmental management evolves, whereas Landrum's (2018) model is a comprehensive integration of micro and macro models.

Although Dunphy et al.'s (2007) phase model covers human and ecological sustainability (social and natural capital), we concentrate here on the environmental aspects, which are characterised through these stages:

- *Rejection*: environment seen as 'free good' to be exploited for economic gain, no responsibility for environmental impact taken. Opposition to outside pressure for environmental action.
- *Non-responsiveness*: more ignorance and lack of awareness than opposition. Environment not considered relevant, taken for granted. Business as usual – financial and technological factors dominant.
- *Compliance*: meet minimum standards to reduce risks from sanctions. Most obvious environmental impacts eliminated to avoid liability/pressure, others ignored.
- *Efficiency*: growing awareness, environment seen as avoidable cost and opportunity to increase efficiency (waste elimination, review of processes), some proactive and systematic approaches.
- *Strategic proactivity*: sustainability important part of business strategy, source of strategic (leadership) opportunity, product redesign, new products and processes replace existing, environmentally damaging ones.
- *The sustaining corporation*: internalisation of ecological sustainability values, environmental best practice right thing to do, promoted externally.

We see the strategic responses to sustainability in terms of value covered previously reflected in the stages: the rejection phase destroys value, the non-responsiveness phase limits values – both form part of the so-called *first wave*. In the *second wave*, compliance conserves value, efficiency and strategic proactivity create value, whilst the *third wave* of the sustainable corporation has the effect of sustainable business as value.

Each phase is also characterised by a predominant attitude, which is the focus of senior executives. These attitudes go from opposition and ignorance in the first wave, to risk, cost and competitive advantage in the next and to transformation in the final stage and wave. This also implies a shift from instrumental to non-instrumental motivation (Dunphy et al. 2007).

While the model can be used to chart possible ways forward, there is no assumption of automatic progression through a step-by-step trajectory – some companies for example are founded on ethical and environmental principles and can be placed in a later stage from the beginning. A company may also leapfrog phases or even regress if it abandons previously established sustainability practices. It is also important to note that a company is not seen as a unitary bloc as its constituent parts may be in different phases.

Similarly to Dunphy et al. (2007), Ormazabal and colleagues (2017) stress the non-linearity of progression from lower to higher stages of environmental management and include feedback and causal loops in their maturity stages framework which aims to be prescriptive rather than simply descriptive. Throughout the six stages, commitment from top management and environmental measures and actions are significant factors:

- *Legal Requirements*: common place to start examining ecological maturity, focus on regulatory environment and compliance. End-of-pipe actions, investment in new equipment as result.
- *Responsibility assignment and training*: to comply with legal requirements, positions need to be assigned to take responsibility for environmental management, and employees must be trained, generally and specifically, with regards to equipment and technology.
- *Systemisation*: good environmental practices are systemised through formalisation of processes, which may be in the form of certification (e.g. ISO 14001).
- *ECO2*: both ecological and economic benefits accumulate, operational savings achieved reinforce senior management and employee commitment as well as implemented processes.
- *Eco-Innovative Products & Services*: proactive approach to eco-designed new products leads to increased market share.
- *Leading Green Company*: environmental practices made public, reputational capital further increases market share, environmental stewardship as competitive basis, sustainability leader in industry.

The dynamic nature of the model provides both a framework for classification and for identifying steps needed for ecological advancement, including nontechnical elements, such as top management commitment.

Landrum (2015, 2018) puts forward a developmental model of corporate sustainability that takes into account how corporations view natural capital compared to other types of capital. As covered in Chapter 2, weak sustainability implies substitutability between different kinds of capital, whilst strong sustainability suggest complementarity instead and a separation between natural and other capital. Moving along the spectrum from very weak and weak to intermediate and then strong and very strong sustainability, the proposed model includes the following stages (Landrum 2015, 2018):

- *Non-participatory*: a reactive stance.
- *Compliance*: defensive stance, focus on risk management, very weak sustainability.
- *Business-centred*: proactive, efficiency-focused approach ("do less bad"), weak to moderate sustainability.
- *Systemic/Integrative*: encompasses a strategic triple bottom line approach ("do more good"), intermediate sustainability.
- *Regenerative*: systemic approach to restore and repair impact of industrial consumerist economy, strong sustainability.
- *Co-evolutionary*: co-evolution of economy and society as part of nature, circular model, very strong sustainability.

In including strong sustainability, this model goes beyond other models that do not encompass such stances (Landrum 2015). In extending the stages to regenerative and co-evolutionary it can enhance how corporations view sustainability and help businesses move beyond the triple bottom line and integrative sustainability. Although Hawken and colleagues (1999) do not use the terms strong and weak sustainability, their proposed natural capitalism strategy of reinvestment in natural capital reflects a strong sustainability stance. Similarly, The Natural Step's systems conditions and sustainability principles echo strong sustainability and the protection of natural capital.

Whichever model one uses, the later (or higher) stages all signal a move from add-on to integrated sustainability, ingrained in organisational culture and behaviours, and a willingness to share expertise in collaboration and partnership with others. Although there is no automatic and linear trajectory, backwards or retrograde evolution is difficult to imagine (Grayson 2012). The question then is, how does a company move along the sustainability spectrum, what drives a company to move from non-responsiveness to compliance, from efficiency to co-evolution? What changes are needed?

ORGANISATIONAL CHANGE AND LEADERSHIP FOR SUSTAINABILITY

Dunphy and colleagues (2007, see also Benn et al. 2014) outline the organisational changes needed to move a company along the sustainability spectrum. Although traditionally government and other external factors have driven companies towards sustainability, internal factors have become more important in the developmental stages. In particular, the role of individuals as change agents is critical as only through their actions can progress on the path to corporate sustainability be achieved. Internal change agents may come in the form of board members, CEOs, executives, managers, supervisors, professionals, whereas external change agents could be politicians, investors, consultants, suppliers, financial analysts, activists and other stakeholders such as the local community, regulatory agencies and consumers. The role of change agents highlights the importance of leadership, to which we turn later.

The change these agents enact can happen in two ways, incremental or transformational (or transformative), which can be characterised as follows (Dunphy et al. 2007; Elkington 2012):

- Incremental: planned, emergent, continuous, ongoing, short time scale, narrow scope.
- Transformational: large-scale, revolutionary, radical changes in strategy, structure, capability, and culture, long time scale, wide scope.

Incremental change can be undertaken with resources internal to an organisation, while transformative change often requires the use of external consultants that are experienced, qualified and competent. The following table gives a brief overview of the steps involved in either type of change:

TABLE 4.3

Steps of change towards corporate sustainability

8 Steps of incremental change	10 Steps of transformational change
Future workshops/search conferences	Know where you are now
Assess current position in relation to sustainability	Develop the vision: the dream organisation
Evaluate type of change programme needed	Identify the gap
Identify change agents	Assess the readiness for change
Pilot new practices and innovations	Set the scene for action
Harness further resources	Secure basic compliance first
Communicate and extend programme	Move beyond compliance
Align organisational systems	Establish performance criteria for "compliance-plus"
	Launch and manage the transformational change programme
	Maintain the rage

Source: Benn et al. 2014; Dunphy et al. 2007.

The choice of change strategy needs to be carefully considered and aligned with a company's overall strategy and its situational context – some circumstances warrant incremental change, whereas others require organisational transformation, and change for change's sake should be avoided. Such change "faddism" is one of the pitfalls of incremental change as is a lack of fit with the external environment. Other limitations include decreased managerial support, a lack of a clear vision and objectives of the change programme and insufficient learning mechanisms.

The advantages of incremental change can help address these pitfalls. Benefits include the power of small wins as a driver for change, a positive culture change, efficiency improvements, the creation of new organisational structures, and capability development of technical and human competencies as well as change competencies.

Dunphy et al. (2007) argue for organisational flexibility regarding the two change approaches, in line with external market conditions, and suggest the development of ambidextrous capabilities to manage both approaches to change. Similarly, depending on fit and suitability, companies can use top-down and bottom-up approaches to initiate change, which for Laszlo and Zhexembayeva (2011) centres around four interdependent areas of focus – getting the right start, building the buy-in, moving from incremental to breakthrough, and staying with it. The importance of buy-in and involvement of key stakeholders is also emphasised by Yoxon (2011), though change does not have to come from the top as leadership is important throughout the company (Blowfield 2013).

Whether top-down or bottom-up, King (2011) outlines that the transition to sustainability requires work at all organisational levels, each with their own timeframes and challenges:

- Level 1 task focus: day-to-day operational activities, time span perspective of three months.
- Level 2 first line supervision or management: still operational, but some more responsibility, at most, time span perspective of one year.
- Level 3 department and system management: wider responsibilities including some operational, but also involvement in corporate planning, budgeting, implementation, monitoring and reporting system and cycle; longer time span to achieve goals, two years.
- Level 4 general management of a division/function: e.g. head of function, for a line of business, a division or department; first level at which management can be comprehensive and tactical; time horizon of five years.
- Level 5 business unit/corporate division management: tactical, comprehensive work and strategic intent; contribute to corporate policies and developing business unit strategies; expansion of time horizon, about ten years.
- Level 6 international level: fully strategic, time horizon of 20 years; level where sustainability has to be the concern of the whole top team of an international organisation.
- Level 7 global level: higher strategic level with time span of up to 50 years and working in relation to global environment; ability to see and preview possible futures, ready and interested in contributing by playing an active part in bringing about desirable futures.

Long-term orientation is a key characteristic of sustainable business models (discussed earlier): whilst incremental change delivers quick wins through low-hanging fruit, true sustainability is a long-haul game.

Time is also emphasised by Elkington (1997), who identifies seven revolutions that are needed for a transition to sustainability, a change from old to new paradigm: Markets change from compliance to competition, values from hard commercial ones to softer triple bottom line values, transparency from closed to open, life cycle technology changes focus from product to function, partnership shifts from subversion to symbiosis, time from wider to longer, and corporate governance from exclusive to inclusive. Based on these revolutions, he suggests 39 steps to move sustainability to the core of corporate policies and practice. For Elkington (1997), leadership is key to the top-down and bottom-up revolution of sustainability.

Leadership has been highlighted as key to address sustainability effectively though there is no agreement as to how unique sustainability leadership is (Blowfield 2013). Whilst there are many definitions and theories of leadership,[10] which we cannot cover here, it is important to keep in mind that leadership can happen at all levels of an organisation, and a leader is generally someone who has followers (Cavagnaro and Curiel 2012; Blowfield 2013). Effective change leaders towards sustainability need a clear vision alongside a clear understanding of their role, relevant knowledge and the right skills needed to implement the change (Dunphy et al. 2007).

Piasecki and colleagues (1999) stress the comprehensiveness of the skills required for corporate environmental leaders, who firstly need to ensure regulatory compliance before they can then go beyond compliance in capturing business opportunities based on innovation and risk-taking, all the while working with a broad range of stakeholders. An important element of sustainability leadership is the creation of opportunities for innovation across the company, whether it is intrapreneurship or entrepreneurship (Blowfield 2013).

BOX 4.10 THE ROLE OF ENTREPRENEURS AND INTRAPRENEURS IN SUSTAINABILITY

Sustainability leadership is not confined to the top level of an organisation or only to large organisations: it can happen at all levels and in SMEs as well, when breakthrough opportunities for new products and new business models are grasped by intrapreneurs and entrepreneurs.

Whilst entrepreneurs generally run their own organisations, intrapreneurs are employees in a larger organisation who are given the freedom to innovate within a firm, making use of an entrepreneurial spirit within it. This may result in the revitalisation and diversification of the business. Such internal venturing through intrapreneurship encourages expansion in-house as opposed to mergers and acquisitions covered in Box 4.11. Intrapreneurs play an important role in moving companies towards environmental sustainability as their knowledge of the organisation means they can successfully navigate corporate culture, process and strategy, communicate goals and build relationships and teams across corporate functions. Nevertheless, intrapreneurs face obstacles such as inertia and incumbency which entrepreneurs do not, as SMEs provide opportunities for flexibility and disruptive innovation conducive for sustainability. Companies where intrapreneurs have influenced and driven corporate sustainability agendas include Walmart, Unilever, and Nike, amongst others.

SMEs are not a homogenous group as they come in different kinds of entrepreneurship, which has implications for sustainability. Some SMEs might see sustainability as an unnecessary cost, whilst others built their business around a core of sustainability, such as for example social enterprises which work on business-led solutions for social aims and reinvest profits for this purpose. Some argue that by their nature SMEs are more open to sustainability than

BOX 4.10 (CONTINUED)

other companies, as they tend to have closer relationships with communities, are often family-based and might feel less obliged to maximise shareholder value.

Whilst social enterprises have received a lot of attention in literature and practice, sustainability entrepreneurs also come in the form of environmental entrepreneurs focused on creating new ventures, for-profit or not-for-profit, that help address environmental challenges through the introduction of new products and services that are environmentally benign. Drivers of environmental entrepreneurship include economic incentives, personal motivations and the institutional context. Entrepreneurs are in a particularly good position to tackle the challenges inherent to environmental concerns – uncertainty, innovation and resource allocation. Environmental entrepreneurs are also called green entrepreneurs or ecopreneurs. Whilst some environmental entrepreneurs are opportunist entrepreneurs who have found a green niche, others, such as ecopreneurs, set up their enterprises based on sustainability principles. Ecopreneurs are therefore crucial change agents in the sustainability transformation.

Examples of companies set up with sustainability embedded from the start by the founding entrepreneurs include the cosmetics manufacturer and retailer Lush, and Seventh Generation, a green household products brand.

Sources: Blowfield 2013; Elkington 2012; Lenox and York 2012; Stead and Stead 2014; Walley and Taylor 2002.

Stead and Stead (2014) emphasise how a culture built on entrepreneurial learning, collaboration and communication can facilitate transformational change. Organisational culture can be loosely described as "the way we do business around here" and provides a collective base of meaning and continuity out of which employees make sense of their work and organisation (Dunphy et al. 2007: 81). Organisational culture needs to change alongside structure, reward systems and job responsibilities for a shift in environmental strategy (Hoffman 2008). Hoffman suggests a staged road map for organisational change management of four general phases and eight critical steps. The four phases are:

1 *Diagnosis*: recognition that change is necessary, strategy development.
2 *Unfreezing*: preparation for change in organisation.
3 *Movement*: implementation of change plan.
4 *Refreezing*: institutionalisation of changes.

In phase 2, the first three critical steps are to establish a sense of urgency for the change, to form a powerful guiding coalition of internal and possibly external stakeholders and to create a vision of goals, plans and timetables. In phase 3, four steps are critical – communicating the vision set out in the previous phase, empowering others to act, i.e. involving all employees, planning for and creating short-term wins, and consolidating improvements whilst producing still more change. The final critical step is in phase 4 and sees the institutionalisation of new approaches, such as in reward structures, selection, socialisation and organisational structure. We will look at some of these elements in more detail in Chapter 13 on HRM.

As organisational change requires unlearning and presents uncertainty, change agents might well experience resistance to change, which can take various forms, such as habitual routines, resource limitations, fear of the unknown and threats to established power bases (Hoffman 2008). It is the task of the change agent to overcome these restrictions in organisational culture and structure. Willard (2008) offers advice to handle the objections that inhibit corporate sustainability commitments. He provides detailed counterarguments and responses to address what he groups as the four main inhibiters, all with different levels of objections – lack of support from senior leaders, fear of backlash, weak business case and mind-set.

One strategy of learning a company can pursue in changing towards sustainability, is through acquisition (or mergers), which is often used to expand a company's scope and size externally rather than internally through intrapreneurship (see Box 4.10). Acquisition allows a firm to expand vertically across the supply chain or horizontally within the industry, or, using diversification, to move into new products and markets (Stead and Stead 2014). Box 4.11 provides some examples of how companies have addressed sustainability through acquisition.

BOX 4.11 ACQUISITIONS AS A ROUTE TO CHANGE TOWARDS SUSTAINABILITY

Strategy was emphasised as a learning process at the start of the chapter, and one avenue companies have pursued in moving towards sustainability is through acquiring companies further along the sustainability spectrum, making the acquiring company in effect learning organisations. For the acquired companies, this offers advantages of scale, marketing and production, whilst for the acquiring firms, it provides a chance to build a sustainability reputation. Prominent examples where mainstream companies have bought sustainable brands include some of the following.

Cadbury's acquisition of Green & Black's led to Cadbury using the sustainability reputation of its acquisition to strengthen its efforts in the rest of the company e.g. launching the Cadbury Cocoa Partnership for investments in sustainable cocoa production.

Unilever had already started its sustainability journey when it took over ice cream maker Ben & Jerry's, but took important lessons from the principles Ben & Jerry's is run on, such as its triple bottom line mission statement.

Coca-Cola's purchase of *innocent* allowed the smoothie maker to continue its mission, which includes pushing harder for better social and environmental standards in business whilst providing scaling opportunities and market access across the world. Innocent boosted the parent company's revenues but at the same time also influenced its approach to sustainability e.g. in sustainable sourcing.

Puma and luxury brand Kering: sports gear company Puma launched an environmental profit and loss account in 2011, becoming one of the first companies in the world to put a monetary value on the ecosystem services it uses. Parent company Kering (previously PPR) had acquired Puma in 2007 and announced in 2011 that the whole group would follow in Puma's footsteps and publish environmental profit and loss accounts, with the methodology being made public to allow other companies to follow suit.

Further examples of large companies acquiring sustainable or green brands include Coca-Cola's acquisition of Honest Tea, maker of organic

BOX 4.11 (CONTINUED)

and healthy drinks, cleaning company Clorox's acquisition of Burt's Bees, and L'Oréal's purchase of The Body Shop. However, in 2017, L'Oréal put The Body Shop up for sale, whilst Cadbury was taken over by Kraft Foods in 2010 (now Mondelez International), leading critics to ask questions about the sustainability commitment of these corporations.

Sources: Blowfield 2013, www.greenbiz.com, www.kering.com, www.bbc.co.uk/news/business-18167345, www.innocentdrinks.co.uk, www.marketingweek.com/2014/02/21/how-innocent-and-coke-have-influenced-each-other/.

Although leadership is a significant element of a company's move towards sustainability, and shifts are indeed often triggered by the arrival of for example a new CEO, there are other drivers for these shifts as well, such as stakeholder pressure, new regulation, economic fluctuation or other staff changes (Dunphy et al. 2007).

When is a business truly sustainable is the question Dyllick and Muff (2016) ask, outlining the shifts that are required to move business sustainability from version 1.0 (Refined Shareholder Value Management) via 2.0 (Managing for the Triple Bottom Line) to 3.0 (True Sustainability). The starting point is business as usual, the current economic paradigm based on shareholder value, which is refined in a first step to business sustainability (1.0). The economic bottom line is then extended to take into account social and environmental concerns for sustainability 2.0. A truly sustainable business in sustainability 3.0 translates sustainability challenges into business opportunities. The authors (2016: 165f) define a truly sustainable business as follows:

> Truly sustainable business shifts its perspective from seeking to minimize its negative impacts to understanding how it can create a significant positive impact in critical and relevant areas for society and the planet. A Business Sustainability 3.0 firm looks first at the external environment within which it operates and then asks itself what it can do to help overcome critical challenges that demand the resources and competencies it has at its disposal.

The challenges that need to be overcome in this roadmap (Table 4.4) to true sustainability include domains of organisational development such as ownership (finance, time horizons), governance (stakeholders, measuring and reporting, transparency), strategy (products and services), and business models but also culture and organisational behaviour encompassing leadership and innovation (Dyllick and Muff 2016; Muff and Dyllick 2014). These all help to achieve the shifts required for true sustainability across a company's concerns (what?), its organisational perspectives (how?) and values created (what for?) – concerns need to broaden from uni-dimensional economic to three-dimensional concerns, value needs to expand and be created beyond shareholders and organisational perspectives need to change to encompass inside-out and outside-in ones. These shifts are summarised by the following table (4.4) classifying business sustainability:

Whichever framework a company uses to implement sustainability in the organisation, communication and collaboration are crucial, as are good leadership and governance (Robertson 2017). This is particularly significant when profound change is needed for the change process towards sustainability, as all aspects of a company are affected – strategy and the values, vision and mission underlying it and operational activities, which all need to be aligned with sustainability principles, with the help of organisational culture (Cavagnaro and Curiel 2012).

TABLE 4.4

Business sustainability – roadmap of key characteristics and shifts

	Concerns (What?)	Organisational perspective (How?)	Values created (What for?)
Business Sustainability 3.0	Sustainability challenges as starting point	Outside-in	For the common good
Business Sustainability 2.0	Three-dimensional	Inside-out	Triple bottom line
Business Sustainability 1.0	Three-dimensional	Inside-out	Shareholder value refined
Business as usual	Economic concerns	Inside-out	Shareholder value

Source: adapted from Dyllick and Muff 2016.

SUMMARY

Having established the scientific and economic foundations, and regulatory principles that underpin environmental sustainability in the previous chapters, in this final chapter of part one of the book we asked, how can a company move towards sustainability? What kind of strategies and business models are available and what kind of change is necessary? The choices available to companies were presented as conventional and sustainability-focused strategy frameworks, incremental versus transformative change, sustainability as an add-on to existing strategy or sustainability as embedded and integrated into corporate strategy.

The choices reflect a theme that ran through the chapter – the changes, shifts and moves that are required to start the journey towards corporate environmental sustainability. From re-thinking the relationship between the economy and the environment from tension to win-win, from viewing sustainability as a cost to viewing it as an opportunity, from inside-out to outside-in approaches, from sustainability as value destruction to value creation, from either-or to both, from offering ownership of products to service provision, from defensive to offensive, from linear to circular, from conventional to natural capitalism, there are plenty of ways to forge a path from business as usual to business sustainability 1.0, 2.0 and even 3.0 where companies shift from minimising their negative environmental impact to creating a positive impact by reinvesting in natural capital.

Beyond these choices and tensions, there are multiple layers and levels that make up strategies and models for business – Valentine (2010) visualises corporate environmental strategies as an onion consisting of an outer macro layer of outside forces, a secondary (external) stakeholder layer, the industry, the firm and lastly, the functional layer, within which strategies are driven by opportunities such as competitive advantage, cost reduction, sustainable consumption and by threats such as risk, legislation and reputation, amongst others, in their quest for corporate environmental sustainability. Having looked at the external, regulatory framework in the previous chapter, we focused on the internal, strategic aspects here and will move on to corporate functions in the third part as environmental sustainability needs implementation across the organisation at a functional level and not just at a higher strategic level.

Piasecki et al. (1999) also stress that corporate environmental strategies require both the visionary facets of strategy and the functional aspects, which means that corporate environmental management is neither an operations nor a staff function but needs integration throughout an organisation.

How companies address environmental sustainability across their functions is the focus of the third part of the book, where the overall strategies covered here are cascaded to the operational level, looking at supply chains, design, production and operations, marketing, information technology and (environmental) management systems, accounting and human resource management.

Before we direct our attention to the inside of companies, we return to the external context again, though with a different angle: so far, we have covered environmental issues and challenges for corporate sustainability rather generally. In the next part, these challenges will become more concrete and focused as we cover three particular challenges in the wider context of environmental issues with which companies are faced – climate change, pollution and waste.

SUGGESTED QUESTIONS AND ACTIVITIES

- Test your sustainability literacy: www.sulitest.org/en/index.html
- Watch Ray Anderson's TED talk on how he turned his company Interface around: www.ted.com/talks/ray_anderson_on_the_business_logic_of_sustainability.html.
- How useful is The Natural Step as a framework for making a company sustainable?
- Case study of IKEA and The Natural Step: http://pdf.wri.org/bell/case_1-56973-250-7_full_version_english.pdf.
- Explore the Natural Capital Coalition website and research how one of the member companies have changed their approach to natural capital: http://naturalcapitalcoalition.org.
- If you are working/have worked for a company, explore and find out in which stage of Dunphy's model it is. Alternatively, use one of the other stage-based frameworks introduced.
- Download the business model canvas: https://strategyzer.com/canvas/business-model-canvas and chart the business model of a company of your choice. If you have a business idea that is based on sustainability, try charting your own business model.
- Watch the following video on how Vigga Svensson switched from manufacturing and selling baby clothes to manufacturing them and hiring them out: www.bbc.co.uk/news/av/magazine-40190007/the-woman-who-wants-you-to-rent-used-baby-clothes – can you find other examples of the service and flow economy which provides functionality rather than ownership?
- Incremental or transformational change? What kind of change would you enact to move a company towards sustainability?
- Check your learning – which part of the chapters were easy to understand and which might you need to revisit? Explain one of the models to a friend to see whether you have understood it.

NOTES

1 Prof Hoffman, University of Michigan, in his Foreword to *Embedded Sustainability*, highlights how nonsensical the question "does it pay to be green?" has become with the shifts in the market, as it is akin to asking "does it pay to innovate?" (Laszlo and Zhexembayeva 2011).

2 There are other schools of strategy thought (such as design, entrepreneurial, cognitive, learning, cultural and political, Laszlo and Zhexembayeva 2011) though these two are the most prevalent.

3 For a re-examination of Porter's Five Forces through a sustainability lens, see Blowfield (2013) and Weybrecht (2009).

4 It should be noted that sustainable here is in the economic and company sense, not as otherwise used in this book.

5 According to Hawken and colleagues (1999: 261), "For all their power and vitality, markets are only tools. They make a good servant but a bad master and a worse religion. They can be used to accomplish many important tasks, but they can't do everything, and it's a dangerous delusion to begin to believe that they can – especially when they threaten to replace ethics or politics". For Anderson (2009) the redesign of commerce for environmental sustainability requires a change of the law.

6 We return to the circular economy in Chapter 7 on waste, 8 on supply chain management and 9 on production and operations.

7 The risk of pursuing eco-efficiency is exemplified in the so-called *rebound* effect, which means that efficiency savings lead to reduced prices, which in turn lead to increased consumption made more affordable and accessible as a result, thus offsetting the efficiency obtained with using more resources for consumption. For more on the rebound effect and its direct and indirect components, and the so-called Jevons paradox, see O'Connor (2015a, 2015b), and Rowson (2013).

8 Boons and Lüdeke-Freund (2013) highlight that that the configuration of the building blocks of the business model can be used for strategy execution. A business model is not a strategy, but rather a result of strategy (see also Casadesus-Masanell and Ricart 2010).

9 Ormazabal and colleagues (2017) provide an overview of the typical environmental maturity stages from the literature whilst Landrum (2018) covers all aspects of sustainability in her review of stage models.

10 Conventional theories of individual leadership include the following: great man, trait, behaviourist, situational leadership, contingency, transactional and transformational theory (Blowfield 2013), whilst definitions of leadership may focus on processes, personality, influence, behaviour, persuasion, power relations and goal achievement, amongst others (Cavagnaro and Curiel 2012).

SOURCES AND FURTHER READING

Anderson, R. C. (1998) *Mid-Course Correction: Toward a Sustainable Enterprise: The Interface Model*, White River Junction, VT: Chelsea Green.

Anderson, R. C. and White, R. (2009) *Confessions of a Radical Industrialist: How Interface Proved that You Can Build a Successful Business Without Destroying the Planet*, London: Random House.

Aragón-Correa, J. A., Hurtado-Torres, N., Sharma, S. and García-Morales, V. J. (2008) Environmental Strategy and Performance in Small Firms: A Resource-Based Perspective, in: *Journal of Environmental Management*, 86(1): 88–103.

Aragón-Correa, J. A. and Sharma, S. (2003) A Contingent Resource-Based View of Proactive Corporate Environmental Strategy, in: *Academy of Management Review*, 28(1): 71–88.

Barney, J. B. (1991) Firm Resources and Sustained Competitive Advantage, in: *Journal of Management*, 17(1): 99–120.

Barney, J. B. (2003) Resource, Capabilities, Core Competencies, Invisible Assets, and Knowledge Assets: Label Proliferation and Theory Development in the Field of Strategic Management, in: Helfat, C. E. (ed.) *The SMS Blackwell Handbook of Organizational Capabilities: Emergence, Development, and Change*, Malden, MA: Blackwell, pp. 422–426.

Benn, S., Dunphy, D. and Griffiths, A. (2014) *Organizational Change for Corporate Sustainability*, London: Routledge, 3rd ed.

Benyus, J. M. (1997) *Biomimicry: Innovation Inspired by Nature*, New York: William Morrow.

Blowfield, Michael (2013) *Business and Sustainability*, Oxford: Oxford University Press.

Bocken, N. M. P., Rana, P. and Short, S. W. (2015) Value Mapping for Sustainable Business Thinking, in: *Journal of Industrial and Production Engineering*, 32(1): 67–81.

Bocken, N. M. P., Short, S. W., Rana, P. and Evans, S. (2014) A Literature and Practice Review to Develop Sustainable Business Model Archetypes, in: *Journal of Cleaner Production*, 65: 42–56.

Boons, F. and Lüdeke-Freund, F. (2013) Business Models for Sustainable Innovation: State-of-the-Art and Steps Towards a Research Agenda, in: *Journal of Cleaner Production*, 45: 9–19.

Bourne, M., Bourne, P. and Ferguson, D. (2012) Strategic Business Performance for Sustainability, in: Grayson, D. and Exter, N. (eds.) *Cranfield on Corporate Sustainability*, Sheffield: Greenleaf Publishing, pp. 80–104.

Casadesus-Masanell, R. and Ricart, J. E. (2010) From Strategy to Business Models and onto Tactics, in: *Long Range Planning*, 43(2–3): 195–215.

Cavagnaro, E. and Curiel, G. (2012) *The Three Levels of Sustainability*, Sheffield: Greenleaf Publishing.

Christensen, C. M. (1997) *The Innovator's Dilemma: When New Technologies Cause Great Firms to Fail*, Boston: Harvard Business Review Press.

Christensen, C. M. and Overdorf, M. (2000) Meeting the Challenge of Disruptive Change, in: *Harvard Business Review*, 78(2): 66–76.

Dunphy, D., Griffiths, A. and Benn, S. (2007) *Organizational Change for Corporate Sustainability: A Guide for Leaders and Change Agents of the Future*, London: Routledge, 2nd ed.

Dyllick, T. and Muff, K. (2016) Clarifying the Meaning of Sustainable Business: Introducing a Typology from Business-as-Usual to True Business Sustainability, in: *Organization & Environment*, 29(2): 156–174.

Elkington, J. (1997) *Cannibals with Forks: The Triple Bottom Line of 21st-Century Business*, Oxford: Capstone.

Elkington, J. (2012) *The Zeronauts: Breaking the Sustainability Barrier*, London: Routledge.

EMF (2013) Towards the Circular Economy: Economic and Business Rationale for an Accelerated Transition, Ellen MacArthur Foundation, Vol. 1, Isle of Wight.

Esty, D. C. and Winston, A. S. (2009) *Green to Gold: How Smart Companies Use Environmental Strategy to Innovate, Create Value and Build Competitive Advantage*, Hoboken, NJ: Wiley.

Everard, M. (2011) Sustainability and Sustainable Development, in: Brady, J., Ebbage, A. and Lunn, R. (eds.) *Environmental Management in Organizations, The IEMA Handbook*, London: Earthscan/ Routledge, 2nd ed., pp. 39–61.

FT (2011) Tesco: Supermarket Chain Puts Its Vehicles and Drivers on Drastic 'F Plan' Diet, *Financial Times*, 20.10.2011, available at www.ft.com/content/141a1602-f294-11e0-931e-00144feab49a (Retrieved, 24.05.2017).

Gauthier, C. and Gilomen, B. (2016) Business Models for Sustainability: Energy Efficiency in Urban Districts, in: *Organization & Environment*, 29(1): 124–144.

Grayson, D. (2012) Overview of Embedding Corporate Sustainability, in: Grayson, D. and Exter, N. (eds.) *Cranfield on Corporate Sustainability*, Sheffield: Greenleaf Publishing, pp. 7–26.

Hahn, T., Figge, F., Pinkse, J. and Preuss, L. (2018) A Paradox Perspective on Corporate Sustainability: Descriptive, Instrumental, and Normative Aspects, in: *Journal of Business Ethics*, 148(2): 235–248.

Hart, S. L. (1995) A Natural-Resource-Based View of the Firm, in: *Academy of Management Review*, 20(4): 986–1014.

Hart, S. L. (1997) Beyond Greening: Strategies for a Sustainable World, in: *Harvard Business Review*, 75(1): 66–76.

Hart, S. L. (2010) *Capitalism at the Crossroads: Next Generation Business Strategies for a Post-Crisis World*, Upper Saddle River, NJ: Pearson, 3rd ed.

Hart, S. L. and Dowell, G. (2011) A Natural-Resource-Based View of the Firm: Fifteen Years After, in: *Journal of Management*, 37(5): 1464–1479.

Hawken, P. (1993) *Ecology of Commerce: A Declaration of Sustainability*, New York: Harper Business.

Hawken, P., Lovins, A. B. and Lovins, L. H. (1999) *Natural Capitalism: The Next Industrial Revolution*, London: Earthscan.

Hawken, P., Lovins, A. B. and Lovins, L. H. (2010) *Natural Capitalism: The Next Industrial Revolution*, 10th Anniversary Edition with New Introduction, London: Earthscan.

Hoffman, A. J. (2008) Strategy Originates Within the Organization, in: Russo, M. V. (ed.) *Environmental management: Readings and Cases*, Los Angeles: Sage, 2nd ed., pp. 151–168.

Hunt, C. B. and Auster, E. R. (1990) Proactive Environmental Management: Avoiding the Toxic Trap, in: *Sloan Management Review*, 31(2): 7–18.

Kim, C. and Mauborgne, R. (2004) Blue Ocean Strategy, in: *Harvard Business Review*, 82(10): 76–84.

King, R. (2011) The Future of the Environmental Profession, in: Brady, J., Ebbage, A. and Lunn, R. (eds.) *Environmental Management in Organizations, The IEMA Handbook*, London: Earthscan/ Routledge, 2nd ed., pp. 543–558.

Kor, Y. Y. and Mahoney, J. T. (2000) Penrose's Resource-Based Approach: The Process and Product of Research Creativity, in: *Journal of Management Studies*, 37(1): 109–139.

Kramer, M. and Kania, J. (2006) Changing the Game: Leading Corporations Switch from Defense to Offense in Solving Global Problems, in: *Stanford Social Innovation Review*, 4(1): 20–27.

Laasch, O. and Conway, R. (2015) *Principles of Responsible Management: Global Sustainability, Responsibility and Ethics*, Stamford, CT: Cengage Learning.

Landrum, N. E. (2015) Evolution of Corporate Sustainability Along the Sustainability Spectrum,

Poster at 2015 Climate Change Conference at Loyola University Chicago, available at www.luc.edu/sustainability/initiatives/climatechange/2015conference/postersession/ (Retrieved, 02.06.2017).

Landrum, N. E. (2018) Stages of Corporate Sustainability: Integrating the Strong Sustainability Worldview, in: *Organization & Environment*, 31(4): 287–313.

Laszlo, C. and Zhexembayeva, N. (2011) *Embedded Sustainability: The Next Big Competitive Advantage*, Sheffield: Greenleaf Publishing.

Lenox, M. and York, J. G. (2012) Environmental Entrepreneurship, in: Bansal, P. and Hoffman, A. J. (eds.) *The Oxford Handbook of Business and the Natural Environment*, Oxford: Oxford University Press, pp. 70–82.

Lovins, A. B., Lovins, L. H. and Hawken, P. (1999) A Road Map for Natural Capitalism, in: *Harvard Business Review*, 77(3): 145–158.

Lüdeke-Freund, F., Carroux, S., Joyce, A., Massad, L. and Breuer, H. (2018) The Sustainable Business Model Pattern Taxonomy – 45 Patterns to Support Sustainability-Oriented Business Model Innovation, in: *Sustainable Production and Consumption*, 15: 145–162.

McDonough, W. and Braungart, M. (1998) The Next Industrial Revolution, in: *Atlantic Monthly*, October: 82–92.

McDonough, W. and Braungart, M. (2002) *Cradle to Cradle: Remaking the Way We Make Things*, New York: North Point Press.

Molthan-Hill, P., Hill, S. and Parkes, C. (2017) A New Framework for Embedding Sustainability into the Business School Curriculum, in: Molthan-Hill, P. (ed.) *The Business Student's Guide to Sustainable Management*, Sheffield: Greenleaf Publishing, 2nd ed., pp. 13–51.

Muff, K. and Dyllick, T. (2014) An Organizational Roadmap to Business Sustainability, Business School Lausanne Working Paper, available at https://ssrn.com/abstract=2442211 (Retrieved, 13.06.2017).

The Nature of Business (2013) Biomimicry for Business? *Blog Post*, available at https://thenatureofbusiness.org/2013/06/19/biomimicry-for-business/ (Retrieved, 14.01.2018).

Nemetz, P. N. (2013) *Business and the Sustainability Challenge: An Integrated Perspective*, London: Routledge.

O'Connor, P. (2015a) Energy Efficiency, Part 1: What's the Point? *Blog Post*, available at http://blog.ucsusa.org/peter-oconnor/energy-efficiency-part-1-whats-the-point-933? (Retrieved, 05.06.2017).

O'Connor, P. (2015b) What is the Rebound Effect? – Energy Efficiency, Part 2, blog post, available at http://blog.ucsusa.org/peter-oconnor/energy-efficiency-what-is-the-rebound-effect-946 (Retrieved, 05.06.2017).

Ormazabal, J., Rich, E., Sarriegi, J. M. and Viles, E. (2017) Environmental Management Evolution Framework: Maturity Stages and Causal Loops, in: *Organization & Environment*, 30(1): 27–50.

Ormazabal, J. and Sarriegi, J. M. (2014) Environmental Management Evolution: Empirical Evidence from Spain and Italy, in: *Business Strategy and the Environment*, 23(2): 73–88.

Osterwalder, A. and Pigneur, Y. (2010) *Business Model Generation*, Hoboken, NJ: Wiley.

Pedersen, E. R. G. (2017) Business Models for Sustainability, webinar, Academy of Business in Society (ABIS), available at www.abis-global.org/resources-publications/abis-webinar-series-2017-business-models-for-sustainability (Retrieved, 07.06.2017).

Piasecki, B. W., Fletcher, K. A. and Mendelson, F. J. (1999) *Environmental Management and Business Strategy: Leadership Skills for the 21st Century*, New York: John Wiley & Sons.

Porter, M. E. (1980) *Competitive Strategy: Techniques for Analyzing Industries and Competitors*, New York: The Free Press.

Prahalad, C. K. and Hamel, G. (1990) The Core Competence of the Corporation, in: *Harvard Business Review*, 68(3): 79–91.

Prahalad, C. K. and Hart, S. L. (2002) The Fortune at the Bottom of the Pyramid, in: *Strategy + Business*, 26(1): 1–14.

Reinhardt, F. L. (1998) Environmental Product Differentiation: Implication for Corporate Strategy, in: *California Management Review*, 40(4): 43–73.

Robertson, M. (2017) *Sustainability Principles and Practice*, London: Routledge, 2nd ed.

Rouse, M. J. and Daellenbach, U. S. (2002) More Thinking on Research Methods for the Resource-Based Perspective, in: *Strategic Management Journal*, 23(10): 963–967.

Rowson, J. (2013) A New Agenda on Climate Change: Facing up to Stealth Denial and Winding Down on Fossil Fuels, Royal Society for the encouragement of Arts, Manufactures and Commerce, available at www.thersa.org/discover/publications-and-articles/reports/a-new-agenda-on-climate-change (Retrieved, 10.07.2017).

Russo, M. V. and Fouts, P. A. (1997) A Resource-Based Perspective on Corporate Environmental Performance and Profitability, in: *Academy of Management Journal*, 40(3): 534–559.

Russo, M. V. and Minto, A. (2012) Competitive Strategy and the Environment: A Field of Inquiry Emerges, in: Bansal, P. and Hoffman, A. J. (eds.) *The Oxford Handbook of Business and the Natural Environment*, Oxford: Oxford University Press, pp. 29–49.

Schaltegger, S., Hansen, E. G. and Lüdeke-Freund, F. (2016) Business Models for Sustainability: Origins, Present Research, and Future Avenues, in: *Organization & Environment*, 29(1): 3–10.

Stead, J. G. and Stead, W. E. (2014) *Sustainable Strategic Management*, Sheffield: Greenleaf Publishing, 2nd ed.

Stubbs, W. and Cocklin, C. (2008) Conceptualizing a "Sustainability Business Model", in: *Organization & Environment*, 21(2): 103–127.

TNS websites Robèrt (2002) The Natural Step, available at www.naturalstep.ca/, www.thenaturalstep.org (Retrieved, 05.11.2019).

Valentine, S. V. (2010) The Green Onion: A Corporate Environmental Strategy Framework, in: *Corporate Social Responsibility and Environmental Management*, 17(5): 284–298.

Van der Byl, C. A. and Slawinski, N. (2015) Embracing Tensions in Corporate Sustainability: A Review of Research from Win-Wins and Trade-Offs to Paradoxes and Beyond, in: *Organization & Environment*, 28(1): 54–79.

von Weizsäcker, E., Lovins, A. B. and Lovins, L. H. (1998) *Factor Four: Doubling Wealth, Halving Resource Use*, London: Earthscan.

Walley, E. E. and Taylor, D. W. (2002) Opportunists, Champions, Mavericks . . . A Typology of Green Entrepreneurs, in: *Greener Management International*, 38: 31–43.

Weybrecht, G. (2009) *The sustainable MBA: The Manager's Guide to Green Business*, Hoboken, NJ: Wiley.

Whitfield, R. and McNett, J. (2014) *A Primer on Sustainability: In the Business Environment*, New York: Business Expert Press.

Willard, R. (2008) Objection-Handling Clinic on Inhibiters to the Next Wave, in: Russo, M. V. (ed.) *Environmental management: Readings and Cases*, Los Angeles: Sage, 2nd ed., pp. 168–190.

Woolley-Barker, T. (2016) 10 Amazing Uses for Biomimicry in Business, *GreenBiz*, 28.12.2016, availableatwww.greenbiz.com/article/10-amazing-uses-biomimicry-business (Retrieved, 14.01.2018).

Yoxon, M. (2011) Engaging with People in Your Organization, in: Brady, J., Ebbage, A. and Lunn, R. (eds.) *Environmental Management in Organizations, The IEMA Handbook*, London: Earthscan/Routledge, 2nd ed., pp. 335–353.

PART II

Environmental Challenges

INTRODUCTION

Type "top 10 environmental issues" or "top 10 sustainability challenges" into any search engine and you will find a collection of topics and themes returned to you, usually with a great degree of overlap. For Esty and Winston (2009) the top 10 environmental issues are the following:

- Climate change
- Energy
- Water
- Biodiversity and land use
- Chemicals, toxics and heavy metals
- Air pollution
- Waste management
- Ozone layer depletion
- Oceans and fisheries
- Deforestation.

Whereas the handbook of the Institute of Environmental Management and Assessment (Brady et al. 2011) covers the following environmental themes:

- Biodiversity and conservation
- Chemicals
- Climate change
- Energy
- Food and agriculture
- Pollution prevention and control
- Transport and greenhouse gases.

Robertson (2017) covers the following issues and solutions in her sustainability book:

- Climate
- Water
- Ecosystems and habitat
- Pollution
- Energy
- Green buildings and sites
- Liveable cities
- Food
- Products
- Waste and recycling.

Rockström and colleagues (2009: 32) categorise nine different planetary boundaries, the crossing of which constitutes "unacceptable human-induced global environmental change":

- Climate change
- Ocean acidification
- Stratospheric ozone
- Global phosphorus and nitrogen cycles
- Atmospheric aerosol loading
- Freshwater use
- Change in land use
- Biodiversity loss
- Chemical pollution.

Cavagnaro and Curiel (2012) deal with the three broad concerns of pollution, resource depletion and disruption of the self-regulating systems keeping Earth fit for life. Blowfield (2013) also identifies three broad changes and challenges: demographic change, ecosystem change and climate change – the most talked about ecosystem change.

There are various ways to categorise and classify environmental sustainability challenges, whether it be by medium – air, water, land – or by the scale of the issue – local, regional, global, or by the triple bottom line – economy, environment, equity or e.g. by sector and the salience of environmental challenges for it. Other suggestions include food, energy and waste or a focus on ecosystem changes, whilst health and poverty feature in more integrated perspectives of sustainability, encompassing social, ecological and economic aspects.

The European Environment Agency (EEA 2016) distinguishes between key environmental themes, their socio-economic dimensions and the holistic perspective as follows, demonstrating the systemic nature of the challenges in table II.1:

TABLE II.1

Classification of environmental themes, wider dimensions and systems perspective

Environmental themes	Socio-economic dimensions	Systemic perspectives
Air pollution	Agriculture	The air and climate system
Biodiversity	Consumption	Green economy
Climate change impacts and adaptation	Energy	Hydrological systems and sustainable water management
Forests	Health and environment	Natural capital and ecosystem services
Freshwater quality	Industry	Urban systems
Land systems	Maritime activities	
Marine environment	Resource efficiency	
Mitigating climate change	Tourism	
Noise	Transport	
Soil		
Waste		

Source: EEA 2016.

Three broad thematic priorities emerge from this categorisation – protecting, conserving and enhancing natural capital, resource efficiency and the low-carbon economy, and safeguarding from environmental risks to health – which can be addressed through an enabling framework across local, regional and global levels, as indicated in Table II.2:

TABLE II.2

Classification of environmental thematic priorities and trends

Protecting, conserving and enhancing natural capital	Resource efficiency and the low-carbon economy	Safeguarding from environmental risks to health
Terrestrial and freshwater biodiversity	Material resource efficiency and material use	Water pollution and related environmental health risks
Land use and soil functions	Waste management	Air pollution and related environmental health risks
Ecological status of freshwater bodies	Greenhouse gas emissions and climate change mitigation	Noise pollution (especially in urban areas)
Water quality and nutrient loading	Energy consumption and fossil fuel use	Urban systems and grey infrastructure
Air pollution and its ecosystem impacts	Transport demand and related environmental impacts	Climate change and related environmental health risks
Marine and coastal biodiversity	Industrial pollution to air, soil and water	Chemicals and related environmental health risks
Climate change impacts on ecosystems	Water use and water quantity stress	

Source: EEA 2016.

This categorisation is based on the EU's Environment Action Programme and could be argued to be specific to the region it is useful wherever we are based, due to its comprehensiveness. Whichever classification is chosen and whichever selection of challenges is discussed, what should be clear from this overview and the first part of the book is that there are no isolated challenges – underlying all issues is often the depletion and unsustainable use of resources. Whilst the initial response to specific environmental challenges was often isolated when they were first discovered and addressed, recent efforts have seen a development to more integrated approaches to corporate decision-making and policymaking, as illustrated in the following table II.3:

TABLE II.3

Evolution of environmental challenges

Characterisation	Specific	Diffuse	Systemic
Key aspects	Linear cause-effect; large (point) sources; often local	Cumulative causes; multiple sources; often regional	Systemic causes; interlinked sources; often global
In the spotlight in	1970s/1980s (and continuing today)	1980s/1990s (and continuing today)	1990s/2000s (and continuing today)
Includes challenges such as	Forest damage due to acid rain; urban wastewater	Transport emissions; eutrophication	Climate change; biodiversity loss
Main policy response	Targeted policies and single-issue instruments	Policy integration and raising public awareness	Coherent policy packages and other systemic approaches

Source: based on EEA 2016.

Throughout the book, we emphasise the systemic nature of environmental sustainability for business, and the interlinkages and interrelationships between the various challenges illustrate these interdependencies. Take the example of water, which is crucial for food and agriculture but is at risk from pollution as well as from desertification and climate change in terms of freshwater, and in terms of seawater is subject to ocean acidification and sea level rises whilst also being a place of overfishing, and that is before considering its role in renewable energies. Or looking at energy, renewable energies provide an alternative to energy from fossil fuels which heavily contribute to climate change in transport, agriculture and heating/cooling requirements, whilst also polluting the air and ground, enabling food production but threatening biodiversity. Fiksel (2012) also highlights unintended consequences such as fertilisers which boost crop yields and ensure food security but threaten water quality through nutrient runoff.

The challenges are thus complex, and it is beyond the scope of the book to cover all comprehensively. We recognise the significance and relevance of all challenges but will focus on three challenges that are deemed to be particularly pertinent for businesses in their quest for environmental sustainability. Whilst we come across some of the other issues again in the book, we direct the interested reader to other sources, of which there are plenty, for finding out about these challenges (what are they?) and what the implications are for businesses (risks, opportunities, responsibilities?). One tool businesses can for example use to manage the complexity of environmental issues is provided by Esty and Winston (2009). The tool is called *AUDIO* analysis which stands for Aspects, Upstream, Downstream, Issues, and Opportunities. It allows companies to identify what aspects of an environmental trend impacts on them before looking at how the issue plays out in their value and supply chain. After this big picture perspective, a more detailed look at what kind of issues emerge from the environmental challenge follows, which is akin to looking at threats and risk, before attention is focused on the opportunities arising from them.

We then turn to the challenges chosen for this book, which comprise climate change, pollution and waste. All three address the thematic priorities of natural capital preservation, resource efficiency and low-carbon economy as well as protection from environmental risks. They are also three challenges that impact heavily on business and industry, in terms of the risks for business arising from these issues, in terms of the role business plays and has played in creating the problems, but also in light of the opportunities to deal with the challenges, internally as well as externally – as emphasised by Esty and Winston (2009: 33):

> Limits can create opportunities.

All three challenges have global as well as local implications, affect water, land and air, and are caused and faced by companies in both developing and industrialised nations. They are all underpinned by unsustainable resource use – whether it is resource depletion and fossil fuels in the case of climate change, too much material in the case of waste, or careless and inefficient use of productive resources in the case of pollution.

In order to provide consistency, similar chapter structures are used for each challenge, starting with a brief overview of what the challenge is, what the concerns are, how it links to other challenges before the business impact is covered, both in terms of cause and effect, i.e. how does business contribute to the problem and how is business affected by it, what are the risks and challenges?. Attention is then turned to how to deal with the challenge by first looking at policy and regulatory responses, which provide the legislative framework for businesses. We then discuss how businesses can use the challenge as an opportunity both internally and externally – how can they address e.g. climate change within a company by minimising impact, and what solutions can they offer on the market, to other businesses, to consumers and others.

Throughout this part, examples of industries, countries and companies, and how they have addressed the challenges, will draw the link between these challenges and concrete actions across local, regional and global levels.

SUGGESTED QUESTIONS AND ACTIVITIES, AND ONLINE SOURCES

- Using an industry or company of your choice, discuss which environmental challenge is the most urgent and/or important one for them to address. Have there been changes over time?
- Are there any environmental challenges you are faced with in your personal life? How are you responding to them?
- Most countries have environment agencies (or environmental protection agencies) that provide information on environmental topics and how they are regulated.
- Research a country of your choice and explore particular environmental challenges with which the country is faced.
- Which environmental challenge is particularly important for the town, region or country in which you live?
- How have policymakers addressed such environmental challenges and at which level? Compare and contrast with another location of your choice.

- Consider which environmental challenge applies to one sector in particular and which one applies to companies regardless of their activities.
- One particular resource that is being increasingly depleted is sand. Using Bendixen and colleagues (2019) as a starting point, research how much sand is mined, what it is used for and what the alternatives are.
- For more material on the breadth of environmental challenges, check out the following:
  - Esty and Winston (2009) provide summaries for each of the 10 environmental challenges listed earlier, covering the substance (what is it?) and business consequences (costs, risks, opportunities).
  - The Millennium Ecosystem Assessment (www.millenniumassessment.org/) explains ecosystem change and their consequences for human well-being whilst The Economics of Ecosystems and Biodiversity (TEEB, www.teebweb.org) looks at the economics of these changes.
  - The United Nations Environment Programme (www.unep.org/) covers a range of environmental issues. Its GEO series (Global Environmental Outlook) assesses environmental trends for air, climate, water, land and biota.
  - The International Energy Agency (www.iea.org/) publishes a world energy outlook annually on the global energy system.
  - Check the World Economic Forum's global risks reports to see how environmental challenges are set in context of other (economic, societal, geopolitical, technological) challenges and what the interconnections are. http://reports.weforum.org/global-risks-report-2020/, http://reports.weforum.org/global-risks-2019/, http://reports.weforum.org/global-risks-2018/, http://reports.weforum.org/global-risks-2017/ are the most recent ones.
  - The World Resources Institute (www.wri.org) works on a number of critical environmental issues providing information for businesses and policymakers.

SOURCES AND FURTHER READING

Bendixen, M., Best, J., Hackney, C. and Lønsmann Iversen, L. (2019) Time Is Running out for Sand, in: *Nature*, 571(7763): 29–31.

Blowfield, M. (2013) *Business and Sustainability*, Oxford: Oxford University Press.

Brady, J., Ebbage, A. and Lunn, R. (2011) *Environmental Management in Organizations, The IEMA Handbook*, London: Earthscan/Routledge, 2nd ed.

Cavagnaro, E. and Curiel, G. (2012) *The Three Levels of Sustainability*, Sheffield: Greenleaf Publishing.

EEA (2016) The European Environment – State and Outlook 2015 Report, European Environment Agency, available at www.eea.europa.eu/soer (Retrieved, 21.06.2017).

Esty, D. C. and Winston, A. S. (2009) *Green to Gold: How Smart Companies Use Environmental Strategy to Innovate, Create Value and Build Competitive Advantage*, Hoboken, NJ: Wiley.

Fiksel, J. (2012) A Systems View of Sustainability: The Triple Value Model, in: *Environmental Development*, 2(1): 138–141.

Robertson, M. (2017) *Sustainability Principles and Practice*, London: Routledge, 2nd ed.

Rockström, J., Steffen, W., Noone, K., Persson, Å., Chapin, F. S. III, Lambin, E., Lenton, T. M., Scheffer, M., Folke, C., Schellnhuber, H., Nykvist, B., De Wit, C. A., Hughes, T., van der Leeuw, S., Rodhe, H., Sörlin, S., Snyder, P. K., Costanza, R., Svedin, U., Falkenmark, M., Karlberg, L., Corell, R. W., Fabry, V. J., Hansen, J., Walker, B., Liverman, D., Richardson, K., Crutzen, P. and Foley, J. (2009) Planetary Boundaries: Exploring the Safe Operating Space for Humanity, in: *Ecology and Society*, 14(2): 32, available at www.ecologyandsociety.org/vol14/iss2/art32/ (Retrieved, 20.03.2017).

CHAPTER 5

Climate change

INTRODUCTION

Climate change is arguably the most significant and serious environmental challenge facing business and society today, dominating environmental policy agendas, corporate sustainability efforts and scientific research as well as featuring heavily in news coverage and public debates. This has resulted in a substantial amount of scholarly, regulatory and practical material on the topic. It is an issue that "changes everything" (Klein 2014: 28). In framing this chapter, we want to emphasise from the outset that the aim here is not to provide a comprehensive coverage of the topic but rather an overview of climate change across several levels and themes – what is it, what characterises business's involvement in terms of cause and impact but also in terms of internal and external responses and how policymakers have addressed it.

CLIMATE CHANGE – WHAT IS IT AND WHY IS IT A CONCERN?

The climate on planet Earth has always changed. In some respects, climate change is therefore nothing new (under the sun), but when we talk about the environmental challenge of climate change, and the related terms of global warming and greenhouse gas emissions, it is the rate of change that is different, and therefore a cause for concern. The speed at which the climate is changing is rapid and unlike the slow changes the planet has experienced before. When we talk about climate, we mean the long-term trends in atmospheric conditions. This is different from the weather, which is about short-term variations in temperature, wind and precipitation (Robertson 2017).

While the term global warming is also used in this context, it is more accurate to use the term climate change, as there are factors involved other than temperature rises, and not all areas of the globe experience warming, though many are subject to disrupted weather patterns and extreme weather events. Other changes associated with climate change are rising sea levels, increased flooding and droughts as a result of changes in precipitation patterns (e.g. rain and snow).

A change in the amount of so-called greenhouse gases (GHGs) in the atmosphere is behind the altered climate. Greenhouse gases make life on earth possible by trapping the heat from the sun, which would otherwise bounce back into space: the so-called greenhouse gas effect was first predicted in 1827 by French mathematician Joseph Fourier and coined by Swedish scientist

Svante Arrhenius in the late 19th century. The gases act like a blanket, insulating the earth's surface. Sunlight on the earth is either absorbed (in the case of darker surfaces) or reflected (in the case of lighter surfaces, such as snow and ice), the so-called albedo effect. GHGs absorb enough of this reflection rising from the surface to stop the world from either freezing or boiling, resulting in a stable global mean temperature of approximately 14°C. Overall, the more GHG in the atmosphere, the warmer the planet's surface, as more heat is trapped instead of radiating into space.

There are six major GHGs: carbon dioxide (CO_2), methane (CH_4), nitrous oxide (N_2O) and three F-gases (fluorinated gases) or halocarbons (hydrofluorocarbons (HFCs), perfluorocarbons (PFCs) and sulphur hexafluoride (SF6)).[1] They are measured in concentration (ppm – part per million or ppb – part per billion[2]) and in terms of their global warming potential (GWP) which is expressed as CO_2e (CO_2 equivalent), i.e. in relation to CO_2, the most abundant of GHGs.[3]

Although GHGs occur naturally, there is scientific consensus that their increase in the atmosphere and the resulting change in climate is anthropogenic, i.e. caused by humans, as e.g. halocarbons were hardly found in nature before the industrial revolution. The international scientific body tasked with comprehensive climate change research, the Intergovernmental Panel on Climate Change (IPCC), puts the certainty of human-induced climate change at 95% – extremely likely. Human activities contributing to climate change include fossil fuel use in transport, agriculture, manufacturing, heating and cooling of buildings, to name but a few. Table 5.1 provides an overview of the major GHGs and the common sources of their emissions:

TABLE 5.1

Greenhouse gases and their sources

GHG	100-year GWP	Abundance 1750 (ppb)	Abundance 2011 (ppb)	Main sources
Carbon dioxide (CO_2)	1	278,000	391,000 (408,000 in 2017)	Deforestation, combustion of hydrocarbons/fossil fuels for energy (oil, gas, coal)
Methane (CH_4)	21	722	1803 (1847 in 2017)	Landfills, agriculture (cattle farming), oil and gas production
Nitrous oxide (N_2O)	296	270	324	Burning of fossil fuels and biomass, released by chemical fertilisers
Hydrofluorocarbons (HFCs)		0	7 (1998)	Chemical by-product, refrigerant fluids
Perfluorocarbons (PFCs)	6k-12k	20	40 (1998)	Manufactured and chemical compounds, used in industrial processes (e.g. aluminium production)
Sulphur hexa-fluoride (SF_6)	22.2k	0	7.28 (ppt) (over 9 in 2017)	Manufactured compound, used in industrial processes (e.g. semiconductors)

Source: Adapted from Lingl et al. 2010; Gore 2011; IPCC 2013, www.esrl.noaa.gov/gmd/ccgg/trends/weekly.html.

The combination of growths in population, the economy and industry based on fossil fuels as observed over the last 150–250 years has brought about the change in climate, primarily through accelerated warming – the hottest years on record were 2015, 2016 and 2017 with 2018 and 2019 not far behind, making the 2010s the hottest decade since records began – which illustrates that underlying this challenge is one of resource depletion and unsustainable use of resources. In 2018, levels of heat-trapping GHG reached a record-high of 407.8 ppm for CO_2, representing 147% of pre-industrial levels, and for methane 259% compared to the year 1750 (WMO 2019).

The changes in the natural environment have been measured mainly through temperature records and proxies such as tree rings, coral reefs, lake sediments and ice cores. The IPCC categorises and measures the changes in the climate system around the atmosphere (temperature, water cycle, air quality), the ocean, the cryosphere (ice), the sea level and the carbon and other biogeochemical cycles. The planet is experiencing three major geophysical changes as a result of climate change and global warming – altered weather patterns, sea level rise (thermal expansion) and ocean acidification – with major consequences and impacts for the environment and socio-economic systems from seemingly modest temperate rises. Table 5.2 outlines some of the major consequences of such increases, which are expected to be in the range of 2.6 to 4.8°C by 2100 if emissions are not cut, or 0.3 to 1.7°C if stringent cuts are made soon.

TABLE 5.2

Consequences of temperature rises

Temperature rise (range +/-0.5°C, e.g. 1°C = 0.5–1.5°C)	Consequences
1°C	Crop yields rise in high-latitude developed countries but fall in lower-latitude developing countries Extensive damage to coral reef systems Disappearance of small glaciers
2°C	Increased number of people at risk from hunger as a result of decline of agricultural yields Irreversible melting of Greenland ice sheet Marine ecosystem at significant risk from ocean acidification
3°C	Significant changes in water availability Loss of ecosystem goods and services with extensive biodiversity loss as a result Increased risk of abrupt, large-scale shifts in climate system (e.g. monsoon patterns, gulf stream)
4°C	Rise in sea levels threatens major world cities, small islands and low-lying coastal areas (London, New York and Tokyo; Florida) Large risks to food security at regional and global levels due to major declines in yields Disappearance of large glaciers

Sources: Blowfield 2013; Stern 2006, www.carbonbrief.org/.

The geophysical changes have wide-ranging impacts, which show the interrelationship to other environmental challenges as well as contradictory trends depending on location for example. We briefly outline a selection, based on a number of sources[4] (e.g. Gore 2011; IPCC 2013; Stern 2006).

- *Water resources*: water supply is affected with increased water availability in high latitudes and moist tropics but decreased availability and increased droughts in mid-latitudes and semi-arid low latitudes. Increased water stress for hundreds of millions of people with increased competition for water. Water quality may also be affected.
- *Coastal areas*: increased risk of damage from floods and storms (related to water resources), erosion of beaches and loss of coastal wetlands.
- *Agriculture and food*: crop yields will change significantly – whilst some changes are positive (increased yields at mid to high latitudes), most are negative (decline in yields) and the impact on water resources will affect agriculture as well with changing irrigation demands. Negative impacts on smallholders, subsistence farmers and fishers.
- *Ecosystems*: coral bleaching and increased coral mortality, many species at risk from extinction (e.g. polar bear), increased species shift (e.g. invasive species) and wildfire risk, loss of forests. Changes in ocean circulation affect marine ecosystems as does increased ocean acidification. Complex and interactive impacts for species and ecosystem from simultaneous warming and ocean acidification. Higher temperatures lead to thawing of permafrost and release of methane, whilst the ability of plants and soils to soak up carbon is negatively affected.
- *Health*: whilst some benefits will occur through reduced winter mortality in higher latitudes, overall the impact here is negative. Increased risks from infectious diseases (e.g. food- and water-borne), malnutrition from diminished food production, greater likelihood of injury, disease and death from more intense heatwaves and fires, but also floods. Consequently, a substantial burden is placed on health services in the long term.

Other risks include extreme weather events and large-scale singular events, such as for example the drying of the Amazon rain forest, changing monsoon patterns in South Asia, the melting of Antarctic and Greenland ice caps and the slowing of the Atlantic thermohaline circulation system. The impacts are thus felt across the globe, but the risks and consequences outlined differ from country to country as they are unevenly distributed. Overall, the impacts are greater for disadvantaged people and communities regardless of their country's level of development, with an increase in displaced people and environmental refugees.

In the introduction to part two of the book we covered a number of environmental challenges, and having introduced the various geophysical changes resulting from climate change and their impact, it is clear that climate change relates to and affects many other challenges. For example, of Esty and Winston's (2009) list of top ten environmental challenges (climate change, energy, water, biodiversity and land use, chemicals, toxics, and heavy metals, air pollution, waste management, ozone layer depletion, oceans and fisheries, deforestation), apart from maybe chemicals, most affect or are affected by climate change. For example, energy use from fossil fuels contributes to climate change, water is affected by global warming (droughts and floods), methane emissions from landfill (waste management) contribute to climate change, whilst oceans and fisheries are affected by ocean acidification and sea level rises.

The relationship between climate change and air pollution warrants a closer look as climate change is often confused with air pollution or is itself called air pollution (see e.g. Anderson 2014). The IPCC (2013) points out that anthropogenic sources of emissions may emit some chemicals that affect air pollution and others that impact on the climate and thus the two are closely linked – the burning of fossil fuels drives climate change and emits air pollutants. Improvements in air quality have effects on climate change, either cooling or warming, depending on which air pollutants are addressed. Broadly speaking, air pollution has local and near-term impacts, whilst the location of the source of GHG emissions is less relevant as the effect is global and long-term. Whereas the major GHGs covered previously are mostly long-lived, air pollutants that contribute to climate change come in the form of so-called short-lived climate pollutants (SLCPs, or short-lived climate-forcing pollutants) and include methane, black carbon, ground-level (tropospheric) ozone and sulphate aerosols (IASS 2017). The relationship is thus complex and is still being studied. Acknowledging the systemic nature of these challenges, we cover them separately, continuing with climate change here, and air pollution in the next chapter, alongside water and ground pollution.

As we have seen, climate change has wide-ranging impacts on the environment and on society, whether it is the risk from floods or droughts threatening livelihoods, including shelter and ability to provide food from subsistence farming or commercial agriculture, or the water scarcity affecting supplies of clean water, or the risk to human health. What are then the impacts on business and industry? We will look at the effects on climate change on business in the next section, starting with the risk to business before looking at how business contributes to climate change.

BUSINESS IMPACT

Climate change affects business in a number of ways, whether it is through physical impacts or market impacts. Whilst there are opportunities from climate change for business, which we address in the latter part of the chapter, we focus first on the impact on business and how business contributes to climate change.

Effect – climate change as a problem for business

The impact of climate change on business comes in various forms, physical and financial. The importance of climate change for industry will increase over time as ecosystems are increasingly affected and regulatory responses are developed by policymakers, changing the natural and legislative environment within which business operates.

The physical impacts outlined earlier all impact on business, depending on their location and sector e.g. the world's key financial centres such as New York, London and Tokyo are all in coastal areas (Stern 2006). A major vulnerability for businesses is global supply chains, which may be subject to increased floods and sea level rises. This poses the risk of broken supply chains as witnessed for example after the floods in Thailand in 2011 which affected supply chains for companies such as Apple, Dell and Intel as production facilities supplying these companies were flooded – 40% of global hard disk drive production is concentrated in Thailand. Droughts in Russia in 2010 and in the US in 2012 destroyed agricultural crops (PWC 2013). In instances of both too much and too little water, higher commodity costs were the results and continue to be.

Such single events and climate change more generally can lead to a contagion of impacts where not only the immediate supplies are disrupted but also entire supply and value chains as well as the infrastructure on which companies rely. For example, prolonged water shortages in India led to a loss of business for Coca-Cola. Furthermore, the social impacts of such single events and long-term trends could be in the form of dislocation of population affecting a company's employees, leading to lost work capacity and reduced labour productivity (Dharmasasmita et al. 2017).

There are thus market impacts in addition to the physical impacts that disrupt and interrupt business as a result of climate change and extreme weather events. Another financial risk that is added to the cost of catastrophic climate events is the rising cost of energy and carbon (NBS 2010), which is felt across all sectors.

Although many of the consequences of climate change are experienced by business across the board, the impacts differ between individual sectors, of which we provide a few examples here:

- *Property and construction industries* are faced with higher insurance costs associated with the increased risk of floods, which may also affect asset values. The insurance industry, which has been faced with the economic costs of extreme natural disasters (which have increased tenfold since the 1950s), was one of the first industries to advocate for action on climate change. In particular, reinsurance companies such as Swiss Re or Munich Re, have raised their voice in the light of increased risk and uncertainty (Esty and Winston 2009). Munich Re has been working on climate change since the 1970s and is part of a group of insurers, ClimateWise, warning of an increasing "protection gap" between the cost of natural disasters linked to climate change and the amount insured. Weather catastrophes are now six times more frequent than in the 1950s and pose a challenge to insurers who argue that public decision-making does not take sufficient account of climate change (Harvey and Watts 2017).
- *Tourism and related industries* are affected through the destruction and loss of attractions (for example in coastal areas or ski resorts) or shorter seasons (skiing) and the increase of tropical diseases affecting the willingness of consumers to travel to areas at risk.
- Sectors providing *transport and infrastructure* have higher maintenance and insurance costs, alongside increased construction costs, all to do with a greater need for protection from storms and floods. This is also an issue faced by coastal *manufacturing and refining activities* as well as sectors dependent on proximity to ports, such as steel mills, petrochemical and energy-related plants (e.g. oil refineries, re-gasification plants) that could be subject to significant costs if they need to be protected from rising sea levels and stronger floods or might need to be relocated inland (Stern 2006). At the same time, some trade routes, such as the North West passage or Russia's Artic sea-lanes could become safer and more reliable as alternatives. The impact on transport and infrastructure may affect other sectors reliant on transport, for example logistics.
- We have already touched upon *agriculture*, which is faced with changing crop yields, and, in most cases, increased irrigation needs and food security risks. This in turn affects *retailers and providers of consumer discretionary goods and services* in terms of security of supply for key consumer staples but also in terms of volatility in earnings for e.g. season-dependent products.

There are thus significant challenges for businesses emanating from climate change directly as well as indirectly, with potential impacts on markets, logistics, processes and finances, as well as premises and people which translate into a range of reputational, environmental, operational, financial, health and safety as well as strategic risks (UKCIP 2010). Although some market and production opportunities emerge, which we discuss in the last part of the chapter, all sectors and all corporate functions are affected negatively by climate change. Similarly, all sectors, directly or indirectly, contribute to climate change, making business part of the problem, which we explore next.

Cause of climate change – business as part of the problem

Table 5.1 has given some indication as to what causes the increase in GHG concentrations in the atmosphere, leading to global warming and climate change. Here we ask specifically how business contributes to climate change, starting with Figure 5.1 depicting which economic sectors contribute how much to global GHG emissions, both directly and indirectly.

As these are fairly general categories, a more detailed breakdown is provided in the flowchart in Figure 5.2 showing activities and industries and the type and volume of GHG they emit.[5]

We need to keep in mind that sectors have different growth rates, and individual sectors might also be substantially different depending on whether they are from developed or developing countries, whilst emissions levels are constantly changing as well (Guardian 2011; WRI 2005). Although there is some element of uncertainty as to the precise distribution, the figure serves to illustrate how every business contributes to climate change. Regardless of sector, all companies rely on energy, and this is where business is part of the problem, which is a growing one as shown in Figure 5.3.

The biggest contributors are industries with carbon-intensive production processes, such as coal, oil and natural gas producers, and cement as well as chemical manufacturers. Other heavy emitters include iron, steel and aluminium industries as well as paper and pulp. It is, however,

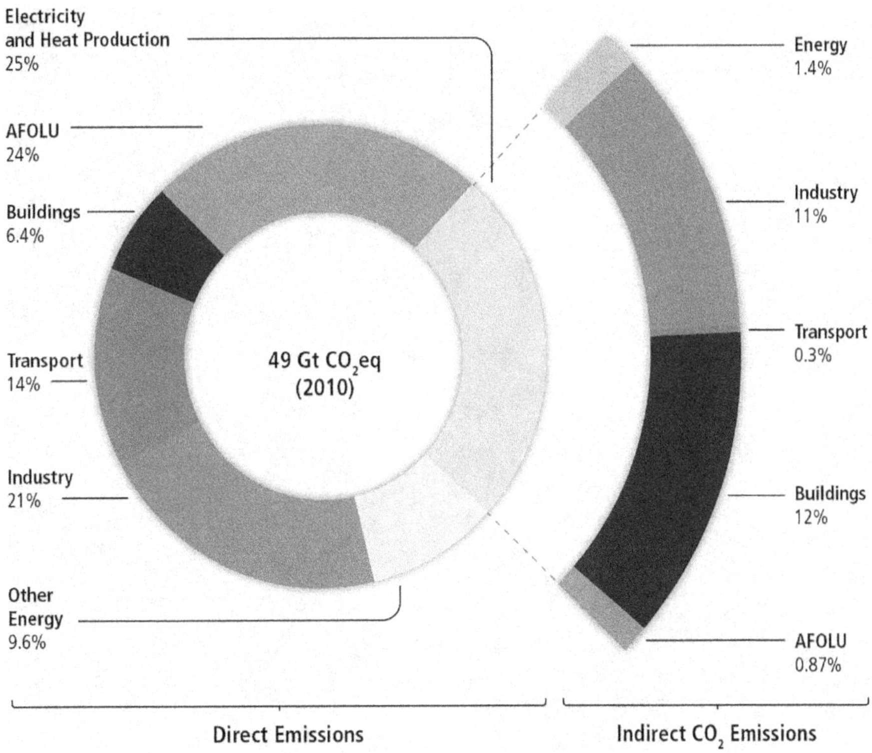

FIGURE 5.1
Greenhouse gas emissions by economic sectors

Source: IPCC, used with permission.

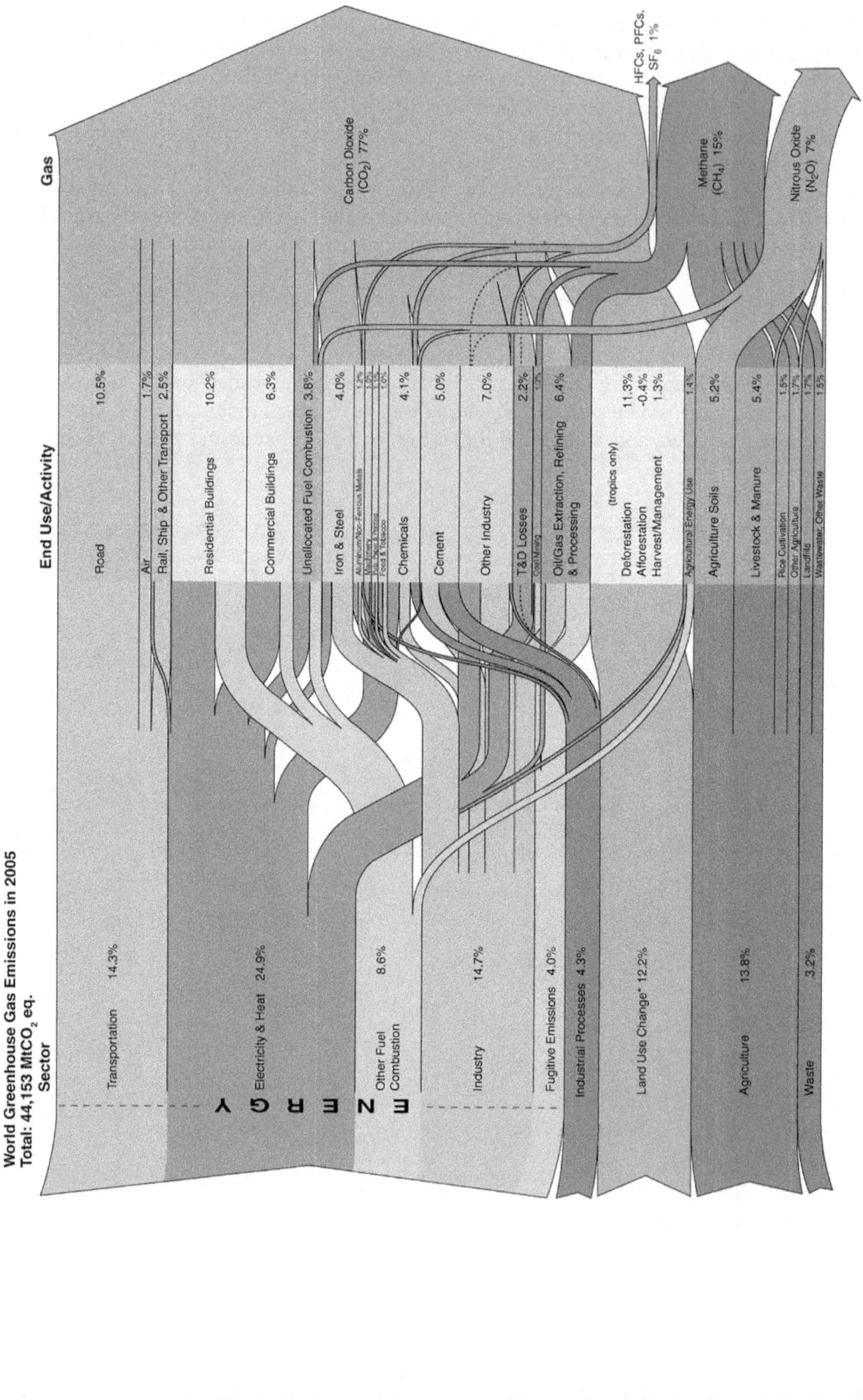

FIGURE 5.2
Detailed breakdown of global GHG emissions

Source: WRI, used with permission.

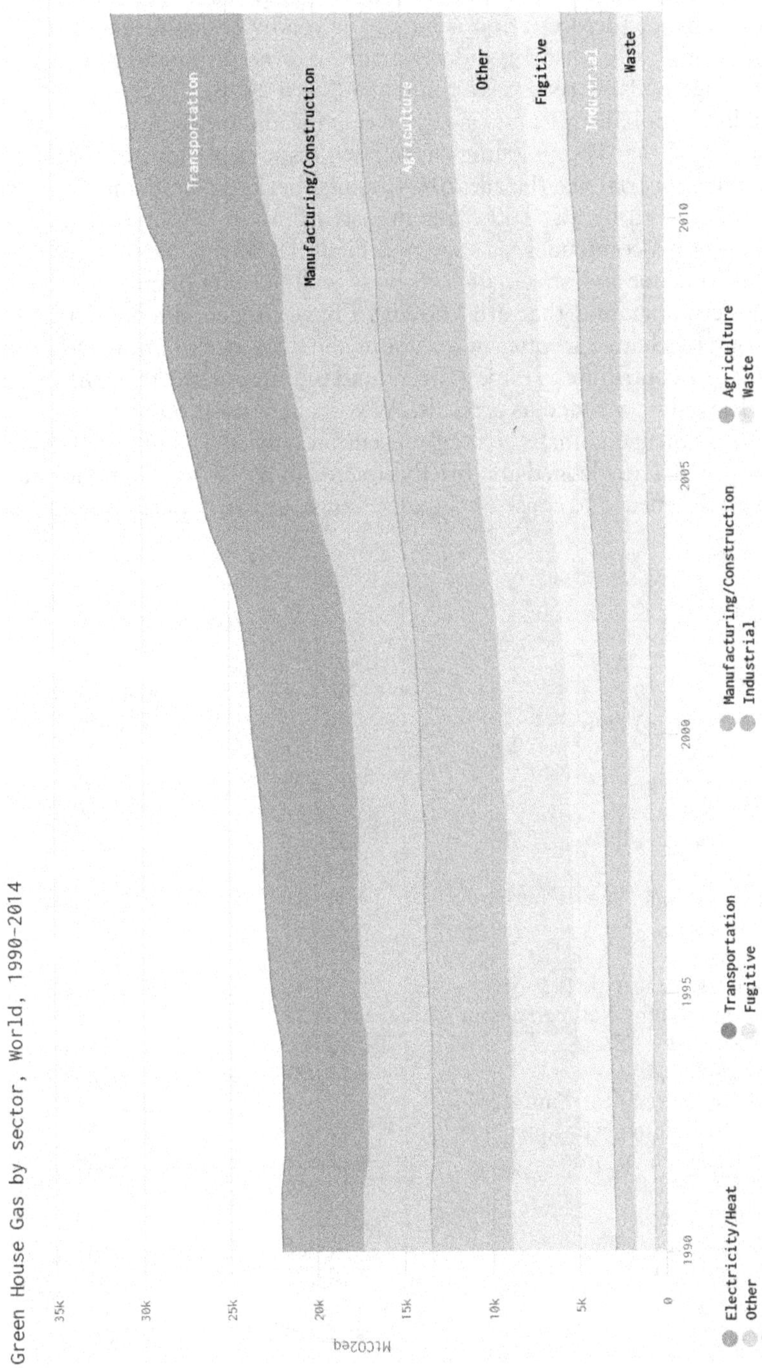

FIGURE 5.3
Historical GHG emissions by sector

Source: TSP, used with permission.

not just industry, but also agriculture, that contributes to GHG emissions, both directly, in the form of fertiliser use, livestock as well as rice and manure management, and indirectly, through emissions from land-use change (deforestation is largely driven by agriculture and releases emissions that forests would otherwise absorb), industry (through the production of fertiliser), and transport (through moving goods by road, water and air) (Stern 2006).

In a study of cumulative global emissions of carbon dioxide and methane by industry between 1751 and 2010 it was found that 90 companies have been responsible for almost two thirds of GHG emissions since the industrial age (Heede 2014), whilst between 1988 and 2015 over 70% of GHG emissions were emitted by just 100 companies (Griffin 2017). Some of these so-called carbon majors are well-known companies, as shown in Figures 5.4a and 5.4b.

The companies here include investor-owned ones as well as state-owned and government-run ones. It also includes companies that are known to have funded climate denial initiatives (Goldenberg 2013). Climate change scepticism questions the existence of anthropogenic climate change, and companies that contribute to it may compound the environmental challenge as action to address it may be delayed or avoided as a result. As we take a position here of accepting the science and the serious challenge climate change presents to society, scepticism and denial by industry and government is acknowledged and briefly covered in Box 5.1 though the overall focus here is on understanding the effect and cause of climate change, and how business and policymakers can address it.

Entity	$MtCO_2e$	% of global
1. Saudi Aramco, Saudi Arabia	59,262	4.38%
2. Chevron, USA	43,345	3.20%
3. Gazprom, Russia	43,230	3.19%
4. ExxonMobil, USA	41,904	3.09%
5. National Iranian Oil Co.	35,658	2.63%
6. BP, UK	34,015	2.51%
7. Royal Dutch Shell, The Netherlands	31,948	2.36%
8. Coal India, India	23,124	1.71%
9. Pemex, Mexico	22,645	1.67%
10. Petroleos de Venezuela (PDVSA)	15,745	1.16%
11. PetroChina / China Natl Petroleum	15,632	1.15%
12. Peabody Energy, USA	15,385	1.14%
13. ConocoPhillips, USA	15,229	1.12%
14. Abu Dhabi, United Arab Emirates	13,840	1.01%
15. Kuwait Petroleum Corp., Kuwait	13,479	1.00%
16. Iraq National Oil Co., Iraq	12,596	0.93%
17. Total SA, France	12,352	0.91%
18. Sonatrach, Algeria	12,302	0.91%
19. BHP Billiton, Australia	9,802	0.72%
20. Petrobras, Brazil	8,676	0.64%
Top Twenty	480,168	35.45%
Global	1,354,388	100.00%

FIGURE 5.4A
Carbon majors

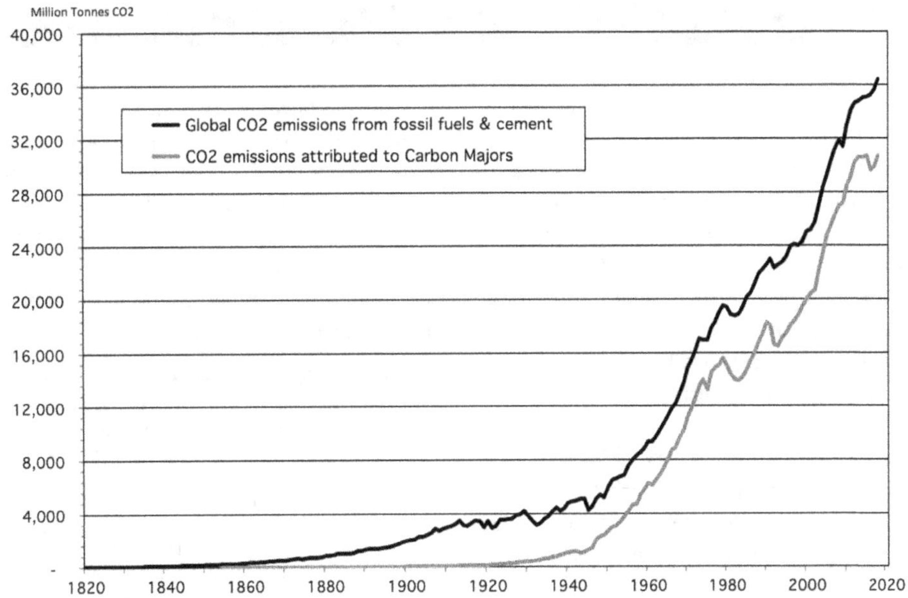

FIGURE 5.4B
(Continued)

BOX 5.1 CLIMATE CHANGE SCEPTICISM

Although the IPCC puts the likelihood of human-induced climate change as extremely probable (95%), and the scientific consensus in peer-reviewed literature on anthropogenic global warming is 90% to 100%, climate change scepticism and denial are still mirrored in public opinion and the media, reflecting a wide gap between the public's understanding of climate change and that of scientists. Notwithstanding the expert consensus, it is important to acknowledge this phenomenon, particularly in the current political environment where US President Donald Trump pulled his country out of the Paris 2015 climate accord (see policy responses to climate change), which commits countries to limiting global warming to 2°C, arguing that it would be detrimental to the US economy, jobs, and competitiveness and would see other countries benefit at the expense of his country. What is particularly telling in this context is that a significant number of US companies, cities and federal states have confirmed their commitment to reducing their GHG emissions in line with the Paris accord. Corporate examples include IBM, GE, Facebook, Apple and Twitter, Goldman Sachs and Microsoft whilst cities include Pittsburgh, New York City, Chicago and Los Angeles, and California, Washington and New York amongst the states.

Reasons often given for scepticism about climate change including the following:

- Faulty science and unreliable models: CO_2 levels may have been equally high in the past which demonstrate that this is part of the natural cycle and not a disaster.
- No consensus: scientists don't agree on climate change.
- Not caused by humans: natural phenomena such as water vapour and volcanoes cause climate change.

BOX 5.1 (CONTINUED)

- Climate change is real, but we can't predict or control what the future will bring.
- Climate change is a hoax, invented to spread fear and raise money for environmental causes.
- Adaptation is better than mitigation: anthropogenic climate change is a fact, but rather than changing the current socio-economic systems the best way to address it is by adapting to it.
- Global warming is a good thing – an argument particularly heard in northern climes.

The hostility towards regulatory initiatives addressing climate change, which have been perceived as attacks on personal liberty and contrary to cultural narratives of humans' role on earth and the faith in markets, seem to originate partly from a fear that such policies would threaten people's lifestyles, particularly in wealthier countries, whilst for industry there is the perceived threat to profitability from climate change policies. Such interpretations of climate change through the lenses of pre-existing ideologies illustrate that climate change is more than just a scientific phenomenon but needs to be addressed through social, psychological, cultural and political perspectives as well.

Sources: Blowfield 2013; Hackett 2011; Klein 2014, www.theguardian.com/environment/climate-consensus-97-per-cent/2017/may/03/is-the-climate-consensus-97-999-or-is-plate-tectonics-a-hoax, www.theguardian.com/technology/2017/jun/02/shortsighted-wrong-apple-facebook-among-tech-giants-to-reject-paris-pullout, www.bbc.co.uk/news/science-environment-40128046, www.nytimes.com/2017/06/01/climate/american-cities-climate-standards.html, see also https://skepticalscience.com and www.edx.org/course/making-sense-climate-science-denial-uqx-denial101x-4.

Beyond carbon-intensive production and manufacturing processes that are exclusively or heavily based on fossil fuels, it is energy for lighting, heating and cooling of office spaces and buildings, use of electronic equipment, transport and logistics for goods and services, and business travel that are major business activities contributing to climate change.

Although more could be said about how business contributes to climate change, what the section has made clear is the large impact of industry on climate change regardless of sector or size, though with the heaviest contribution from carbon majors in fossil fuel extracting industries. Before we look at how business can address climate change internally and externally, the regulatory responses are covered as these influence how business operates and often shape the direction of travel.

REGULATORY RESPONSES TO CLIMATE CHANGE

Both corporate and regulatory responses to climate change can be set within the broader context of general responses to climate change – mitigation, adaptation and geoengineering, though most responses focus on the categories of mitigation and adaptation. Mitigation consists of preventive measures aimed at reducing the extent as well as delaying the timing of climate change by stabilising or lowering GHG emissions, whilst adaptation comprises actions aimed at reducing the magnitude or risk of damages resulting from climate change by dealing with the consequences and adapting to new circumstances (Harris and Roach 2018). Geoengineering involves large-scale efforts that counteract changes in temperature and oceanic chemistry involving aspects of both mitigation and adaptation. Broad approaches here include increasing the albedo effect through solar radiation management (e.g. sulphur reflection) and iron fertilisation. Geoengineering is further explained in Box 5.2, whilst we concentrate on mitigation and adaptation here.

BOX 5.2 GEOENGINEERING

Geoengineering can be defined as deliberate, large-scale efforts manipulating the planet to counteract effects of climate change and global warming. It broadly includes technologies that either remove CO_2 from the atmosphere or reduce global temperatures by reflecting sunlight into space:

1 Carbon Dioxide Removal (CDR) techniques – these address the root cause of climate change (rising CO_2 concentrations), thus techniques have relatively low uncertainties and risks, but work slowly to reduce global temperatures and thus would need to be implemented on a global scale to be of significant impact. Also called carbon geoengineering or decarbonisation techniques. These include some of the following:
   - a Ambient air capture: removing CO_2 directly from the air and storing it elsewhere (carbon capture and storage, CCS).
   - b Ocean fertilisation: adding nutrients, such as for example iron, to increase primary production (e.g. phytoplankton blooms) which captures atmospheric CO_2.
   - c Enhanced weathering: large-scale exposure of minerals to react with CO_2 and storing resulting compound in soil or ocean.
   - d Ocean alkaline enhancement: rocks such as limestone or silicates are ground up, dispersed and dissolved in the ocean to increase carbon storage ability and ameliorate carbon acidification.
   - e Approaches such as afforestation, biochar, and bio-energy with carbon capture and sequestration.

2 Solar Radiation Management (SRM) techniques – these reflect a small percentage of the sun's light and heat back into space. Methods act quickly but are costly and reduce some but not all effects of climate change, while possibly creating other problems as side effects. Do not address root cause as CO_2 levels are unaffected. Also called solar geoengineering. Some techniques include the following:
   - a Albedo enhancement: either on land through surface measures such as white rooftops and roads or land use changes, or through increased cloud albedo.
   - b Space reflectors or orbital sunshades: space-based mirrors reflect sun rays before reaching the Earth.
   - c Stratospheric aerosols: increased stratospheric albedo by stimulating effects of volcanic eruptions through e.g. sulphur particular or sulphur aerosol. Also called sulphur reflection.

Geoengineering comes broadly under mitigation technologies, which it can boost by augmenting reduction efforts though it should not be seen as a replacement for mitigation and adaptation. Geoengineering approaches have in common in assuming that Earth behaves like a machine and can be treated like a linear system, whereas ecosystems are complex and nonlinear, which makes geoengineering's consequences challenging to predict. The difficulty of both CDR and SRM technologies lie in their lack of effectiveness at affordable cost – geoengineering is vastly costly – and in the acceptability of potential side effects and unintended consequences (e.g. how long- or short-lived is the cooling effect of sulphur aerosols). The use of geoengineering has therefore ethical implications, warranting questions on whether such an active manipulation of the environment and intervention in the climate is acceptable even if it is feasible, or whether the potential benefits outweigh the risks, or whether such a severe challenge warrants the suspension of the precautionary principle and an accelerated use of such technologies. Klein (2014) provides a thorough discussion of the ethics and wider risks of climate engineering, whilst Vince (2014)

BOX 5.2 (CONTINUED)

offers some low-tech examples of geoengineering in developing countries, where whitewashing a mountain in Peru and creating artificial glaciers in the Himalayas increase the albedo effect.

As research into geoengineering is still in its early stages and success is not guaranteed, the so-called Oxford Principles propose a set of guiding principles for the governance of geoengineering which includes the regulation of geoengineering, the involvement of the public, disclosure of research and independent assessment of impacts.

Sources: Stilgoe 2018; Oxford Geoengineering Programme 2017; Whitfield and McNett 2014; Vince 2014; Robertson 2017; Klein 2014; Blowfield 2013; Gore 2011.

Scenarios and policy framework

In order to respond appropriately to climate change, scenarios are used which forecast the change in emissions and increase in temperature and their consequences based on different technological and behavioural options. The scenarios use representative concentration pathways (RCP) which are identified by radiative forcing as well as GHG concentration levels measured in CO_2 equivalent (CO_2e) and temperature increases (IPCC 2014a), as shown in Figure 5.5 and Table 5.3.

The IPCC uses baseline scenarios as well as pathways based on mitigation and adaptation. The differences can be seen in the so-called stabilisation wedges where the actions taken to address climate change result in the stabilisation of GHG emissions. Beyond emissions, the various scenarios may also take into account population growth, world GDP growth and the convergence between rich and poor countries (Stern 2006). The scenarios may be used to calculate the cost of climate change in itself as well as the cost of responding to it. The Stern Review on the economics of climate change provided a comprehensive account of the costs of both inaction and action over the next century and beyond, concluding in essence that the benefits of strong, early action outweigh the cost (Stern 2006).

In more detail, the review looked at the cost of inaction, which in effect means business as usual – the BAU scenario. This could lead the stock of GHG to treble by the end of the century and risk a temperature rise of 5°C. The cost of this most severe scenario could lead to a 5–20% reduction in consumption per head and thus to reduced welfare, whilst a temperature rise of 3°C could lead to a permanent loss of up to 3% in GDP. With regard to the feasibility and costs of stabilising GHG by the middle of the century, the review concluded that this would cost 1% of GDP, whilst the benefits (costs avoided) would be 3% of GDP, making mitigation the only economically rational choice. The challenge of such investments is that the costs have to be borne now whilst the benefits fall far in the future (recall the discussion in Box 3.5 on discount rates), but Stern (2006) concluded that although the costs are significant, they are manageable.

In addressing climate change, which the review calls "the greatest market failure the world has ever seen" (Stern 2006: xviii), three policy elements are required – carbon pricing (whether through tax, trading or regulation), innovation and low-carbon technologies (support and deployment), and the removal of barriers to behavioural change (e.g. with regard to energy efficiency and individual action).

Further ways of calculating and illustrating the need for action is in the form of a carbon budget, which is the estimated amount of carbon dioxide (and its equivalents) that can be emitted while keeping global temperature rise below 2°C compared to 1850. According to the IPCC Fifth Assessment Report (AR5), by 2011, the world had used up 52% of its carbon budget and under a high emissions, fossil fuel intensive scenario, where global CO_2 emissions keep growing at current rates, the remaining carbon budget would be spent by 2045. Under a low emissions scenario that

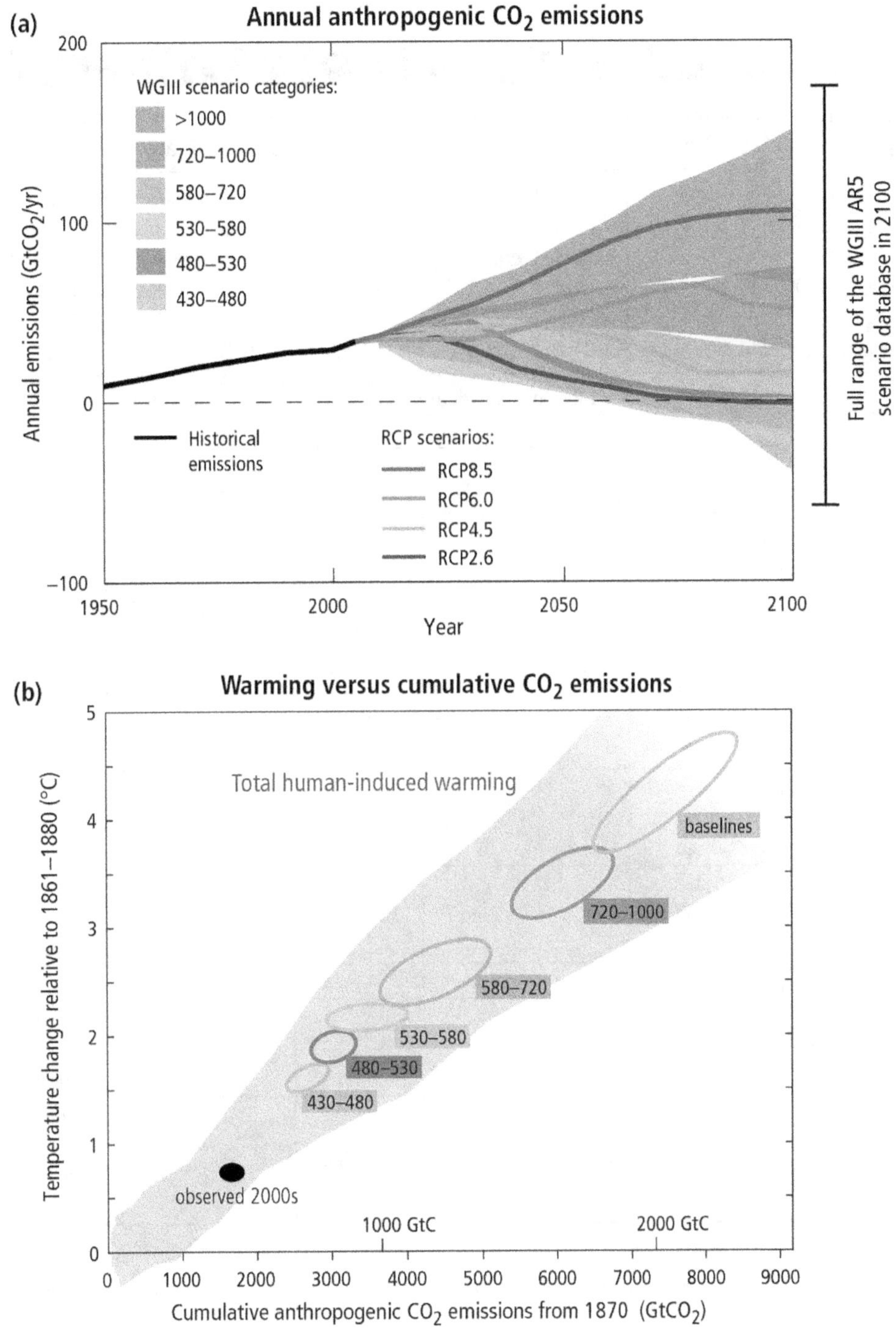

FIGURE 5.5
Scenarios of GHG emission pathways

Source: IPCC, used with permission.

TABLE 5.3

SPM.1 | Key characteristics of the scenarios collected and assessed for WGIIIAR5. For all parameters the 10th to 90th percentile of the scenarios is shown[a]. {Table 3.1}

<table>
<tr><th rowspan="2">CO_2-eq concentrations in 2100 (ppm CO_2-eq)[f] Category label (cone, range)</th><th rowspan="2">Subcategories</th><th rowspan="2">Relative position of the RCPs[d]</th><th colspan="2">Change in CO_2-eq emissions compared to 2010 (in %)[c]</th><th colspan="4">Likelihood of staying below a specific temperature level over the 21st century (relative to 1850–1900)[d*e]</th></tr>
<tr><th>2050</th><th>2100</th><th>1.5°C</th><th>2°C</th><th>3°C</th><th>4°C</th></tr>
<tr><td><430</td><td colspan="8">Only a limited number of individual model studies have explored levels below 430 ppm CO_2-eq i</td></tr>
<tr><td>450 (430 to 480)</td><td>Total range a-s</td><td>RCP2.6</td><td>−72 to −41</td><td>−118 to −78</td><td>More unlikely than likely</td><td>Likely</td><td rowspan="6">Likely</td><td rowspan="8">Likely</td></tr>
<tr><td rowspan="2">500 (480 to 530)</td><td>No overshoot of 530 ppm CO_2-eq</td><td rowspan="4"></td><td>−57 to −42</td><td>−107 to −73</td><td rowspan="6">Unlikely</td><td>More likely than not</td></tr>
<tr><td>Overshoot of 530 ppm CO_2-eq</td><td>−55 to −25</td><td>−114 to −90</td><td>About as likely as not</td></tr>
<tr><td rowspan="2">550 (530 to 580)</td><td>No overshoot of 580 ppm CO_2-eq</td><td>−47 to −19</td><td>−81 to −59</td><td rowspan="3">More unlikely than likely'</td></tr>
<tr><td>Overshoot of 580 ppm CO_2-eq</td><td>−16 to 7</td><td>−183 to −86</td></tr>
<tr><td>(580 to 650)</td><td>Total range</td><td rowspan="2">RCP4.5</td><td>−38 to 24</td><td>−134 to −50</td></tr>
<tr><td>(650 to 720)</td><td>Total range</td><td>−11 to 17</td><td>−54 to −21</td><td rowspan="2">Unlikely</td><td>More likely than not</td></tr>
<tr><td>(720 to 1000)[b]</td><td>Total range</td><td>RCP6.0</td><td>18 to 54</td><td>−7 to 72</td><td rowspan="2">Unlikely[h]</td><td>More unlikely than likely</td></tr>
<tr><td>>1000 ^</td><td>Total range</td><td>RCP8.5</td><td>52 to 95</td><td>74 to 178</td><td>Unlikely[h]</td><td>Unlikely</td><td>More unlikely than likely</td></tr>
</table>

Source: IPCC, used with permission.

involves transformative change, global emissions have to peak by 2020 and then need to decline for the 2°C target to be met cost-effectively (WRI 2014).

It is this 2°C target that was agreed in the Paris Climate Accord in 2015 with a view to further limiting temperate rises to 1.5°C. This continues the global efforts to curb climate change that started in 1992 with the UN Framework Convention on Climate Change (UNFCCC) and carried on with the Kyoto Protocol in 1997, which came into force in 2005 as the first comprehensive international agreement on climate change. Efforts for a binding follow-up agreement failed in 2009 at the Copenhagen Conference of Parties (COP-15), and thus the Paris (COP-21) agreement consists of nationally determined contributions (NDCs) of the 195 signatory nations, based on voluntary planned reductions in GHG emissions. Although criticisms have been levelled at the agreement for not going far enough – e.g. most pledges do not go beyond 2030 – the agreement entered into force within a year, unusually speedy for an international accord, and includes provisions to step up commitments over time as well as arrangements for financial and technical support for developing countries (Harris and Roach 2018).

The comprehensive success of the agreement is reflected in the fact that only two countries, Nicaragua and Syria,[6] were initially outside the agreement, but have since been joined by the United States, which pulled out under newly elected president Trump, a move widely condemned by business and industry in the US and beyond. In 2017, Nicaragua signed the agreement.

The Paris agreement provides an overall, internationally agreed, global framework. Implementation takes place at regional, national and local levels and may take different forms, voluntary or compulsory efforts, command and control approaches or market-based regulations, using incentives. We will now look at the general policy options available, providing specific examples, where appropriate, of the legislative framework for business with regard to climate change.

Policies and pathways for adaptation and mitigation

The main approaches to address climate change are mitigation and adaptation, which are complementary and may be set in a wider context of four policy opportunities (EEA 2016):

- Mitigate – reducing environmental pressures or offsetting harmful effects
- Adapt – anticipation of adverse effects as some environmental change is inevitable
- Avoid – precautionary principle (see Chapter 3) is applied to avoid potential harm in situations that are complex and uncertain
- Restore – remediation of environmental degradation (akin to investment in natural capital covered in Chapter 4).

The difference between preventive and adaptive strategies is the attitude to change – mitigation prevents changes from happening in the first place, whereas adaptation means adjusting to the consequences of climate change. In reality, both are needed to address climate change in a cost-effective and sustainable way (Stern 2006) e.g. by developing climate-resilient pathways (IPCC 2014b).

Adaption includes for examples measures such as building dykes and seawalls to protect against rising sea levels and floods, building dams to store water for use during longer and drier summers, shifting agricultural cultivation patterns to adapt to changing weather conditions, and the creation of institutions for mobilising resources (financial, human, material) to respond to climate-related disasters (Harris and Roach 2018). Other adaptive measures might include better storm run-off systems, biological reserves to protect endangered species, new disease control systems, development of improved crops, as well as groundwater irrigation systems (Gore 2011). The IPCC (2014b) emphasises that adaptation is place- and context-specific, which means there is no single approach to reducing climate risks. As vulnerability to the impacts of climate change is directly linked to incomes and technological development, the burden of adaptation is expected to fall most heavily on poorer countries (Gore 2011) and thus needs to reduce exposure and vulnerability in addition to incremental and transformational adjustments.

Mitigation includes preventive measures to reduce GHG emissions in order to slow the rate of climate change. Such measures include efficiency-increasing and decarbonising measures from GHG-producing activities, replacing GHG-intensive energy sources with lower-emissions alternatives and removing CO_2 from the atmosphere through enhancing natural carbon sinks, which store CO_2 (e.g. forests, soil, reservoirs, glacial streams, oceans).

There are multiple so-called mitigation pathways that may limit warming to below 2°C (IPCC 2014c), all of which require substantial reductions of GHG emissions, up to 80% below the annual levels of 1990. This would result in a level of GHG emissions that balances the planet's natural capacity to remove GHG from the atmosphere (Stern 2006). The required stabilisation can be illustrated through so-called wedges which bring the BAU scenario (which doubles emissions by 2050) down to sufficiently low, flat emissions levels (Harris and Roach 2018). These climate stabilisation wedges provide a quantitative tool to measure how emissions can be reduced using existing technologies. Each wedge represents one gigaton of CO_2. Whilst the original paper of Pacala and Socolow in 2004 included seven wedges, developments since then have increased the number needed to nine. The wedges can be broadly categorised into increased efficiency, energy supply-side shifts and carbon storage (Harris and Roach 2018), overall consisting of 15 strategies, which can be seen in Table 5.4.

TABLE 5.4

Strategies in stabilisation wedges

Category		Options	Brief comment on strategy
Efficiency and Conservation	1	Improved fuel economy	Transport efficiency
	2	Reduced reliance on cars	Transport conservation (fewer miles travelled)
	3	More efficient buildings	Heating, cooling, lighting, refrigeration
	4	Improved power plant efficiency	Efficient electricity production
Decarbonisation of electricity and fuels – fossil fuels and renewables	5	Substituting natural gas for coal	Fuel switching
	6	Storage of carbon captured in power plants	Fossil-based electricity with carbon capture and storage (CCS)
	7	Storage of carbon captured in hydrogen plants	Fossil-based hydrogen fuel with CCS Coal synfuels with CCS
	8	Storage of carbon captured in synfuel plants	Nuclear fission Onshore and offshore
	9	Nuclear electricity	Photovoltaic power
	10	Wind electricity	Wind hydrogen for vehicle fuel
	11	Solar electricity	(e.g. ethanol)
	12	Renewable hydrogen	
	13	Biofuels	
Natural sinks (bio-storage)	14	Forest management	Reduced deforestation, increased reforestation
	15	Agricultural soils management	(e.g. conservation tillage, erosion control)

Sources: Pacala and Socolow 2004; CMI 2015.

This demonstrates that currently available technologies already provide a significant basis from which to address climate change. These can be combined with carbon pricing, innovation and behavioural change, as suggested by Stern (2006).

In terms of policies that can encourage, direct and force mitigation, three types of regulatory responses are available to policymakers, which we already encountered in Chapter 3 (Gore 2011), of which the second and third are forms of market-based regulations, using economic incentives:

- Command and control mechanisms, which may mandate levels of emissions or technology but may not constitute the most cost-effective reductions.
- Taxation, which sets a price on emissions providing revenue and price certainty but may not ensure envisaged reduction levels.
- Cap-and-trade, where maximum emission levels are set and operators are allowed to trade permits. A variant are offsets where external reductions are purchased from a third party.

Table 5.5 provides an overview of the economic policy instruments that promote renewable energy, energy efficiency and energy conservation.

TABLE 5.5

Policy instruments

Instrument	Explanation/Comment	Example
Cap-and-trade	Tradable permit system for carbon emissions. Complex mechanism. Option of phasing in stricter rules over time.	European Union's Emissions Trading Scheme
Carbon tax	Per-unit tax. Simple mechanism. Uncertain reductions and political difficulties raising tax over time.	British Columbia's carbon tax
Direct subsidies	Promotes R&D in renewable energy or energy efficiency, and commercialisation of alternative technologies. Opposite of tax.	China, UK, US
Feed-in tariffs	Guarantees renewable energy producers grid access and payment to recover costs.	Commonly used in European countries (e.g. Spain and Germany)
Loan programmes	Provides access to credit where market fails, supports capital investment in renewable energy or energy efficiency.	EU
Rebates	Promotes household investment in renewable energy or energy efficiency.	Canada, France
Renewable Portfolio Standards	Setting minimum percentage of electricity generated from renewable sources for utilities.	California
Smart Grid	Can adjust energy demand to supply fluctuation and therefore facilitate increased utilisation of renewable energies.	Austin, Texas; Italy; China
Standards	Energy efficiency standards for appliances, automobiles and buildings.	EnergyStar programme (US), EU energy label
Tax credits (indirect subsidies)	Create incentives similar to rebates or subsidies.	US
Urban planning	Transit-oriented development, clustering of housing, shopping, commerce to reduce need for private car commuting.	Curitiba, Brazil

Sources: Hackett 2011; Harris and Roach 2018.

The challenge for policymakers in addressing climate change can be seen as three-fold – path dependence, behaviour and urgency. Path dependence refers to the limitation of today's decisions by past policy and investment decisions, where the circumstances under which they were made may no longer be relevant (Hackett 2011). For climate change, the challenge is the transition to a low- or zero-carbon (post-carbon) economy in a world where most infrastructure and technology are all built around and based on fossil fuels (e.g. cars running on petrol from petrol stations). The behavioural challenge sits in the dual dilemma of collective versus individual interest, and long-term versus short-term perspectives – whilst it is in society's interest to combat climate change now rather than later, individuals might benefit from acting in another way as climate change and action against it might not have immediate local (and personal) effects. Simply educating people or using fear or guilt is not sufficient, and whilst human behaviour is determined by internal factors, external factors need to be taken into account as well to effect a change in behaviour (Abraham 2017, based on Amel et al. 2017). Whilst the science case is well made, the case for behavioural change needs to be made in the cultural and social sphere, using both individual and collective action, with the former playing a critical role in inspiring people not just to approve of sustainable behaviour but to act in sustainable ways.[7] The third part of the challenge is the urgency in action needed to get GHG emissions 80% below 1990 levels by 2050. Whilst the transition to low-carbon energy driven by technology is already under way, scientists have stated that a peaking of emissions should not be delayed beyond 2020 – a year declared a climate turning point – in order to avoid unsurmountable difficulties in achieving the temperature goals agreed in Paris in 2015, as shown in Figure 5.6 (Figueres et al. 2017).

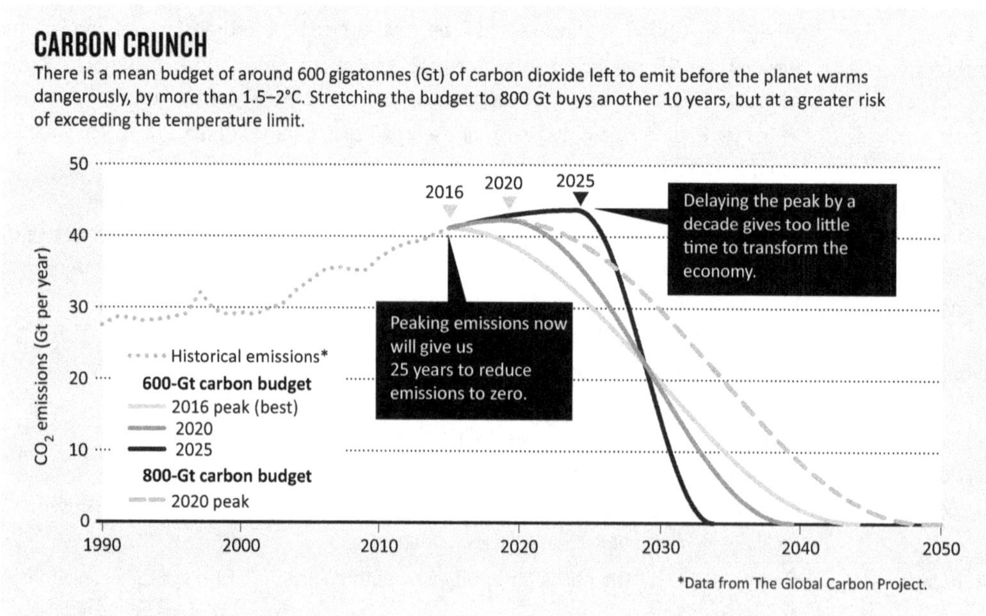

FIGURE 5.6
Carbon crunch

Source: Figueres, *Nature*, used with permission.

In order to level GHG emissions by 2020 and then to lower them continually, six milestones can help to prioritise action (Figueres et al. 2017) – these are supported by the World Business Council for Sustainable Development and underpinned by the collaborative campaign Mission 2020 (www.mission2020.global):

- Energy: renewable energy to supply 30% of electricity and no new power plants based on coal beyond 2020.
- Infrastructure: full decarbonisation of buildings and infrastructure by 2050.
- Transport: increase sales of electric vehicles to 15% of all new cars, increase fuel efficiencies for heavy-duty vehicles by 20%, decrease GHG emissions from aviation by 20% per kilometre travelled, doubling of public transport.
- Land: cut net emissions from deforestation to zero while using afforestation and reforestation to create carbon sinks by 2030, alongside sustainable agricultural practices.
- Industry: carbon-intensive industries (e.g. steel and iron, cement, chemicals, gas and oil) halve emissions before 2050 by cuts and efficiencies using science-based targets.
- Finance: annual investment in climate action from public ($800 billion) and private ($200 billion) resources, disclosure of climate-related financial risks, elimination of fossil fuel subsidies, implementation of carbon pricing mechanisms in all major economies.

Although the milestones amount to a "monumental task" (Figueres et al. 2017: 593), the targets are necessary, desirable and achievable as GHG emissions are already decoupling from production and consumption (e.g. population growth and per capita emissions, and growth of GDP and emissions).

The task is at the same time simple and complex – simple as in "do something or we're doomed", but complex as soon as the question is asked, what needs to be done? (Rowson 2014). In order to capture the complexity inherent in tackling climate change, Rowson (2014) suggests seven dimensions for action on climate change:

- Science: provides facts as objective reference point and authoritative agreement.
- Law: acts as powerful constraint and provides agreement at international level for mitigation e.g. on global carbon budget and at national level e.g. implementation of carbon taxes; provides rules.
- Money: capitalism as main financial operating system, makes redirection of capital possible, divesting from fossil fuels, investing in renewables, provides resources.
- Technology: provides tools and innovative ways of creating, storing and transporting energy.
- Democracy: provides institutions, making collective decisions possible for biggest collective action problem facing societies; policymakers can provide appropriate regulation and market signals based on political will in society.
- Culture: provides ideas and opportunities to frame the importance of climate change relative to other priorities.
- Behaviour: although choices are shaped by the first six dimensions, what is actually done and how people behave individually and collectively is what matters.

It is particularly the last dimension that is of importance, as although science has created the moral imperative to act, which is generally accepted, this is diluted by humanity's inability to act responsibly and collectively – a situation Rowson (2013) calls stealth denial. To address this, Rowson suggests connecting people to solutions, and solutions to problems, thus focusing on communication and on solutions that change behaviour based on consensus, which can be broadly summarised as reducing energy demand by improving efficiency and reducing consumption, and substituting supply.

All seven dimensions suggested previously interact with business – whilst science provides the foundation of facts, the law provides the regulatory framework and thus planning certainty for businesses within which they can operate. The business case for climate change policies lies not only in the regulatory certainty but also in the economic opportunities arising from climate solutions such as clean energy and technology, and in the reputational benefits of supporting public policies that combat climate change (Pew 2010). This reflects the monetary, technological, democratic and cultural dimensions discussed earlier, which alongside science and law help to change and shape individual, collective and corporate behaviour. So what can corporate behaviour set out to do in terms of climate change? What solutions can business bring to the table to address this environmental challenge through compliance and by going beyond it?

BUSINESS RESPONSES TO CLIMATE CHANGE: INTERNAL AND EXTERNAL OPPORTUNITIES

The stabilisation wedges and policy instruments presented earlier already introduced some of the measures available to policymakers and businesses. Here we first look at how businesses can reduce their GHG emissions internally before we turn our attention to the external opportunities business is presented with through climate change. Both of these aspects are situated within the broader context of using innovation and technology (including geoengineering) and carbon pricing, as well as behavioural changes to mitigate against and adapt to climate change.

Addressing climate change involves the challenge of decoupling emissions from revenues for companies to ensure that they are financially as well as environmentally sustainable – reflecting the wider challenge of decoupling emissions from economic and population growth (i.e. decoupling production and consumption, see also Chapter 10) as touched upon earlier. Furthermore, businesses need to take into account both risks and opportunities in addressing climate change, where risks include for example the rising costs of energy, supply chain impacts, effects on infrastructure and natural resources, costs from extreme events and social dislocation, whilst opportunities include efficiency in energy and operations, competitive differentiation through new products and services, involvement in climate-related government programmes and reducing losses from catastrophic weather events (NBS 2010).

Internal opportunities – how to minimise impact within a business

An important step for business is to incorporate climate change into strategic planning, which can then support the development of mitigation and adaptation actions, and the measurement, reporting and management of GHG emissions. The latter can be summarised as carbon management, which is a systematic and integrated approach to understand and respond to climate change at an organisational level, which involves measuring the carbon footprint of the company's GHG sources and sinks, and developing an approach to reduce emissions (Gore 2011).

Broadly speaking, the following elements are involved in how businesses can manage their GHGs (Robertson 2017):

- Making a plan: setting specific and measurable goals (e.g. reduce GHG emissions by 5% per year or become climate neutral by 2025 rather than a vague aspiration to become climate-friendly). This is underpinned by the resources available.

- Measure and assess: conducting a GHG emissions inventory that involves setting boundaries, collecting data, calculating emissions from data. Quality control and monitoring are important in the iterative process of this and the final element.
- Reduce emissions: decide on strategies and implement action plan. If climate neutrality is the goal but cannot be achieved on-site, offsetting may be used to bring total net emissions to zero.

This is all underpinned by communication, which should be clear and consistent regarding information and responsibility while avoiding greenwashing. These efforts are strengthened by setting science-based targets which are in line with the decarbonisation efforts required to keep global temperature increases below 2°C, ideally 1.5°C (Science Based Targets 2017). Setting such externally driven targets will build credibility, reputation, as well as business resilience, and increase competitiveness by driving innovation and transforming business practices. Science-based targets are different from incremental targets, which might only lead to the exploitation of low-hanging fruit rather than to a long-term vision and commitments (ibid.).

We will now look at the measurement and reductions of GHG emissions in more detail.

Measuring emissions

A widely recognised method of measuring and reporting corporations' climate impact is the Greenhouse Gas Protocol (GGP), an initiative jointly developed by the World Resources Institute (WRI) and the World Business Council for Sustainable Development (WBCSD). The ISO 14064 standard was subsequently developed for quantifying GHG emissions. For both standards, companies need to establish the boundaries of the organisation and to decide which sources of emissions are to be included. The GGP includes detailed guidance and a range of calculation tools across sectors and within specific sectors, as well as for countries and cities and, in particular, a corporate accounting and reporting standard that guides companies through the entire inventory process, though it can be used by other organisations as well (GGP 2017).

After the first step of defining organisational and operational boundaries, the scope of data collection needs to be decided on, i.e. which emissions sources need to be included. The GGP has three categories known as Scope 1, Scope 2 and Scope 3, which capture direct and indirect emissions (the following is based on Dharmasasmita et al. 2017; GGP 2017; Robertson 2017):

- Scope 1: direct emissions from sources owned or controlled by the organisation; these include the processes over which the company has the most control, such as heat and electricity generation, emissions from corporate vehicles and manufacturing processes as well as fugitive emissions.
- Scope 2: indirect emissions over which the company has some control, such as purchased electricity or heat.
- Scope 3: indirect emissions from processes outside the control of the organisation, such as employee vehicles, outsourced logistics, extraction and production of materials purchased, emissions from the end use and disposal of the goods produced by the company.

All Scope 1 and Scope 2 emissions should be part of an inventory, though in order for GHG emissions to be managed effectively, all three categories should be examined by a company in order to identify risks and opportunities. Scope 3 is captured through the corporate value chain standard,

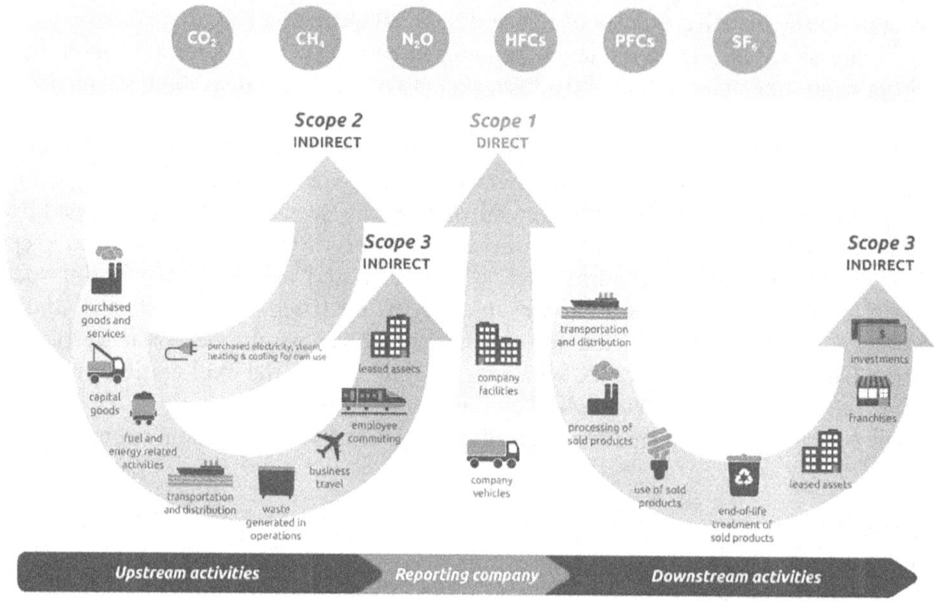

FIGURE 5.7
GGP scope and emissions across value chain

Source: GGP, used with permission.

which allows companies to analyse their supply chain, where often most of the carbon footprint of products may be found. Figure 5.7 shows emissions across the different scopes.

After identifying the sources of emissions (e.g. electricity, business travel, freight transport, outsourced services), the company needs to identify where emissions data can be found (e.g. through utility bills, purchase or expense records, shipping invoices). This then enables the identification of the various data units (KWh, MWh, km or miles flown, fuel type) and the calculation of emissions. The universal formula is: (activity data) x (emission factors) = GHG emissions, which may be expressed as CO_2 equivalent (CO_2e). Although the process is complex and time-consuming, requiring accuracy and good communication between different departments, it is crucial for effectively managing GHG emissions as it allows a company to identify where and how reductions are possible. These may include reductions in material use, water, waste and energy, identification of cost savings and of risks across the supply chain (if Scope 3 is calculated) as well as of opportunities for innovation from low carbon product design (Carbon Trust 2017).

Having measured and calculated emissions, what then are strategies available to companies for reducing emissions?

Reducing emissions

Several approaches are possible here (Robertson 2017) – a company could start with projects that are easy and quick to do at low cost. A two-stage approach might involve a pilot project as a smaller-scale test for a larger project, as not every project will be successful and a test project provides the opportunity to collect data, analyse and adjust. The visibility and appeal of a project also play a role in the decision-making process, for example as part of a new construction or renovation.

Reduction strategies evolve around energy use, transportation, buildings, operations and purchasing, forests (and farms and soil), and carbon sequestration, which can then be complemented by offsets, emissions trading and adaptation (Robertson 2017). As illustrated in Figures 5.1 and 5.2, energy for electricity and heat is responsible for indirect emissions and is a major source of GHG emissions generated from fossil fuels. In the following, we provide suggestions for each area, drawing on various sources (e.g. Robertson 2017; EPA 2017; IPCC 2014d).

Energy use

One approach here that is related to eco-efficiency introduced in Chapter 3 is *energy efficiency*, which aims to achieve more (or the same) with less energy. Energy efficiency is considered the lowest-cost alternative to fossil fuels and easy to achieve as saving fuel is cheaper than buying it in the first place. Similar to GHG management in general, energy efficiency can be measured, then goals set for it to improve performance, which is followed up by audits. Energy efficiency can be increased both at the end-use in companies by reducing energy demand as well as at the source in power plants by using advanced technologies or replacing fuels that burn more efficiently.

Another approach to tackling GHG emissions from energy is to *switch power* from fossil fuels to renewable energies from two types of sources:

- Distributed power sources (produced on the customer side of the meter, "local generation"): solar photovoltaic panels, geothermal power, combined heat and power (CHP, cogeneration), and biomass.
- Centralised power sources (produced on the utility side of the meter): hydroelectric power, concentrating solar power, wind power, wave and tidal energy as well as biogas or methane from landfills.

Another form of fuel switching is the use of nuclear processes to generate energy, though this comes with other uncertainties and risks, such as the long-term storage of nuclear waste. It is therefore probably best viewed as a transitional form of alternative energy.

Energy is not just a cost to be managed, but provides opportunities to reduce risk, build resilience and create new value, and therefore needs commitment at the highest level as well as integration into corporate strategy and operations (Winston et al. 2017). This is underpinned by tracking energy at all levels, tapping into new technologies and engaging key stakeholders.

Transportation

Transportation in companies includes the movement of people and products by road, air, railway, or shipping locally, nationally and internationally.

Options to reduce emissions here include the following:

- fuel switching to fuels that emit less CO_2,
- improvements to fuel efficiency with advanced design, materials and technologies,
- improvements of operating practices,
- reductions of travel demand.

Alternative energies for transportation include natural gas, electricity, biofuels, hydrogen and fuel cells. In particular, electric vehicles fuelled by renewable energies are an increasingly popular choice for commuting and short-distance logistics, especially in light of major cities such as Paris and Mexico City planning to ban automobiles running on diesel by 2025 and other cities (and countries) considering similar bans of fossil fuel-based vehicles.

Companies can offer various structures to employees when it comes to addressing transport emissions, from subsidising season tickets for public transport, offering tax reliefs for bicycle purchases and improved facilities for cyclists (e.g. storage and showers), encouraging ride sharing, allowing flexible working (working from home), charging variable parking fees according to vehicle CO_2 emissions, to providing parking spaces with charging points for electric cars. Furthermore, using tele- or videoconferencing cuts the amount of business travel needed and associated emissions – Polycom and Zoom are two such providers.

For tackling transport emissions from logistics rather than personal transportation, companies also have a range of options available beyond alternative fuels and hybrid vehicles. Reducing aerodynamic resistance through altering the shape and design of vehicles as well as cutting the weight of vehicles through using different materials improve fuel efficiency. Optimising vehicle load and transport routes have been adopted by various companies and can even be offered to consumers – for example furniture maker Warren Evans of London gave customers the option of choosing the greenest option of delivery, which involves careful planning and timing of deliveries to minimise miles travelled. Less technology-intensive options include improved driving practices (e.g. training drivers to drive more efficiently and to reduce idling) and vehicle maintenance (e.g. tyre pressure), whilst some companies limit the speed of their vehicles to save on fuel. Chapter 8 on supply chain management covers transportation as well.

Buildings

Buildings use vast amounts of energy, whether for cooling, heating, electricity consumption for lighting, electronic equipment, catering and so on. Addressing GHG emissions in existing buildings includes options covered earlier in energy use. In designing new buildings, more options for energy efficiencies become available through "green building" techniques such as better insulation and ventilation, passive heating and lighting making use of sunlight, and purchasing energy-efficient appliances and electronics. Existing buildings may be retrofitted to improve energy performance.

Operations and purchasing

This covers a range of activities and will differ according to the sector in which companies operate. For example, HFC refrigerants are used in air conditioning, heat pumps and refrigeration (which are emitting CO_2 through their energy use). HFCs have a high GWP, so regular maintenance is therefore needed to detect leaks whilst making efforts to find refrigerants with lower GWP.

Every company is involved in sourcing and using supplies and materials for which it has purchasing decisions to make and for which GHG implications can be taken into account. For example, paper use is linked to climate change in that trees store carbon, which is released when trees are felled and paper is produced. Changing paper use and purchasing post-consumer recycled paper offers opportunities for reducing emissions for companies regardless of sector. Looking at manufacturing and construction in particular, large quantities of materials are required. Here companies can choose to use recycled or renewable materials to avoid using raw or virgin material, which is produced using fossil fuels (e.g. scrap steel or aluminium). Using recycled materials also affects waste management, reducing the material going to landfill or incinerators where methane and other GHGs are emitted. Purchasing policies offer companies a formalised way to stipulate recycled or local content. We return to purchasing in Chapter 8 and operations in Chapter 9.

Other strategies for reducing emissions

In addition to these direct strategies, companies can use other, more indirect, strategies to reduce their GHG emissions. These include forests, farms and soil, as well as carbon sequestration, which may be complemented by offsets, emissions trading and adaptation (Robertson 2017); we will briefly explain each. Whilst cars and power plants are carbon sources, forests are carbon sinks, which means they store or sequester carbon. The same goes for soil and agricultural practices, which can be used to sequester carbon in the soil. In contrast to these natural processes which can be used to enhance GHG reduction efforts, carbon capture and sequestration (or storage, CCS) is a form of geoengineering technology that captures or removes carbon dioxide from industrial processes and sequesters or stores it underground in geological formations (Robertson 2017). There are considerable trade-offs, costs and risks associated with CCS (see e.g. Stern 2006), and it is therefore not a permanent solution, particularly as it allows the continued use of (e.g. coal power plants) rather than encouraging alternatives, and there are questions about its commercial viability (Nemetz 2013).

Two market-based strategies using carbon credits and carbon pricing are available for companies to further reduce their emissions to net zero once other options have been exhausted. *Offsets* allow the quantification of emissions reductions created by one company, which is then bought by another enterprise, for which this is a less expensive form of emissions reduction than internal efforts (Robertson 2017). As offsets can be bought by companies and individuals anywhere, they illustrate the global nature of climate change. Offsets may be generated by various GHG reduction activities (e.g. wind farms, methane-capture facilities or forests). As they are intangible products and available worldwide, third-party certification and verification schemes ensure real reductions and the additionality of reductions beyond business as usual (Gore 2011). Nevertheless, offsetting has been criticised for resembling the practice of selling indulgences in the Middle Ages for the absolution of sins, since buying offsets allows continued carbon emitting behaviour of individuals and companies, leading possibly to complacency (Clark 2011, see also Klein 2014).

Emissions trading could be called the inverse of offsetting (Robertson 2017). Known also as cap-and-trade systems, a maximum level for emissions is set, within which companies buy and sell permits (or allowances) to emit (see also Chapter 3 for a reminder of how cap-and-trade works). The EU's Emissions Trading Scheme (ETS) was one of the first carbon markets, starting in 2005 with a generous initial allocation of permits to ease the transition (Hackett 2011). During the global economic crisis, prices fell due to excess permits, which in effect has made it cheaper for companies to emit GHGs rather than change their behaviour. In a report on carbon pricing, leading economists (among them Nobel laureate Joseph Stiglitz and Nicholas Stern of the Stern Review on the Economics of Climate Change) have argued that the price of carbon needs to rise significantly in order to incentivise businesses to lower their emissions even in the case of cheap fossil fuels, and the price should be in the form of a carbon tax (Inman 2017).

Whilst both offsetting and carbon trading are external tools to reduce emissions, an increasing number of companies are using internal carbon pricing to make production more efficient and to encourage innovation (CDP 2016). It is one approach to internalise the external costs of GHG emissions, which helps to identify risks and opportunities as well as to provide incentives to find more ways to reduce carbon emissions. One retailer to use an internal carbon price in its climate change action programme is the Carrefour Group.

Adaptation is the second response to climate change after mitigation, which attempts to lower GHG emissions. Adaptation is a recognition that some temperature increases are inevitable as a consequence of past emissions, and that this requires adjusting to the new circumstances of a changed climate and planet. This is not a reduction strategy per se and will therefore not be discussed any further – though an example of businesses adapting to climate change might be in buying flood insurance or constructing office and factory buildings that are better able to withstand extreme weather events such as floods.

Interim Summary

Whichever sector or activity we look at, staff involvement and engagement are crucial, for which training and awareness raising programmes can be developed. As this also concerns the role of human resource management, we will return to this in Chapter 13. The commitment of staff members is particularly important in the case of climate change, which requires long-term solutions, as the present-time perspectives and low tolerance for uncertainty of individuals can interact with and reinforce short-termism and uncertainty avoidance at organisational and institutional levels, leading to overall inaction on climate change (Slawinski et al. 2017).

Whilst the overall emphasis for companies addressing climate change is on reducing emissions, which also involves decoupling them from revenues, and on building resilience, challenges differ across sectors, as some are more energy-intensive or transport-intensive than others.[8] For sectors such as transport and infrastructure, priorities will be the use of alternative fuels, while for other sectors, such as food and drinks, pharmaceuticals, IT equipment and mining, water plays an important role as production input, which is affected by climate change in various ways (e.g. flooding, droughts, changing precipitation) (Peace and Maher 2015). Extraction industries and fossil fuel producers have been urged to "keep it in the ground", as it is estimated that within the global carbon budget (see earlier discussion) only between one-fifth and one-third of proven oil, coal and gas reserves may be burned, and the remaining fossil fuels should be kept under ground in order to keep global temperature rises below 2°C[9] (Clark 2015).

As there are a variety of approaches to reduce emissions and to build resilience, there is no one correct path to follow, though Peace and Maher (2015) stress the importance of incorporating climate change into strategy and risk management. Companies need to assess their vulnerability, identify where their risks lie, track their carbon footprint and report on it, engage and communicate with internal and external stakeholders, and integrate climate risks and actions into their long-term planning.

Climate change not only offers efficiency opportunities to companies internally, as discussed in this part of the chapter, but also market opportunities externally, which we turn to now.

External opportunities – what can business offer to consumers and other companies?

Whilst addressing the business impact on the climate internally could be seen through the lens of business as the problem, looking at the opportunities climate change offers to companies allows a perspective of business as the solution. We focus here on clean technologies and behavioural change, which are two important mechanisms to address climate change alongside carbon pricing.

Cleantech – clean technologies

Clean technologies (abbreviated as cleantech, Elkington 2012) are not closely defined, though Pernick and Wilder (2012: 3), both of research firm Clean Edge, describe clean technology as

> any product, service, or process that harnesses renewable materials and energy sources, reduces the use of non-renewable natural resources, and cuts or eliminates emissions and wastes.

Clean technologies are not limited to one particular technology, and with their dynamic and potentially disruptive nature can become a central element of competitiveness for companies (Pernick and Wilder 2012). Cleantech includes a wide range of technologies, amongst them green buildings, high-efficiency lighting, hybrid vehicles and solar and wind power, that can help address climate change, whilst other clean technologies deal with other environmental challenges. Overall, cleantech is a fast growing sector (Elkington 2012; Carroll 2017) that features innovative start-ups as well as established companies as such technologies have become cheaper and smarter and as support has increased in the form of venture capital, dedicated incubators and hubs as well as conferences, exhibitions and other events rewarding and showcasing low-carbon technologies. Examples of hubs include New York City's Urban Future Lab, a hub for smart cities, clean energy and smart grid technology, and Los Angeles Cleantech Incubator (LACI), whilst Impact Hubs can be found across the globe, having expanded to a diverse network of over 15,000 members in over 80 locations. In Europe, Climate-KIC is a climate innovation initiative that includes an incubator, a platform for cleanteach ideas and a network of entrepreneurs. Conferences and exhibitions include e.g. Cleantech Innovate in the UK.

A lot of investment has been made into clean technologies across a range of sectors. Whilst it is not possible to provide a comprehensive discussion here, the following offers some snapshots of examples of such technologies that help address climate by reducing society's and industry's dependence on fossil fuels and by helping to change behaviour. Table 5.6 provides an overview of clean technologies in terms of patents as well as current topics in cleantech applicable across business activities.

TABLE 5.6

Overview of clean technologies

Trends in sustainable innovation	Higher shares of clean tech patents	Lower shares of clean tech patents
Fleet Tracking	Transportation	Solar
Virtual power plant	Advanced green materials	Air
Internet of things	Energy efficiency	Bioenergy
Robotics	Energy storage	Water and wastewater
Blockchain		Conventional fuel
Autonomous vehicles		Wind
Drones		Recycling and waste
Smart buildings		Hydro and marine power
3D printing		Nuclear
		Geothermal

Sources: www.cleantech.com, Saha and Muro (2017).

Buildings

Hive is a cleantech innovation by British Gas and offers smart home solutions regarding for example heating and light with the help of sensors, plugs and accessories, and a dedicated app that allows remote control of any appliances. It can be used residentially as well as for businesses and does not require energy to be supplied by any specific provider.

Advanced green materials – responsible for a large share of clean tech patents – can be used in constructing new buildings, whether it is insulation materials, glass or bricks, whilst architects can design and plan buildings in ways that are energy and water efficient e.g. through use of geothermal and solar power and heating, and better storm water management. An example of zero energy buildings is the Passivhaus standard, which uses passive measures such as strong insulation minimising thermal bridges, high levels of airtightness, mechanical ventilation and high performance windows (Passivhaus Trust 2017).

Transportation

Transportation features the highest shares of patents when it comes to clean technologies (Saha and Muro 2017). Toyota and Honda were early entrants in the market for hybrid vehicles, with all other manufacturers following suit eventually. Volvo, owned by Chinese car manufacturer Geely, announced in 2017 that from 2019 all its cars would have an electric power train and thus be either hybrid cars or fully electric vehicles, making Volvo the first major car manufacturer to "go all electric" (Tovey 2017). Niche manufacturer Tesla already produces electric vehicles only. These innovations all help consumers reduce emissions whilst also ensuring regulatory compliance for producers' CO_2 emissions across their fleets. A challenge in this area – beyond so-called range anxiety – is how the electricity used to power these cars is generated – by renewable energy or by fossil fuels. Another challenge is the development of the cost of batteries (Spector 2017). Although the share of electric cars is still small in most markets (2% at the most, with the exception of Norway, Milne 2017; Kommenda 2019), it is a fast growing market, surpassing 2 million vehicles in 2016, with estimates that within two decades all cars sold in Europe for example could be electric, helped e.g. by the UK's and France's commitment to ban all sales of new petrol and diesel cars by 2040 (IEA 2017; Vaughan 2017). A further challenge for the car industry is the highly profitable and quickly growing SUV segment of sports utility vehicles, which was the second biggest contributor to a rise in global CO_2 emissions from 2010 to 2018 after the power sector (Cozzi and Petropoulos 2019). Other alternatives in transport include vehicles run with hydrogen fuel cells, though infrastructure here is less developed than for electric cars.

A further and more recent area of innovation in transport are driverless (i.e. self-driving or autonomous) vehicles designed for optimising safety, routing and environmental impact. This is a competitive market that sees not just traditional vehicle manufacturers but also high-tech competitors such as Google and Apple invest in research and development. Other developments here include ride and vehicle sharing, fleet-tracking apps and transportation software (Muro 2017). One example is all-electric vanpool provider Green Commuter, which combines car sharing, vanpooling and fleet replacement, though it currently operates only in California.

Energy efficiency

Energy efficiency and storage are among the top four when it comes to patents for clean technologies. Energy efficiency includes for examples technologies related to heating, ventilation, and air conditioning (HVAC), water heating, appliances, lighting, windows and building envelope, sensors, controls and smart meters (Saha and Muro 2017). Air conditioning represents a particular paradox – as the world gets warmer, more air conditioners are used, which in turn contributes to global warming (Buranyi 2019). Energy storage is particularly relevant for supporting renewable energies such as solar and wind that naturally fluctuate and are thus intermittent in their capacity to supply energy.

The market for energy efficiency and management is a growing one, as is the retail market for energy-efficient consumer goods. One example of a company helping other firms to cut energy costs and emissions is the UK-based firm Energy Technology & Control (ETC) through a new technology that electronically regulates the amount of fuel and air used in combustion boilers (company website, www.energytechnologycontrol.com). Another example is US-based Ice Energy, which produces air conditioning units that can store energy and cool homes without using electricity (Carroll 2017, company website, www.ice-energy.com/).

For countries whose energy and transportation infrastructures are still developing, the necessity to address climate change provides incentives and opportunities to leapfrog to clean technology and avoid investments into fossil fuel-based alternatives (Hart 2010). Not too long ago, China built a new coal power plant every day, but its coal consumption has been declining for several years now and the country is today a global leader in renewable energy with twice the capacity than the United States in 2017, having leapfrogged Germany and Japan as well (Christie 2017).

Other areas: Geoengineering

Geoengineering was explained in Box 5.2. It is part of so-called negative emissions technologies, a term that covers a range of low and high-tech solutions from reforestation projects to fertilising the ocean with iron fillings or seeding the stratosphere with sulphates (Nogrady 2017). Whilst reforestation (i.e. planting more trees) can address deforestation and land degradation, the scale needed to address climate change is substantial, which has led companies and researchers to look for other solutions, such as for example biochar and enhanced weathering.

Although geoengineering is not considered an established and effective way of addressing climate change, as it comes attached with considerable uncertainties, ethical challenges and high expenses, there are examples of companies, in particular start-ups supported by venture capital, that have invested in geoengineering research, such as for example Global Thermostat and Carbon Engineering working on CDR techniques or StratoShield which would use SRM techniques but remains at a conceptual stage (Brogan 2016). Asphalt coating manufacturer GuardTop developed CoolSeal, a reflective street surface that is lighter in colour and can reduce surface temperatures by ca. 5°C (10°F). After two years of testing, it is now used in Los Angeles, a city that has seen temperatures rise by more than average due to the vast amounts of asphalt creating heat island effects through roads, parking lots and roofs, for example (Bartholomew 2017; Carroll 2017).

Companies and behavioural change

Wynes and Nicholas (2017) looked at what are the most effective individual actions consumers can take to reduce their GHG emissions and compared it with advice from governments and educational institutions. Widely recommended strategies such as recycling or changing lightbulbs have only a low to moderate impact, whilst efforts to purchase green energy and to reduce the effects of driving by buying a more efficient car have a high impact and are perceived to have great effect, albeit not as great as the four actions that would substantially decrease individuals' carbon footprints. These four actions are living car-free, avoiding air travel, eating a plant-based diet, and having smaller families, i.e. fewer children. The authors acknowledge that these are very personal and individual choices in terms of lifestyle, and propose empowering individuals to concentrate on changing those behaviours most effective at reducing personal GHG emissions. Despite cultural and structural barriers, Wynes and Nicholas (2017) assert that "serious behavioural change is possible".

In the context of the chapter here, we remind ourselves that just 100 companies are responsible for over 70% of global GHG emissions, whilst households are responsible for 20% of emissions in the EU for example (Griffin 2017; Eurostat 2016). Whilst corporate actions to address climate change are undoubtedly the most significant as they have the highest impact, the challenge requires actions from all stakeholders, also from consumers. We therefore ask the question: what role is there for companies in affecting behavioural change?

The environmental consequences of *family size* are rarely discussed, and the recommendation of having fewer children is a sensitive and potentially contentious one, with personal choices and preferences possibly at odds with the common good, influenced by cultural, religious and political factors and debates about global overpopulation. The role of companies to affect behavioural change is therefore limited, and it is not clear what they could offer beyond contraceptive products and services in this area.

Looking at *plant-based diets*, the market for vegetarian and vegan food, especially for meat substitutes, has experienced phenomenal growth. For example, Quorn Foods reported growth rates of its meat replacement products made of fungus of 19% in 2017, with the US and Australia growing by 40% and 35% respectively (Nilsson 2017). The British company reports that the carbon footprint of its products are 90% lower than beef and 70% lower than chicken, which has been third-party certified by The Carbon Trust (Quorn 2019).

The firm and others like it (e.g. Beyond Meat, Linda McCartney's Foods or Fry's) have benefitted from the rise in "flexitarian" or partially meat-free diets (Nilsson 2017). Campaigns such as Meat Free Monday (launched by Paul, Mary and Stella McCartney in 2009) or World Meat Free Day have helped to raise awareness of the climate impact of agriculture, in particular meat. Market research conducted by Mintel (2016) in its global food and drinks report found a 25% and 250% increase in vegetarian and vegan claims for new product launches. The overall share is still low, but vegetarian launches now make up 11% of global food and drinks launches, up from 8%, whilst vegan launches are up from 1% to 4%, with numbers still increasing – in the UK the latter accounted for 18% of new food products in 2019.

The latest development is research into lab-based meat (also called animal-free meat or synthetic or alternative meats) which aims to resemble the texture, taste and aroma of meat, thus making it easier for consumers to make the switch. Some of these are entirely plant based, such as the burger made by Impossible Foods, whilst others, such as Mosa Meat and Memphis Meat, are developing cultured meat in labs, using stem cells from cows (Challen 2015, see also Kobie 2017a).

For *avoiding air travel*, there are already recommendations available for individual action, such as taking only direct and daytime flights (avoiding stopovers), combining trips to minimise the number of flights, flying economy, packing light and purchasing carbon offsets. Although companies can make carbon offsets for their business, these recommendations still involve air travel, so what else can industry offer? Tele- or videoconferencing was mentioned previously, but what else can be done if travel is unavoidable?

High-speed trains have already eliminated the need for air travel, using both dedicated high-speed lines as well as upgraded classic lines. Examples include the Eurostar travelling between the UK (e.g. London), France (e.g. Paris), Belgium (Brussels) and the Netherlands (Amsterdam), TGV in France, ICE in Germany, AVE in Spain, THSR in Taiwan whilst China's bullet train network features the fastest as well as the longest high-speed train service in the world (Railway Technology website, China Daily 2017). In the US, the planned California High Speed Rail between San Francisco and Los Angeles is expected to divert more than a third of air travel between the two cities. The Shinkansen in Japan has about a 75% market share of the air transportation market for journeys of about 2.5 hours (Railway Technology website). China also boasts the world's first high-speed magnetic levitation line, the Shanghai Maglev, connecting the airport with the city.

An even faster alternative is being developed with Hyperloop, a mode of transport that uses pods travelling through low-pressure (near-vacuum) tubes. Whilst the idea of travelling through frictionless vacuum tubes is a century old, it is entrepreneur Elon Musk (known for his ventures PayPal, Tesla and SpaceX) who is behind the current developments. Making the design open source and holding initial competitions have spurned collaborative efforts into developing the technology – for example Hyperloop Transportation Technologies draws on over 600 team members across the world from R&D and commercial partners, whilst Hyperloop One was the first company to test a full-scale Hyperloop in 2017 involving all combined components (Kobie 2017b). Both companies have proposed routes across the world with planned speeds of up to 700 mph. Building these would require less ground space than high-speed railway lines, run on less energy and thus provide savings of both cost and time (Field 2017).

The final strategy recommended by Wynes and Nicholas (2017) for individual action is *living car-free*. As for the previous action, there are recommendations available for changing individual behaviour, such as using public transportation, car sharing, ride sharing, cycling and walking, and choosing living location to match transportation choice. Here, opportunities exist for companies providing public transport, car clubs (e.g. ZipCar), peer-to-peer car sharing (e.g. GetAround or Turo), ride-sharing apps (e.g. Lyft) as well as construction companies building houses in more convenient locations, and bicycle manufacturers. Many major cities and increasingly smaller towns offer self-service cycle hire schemes since the launch of the world's first large-scale bike-sharing scheme – Velib in Paris in 2007. More recent developments include dockless hire bikes that can be located and unlocked via smartphone apps, a move away from less flexible docking stations where users rely on fixed locations from which to pick up bikes. Examples include Singapore-based oBike which has expanded to Malaysia, Australia, the UK, Switzerland, the Netherlands and Taiwan, while ofo and Mobike, both from China, have also started offering services in selected cities abroad (Collinson 2017).

Beyond transportation and location there are other ways in which companies can help consumers to become car-free. Retailers have adapted to this – instead of out-of-town big hypermarkets that require personal transportation, supermarkets have for example opened smaller shops closer to where people live to enable smaller, more frequent shops, whilst also offering online shopping and home deliveries. Vehicle ownership rates are already declining in the US, and younger people are less likely to own a car and more likely to prefer accessing mobility as a service, particularly in urban environments where local governments and city planners talk about "peak car" and a growing appetite for car alternatives (Bouton et al. 2015; Moss 2015).

The transition movement advocates localisation – an important element of car-free living – and collaboration of multiple stakeholders in local communities, involving companies, local authorities and individual citizens and consumers. The Transition Network was founded in 2005 to address climate change and peak oil by using a model of localisation that uses for example backcasting – asking what the future might look like and then working backwards to get there (see Chapter 4 on The Natural Step for a reminder). The aim is to build resilience in a low-carbon world. The network is active in over 50 countries with thousands of initiatives across schools, universities, towns and cities, and very much built on optimism and opportunity rather than on threat (Transition Network 2017; Flintoff 2013; Robertson 2017). Initiatives include locally produced power (e.g. in the form of combined heat and power (CHP)), and support for entrepreneurship. Examples of cities include Växjö in Sweden, which aims to be fossil-free by 2030, whilst Copenhagen aims to be carbon neutral by 2025. Transition towns in the UK include Totnes and Bristol, which use their own local currencies to circulate money in the local economy. In Bristol, several hundred businesses take part in the scheme, which allows transactions in cash, by text and online for thousands of users (https://bristolpound.org). The role of cities in addressing climate change can for example be seen in the response of cities to the announcement that the US federal government would pull the country out of the Paris climate accord – a response very much supported by businesses (see also Box 5.1).

We can conclude here that while there is an important role for companies in affecting behavioural change in individuals, it is collaboration that plays an even more significant role as climate change cannot be addressed in isolation but requires the commitment and initiative of all levels of state, market and civil society across national and local governments, cities, companies, citizen initiatives, consumers and so on. It can for example be seen in the Hyperloop initiative or in below50, a global collaboration bringing together the entire supply chain for sustainable fuels (WBCSD 2017), whilst the Network for Business Sustainability (NBS 2013) recommends that innovation is built on collaboration, sharing and multiplication of impact. Hargadon (2011) also found that collaboration and dialogue with policymakers is crucial for low-carbon innovations which companies need to commit to in order to succeed in the long term.

SUMMARY

Climate change is arguably one of the biggest and most encompassing challenges facing business and society, requiring urgent action of mitigation and adaptation, using innovative technologies, carbon pricing and behavioural change. Companies can use strategy, design and technology to address it in conjunction with policy. *The Economist* encourages policymakers to target zero emissions rather than 100% renewable energies to fight global warming instead of favouring particular technologies (Economist 2017). Such an approach also provides greater incentives for companies for whom acting on climate change is a *no regrets* strategy, based on for example the efficiency savings and resilience achievable in addition to business opportunities.

Following an introduction to what climate change is, we looked at the impacts of climate change – more generally as well as more specifically on business and how business contributes to and causes climate change. We then covered regulatory responses, which provide the policy environment within which business addresses the problem. We looked at what companies can do to reduce their GHG emissions internally and what they can offer externally on the market, to other companies and consumers through clean technologies and facilitating behavioural change.

Looking ahead to part three of the book, climate change is covered as well – through the lens of corporate functions in addressing environmental challenges. Chapter 8 on supply chain management is relevant as for many companies most of their carbon impact is through the value or supply chain (Scope 3 emissions), whilst Chapter 9 on production and operations includes for example lean manufacturing and design for environment which enables companies to research and develop low or zero carbon materials and processes. Chapter 10 on marketing covers the role of consumers, whilst Chapter 11 on environmental management systems and Chapter 12 on accounting discuss the role of monitoring, controlling, measuring and reporting in order to manage the carbon footprint. Chapter 13 on human resources covers the importance of leadership and change management as well as the role of sustainability-based incentives in addressing environmental challenges.

Three integrative case studies in the concluding part of the book allow for a broader discussion of climate change in the context of specific industries, companies and other organisations.

SUGGESTED QUESTIONS AND ACTIVITIES

- Use a carbon footprint calculator to measure your personal impact: www.carbonfootprint.com/calculator.aspx or for a broader measure www.footprintcalculator.org/. For any US residents, https://www3.epa.gov/carbon-footprint-calculator/ is available – for other countries, research whether your government offers a carbon footprint calculator.
- Visit The Shift Project's webpage and research how much GHG emissions a country or a sector of your choice emit: https://theshiftdataportal.org/.
- Visit the GGP website to explore what calculation tools are available for different sectors: www.ghgprotocol.org/calculation-tools
- Learn more about setting science-based targets for companies: http://sciencebasedtargets.org/
- Choose one of the frameworks, models, or strategies from Chapter 4 and use it to plan a company's approach to climate change.
- Use the following two pages to research particular companies and their contributions to climate change: www.theguardian.com/environment/interactive/2013/nov/20/which-fossil-fuel-companies-responsible-climate-change-interactive and http://climateaccountability.org/carbonmajors.html.
- How would you debate with a climate change sceptic? Check out this MOOC (massive online open course) for more on climate science denial: www.edx.org/course/making-sense-of-climate-science-denial.
- Harvard Business School's 900 MBA students each had to research a company that is affected by climate change in one way or another. Their assignments have been made public – browse the site for some examples and find out about any more recent developments: https://rctom.hbs.org/assignment/climate-change-challenge/
- Play the climate stabilisation wedges game: https://cmi.princeton.edu/wedges/game.
- Debate the following motions, preparing for both agreement and disagreement on the following statements:
  - Geoengineering offers a reliable approach to addressing climate change.
  - Trying to regulate climate change is like rearranging deckchairs on the Titanic.
  - GHG management is anti-capitalist.
- Choose one form of renewable energy and use the REN21 Network's global status report on renewables to analyse its current state: www.ren21.net/reports/global-status-report/.
- Visit the homepage of one of the incubators and hubs or conferences mentioned in the chapter and see what the latest innovations are in cleantech regarding climate change: https://laincubator.org/, https://impacthub.net/, http://ufl.nyc/, www.climate-kic.org, www.cleantechinnovate.com/.
- Find out how far Hyperloop has been developed – full-scale tests were first conducted in 2017; when will it be commercially available to consumers?
- Go to https://transitionnetwork.org/transition-near-me/ and see if there are any transition movement initiatives near you.

NOTES

1 Other halocarbons include CFCs (chlorofluorocarbons) and HCFCs (Hackett 2011). There are 13 HFCs and seven PFCs, which all exhibit different global warming potentials, expressed in CO_2 equivalents, CO_2e (Nemetz 2013).

2 ppm (parts per million) or ppb (parts per billion, 1 billion = 1,000 million) is the ratio of the number of gas molecules to the total number of molecules of dry air (IPCC 2013), for some gases even ppt (parts per trillion) is used.

3 Technically, water vapour is the most common GHG in the atmosphere, but as human activities have only a very small impact, it is not generally included in discussions on anthropogenic climate change.
4 The attentive reader will notice fewer references in the first part of this chapter – a deliberate decision to provide greater readability for a complex topic. Sources (listed in the back) range from sustainability textbooks to the IPCC and various climate change think tanks and media outlets as well as company websites for selected examples.
5 Although more up-to-date statistics exist, the graph has been selected for its useful comprehensiveness.
6 For Nicaragua the agreement did not go far enough, whilst war-torn Syria did not participate in the process.
7 Whilst most people approve of sustainability, in reality, they behave in unsustainable ways (Amel et al. 2017).
8 The following provide simple snapshots. For a more detailed view across sectors of the top 100 GHG-emitting companies, see Lubin and colleagues (2017).
9 In the US, Democrats even proposed a symbolic "Keep it in the Ground" bill to prohibit new fossil fuel extraction on public lands even though it had little chance of being passed by Republican lawmakers.

SOURCES AND FURTHER READING

Abraham, J. (2017) Study: Inspiring Action on Climate Change Is More Complex Than You Might Think, *The Guardian*, 19.05.2017, available at www.theguardian.com/environment/climate-consensus-97-per-cent/2017/may/19/study-inspiring-action-on-climate-change-is-more-complex-than-you-might-think (Retrieved, 10.07.2017).

Amel, E., Manning, C., Scott, B. and Koger, S. (2017) Beyond the Roots of Human Inaction: Fostering Collective Effort Toward Ecosystem Conservation, in: *Science*, 356(6335): 275–279.

Anderson, D. A. (2014) *Environmental Economics and Natural Resource Management*, London: Routledge, 4th ed.

Bartholomew, D. (2017) 'Cool Pavement' to Cut Urban Street Heat Gets First California Tryout in Canoga Park, *Los Angeles Daily News*, 22.05.2017, available at www.dailynews.com/environment-and-nature/20170520/cool-pavement-to-cut-urban-street-heat-gets-first-california-tryout-in-canoga-park (Retrieved, 26.07.2017).

Blowfield, M. (2013) *Business and Sustainability*, Oxford: Oxford University Press.

Bouton, S, Knupfer, S. M., Mihov, I. and Swartz, S. (2015) Urban Mobility at a Tipping Point, McKinsey Insights, available at www.mckinsey.com/business-functions/sustainability-and-resource-productivity/our-insights/urban-mobility-at-a-tipping-point (Retrieved, 27.07.2017).

Brogan, J. (2016) What's the Deal with Geoengineering? available at www.slate.com/articles/technology/future_tense/2016/01/can_we_stop_climate_change_by_tinkering_with_the_atmosphere.html (Retrieved, 25.07.2017).

Buranyi, S. (2019) The Air Conditioning Trap: How Cold Air Is Heating the World, *The Guardian*, 29.08.2019, available at www.theguardian.com/environment/2019/aug/29/the-air-conditioning-trap-how-cold-air-is-heating-the-world (Retrieved, 04.12.2019).

Carbon Trust (2017) The Carbon Trust, website, available at www.carbontrust.com (Retrieved, 17.07.2017).

Carroll, R. (2017) 'Grow Food on Mars': LA Startups Tackle Climate Change with Inventive Solutions, *The Guardian*, 18.07.2017, available at www.theguardian.com/us-news/2017/jul/18/climate-change-los-angeles-tech-startups (Retrieved, 24.07.2017).

CDP (2016) Out of the Starting Blocks: Tracking Progress on Corporate Climate Action, Carbon Disclosure Project, available at www.cdp.net/en/research/global-reports/tracking-climate-progress-2016 (Retrieved, 21.07.2017).

Challen, J. (2015) Lab-Grown Burgers Edge Closer to Menus, Future Agriculture Portal, available at http://futureag.info/news/lab-grown-burgers-edge-closer-to-menus/ (Retrieved, 26.07.2017).

Christie, S. (2017) Panda-Shaped Solar Farm Built by Chinese Energy Firm Nearing Completion, *The Telegraph*, 25.07.2017, available at www.telegraph.co.uk/business/2017/07/25/chinese-energy-firm-builds-solar-farm-shaped-like-panda/ (Retrieved, 25.07.2017).

Clark, D. (2011) A Complete Guide to Carbon Offsetting, *The Guardian*, 16.09.2011, available at www.theguardian.com/environment/2011/sep/16/carbon-offset-projects-carbon-emissions (Retrieved, 20.07.2017).

Clark, D. (2015) How Much of the World's Fossil Fuel Can We Burn? *The Guardian*, 25.03.2015, available at www.theguardian.com/environment/keep-it-in-the-ground-blog/2015/mar/25/what-numbers-tell-about-how-much-fossil-fuel-reserves-cant-burn (Retrieved, 21.07.2017).

CMI (2015) Carbon Mitigation Initiative, available at http://cmi.princeton.edu/wedges (Retrieved, 07.07.2017).

Collinson, P. (2017) On Your Bike: The Best and the Worst of City Cycle Schemes, *The Guardian*, 25.02.2017, available at www.theguardian.com/money/2017/feb/25/best-and-worst-city-cycle-schemes-bike-sharing-london (Retrieved, 27.07.2017).

Cozzi, L. and Petropoulos, A. (2019) Commentary: Growing Preference for SUVs Challenges Emissions Reductions in Passenger Car Market, *International Energy Agency*, 15.10.2019, available at www.iea.org/newsroom/news/2019/october/growing-preference-for-suvs-challenges-emissions-reductions-in-passenger-car-mark.html (Retrieved, 03.12.2019).

Dharmasasmita, A., Kennedy, E., Puntha, H. and Holmes, R. (2017) Climate Change and Greenhouse Gas Management, in: Molthan-Hill, P. (ed.) *The Business Student's Guide to Sustainable Management*, Sheffield: Greenleaf Publishing, 2nd ed., pp. 412–462.

Dunphy, D., Griffiths, A. and Benn, S. (2007) *Organizational Change for Corporate Sustainability: A Guide for Leaders and Change Agents of the Future*, London: Routledge, 2nd ed.

EEA (2016) The European Environment – State and Outlook 2015 Report, European Environment Agency, available at www.eea.europa.eu/soer (Retrieved, 21.06.2017).

Economist (2017) A Green Red Herring: Better to Target Zero Emissions than 100% Renewable Energy, *The Economist*, 15.07.2017, available at www.economist.com/news/leaders/21725001-goal-after-all-curb-global-warming-not-favour-particular-technologies-better (Retrieved 25.07.2017).

Elkington, J. (2012) *The Zeronauts: Breaking the Sustainability Barrier*, London: Routledge.

EPA (2017) Sources of Greenhouse Gas Emissions, United States Environmental Protection Agency, available at www.epa.gov/ghgemissions/sources-greenhouse-gas-emissions (Retrieved, 20.07.2017).

Esty, D. C. and Winston, A. S. (2009) *Green to Gold: How Smart Companies Use Environmental Strategy to Innovate, Create Value and Build Competitive Advantage*, Hoboken, NJ: Wiley.

Eurostat (2016) Greenhouse Gas Emissions by Industries and Households, available at http://ec.europa.eu/eurostat/statistics-explained/index.php/Greenhouse_gas_emissions_by_industries_and_households (Retrieved, 26.07.2017).

Field, M. (2017) What Is Hyperloop and Will it be the Future of Transport? *The Telegraph*, 18.07.2017, available at www.telegraph.co.uk/technology/0/hyperloop-will-future-transport/ (Retrieved, 26.07.2017).

Figueres, C., Schellnhuber, H. J., Whiteman, G., Rockström, J., Hobley, A. and Rahmstorf, S. (2017) Three Years to Safeguard Our Climate, in: *Nature*, 545(7660): 593–595.

Flintoff, J.-P. (2013) Local, Self-Sufficient, Optimistic: Are Transition Towns the Way Forward? *The Guardian*, 15.06.2013, available at www.theguardian.com/environment/2013/jun/15/transition-towns-way-forward (Retrieved, 27.07.2017).

GGP (2017) Greenhouse Gas Protocol, website, available at www.ghgprotocol.org (Retrieved, 13.07.2017).

Goldenberg, S. (2013) Just 90 Companies Caused Two-Thirds of Man-Made Global Warming Emissions, *The Guardian*, 20.11.2013, available at www.theguardian.com/environment/2013/nov/20/90-companies-man-made-global-warming-emissions-climate-change (Retrieved, 05.07.2017).

Gore, J. (2011) Climate Change, in: Brady, J., Ebbage, A. and Lunn, R. (eds.) *Environmental Management in Organizations, The IEMA Handbook*, London: Earthscan/Routledge, 2nd ed., pp. 169–183.

Griffin, P. (2017) The Carbon Majors Database: CDP Carbon Majors Report 2017, Carbon Disclosure Project, available at www.cdp.net/en/reports/downloads/2327 (Retrieved 26.07.2017).

Guardian (2011) Which Industries and Activities Emit the Most Carbon? available at www.theguardian.com/environment/2011/apr/28/industries-sectors-carbon-emissions (Retrieved, 05.07.2017).

Hackett, C. (2011) *Environmental and Natural Resource Economics*, London: Sharpe, 4th ed.

Hargadon, A. (2011) The Business of Innovating: Bringing Low-Carbon Solutions to Market, Center for Climate and Energy Solutions, available at www.c2es.org/docUploads/LowCarbonInnovation.pdf (Retrieved, 27.07.2017).

Harris, J. M. and Roach, B. (2018) *Environmental and Natural Resource Economics: A Contemporary Approach*, New York, NY: Routledge, 4th ed.

Hart, S. L. (2010) *Capitalism at the Crossroads: Next Generation Business Strategies for a Post-Crisis World*, Upper Saddle River, NJ: Pearson, 3rd ed.

Harvey, F. and Watts, J. (2017) Insurers not Surprised by Harvey Costs After Years of Warning on Climate Risks, *The Guardian*, print edition, 02.09.2017.

Heede, R. (2014) Tracing Anthropogenic Carbon Dioxide and Methane Emissions to Fossil Fuel and Cement Producers, 1854–2010, in: *Climatic Change*, 122(1–2): 229–241.

Heede, R. (2019) Carbon Majors, available at http://climateaccountability.org/carbonmajors.html (Retrieved, 24.02.2020).

IASS (2017) Air Pollution and Climate Change, Addressing the Interlinkages Between Greenhouse Gases, Climate Change and Air Quality, IASS Potsdam, Institute for Advanced Sustainability Studies, available at www.iass-potsdam.de/en/content/air-pollution-and-climate-change (Retrieved, 27.06.2017).

IEA (2017) Global EV Outlook 2017, International Energy Agency, available at www.iea.org/media/topics/transport/Global_EV_Outlook_2017_Leaflet.pdf (Retrieved, 25.07.2017).

Inman, P. (2017) Sky-High Carbon Tax Needed to Avoid Climate Catastrophe, Say Experts, *The Guardian*, 29.05.2017, available at www.theguardian.com/environment/2017/may/29/sky-high-carbon-tax-needed-to-avoid-catastrophic-global-warming-say-experts (Retrieved, 20.07.2017).

IPCC (2013) *Climate Change 2013: The Physical Science Basis*, Contribution of Working Group I to the Fifth Assessment Report of the Intergovernmental Panel on Climate Change [Stocker, T. F., D. Qin, G.-K. Plattner, M. Tignor, S. K. Allen, J. Boschung, A. Nauels, Y. Xia, V. Bex and P. M. Midgley (eds.)]. Cambridge University Press, Cambridge, United Kingdom and New York, NY, USA.

IPCC (2014a) Summary for Policymakers, in: *Climate Change 2014, Mitigation of Climate Change*, Contribution of Working Group III to the Fifth Assessment Report of the Intergovernmental Panel on Climate Change [Edenhofer, O., R. Pichs-Madruga, Y. Sokona, E. Farahani, S. Kadner, K. Seyboth, A. Adler, I. Baum, S. Brunner, P. Eickemeier, B. Kriemann, J. Savolainen, S. Schlömer, C. von Stechow, T. Zwickel and J. C. Minx (eds.)]. Cambridge University Press, Cambridge, United Kingdom and New York, NY, USA.

IPCC (2014b) Summary for Policymakers, in: *Climate Change 2014, Impacts, Adaptation, and Vulnerability. Part A: Global and Sectoral Aspects*. Contribution of Working Group II to the Fifth Assessment Report of the Intergovernmental Panel on Climate Change [[Field, C. B., V. R. Barros, D. J. Dokken, K. J. Mach, M. D. Mastrandrea, T. E. Bilir, M. Chatterjee, K. L. Ebi, Y. O. Estrada, R. C. Genova, B. Girma, E. S. Kissel, A. N. Levy, S. MacCracken, P. R. Mastrandrea, and L. L. White (eds.)]. Cambridge University Press, Cambridge, United Kingdom and New York, NY, USA.

IPCC (2014c) Summary for Policymakers, in: *Climate Change 2014, Synthesis Report*, Contribution of Working Groups I, II and III to the Fifth Assessment Report of the Intergovernmental Panel on Climate Change [Core Writing Team, R. K. Pachauri and L. A. Meyer (eds.)]. IPCC, Geneva, Switzerland.

IPCC (2014d) *Climate Change 2014, Mitigation of Climate Change*, Contribution of Working Group III to the Fifth Assessment Report of the Intergovernmental Panel on Climate Change [Edenhofer, O., R. Pichs-Madruga, Y. Sokona, E. Farahani, S. Kadner, K. Seyboth, A. Adler, I. Baum, S. Brunner, P. Eickemeier, B. Kriemann, J. Savolainen, S. Schlömer, C. von Stechow, T. Zwickel and J. C. Minx (eds.)]. Cambridge University Press, Cambridge, United Kingdom and New York, NY, USA.

Klein, N. (2014) *This Changes Everything*, London: Penguin Books.

Kobie, N. (2017a) Meatless Meat: The Startups Ditching Animals from Meat, *Wired Magazine*, available at www.wired.co.uk/article/lab-grown-plant-based-meat-substitutes-impossible-foods-beyond-meat (Retrieved, 24.01.2018).

Kobie, N. (2017b) Hyperloop One Hits 309km/h in Latest Test, *Wired Magazine*, available at www.wired.co.uk/article/hyperloop-one-test-phase-2 (Retrieved, 24.01.2018).

Kommenda, N. (2019) Demand for SUVs 'Second Biggest Cause of CO_2 Rise', *The Guardian*, print edition, 26.10.2019.

Lingl, P., Carlson, D. and The David Suzuki Foundation (2010) *Doing Business in a New Climate: A Guide to Measuring, Reducing, and Offsetting Greenhouse Gas Emissions*, London: Earthscan.

Lubin, D., Moorhead, J. and Nixon, T. (2017) Global 100 Greenhouse Gas Performance: New Pathways for Growth and Leadership, Report, CDP and Thomson Reuters, available at http://sustainability.thomsonreuters.com/wp-content/uploads/2012/08/Global-100-Greenhouse-Gas-Performance-New-Pathways-for-Growth-and-Leadership-2017.pdf (Retrieved, 21.07.2017).

Milne, R. (2017) Reality of Subsidies Drives Norway's Electric Car Dream, *Financial Times*, 14.06.2017, available at www.ft.com/content/84e54440-3bc4-11e7-821a-6027b8a20f23 (Retrieved, 27.07.2017).

Mintel (2016) Global Food and Drink Trends 2017, Report, available at www.mintel.com/global-food-and-drink-trends (Retrieved, 26.07.2017).

Moss, S. (2015) End of the Car Age: How Cities Are Outgrowing the Automobile, *The Guardian*, 28.04.2015, available at www.theguardian.com/cities/2015/apr/28/end-of-the-car-age-how-cities-outgrew-the-automobile (Retrieved, 27.07.2017).

Muro, M. (2017) What Patents Reveal About Who's Behind Clean-Technology Innovation, *Wall Street Journal*, 20.06.2017, available at https://blogs.wsj.com/experts/2017/06/20/what-patents-reveal-about-whos-behind-clean-technology-innovation/ (Retrieved, 25.07.2017).

NBS (2010) Primer: Climate Change, Network for Business Sustainability, available at http://nbs.net/knowledge/climate-change/primer/ (Retrieved, 28.06.2017).

NBS (2013) Climate Change Innovation: 5 Recommendations, Network for Business Sustainability, available at http://nbs.net/knowledge/climate-change-innovation-5-recommendations/ (Retrieved, 27.07.2017).

Nemetz, P. N. (2013) *Business and the Sustainability Challenge: An Integrated Perspective*, London: Routledge.

Nilsson, P. (2017) Quorn's Turnover Grows on Rise of Partially Meat-Free Diet, *Financial Times*, 24.07.2017, available at www.ft.com/content/59a9d546-7069-11e7-93ff-99f383b09ff9 (Retrieved, 26.07.2017).

Nogrady, B. (2017) Negative Emissions Tech: Can More Trees, Carbon Capture or Biochar Solve our CO_2 Problem? *The Guardian*, 05.05.2017, available at www.theguardian.com/sustainable-business/2017/may/05/negative-emissions-tech-can-more-trees-carbon-capture-or-biochar-solve-our-co2-problem (Retrieved, 25.07.2017).

Oxford Geoengineering Programme (2017) What Is Geoengineering? and Oxford Principles, University of Oxford, available at www.geoengineering.ox.ac.uk/what-is-geoengineering/what-is-geoengineering/ and www.geoengineering.ox.ac.uk/oxford-principles/principles/ (Retrieved, 12.07.2017).

Pacala, S. and Socolow, R. H. (2004) Stabilization Wedges: Solving the Climate Problem for the Next 50 Years with Current Technologies, in: *Science*, 305(5686): 968–972.

Passivhaus Trust (2017) What Is Passivhaus? *The Passivhaus Trust*, available at www.passivhaus-trust.org.uk/what_is_passivhaus.php (Retrieved, 25.07.2017).

Peace, J. and Maher, K. (2015) Weathering the Next Storm: A Closer Look at Business Resilience, Report, C2ES, Center for Climate and Energy Solutions, available at www.c2es.org/publications/weathering-next-storm-closer-look-business-resilience (Retrieved, 21.07.2017).

Pernick, R. and Wilder, C. (2012) *Clean Tech Nation: How the U.S. Can Lead in the New Global Economy*, New York: Harper Business.

Pew (2010) In Brief: The Business Case for Climate Legislation, Pew Center on Global Climate Change (now C2ES – Center for Climate and Energy Solutions), available at www.c2es.org/publications/business-case-for-climate-legislation (Retrieved, 10.07.2017).

PWC (2013) Business-Not-as-Usual: Tackling the Impact of Climate Change on Supply Chain Risk, by R. Gledhill, D. Hamza-Goodacre and L. Ping Low, available at www.pwc.com/gx/en/governance-risk-compliance-consulting-services/resilience/publications/pdfs/issue3/business_not_as_usual.pdf (Retrieved, 28.06.2017).

Quorn (2019) Sustainable Development Report 2019, Quorn, available at www.quorn.co.uk/files/content/Sustainable-Development-Report2019.pdf (Retrieved, 05.11.2019).

Robertson, M. (2017) *Sustainability Principles and Practice*, London: Routledge, 2nd ed.

Rowson, J. (2014) The Seven Dimensions for Action on Climate Change, *The Guardian*, 14.02.2014, available at www.theguardian.com/sustainable-business/behavioural-insights/2014/feb/14/seven-dimensions-action-climate-change (Retrieved, 10.07.2017).

Rowson, J. (2013) A New Agenda on Climate Change: Facing up to Stealth Denial and Winding Down on Fossil Fuels, Royal Society for the Encouragement of Arts, Manufactures and Commerce, available at www.thersa.org/discover/publications-and-articles/reports/a-new-agenda-on-climate-change (Retrieved, 10.07.2017).

Saha, D. and Muro, M. (2017) Patenting Invention: Clean Energy Innovation Trends and Priorities for the Trump Administration and Congress, Report, The Brookings Institution, available at www.brookings.edu/research/patenting-invention-clean-energy-innovation-trends-and-priorities-for-the-trump-administration-and-congress/ (Retrieved, 25.07.2017).

Science Based Targets (2017) Science Based Targets Initiative, website, available at http://sciencebasedtargets.org (Retrieved, 17.07.2017).

Slawinski, N., Pinkse, J., Busch, T. and Banerjee, S. B. (2017) The Role of Short-Termism and Uncertainty Avoidance in Organizational Inaction on Climate Change: A Multi-Level Framework, in: *Business and Society*, 56(2): 253–282.

Spector, J. (2017) Study: We're Still Underestimating Battery Cost Improvements, 17.08.2017, available at www.greentechmedia.com/articles/read/were-still-underestimating-cost-improvements-for-batteries (Retrieved, 04.09.2017).

Stern, N. (2006) *The Economics of Climate Change: The Stern Review*, Cambridge, UK: Cambridge University Press.

Stilgoe, J. (2018) Geoengineering, in: Castree, N., Hulme, M. and Proctor, J. D. (eds.) *Companion to Environmental Studies*, London: Routledge, pp. 679–683.

Tovey, A. (2017) Volvo Becomes the First Major Car Manufacturer to Go All Electric, *The Telegraph*, 05.07.2017, available at www.telegraph.co.uk/business/2017/07/05/volvo-becomes-first-major-car-manufacturer-go-electric/ (Retrieved, 25.07.2017).

Transition Network (2017) Transition Network webpages, available at https://transitionnetwork.org (Retrieved, 27.07.2017).

UKCIP (2010) A Changing Climate for Business, UK Climate Impacts Programme, 3rd ed., available at www.ukcip.org.uk/wp-content/PDFs/UKCIP_Business.pdf (Retrieved, 29.06.2017).

Vaughan, A. (2017) Electric Cars to Account for All New Vehicle Sales in Europe by 2035, *The Guardian*, 13.07.2017, available at www.theguardian.com/environment/2017/jul/13/electric-cars-to-account-for-all-new-vehicle-sales-in-europe-by-2035 (Retrieved, 25.07.2017).

Vince, G. (2014) Saving the World One Rock at a Time, *The Guardian Weekend*, print edition, 28.06.2014.

WBCSD (2017) Below50, World Business Council for Sustainable Development, available at www.wbcsd.org/Clusters/Climate-Energy/below50 (Retrieved, 27.07.2017).

Whitfield, R. and McNett, J. (2014) *A Primer on Sustainability: In the Business Environment*, New York: Business Expert Press.

Winston, A., Favaloro, G. and Healy, T. (2017) Energy Strategy for the C-Suite, in: *Harvard Business Review*, 95(1): 138–146.

WMO (2019) Greenhouse Gas Concentrations in Atmosphere Reach Yet Another High, World Meteorological Organization, Press Release, available at https://public.wmo.int/en/media/press-release/greenhouse-gas-concentrations-atmosphere-reach-yet-another-high (Retrieved, 25.11.2019).

WRI (2005) World Greenhouse Gas Emissions: 2005, World Resources Institute, available at www.wri.org/resources/charts-graphs/world-greenhouse-gas-emissions-2005 (Retrieved, 05.07.2017).

WRI (2014) Visualizing the Global Carbon Budget, World Resources Institute, available at www.wri.org/blog/2014/03/visualizing-global-carbon-budget (Retrieved, 06.07.2017).

Wynes, S. and Nicholas, K. A. (2017) The Climate Mitigation Gap: Education and Government Recommendations Miss the Most Effective Individual Actions, in: *Environmental Research Letters*, 12(7): 074024.

CHAPTER 6

Pollution

INTRODUCTION

We have frequently used the term pollution and related words in the previous chapters. Before we look into detail at this environmental challenge and business's role in it, it is worth asking, what do we really mean by this, as we all have preconceived notions of what pollution is, what causes it and what can be done to remedy or prevent it. We will therefore look at what constitutes pollution, how can it be classified, and what are the different kinds of pollution. As in the previous chapter, we will then discuss the business impact in terms of cause and effect before addressing the regulatory response to pollution. The final part of the chapter looks at the internal and external opportunities for businesses to address pollution.

As with climate change and waste, also underlying this challenge is the issue of depletion and irresponsible or inefficient use of resources. Whilst we treat these three challenges in separate chapters, there are overlaps and interconnections as already alluded to in the introduction to this part of the book. For example, in food production, the use of fertilisers leads to agricultural runoff, which can be considered waste as well as a source of water pollution. Another example is waste on landfill sites which can lead to both water and ground pollution through leakages if not sufficiently secured, while waste also emits methane, a greenhouse gas contributing to climate change. Although climate change is considered by some as a form of air pollution (see e.g. Anderson 2014), waste, in particular hazardous waste, can also lead to pollution. Friedman (2011) highlights how waste – if seen in its multiple forms – can include all materials released into the environment and therefore minimisation of multimedia waste includes reductions of air emissions, effluences and hazardous and non-hazardous wastes. One particular example that has garnered attention in 2017 and 2018 is plastic pollution when soil samples and tap water around the world as well as bottled beer were found to be contaminated with microplastic fibres which cannot be fully filtered with current treatment systems (see e.g. Carrington 2017). Plastic pollution will be covered as a waste issue in the next chapter.

It is important for business to be at least aware of the overlaps, as addressing one environmental challenge could help with another, whereas in other cases there could be unintended side effects and trade-offs with seemingly unrelated environmental challenges.

POLLUTION – WHAT IS IT AND WHY IS IT A CONCERN?

Having drawn the link to other environmental challenges, we can now ask, what is pollution? Very broadly, it is the degradation of the environment (Cavagnaro and Curiel 2012). Robertson (2017: 133) simply defines it as "an unwanted substance that is harmful" whilst *The Dictionary of Global Sustainability* (Green 2013: 103) refers to pollution as

> generally, the presence of a substance in the environment that because of its chemical composition or quantity prevents the functioning of natural processes and produces undesirable environmental and health effects.

Pollution is not limited to chemical pollution, though – the Clean Water Act in the United States defines pollution as human-made or human-induced "alteration of the physical, biological, chemical, and radiological integrity of water" to which we can add other media beyond water, such as air and soil. Similarly to climate change, where we distinguish between naturally occurring greenhouse gases, and anthropogenic and synthetic ones, we can distinguish between *natural sources* of pollution, such as volcanic eruptions, *organic sources* such as sewage (a natural by-product of life that only becomes pollution when it accumulates beyond what nature can absorb), as well as pollution from *human activities*, such as synthetic chemicals (Robertson 2017), also called anthropogenic pollutants (Anderson 2014).

The planet has a natural capacity to absorb pollution through so-called sinks. This capacity is limited (recall the carrying capacity discussed in Chapter 2), thus pollution acts as a warning that we have surpassed earth's ability to absorb the output of human activities through breaking it down (Cavagnaro and Curiel 2012). In the past, it was considered normal business practice to discharge effluents in air, water or on land. It was only in the 1960s, after Rachel Carson's *Silent Spring* alerted society to the detrimental effects of DDT (dichloro-diphenyl-trichloroethane), that a better understanding started to develop about how even small amounts of pollutants can accumulate and reach dangerous levels. DDT and other chemicals, such as fertilisers and pesticides, are examples of human-synthesised chemicals initially developed to improve human life that have then resulted in unintended and unpredicted consequences of destroying rather than saving life (Cavagnaro and Curiel 2012). Although some synthetic chemicals can be broken down by natural processes or organisms, many cannot, as they have never before existed. As a consequence, the means to break them down naturally have not evolved. Such chemicals are also called *xenobiotic* as they are foreign to planetary life (Robertson 2017). As natural processes can accommodate and absorb some pollution, the toxicity depends on the dose.

Pollution can be classified in a number of ways. We introduce some classifications to develop a better understanding of this challenge before focusing on three types of pollution by location: air, water, and soil. Knowledge of what constitutes pollution is important for businesses and policymaking in order to deal with it appropriately – some of the following might be technical, but it provides a useful introduction to matters that are complex and are often dealt with in even more specialist and technical terms.

Anderson (2014) distinguishes between stock, fund and flow pollutants, dependent on the absorption capacity of the planet where environmental sinks can transform, dilute, disperse or absorb pollutants. *Stock pollutants* cannot be absorbed or can only be absorbed to a limited extent – they quickly exceed the absorptive capacity of sinks and accumulate in the environment. Examples include DDT, dioxin and polychlorinated biphenyls (PCBs). *Fund pollutants*, such as CO_2 and organic waste, can be absorbed by nature to some extent, whilst *flow pollutants* may cause initial damage before being dissipated. Examples here include biodegradable litter, light, noise and heat pollution, and smog.

Whilst we can broadly differentiate between natural and anthropogenic pollutants, within human activities as sources of pollution we can distinguish whether pollution comes from *stationary sources*, such as factories or power plants, or from *mobile sources*, such as motor vehicles or ships (Anderson 2014). We can further differentiate between the identifiability of sources – any identifiable source is a *point source* (e.g. a pipe or a leaking tank) whereas diffuse sources, where no single point of origin may be identified, are *nonpoint sources*. Runoff of storm water (e.g. rain, snow) carries pollutants (e.g. fertilisers, oil, salt, bacteria from waste, all from varied sources such as agriculture, mining, construction, urban activities) and then flows into waterways (Green 2013; Anderson 2014). Another type of source is *radioactive* pollution which occurs naturally and as a by-product of human activity (Robertson 2017).

Another way to categorise pollutants is not by where they originate but by their range and density (Anderson 2014; Nemetz 2013). *Uniformly distributed* pollutants (also called uniformly mixed) act globally, which means they cause the same damage regardless of where they are emitted – for example ozone-destroying chemicals or greenhouse gases.[1] *Nonuniformly distributed* pollutants (or nonuniformly mixed) lead to some damage elsewhere but have relatively severe effects on the local area – for example runoffs from fertilisers contaminating local water sources. *Concentrated* pollutants primarily cause damage within a local area, with so-called *hotspots* featuring particularly high levels of pollutant accumulation.

Pollutants can be characterised by their content, whether they are *organic pollutants* or *inorganic pollutants* (Robertson 2017). Organic pollutants are synthetic chemical compounds containing carbon and often hydrogen atoms. All fossil fuels are hydrocarbons, which are also contained in their derivatives (e.g. benzene), pesticides (e.g. DDT) and synthetic plastics (e.g. PCBs). One particular form of organic pollutants is volatile organic compounds (VOCs) that vaporise at room temperature. Whilst they can be found in nature (e.g. CFCs – chlorofluorocarbons), synthetic VOCs are of concern, as some are carcinogenic. End-use examples containing organic pollutants include pharmaceuticals, food additives, solvent-containing cleaning products, cosmetics, formaldehyde-containing paint, adhesives and carpets, to name a few. Inorganic pollutants include salts (such as halogens, sodium, nitrate) and heavy metals, including mercury, lead, arsenic, cadmium and chromium. Salts are of concern as they are highly mobile in water, and persistent, whilst many heavy metals are toxic and persistent as well as cumulative.

Two particularly hazardous types of pollutants can be categorised by their behaviour: *endocrine disruptors* and *persistent organic pollutants* (Robertson 2017). Persistent organic pollutants (or POPs) are substances that have three critical properties, making them especially problematic – they persist in the environment, they bioaccumulate in the food web and are subject to long-range transport (Nemetz 2013) – travelling to the most remote regions of the planet (Anderson 2014). Bioaccumulation occurs when toxic chemicals gradually accumulate in fish or animals as they grow and age. A related concept is biomagnification which refers to the increasing concentration of toxics through the food chain (Nemetz 2013). These two characteristics are shared by some heavy metals (Robertson 2017). Most POPs are produced intentionally (though some occur as by-products) and are used as solvents and ingredients in pharmaceutical production and industrial processes. Many POPs are pesticides – designed to kill some form of living organism – and include for example insecticides, rodenticides, herbicides, fungicides, growth regulators, defoliants and desiccants, as e.g. defined under US law (Robertson 2017). Endocrine disruptors can interfere with human hormonal systems and are sometimes therefore referred to as hormone mimics, oestrogen disruptors, pseudo-oestrogens or environmental hormones (Nemetz 2013; Robertson 2017). They are globally pervasive chemicals that affect body functions and processes such as for example the thyroid, immune, metabolic and cardiovascular systems, reproduction and neurodevelopmental processes.

Endocrine disruptors can also be classified as a *type of toxin*. Other toxins include allergens (triggering a response when a response is not necessary – e.g. formaldehyde), mutagens (damaging genetic material) which include also carcinogens (causing cancer) and teratogens (causing abnormalities in embryos – e.g. lead and alcohol), and neurotoxins (causing nerve damage e.g. mercury and pesticides).

Pollutants are furthermore characterised by their pathways – how they move through the environment ("transport") from a source to a sink, and how they change over time ("fate") (Robertson 2017; Green 2013). They differ in terms of the speed and paths in which they move, with the greatest risk coming from pollutants and contaminants that are highly mobile, persistent and toxic. Transport of pollutants is by air flows or through water, which are also the two main locations where pollution occurs. We will now discuss these two as well as soil and land as the third major location of pollution. Although we discuss them separately, pollutants are subject to interrelationships between toxic substances that can act in synergistic and nonlinear ways (Costanza et al. 1997), which may render the boundary between air, water and land pollution blurry at times.

Air pollution

Air pollution has been found to be a significant contributor to human illness and death, and to harmful environmental effects (Nemetz 2013; Green 2013). As the entire biosphere is surrounded by air, airborne pollutants are the most likely cause of health problems (Robertson 2017). Other effects of air pollution include acidification (change of the chemical balance in the natural environment affecting soils and waters), eutrophication (excess of nutrients that threatens biodiversity e.g. algal blooms) and physical damage (e.g. to infrastructure by corroding buildings).

The Clean Air Act in the US identifies six major air pollutants, also known as criteria pollutants. These are particulate matter (or particulates), carbon monoxide, sulphur oxides, nitrogen oxides, lead and ground-level ozone. The first five are primary pollutants, directly released from their sources, whilst ground-level ozone is a secondary pollutant only forming as a result of chemical reactions between primary pollutants and other constituents in the air (Robertson 2017). VOCs (volatile organic compounds) are a key component of this formation, which is why they are included as a main air pollutant in the EU's list of main air pollutants, as is ammonia as a building block for secondary particulate matter, and methane which contributes to ozone formation through reactions with nitrogen dioxide. The World Health Organization lists particulate matter, ozone, nitrogen dioxide and sulphur dioxide as key air pollutants.

All six criteria pollutants result from the burning of fossil fuels, which is thus a commonality of source and cause that air pollution shares with climate change. The following table (6.1) gives an overview of the major air pollutants, their sources and effects. It also refers to control methods, anticipating some of the regulatory and business responses covered later in the chapter.

Other air pollutants that may cause severe damage to health and the environment include heavy metals other than lead (such as mercury, arsenic, cadmium and nickel), and polycyclic aromatic hydrocarbons (PAHs, such as benzo[a]pyrene). Unlike the six criteria pollutants in the US that originate mostly from fossil fuel burning, the three pollutants included by EU air quality standards often result from agricultural uses: ammonia (NH_3), methane (CH_4), and VOCs (European Commission 2016a). As methane is both a greenhouse gas and an air pollutant, we can see again the overlap between air pollution and climate change. In addition to criteria pollutants, the Clean Air Act identifies hazardous air pollutants (HAPs), also called toxic air pollutants or air toxics, in a list of over 180 carcinogens, mutagens, neurotoxins, endocrine disruptors and other

TABLE 6.1

Overview of major air pollutants

Pollutant	Characteristic	Sources	Effects	Control methods
Particulate matter (PM_{10}, $PM_{2.5}$)	Airborne particles (aerosols), characterised by size not composition – less than 10 μm (micrometres) and 2.5 μm. Solid or liquid.	For example soot, smoke, dust, metals, acids from *primary particles* (construction sites, incomplete combustion, metal smelting, unpaved roads) and from *secondary particles* (vehicles, power plants, industrial processes, incinerators, wood burning).	Heart and lung problems, toxic materials attach to fine particles and are carried into respiratory system, impaired or reduced visibility.	Filters (e.g. particulate filter for diesel vehicles) Scrubbers Reduction in fuel combustion
Sulphur oxides (SO_2 or "sox")	Colourless gas with strong, distinctive smell.	90% from fossil fuel burning, especially coal, also oil refineries, ore smelters, cement production, paper mills, diesel fuel combustion. Natural sources (volcanoes). SO_2 oxidises to form SO_3 (sulphur trioxide), which reacts with water vapour or rain to form sulphuric acid – component of acid rain.	Corrosive substance, damaging to plants and animals – heart and lung problems, second only to tobacco in causing health problems related to air pollution. Can damage marble, iron, steel, crops and vegetation. Acidification of soils and inland waters.	Low-sulphur fuel Energy conservation (reduces power plant emissions) Pollution control equipment
Nitrous oxides (NOX/NO_x)	Include nitric oxide (NO), nitrogen dioxide (NO_2 – reddish-brown gas), nitrous oxide (N_2O – GHG).	NO from fossil fuel burning, oxidised NO results in NO_2, used as indicator of other NOx present. Further oxidisation results in NO_3, which reacts with water vapour or rain to form nitric acid – component of acid rain. Fuel combustion in cars and industrial sources (heating, power).	NOX damaging to plants and animals – lungs and mucous membranes. NO_2 30–100% higher near major roads – affects low-income people disproportionately.	Focus on NO_2: Exhaust gas recirculation and reduction of combustion temperature Energy conservation Pollution control equipment
Carbon monoxide (CO)	Odourless, colourless gas.	Incomplete combustion of fossil fuels in vehicles and other machinery.	Impaired delivery of oxygen to vital organs. Can cause dizziness, unconsciousness or death. Heart damage. Reduces Earth system ability to regulate methane levels (GHG).	Transport planning Testing of vehicle emissions Efficient combustion techniques Energy conservation
Lead (Pb)	Toxic heavy metal.	Smelters, waste incinerators, coal-fired power plants, lead-acid battery production, leaded petrol.	Toxic to nervous system and other body function, behavioural problems, learning deficits, cardiovascular effects.	Phase-out of leaded fuel Pollution control equipment
Ozone (O_3)	Secondary pollutant; colourless or bluish gas (smog), formed from VOCs and NOX in sunlight. Gives urban smog characteristic odour.	Fuel combustion in motor vehicles, fuel storage and transport, solvents and paint, landfills.	Harms human health (respiratory problems) and sensitive construction materials (metals, paints) as well as rubbers and textiles, reduces crop yields.	Pollution control equipment Testing of vehicle emissions Evaporative controls Low-VOC solvents.

Sources: Robertson 2017; Anderson 2014, http://ec.europa.eu/environment/pubs/pdf/factsheets/air/en.pdf, www.epa.gov/criteria-air-pollutants, www.who.int/mediacentre/factsheets/fs313/en/.

toxic compounds (Anderson 2014; Robertson 2017). HAPs are less pervasive and found in fewer quantities than criteria pollutants but are nevertheless dangerous as they accumulate and persist in the environment. Health risks include breathing contaminated air, as well as eating and drinking contaminated food and water (EPA 2017a).

In the case of ozone, it is worth highlighting that it can be damaging or protective, depending on where it is found in the atmosphere. Stratospheric ozone, found in the upper atmosphere, forms the ozone layer when solar radiation strikes oxygen molecules. This shields the Earth from ultraviolet radiation. Tropospheric ozone, found in the lower atmosphere, near the ground, is considered a pollutant (Robertson 2017). The ozone layer has been subject to depletion resulting in a hole as a consequence of CFCs (e.g. used in refrigeration) which were banned under the Montreal Protocol.

Acid rain could be interpreted as air pollution that ends up as water pollution as well as soil pollution. Acid deposition is a by-product from the burning of fossil fuels released by power plants and vehicle exhausts. Emissions from SO_2 and NOx (sulphur dioxide and nitrogen oxide) in the form of aerosols react with water droplets, oxygen and other oxidants in the atmosphere, forming solutions of sulphuric and nitric acid (Hackett 2011). Precipitation returns these acids back to the ground as acid deposition, mostly in the form of acid rain. Such acid depositions lower the pH levels of soils, streams and other water bodies and damage these terrestrial and aquatic ecosystems, reducing their biotic productivity through acidification.[2] Acid rain also damages buildings and sculptures (Robertson 2017).

So far we have focused on outdoor air, also called ambient air. In 2012, the World Health Organization estimated that outdoor pollution is responsible for 3 million premature deaths worldwide annually. In addition, indoor and household air pollution is thought to be responsible for over 4 million premature deaths, largely in low- and middle-income countries where around 3 billion people still use solid fuels such as biomass and coal to cook and to heat their homes (WHO 2016). The inefficient use of solid fuels for cooking releases particulate matter (soot) which can cause pneumonia, stroke and lung cancer, amongst others. Other sources of indoor air pollution include tobacco smoke, pesticides, cleaning agents and gases released from building materials and furnishings (e.g. floors and carpets) treated with chemicals as well as outdoor pollution entering buildings (EEA 2017; EPA 2017a). One health effect of chemical releases from paints, furniture and carpets is so-called sick building syndrome (Esty and Winston 2009).

One particular form of indoor air pollution is contaminated air in aircraft caused by a so-called fume event, which can lead to aerotoxic syndrome and organophosphate poisoning, which may affect cabin crew and frequent travellers (Turner 2015; Leahy 2017).

Two forms of flow pollutants (pollutants that dissipate and do not built up in the environment) that we could categorise as a variant of air pollution are noise and light, which are briefly covered in Boxes 6.1 and 6.2.

BOX 6.1 SPOTLIGHT ON LIGHT POLLUTION

Pollution through light can affect both wildlife and human health. We can distinguish between three types of light pollution: light trespass, astronomical and ecological light pollution. Light trespass is light that is unwanted and infringes on property boundaries. Astronomical light pollution restricts the usage of telescopes and the ability to observe night sky features – e.g. the once-dominant Milky Way galaxy is often no longer visible. Ecological light pollution disrupts the functioning

BOX 6.1 (CONTINUED)

of ecosystems, including plants and animals, by interfering with so-called circadian rhythms – the daily regular cycles of activity – of feeding, reproducing and migrating. The seasonal and daily differences of dark and light are vital for both plant and wildlife as signals for reproduction and growth. Artificial light has been shown to impact animals such as insects, birds, sea turtles, amphibians, fish and zooplankton. Artificial night lighting has also been shown to contribute to a range of health problems in humans. Light pollution is a side effect (or negative externality) of industrial civilisation and comes from exterior and interior lighting of buildings and commercial properties, advertising, streetlights and lit sporting venues, to name a few. These cause glare, skyglow and clutter.

A particular form of light pollution is the use of kerosene lights in developing countries, which emit toxic black smoke and cause hazardous indoor air pollution whilst also being very costly.

Overall, excess light that is spilled into the atmosphere is also a form of wasted energy and thus a source of environmental degradation and unnecessary costs.

Sources: Robertson 2017; Anderson 2014, www.darksky.org/light-pollution/, https://solar-aid.org.

BOX 6.2 SPOTLIGHT ON NOISE POLLUTION

Air pollution as covered elsewhere in the chapter is characterised by the presence of contaminants in the air. Noise pollution, in contrast, changes the properties of air – pressure is modified locally, causing sound pressure waves. Unwanted or harmful sounds may come from industrial noise, environmental noise, occupational noise, construction and transport noise (road, rail, air) as well as entertainment and neighbourhood noise. Excessive sound levels from human activity affects both wildlife and humans. In humans, excess noise can lead to temporary or permanent hearing loss, interfere with sleep quality and cognitive abilities, and result in emotional disturbances and mental health problems such as increased anxiety and stress. It may also affect property values. Noise is particularly a problem in urban areas, though not restricted to them.

Animals use sounds to mate and locate other members of their species, to prey or to escape predators. Noise pollution can mask quieter sounds animals use as cues, and also lead to other detrimental effects, such as changed heart rates, and repercussions from the stress caused on digestion, reproduction and thyroid functions. Bioacoustic research of ecosystem sounds has shown that natural soundscapes are changing to such a degree that some local species' extinction is taking place.

Noise pollution is not confined to the air – it is also a problem underwater, caused by ships and tankers, oil exploration activities, military sonar, explosives and dredging. The damage to human health here might be limited but causes harm to marine life, leading to hearing loss, swimming off course, habit abandoning, behavioural change, or drowning out other sounds from mates, calves, predators – so-called masking.

Sources: Robertson 2017; Anderson 2014; Green 2013; Pullen 2011, http://ec.europa.eu/environment/noise/index_en.htm.

Water pollution

Water is essential to life, and for business an important input to industrial and agricultural processes. Water pollution in the form of objectionable or harmful material – whether solid, liquid or gaseous – changes the chemical, physical, biological and radiological condition (or integrity) of water and makes it unfit for use (EEA 2017). Water is a complex environmental topic as it comprises a range of issues of both quality and quantity, such as pollution, droughts, floods, access and distribution, sanitation, waste, as well as various types of water – surface water, groundwater, coastal water, inland water, fresh water, salt water, drinking water, bathing water. It overlaps with other environmental challenges (e.g. with waste in the case of wastewater or with climate change). In terms of access and distribution, water stress and water scarcity are also a source of international conflict and (social) injustice due to a mismatch of supply and demand, aggravated by overpopulation and overconsumption, in the context of the tragedy of the commons. All of these complex issues affect businesses, which are increasingly scrutinised for how they use and pollute water. We focus here on the issue of pollution and advise the reader to consult other sources to research water beyond pollution.[3]

In the past, the approach to water pollution has often been dilution – flippantly put as "dilution is the solution to pollution" (Robertson 2017: 93). Its ability to dissolve and carry other substances is water's most significant chemical property, but also the one that makes its contamination easy and renders the clean-up process difficult and expensive. In particular, aquatic pollutants from *diffuse sources* have no clearly defined origin, resulting from widespread activities. *Point source* pollutants can be traced back to a single location or source, making them easier to control and to prevent pollution in the first place (McIntyre et al. 2011).

Water pollution results from activities such as industrial processes, agriculture, transportation, mining, urban drainage and disposal of domestic sewage as well as accidental or deliberate pollution incidents. Although some pollutants are synthetic, many are natural but occur in dangerously high concentrations. Such pollution results in various impacts, such as on human health as a consequence of detrimental effects on water quality, but also on recreational activities as well as in damage to commerce and agriculture, and to aquatic ecosystems, whether freshwater, coastal or marine areas (McIntyre et al. 2011).

Categories of water pollutants include conventional pollutants, priority pollutants and nonconventional pollutants, a distinction made by the Clean Water Act in the US, whereas the EU's water framework directive focuses on priority substances (Robertson 2017). Priority pollutants, or priority hazardous substances, are individual chemicals such as heavy metals, plant protection products (pesticides), PCB and PAHs. Conventional pollutants are compounds, i.e. not individual chemicals, and include biochemical oxygen demand (or oxygen-demanding wastes), suspended solids, faecal coliform bacteria, oil and grease (Robertson 2017). Other general types of water pollutants include pathogenic and microbiological organisms, plant nutrients, synthetic organic chemicals, inorganic chemicals, radioactive substances, sediments, oil, heat (thermal energy) and colour as well as pH and salinity (McIntyre et al. 2011). The latter pollutants are also called nonconventional pollutants which are less well understood or not statutorily listed (Green 2013).

Other forms of water pollution include thermal pollution where water temperature changes as a result of industrial processes, whilst eutrophication results from agricultural processes (fertiliser runoff and manure) and leads to the build-up of organic matter and nutrients in waterways. A consequence is algal blooms, which deplete water of oxygen as they decompose, leading to dead zones of water unfit for aquatic life (Robertson 2017). Water pollution can also be of aesthetic form where discolouration and odours change its properties and its fitness for purpose (e.g. for recreational uses) (Pullen 2011).

Land pollution

Land pollution is also called ground or soil pollution, whilst land contamination is also used as a term. Although the terms land, ground and soil are often used interchangeably, land can mean more than just soil or ground and may also include water beneath, and structures and buildings above, i.e. land and its constituents (Witherington 2011; EEA 2017). Soil, in contrast, can be more specifically characterised – it is the thin layer covering large surface areas of land, a complex ecosystem providing fertility, decomposition, storage of rain, nutrients and carbon, amongst others (Robertson 2017; Green 2013). Similar to water, soil and land are a complex environmental topic, where issues such as land degradation, deforestation, soil erosion and loss of biodiversity from changed land use interact with each other and with further environmental challenges, such as e.g. climate change and water. While we focus on pollution here, it is appropriate to highlight the distinction from contamination – a term more frequently applied to land. *The Dictionary of Global Sustainability* defines the two terms as follows (Green 2013: 103, 33):

- Pollution: "Generally, the presence of a substance in the environment that because of its chemical composition or quantity prevents the functioning of natural processes and produces undesirable environmental and health effects".
- Contamination: "Introduction into water, air, and soil of microorganisms, chemicals, toxic substances, wastes, or wastewater in a concentration that makes the medium unfit for its next intended use".

Pollution is more general and serious over time, while contamination is more serious and more specific, which is also reflected in the European Environment Agency's definition (EEA 2017) of a contaminated site: "A location where as a result of human activity an unacceptable hazard to human health and ecosystems exists. Local contamination (contaminated sites) is a problem in restricted areas (or sites) around the source, where there is a direct link to the source of contamination".

In some jurisdictions, the term contaminated land has specific legal meaning – e.g. in the UK this term is applied only to land where there is "significant harm" or "significant possibility of significant harm" i.e. a threat to health and safety, property or groundwater. The term contamination is not used in the absence of (possible) significant harm even if contaminants can be found on site (Witherington 2011).

Releases from industrial processes in the form of spillages, leakages or airborne emissions (deposited back on the ground), or deposits (accidental or deliberate) of material on the ground lead to land contamination. Soil polluting activities include waste disposal, mining, oil drilling and agriculture (Robertson 2017; Witherington 2011). Waste could be industrial or domestic and municipal waste, while agricultural sources include chemicals such as fertilisers, sewage sludge and pesticides (Pullen 2011). In industrialised countries, land contamination also results from the industrial heritage of 200 years of industrialisation. Abandoned former industrial sites are known as brownfields where the presence of pollutants creates challenges for reuse or redevelopment (Robertson 2017).

A particular difficulty with land pollution is that contamination may not be limited to the soil as leachates and runoff may end up in groundwater, the pollution of which is notoriously difficult to remedy. Furthermore, soil contaminants are less homogenously dispersed compared to air and water pollutants and may thus be more difficult to detect (Pullen 2011). Frequent soil contaminants are heavy metals such as lead, mercury, arsenic for example, mineral oil and persistent organic pollutants (POPs) such as dioxins that do not break down easily (SCU 2013). Beyond affecting food and water supplies, soil pollution affects human health and ecosystems but also affects e.g. land and property values. Compared to water and air pollution, soil contamination is less well understood and subject to ongoing research (SCU 2013).

BUSINESS IMPACT

It is already clear from discussing the various types, locations, contents, ranges and particularly sources of pollution that businesses are a large part of the problem. We will briefly cover business as the cause of pollution before looking at how businesses themselves are affected by it.

Cause of pollution – business as part of the problem

For air pollution, the sources referred to in Table 6.1 and elsewhere already indicate how business contributes to the problem. Fossil fuel burning in transport as well as industrial processes, particularly in heavy industry, such as iron and steel, electricity supply and large combustion processes, are significant sources of air pollution with coal-fired electric power generation one of the largest alongside construction and agriculture[4] (Nemetz 2013). For example, with regard to air pollution from fine particulate matter, emissions from farming are greater than all other anthropogenic sources in the US, Europe, Russia and China, as nitrogen compounds from fertilisers and animal waste combine with pollutants from combustion in transport and industry to form solid particles (aerosols, $PM_{2.5}$), causing lung or heart diseases (Bauer et al. 2016).

Broadly speaking, the biggest sources of water pollution are water companies, steel works and bulk chemical producers to which we can add agriculture, which for example in the US is the largest source of diffuse and anonymous water pollution (Bigg 2011; Hawken et al. 1999). This is perhaps not surprising considering that agriculture is responsible for 70% of water use and industry for 20%, with the remaining 10% from domestic use. More specifically, metal, electronics as well as pulp and paper industries are heavy water users, whilst mining companies, chemical and petroleum plants and machinery producers are not only major industrial users of water but also heavy polluters (Robertson 2017). Textile manufacturing also contributes to water pollution.

As with air and water pollution, agriculture is also a large contributor to soil pollution as agricultural pollution includes both solid and liquid wastes from fertiliser runoff, animal manure and crop residues as well as erosion and dust from ploughing (EEA 2017). Other major industries that are a cause of land pollution include gas and steel works, oil drilling and production (e.g. fracking) as well as oil storage and transport, and mining. Extractive industries in particular use large quantities of soil and rock for processing with the use of chemicals to obtain metals which results in mine tailings and tailing ponds (Robertson 2017; Nemetz 2013). In addition to day-to-day operations, industrial accidents can pollute vast areas of land.

BOX 6.3 SPOTLIGHT ON POLLUTION IN CHINA

In the country's push for industrialisation and economic growth, the environment in China has experienced high demands on natural resources and much degradation. At the height of its growth, China was adding one coal-fired power plant per week to its energy network. Growing awareness of coal's dual role in air pollution and climate change has seen a decline in the number of such plants being planned and built, and investments made into renewable energies. China is now a major exporter of solar panels and wind turbines though Chinese companies are also building large coal power plants abroad. The smoggy air in China is notorious and has led to government action to combat air pollution and provide cleaner air. Water pollution, however, is continuing to be a major concern. Surface water that is fit for human contact or suitable as a drinking water source is

BOX 6.3 (CONTINUED)

a scarce resource – for example in Beijing 40% of water is so polluted that it essentially serves no function, whilst in the port city of Tianjin only 5% of water is usable for drinking. Heavy industry is a problem, illegally releasing toxic and chemical wastes into rivers, making 80% of groundwater from major river basins unsuitable for human contact. The health of citizens, of towns and villages along major polluted rivers suffers as a result. Polluters include paper and pulp mills, dyeing and tanning plants, and chemical factories. Soil pollution is equally a problem with thousands of brownfields seriously polluted with heavy metals and pesticides – it is estimated that a third of China's brownfields and a sixth of the country's soils exceed national pollution standards. Health problems ranging from nose bleeds to cancer emerged when industries moved from urban areas to suburban and rural locations, and brownfield sites were built on without sufficient remediation of the pollution.

Sources: Economy 2004; Yang 2016; Economist 2017; Tabuchi 2017; Tingting 2017.

EFFECT – POLLUTION AS A PROBLEM FOR BUSINESS

Although businesses are clearly a part of the problem of pollution, at the same time they are also affected by it in various ways, such as lost work days due to illness, regulatory costs, impact on their reputation, polluted inputs (e.g. water) to their production processes, potential legal liabilities and costs which all pose various risks and challenges.

In a study for the UK government, Birchby and colleagues (2014) found impacts on productivity from air pollution through temporary or permanent withdrawal from work, leading to missed days at work, reduced productivity of employees not absent from work through illness as well as premature deaths removing labour from employment. Other findings include the impact on outputs from commercial crops (reduced yield) and the corrosion of buildings and other fixed productive assets through acid rain.

The severity of air pollution is also highlighted in a World Bank report – it is associated with one in ten premature deaths and the fourth leading fatal health risk after metabolic risks, dietary risks and tobacco smoke (World Bank 2016). In lower income countries it is the third leading health risk. The health risks translate into economic burdens, which in 2013 were estimated to cost the global economy $225 billion in lost labour income. Businesses thus lose productive labour in this way. Furthermore, air pollution also reduces cities' competitiveness as they are less able to attract skilled workers, affecting companies in turn (World Bank 2016). Other health consequences of pollution that affect business include the impacts on cognitive functions, including self-control and cooperation (Phillips 2017).

Outdoor air pollution not only affects outdoor workers, such as agricultural workers in the case of ozone pollution (Graff Zivin and Neidell 2012), it also affects the productivity of indoor workers. Whilst some pollutants do not travel indoors – such as ozone – particulate matter can easily penetrate buildings. This is especially the case for $PM_{2.5}$, which was found to affect productivity level of workers at a pear-packing facility at levels below air quality standards in the US and of call-centre workers in China (Chang et al. 2016a, 2016b). $PM_{2.5}$ pollution has also been found to influence investor behaviour at the New York Stock Exchange where increased levels of pollution lowered stock performance and the risk tolerance of traders, thus affecting the efficient operation of the stock market (Heyes et al. 2016, see also Berinato 2017).

Whilst agriculture is to a large extent responsible for water pollution, it is also heavily affected by it. Crops can be contaminated by the use of wastewater and polluted surface or groundwater which may lead to diseases being transmitted to farm workers and consumers (FAO 1996). In 2017, millions of eggs that had been contaminated with the insecticide fipronil had to be recalled from supermarkets in a dozen European countries. Fipronil is a pesticide used to de-flea household pets but also to treat red lice in poultry, which is how it entered the food chain despite being banned for use on animals destined for human consumption (BBC 2017). Such incidences lead to increased costs for businesses from the loss of trust and reputation, which harm consumer confidence, as well as from recall costs and reduced demand and sales, a rise in prices and potential legal liabilities. In the fipronil case, farms were temporarily shut down. In the same year, high street chain Primark had to recall men's flip-flops in their thousands as they were found to contain the chemical chrysene, a restricted substance that can cause cancer (Slawson 2017). Other examples of the costs of pollution include the law suits facing Monsanto for its use of PCBs even after it had been banned, and BP's costs for paying the damages following its Deepwater Horizon disaster.

Regulatory costs for complying with pollution-preventing or pollution-abating policies also affect companies, though one could also argue that whilst this might be a cost in the short term, in the long term it protects companies from facing even higher costs in the future e.g. in the form of clean-up costs, fines for breaching limits, reputational damages or working days lost from ill employees.

REGULATORY RESPONSES TO POLLUTION

Having focused on what pollution is and what the relationship to business is in terms of cause and effect, we briefly cover regulatory responses before looking at business responses to and opportunities from pollution as policymaking provides the reference framework within which businesses operate.

Early efforts to address pollution can be found in the 19th century – for example in London a cholera epidemic was traced back to a contaminated water pump in 1855, but clean water was not viewed as an important public necessity until the 1880s. Air pollution came to attention in the 1950s when deadly smog events affected London, New York and Los Angeles (Post 2012).

Regulating pollutants poses a challenge for policymakers as some pollutants have only local effects whereas others have local and regional or even global effects, depending on their range of transport (Nemetz 2013). Another challenge is that zero pollution might not be feasible. Anderson (2014) emphasises that efficiency concerns command that the optimal quantity of pollution does not equal zero. Nemetz (2013) argues similarly that it might be neither feasible nor economically desirable to reduce (air) pollution to zero, recommending the use of cost-benefit analysis to determine optimal reduction levels.

The regulatory responses to pollution are broadly set between remediation and prevention, which can also be seen in the wider context of the policy options of mitigating, adapting, avoiding and restoring (EEA 2016). Another term used for addressing this environmental challenge is pollution prevention and control. Whilst the precautionary principle guides the prevention of pollution, the polluter pays principle applies to both remediation and prevention (see Chapter 3 for more on these principles).

Regulatory tools available to achieve pollution prevention and control include permits, (environmental) taxes, trading schemes, negotiated or voluntary agreements and environmental management systems (Bigg 2011). Permit-based regimes can be costly to administer compared with other instruments but allow for environmental objectives to be delivered quickly and widely, in particular localised issues of air, water and groundwater pollution. Such direct regulation benefits from other regulatory instruments such as taxes, trading and voluntary agreements. Taxes complement conventional policy tools and send effective price signals to achieve behavioural changes and encourage innovation. Trading of pollutant emissions provides a low-cost tool to achieve environmental goals, giving companies incentives and options for how to lower their environmental impact (Bigg 2011). Trading is particularly suitable for transboundary pollution that has not just local but also international and global effects.

Voluntary or negotiated agreements and environmental management systems (EMSs) might have the input of policymakers but are overall lighter in regulatory touch. Agreements benefit from higher commitment of companies and early buy-in before regulation, though work best for industries with a small number of larger players and with effective penalties in place (Bigg 2011). EMSs (see also Chapter 11) allow industry to take responsibility for and to actively manage their environmental impact.

Air pollution

Air pollution was an early focus of environmental policy and continues to play a major role in environmental regulations. Regulations are often complementary, legislating both for emissions at source and for the air quality in the receiving environment. Source-specific legislation aims to limit emissions from particular sectors and major emitters, such as for example combustion plants, motor vehicles and ships. For stationary sources it is comparatively easy to monitor the output and hence to legislate for emissions, whilst for mobile sources, legislating for the input (i.e. the type of fuel) is more feasible as the emissions are harder to account for (Anderson 2014). An example here is the ban of lead for use in petrol or the introduction of low sulphur fuels.

Air quality standards focus on the local concentration of air pollutants. Depending on which jurisdiction, such standards stipulate various averages, ranging from one- and three-hour averages to eight- and 24-hour ones as well as annual arithmetic means and rolling three-month averages. They are either expressed as $\mu g/m^3$, or as ppm or ppb (parts per million/billion). Some pollutants are permitted to exceed their target or limit value[5] each year for a limited number of times.

Economic instruments suitable for regulating air pollution include tradeable emission permits, emission charges, technology subsidies as well as low-interest loans and deposit refund systems (Nemetz 2013).

Water pollution

Unlike the air, water is subject to anything being spilled or dumped into it, with nonpoint sources the major cause of water pollution (Anderson 2014). Similar to air pollution, regulations for water pollution address the source of pollution by regulating water pollutants as well as the receiving environment by setting water quality standards. Water pollutants that are regulated include individual as well as groups of pollutants, such as priority substances and priority hazardous substances which are priority substances that are persistent, bio-accumulative and toxic (European Commission 2016b).

Water quality standards stipulate annual average concentrations and maximum allowable concentrations (or maximum contaminant level) of pollutants and may be expressed as µg/l, mg/l or number/100 ml, depending on jurisdiction, substance, polluting sector (e.g. industry or agriculture) and type of water (e.g. drinking, bathing, surface, river, coastal, used).

Suitable economic instruments for addressing water pollution include tradeable effluent permits, effluent charges, water treatment fees and low-interest loans (Nemetz 2013). Economic instruments and uniform emission limits are not the only possible responses to water pollution, however. Ecosystems themselves have a natural ability to deal with pollution, as shown in a study of the re-introduction of Eurasian beavers in Devon, a region in the UK. The beavers and the dams they built changed the structure and function of the local ecosystem, increasing water storage, reducing flood risks and improving water quality – pollutants from nearby agricultural land such as phosphates and fertilisers containing nitrogen were filtered out (Puttock et al. 2017). Nature-based solutions are also used in treating wastewater where living organisms (e.g. bacteria) break down organic and chemical wastes (Robertson 2017).

Soil pollution

Regulation addressing soil pollution tends to be less comprehensive and coherent compared to air and water pollution. For example, in the EU both water and air pollution are addressed through a set of union-wide rules, whilst soil protection may only be indirectly covered through policies on agriculture, waste, water, chemicals and industrial pollution (European Commission 2016c). Few countries have specific regulations regarding soil pollution, and on an international level there is no direct soil legislation (Davies 2017). In the US, there are programmes for cleaning up contaminated sites such as the Superfund programme, and for land revitalisation of brownfields, for example. As the impact of soil pollution generally tends to be more localised and more contained, legislation is also more focused on a more local and regional level as well as on remediation of contaminated legacy sites, which can be costly (FAO and ITPS 2015).

For land and soil pollution, economic instruments available to policymakers include land rights and use rights, property and land use taxes, soil conservation incentives such as loans, and local reclamation bonds (Nemetz 2013).

Other policies and approaches

There is not much established legislation concerning *light pollution*,[6] although a growing number of countries regulate it (Morgan-Taylor 2015). Light pollution is often addressed at a local level to take into account different needs and interests of diverse communities (Anderson 2014). Voluntary efforts to address light pollution can be found for example in the protection of sites from astronomic light pollution – e.g. the International Dark-Sky Association designates and awards places that protect and preserve the night sky with Dark Sky Places certification.

Noise pollution is legislated on the basis of exposure and where it comes from – the source. Environmental noise in the EU is regulated at the source and applies to road traffic noise (motor vehicles and cycles as well as tyres), aircraft noise (level and operating times), railway noise and outdoor equipment (European Commission 2016d). Regulatory requirements include noise maps and noise abatement measures such as e.g. isolating and damping vibration sources, providing noise barriers and replacing components. In the US, the Noise Control Act regulates major sources of noise, but cutting of funding for the Office of Noise Abatement and Control within the Environmental Protection Agency has meant limited implementation and regulatory responsibilities passed to state and local levels (Anderson 2014).

A general approach to regulating pollution that incorporates incentive-based legislation as well as command and control standards is advocated by Costanza and colleagues (1997). In the first zone ("property rights zone") where levels of emission and concentration of pollutants are so low that no damage can be measured or detected, and pollutants do not accumulate, emitters would be permitted to release pollutants without any charge as the receiving environment has the assimilative capacity to deal with the emissions. In the next zone ("incentive zone"), damage to the environment is measurable and productivity reduced, and emitters are charged a pollution tax which provides them with an economic incentive to reduce pollution. When levels of pollutant emissions and concentrations reach a non-sustainable stage, causing long-term damage, the third level of policy concern is reached ("regulatory zone") where the pollute-and-pay approach is superseded by regulatory prohibition of any further emissions. Tradable permits could be used instead of charges in the second zone, the number of which could be limited. Overall, such an approach combines the best of both approaches – direct regulation and economic incentives (Costanza et al. 1997).

We briefly mentioned pollution havens in Chapter 3 – it is worth returning briefly to the hypothesis in the context of pollution regulation as this is a term often bandied about. Stricter legislation in conjunction with freer trade is thought to influence location decisions of multinational enterprises in pollution-intensive industries which shift from countries with stringent regulations to countries with weaker ones that might use these deliberately to attract foreign direct investment. In reality, there is little evidence to support the pollution-haven hypothesis as other factors come into play in international trade as well (Hackett 2011; Copeland and Taylor 2004).

In the following, selective examples of regulatory and voluntary pollution prevention and control efforts will be given to illustrate the legislative framework across different levels within which businesses operate.

Global responses to transboundary pollution

Transboundary pollution refers to pollution generated in one country and transmitted to others (EEA 2017). Addressing pollutants that are persistent and can be transported long distances requires international cooperation, which the United Nations Environment Programme (UNEP) can facilitate by hosting global conventions. Examples include the Rotterdam Convention which was signed in 1998 and entered into force in 2004 to cover pesticides and industrial chemicals that are hazardous to the environment and human health and the Stockholm convention of 2001 which banned or restricted the production of persistent organic pollutants (POPs) and also entered into force in 2004. Together with the Basel Convention on transboundary movements of hazardous wastes and their disposal, these multilateral environmental agreements share synergies and common objectives in addressing hazardous pollution. The Minamata Convention on Mercury, signed in 2013, entered into force in 2017 and addresses the adverse effects of mercury which is subject to long-range atmospheric transport, is persistent and bio-accumulates (UNEP 2017).

The International Convention for the Prevention of Pollution from Ships (MARPOL) addresses accidental pollution and pollution from routine operations. It has mostly addressed water pollution since entering into force in 1983, but a 2005 protocol also addressed air pollution prevention. It requires for example that new tankers are double-hulled (Anderson 2014).

Regional responses to pollution

On a regional level, the United Nations Economic Commission for Europe (UNECE) hosts the Convention on Long-range Transboundary Air Pollution (CLRTAP) which was signed in 1979 before entering into force in the early 1980s. Fifty-one mostly European nations are signatories of the convention, which is also supported by the European Union.

A proposal for a soil framework directive in the EU was withdrawn in 2014, but for air and water substantive legislative frameworks are in place (previously also known as Integrated Pollution Prevention and Control). These are regularly reviewed and updated to take into account better scientific understanding of the matter and the development of innovative pollution control technologies. More than 30% of citizens in the EU are exposed to air pollutants above EU standards – air pollution is the biggest environmental cause of premature deaths in the region, which number more than 400,000 per year (European Commission 2017).

In the far east, Clean Air Asia (formerly Clean Air Initiative for Asian Cities) was set up with help of the World Bank and others to work with governments, NGOs and the private sector for better air quality in a region that is urbanising rapidly and suffering from air pollution as a result of accelerated industrialisation and motorisation alongside poor urban planning (Clean Air Asia 2017).

Regional responses to marine pollution in the North-East Atlantic include the Oslo Convention for the Prevention of Marine Polluting by Dumping from Ships and Aircraft and the Paris Convention for the Prevention of Marine Pollution from Land-based Sources, now a single entity in the OSPAR Convention (Anderson 2014).

National responses to pollution

In the US, the Oil Pollution Act of 1990 required oil tankers to be double-hulled by 2015 to protect from oil spills and water pollution, paralleling MARPOL efforts (Anderson 2014). Applying to all forms of pollution, the Toxic Release Inventory (TRI) adopted in 1989 was an innovative approach by the US government requiring major industrial plants to disclose pollutants (names and quantities) released to air, water and soil (Nemetz 2013). The provision of such information to the public has nudged major emitters to lower their emissions voluntarily, which has helped avoid the threat of more stringent regulation (Hackett 2011; Nemetz 2013).

China's problems with pollution were highlighted in Box 6.3. In recent years, efforts have been made to address pollution, though industries that were poorly regulated in the past and lack of local enforcement have posed challenges (Economist 2017). In 2017, the country launched 8,000 projects to clean up water (Reuters 2017). Whilst closing factories during smog is only a short-term measure, Chinese authorities are recognising that reducing pollution is a long-term process and have for example included $PM_{2.5}$ in the current Five-Year Plan (Batmanian 2016) and announced a crack-down on emissions from ozone and PM in 2017 (Date 2017).

Local responses to pollution

In preparation for the London Olympics in 2012, the site of the games, a former industrial area, had to be decontaminated. The soil was washed and treated with bacteria, to the cost of £3,000 ($3,900) per cubic meter[7] (Economist 2017). This was the largest-ever soil washing operation in the UK at the time, with six "soil hospitals" cleaning most of the million cubic meters of soil contaminated with petrol, oil, tar, arsenic, cyanide and lead (Elkington 2012). The previous and subsequent Olympics in Beijing 2008 and Rio de Janeiro 2016 were also faced with the challenge of controlling air and water pollution ahead of hosting the games, featuring heavily in the news.

Other major cities such as Oslo, Singapore, Milan and London have introduced congestion charging systems to reduce air pollution from motor vehicles in their centres, whilst other places such as Jakarta and Bogota have started regular car-free days or have established car-free zones. Delhi has used a car rationing system using odd and even numbers for days and car registration plates to deal with hazardous air pollution.

BUSINESS RESPONSES TO POLLUTION: INTERNAL AND EXTERNAL OPPORTUNITIES

In the introduction to part two of the book, we introduced Esty and Winston's (2009) AUDIO tool which companies can use to analyse a particular environmental challenge. By looking at the aspects of the challenge from an upstream and downstream perspective, businesses can identify relevant issues (or vulnerabilities and risk) and opportunities arising from the challenge. For example, concerning air pollution retailers are faced with air emissions from their own facilities, but also with emissions at supplier sites and from energy sources as well as with emissions from products. Tightening regulatory controls can be a vulnerability which the business can for example address by increasing efficiency which reduces emissions as well as costs, and by considering the use of pollution abatement or control technologies.

Opportunities to respond to pollution lie in both the prevention and control of it. Whilst not all pollution might be preventable, as zero pollution is often not feasible, controlling pollution should always be possible. Nevertheless, pollution does not have to be inevitable, since it can be a symptom of poor design and inefficiency, as Robertson (2017) emphasises (see also Porter and van der Linde 1995).

Control methods for air pollution include filters, scrubbers, reduced fuel combustion and use of low-sulphur and lead-free fuel, energy conservation, pollution control equipment, transport planning, testing of emissions, efficient combustion techniques (e.g. reduction of temperatures), evaporative controls and exhaust gas recirculation (Anderson 2014). Other technologies include noise mufflers, waste-water treatment facilities and composting of waste (EEA 2017).

One issue with pollution control lies in abatement measures such as end-of-pipe solutions, which may be perceived as simply shifting the problem from one location to another rather than addressing it – e.g. building higher chimneys simply brought smog further away rather than removing already-formed contaminants from the air (Cavagnaro and Curiel 2012). It is a problem 3M faced which then used its *Pollution Prevention Pays* programme to stop pollution from becoming a problem in the first place (Esty and Winston 2009). This can be seen as an internal approach to minimising pollution, which we will cover in the next part of the chapter before we move on to opportunities arising for businesses from offering pollution solutions to other companies and consumers. The following list provides examples of technologies and businesses preventing and controlling pollution. This should be seen as a starting point rather than an exhaustive list of how companies address this particular environmental challenge. Weybrecht (2014) differentiates between four types of technologies:

- remediation technologies (e.g. soil and water clean-up methods) that treat and remediate environmental problems after their occurrence and are therefore often expensive;
- abatement technologies that treat or capture pollutants before they are emitted into the environment through chemical, physical or biological mechanisms, and that are often capital-, energy- and waste-intensive;
- pollution prevention technologies that comprise alternative or improved processes in industry and agriculture as well as alternative products, both of which reduce or avoid pollution;
- sustainable technologies that focus on eliminating any ecological impact and on achieving resource efficiency.

Remediation as the stopping and reversing of environmental damage involves a number of steps, starting with understanding and characterising the type, source and location of pollution before deciding on containment, removal or treatment (Robertson 2017). Containment is in the form of physical or hydraulic barriers, whilst treatment can be physical or biological, on site or off site. However, not even the best remediation is optimal, and the more sustainable option is to prevent pollution from happening in the first place.

Internal opportunities – how to minimise impact within a business

It could be argued that businesses that manage their environmental impacts and therefore polluting activities well are almost always stronger and more competitive as they tend to be better managed but are also more efficient in their resource use (Bigg 2011). Reducing emissions voluntarily has benefits for companies in terms of reducing costs, pre-empting liability and improving public relations (Friedman 2011). Developing voluntary pollution prevention programmes may not only be cheaper than waiting for regulation, but it might also be easier. Bigg (2011) likens voluntary efforts in organisations to observing a car's speedometer, which allows effective monitoring of speed as compared to radar speed guns. Arguably, voluntary anti-pollution efforts give companies more control over pollution prevention than regulation, though the latter ensures a level playing field. Whichever approach – voluntary or mandatory – is used to minimise pollution, companies are likely to benefit from improved productivity, increased product yields, energy conservation and reduced costs regarding pollution treatment and potential liability (Friedman 2011). The financial benefits of addressing pollution can for example be seen in 3M's Pollution Prevention Pays programme and Dow Chemicals' initiative Waste Reduction Always Pays (WRAP).

Pollution prevention technologies include for example paper production that eliminates chlorine bleaching in terms of processes (Weybrecht 2014). By changing its processes and tools through a modified wet processing tool, chipmaker AMD reduced its use of both chemicals and water (Esty and Winston 2009).

Other more general pollution prevention activities that companies can use comprise inventory management, including purchasing procedures, process modifications, good operating practices, substitution of materials, redesign of products and in-process recycling as well as water and energy conservation (EEA 2017; EPA 2017b). We will cover some of these in more depth in Chapters 8 and 9 when we discuss supply chain management and design for environment.

Air pollution control devices clean emissions at the source where they are generated, such as incinerators, smokestacks or automotive exhaust systems (Green 2013) and include scrubbers and catalytic converters. Scrubbing systems can effectively remove a large percentage of SO_2 and NO_x emissions from coal-fired plants (Anderson 2014).

BOX 6.4 SOLUTIONS TO LIGHT POLLUTION

Light pollution was introduced in Box 6.1 as excess light spilling into the air, which is a form of wasted energy and costs. Addressing light pollution represents cost savings for companies e.g. by using light only where and when necessary (e.g. moving from all-night illumination of a site to only illuminating facilities in use and eliminating unnecessary fixtures or using motion sensors for turning on lights), by controlling the direction and spread of light through using fully shielded lighting that points downward and inward rather than outward and upward, by reducing bulb wattages and by switching the type of light and bulbs (e.g. using less blue light and more warm light, moving from incandescent to fluorescent bulbs).

Sources: Robertson 2017; Anderson 2014, www.darksky.org.

In the course of the chapter we identified agriculture as a major polluter of air, land and ground. Efforts to cut its impact include reducing its use of water and fertilisers. Switching to organic farming involves a major shift to inputs and processes as well as outputs, such as restricting the use of artificial chemicals and fertilisers common in intensive, conventional agriculture (Hughes 2011). Even without a transition to organic practices, pollution can be prevented by adopting less harmful pesticides and by cultivating naturally pest-resistant crop strains as well as protecting sensitive areas (EPA 2017b). Furthermore, the storage and management of animal waste is another area where groundwater pollution through bacteria and nitrogen can be prevented. Agricultural companies have been pushed by investors to address water pollution as part of wider water risk concerns (Clancy 2016).

Pollution abatement in the case of oil pollution at sea comes in various forms – under MARPOL (see earlier discussion) all ships are required to be capable of retaining the residue of tank cleaning operations with the help of an empty cargo tank, also called "slop" tank. This is a "load-on-top" system; alternatives are the spraying of oil dispersive and blasting sawdust and straw as a quasi-blotting paper on water and land (EEA 2017). British Petroleum (BP) developed the Vikoma System for containing oil spills at sea, which uses seabooms to trap and then skim the oil. Following a management buy-out, Vikoma International is now an independent company offering a range of oil spill solutions and wastewater treatments. The example of Vikoma also demonstrates how addressing pollution internally can lead to solutions offered externally as well, making the distinction between internal and external responses and opportunities somewhat redundant.

Before we move on to the external side, we want to provide two examples of what we referred to in the introduction – how addressing one environmental challenge could either help address another one or could have unintended trade-offs and side effects. British DIY retailer B&Q had to pay compensation to hundreds of customers after the paint they sold from US-based firm Valspar started to smell like rotting animals or cat's urine, leaving consumers with uninhabitable rooms and the costs of repainting. The ammonia-type odour developed in the products after Valspar had removed a preservative additive from its formula which made the paint more susceptible to bacterial contamination. Tighter regulations on chemicals have restricted the type of solvents and additives manufacturers can use. Valspar has since re-added the preservative (Ward 2017).

Chemical leasing addresses both waste and pollution, generating synergies not only between two environmental challenges but also between suppliers and buyers (or users) of chemicals. It shifts the provision of chemicals from buying a product in increased quantities to buying a service (Eacott 2011). In providing a service rather than a product, chemical manufacturers are incentivised to reduce the cost per service and thus their energy and resource use, generating environmental and economic benefits simultaneously (Stringer 2014). Examples of companies using chemical leasing include for example Carlsberg and Coca-Cola which contracted Eco-Lab to help improve its use of chemicals in lubricating, cleaning and washing. Industrial cleaning is an area where the benefits of chemical leasing are obvious (Stringer 2014). Chemical leasing has been criticised for not necessarily removing hazardous substances – something so-called green chemistry attempts to do. Nevertheless, it is one approach to pollution and waste prevention that attempts to address these environmental challenges both internally and externally.

External opportunities – what solutions can be offered to customers and consumers

Dealing with pollution through remediation, abatement, control and prevention provides businesses with opportunities to offer solutions to other companies and to consumers.

Earlier, we mentioned Vikoma, initially an in-house project within BP, which was developed following the UK's largest oil spill in 1967. The company now offers oil spill control systems to the oil and gas industry both onshore and offshore, including ports and terminals, inland waters and arctic environments. Solutions offered include booms, skimmers, tanks and storage, beach and land cleaning, dispersants, absorbents and ancillary equipment. In addition to selling equipment, the company also offers training courses to first responders, engineers and operators. Competitors of Vikoma include for example Bulbeck Envirosolutions Pty Ltd from Australia, Cornelsen Umwelttechnologie GmbH from Germany, DESMI A/S from Denmark and MAVI DENIZ Environmental Protection Services Co. from Turkey, to name just a few (see www.environmental-expert.com or www.ukspill.org for more).

For oil-polluted soil, fungi have been found to be an effective treatment for cleaning up the contaminated land, a form of so-called bioremediation. While composting, landfill and incineration can be used to clean up polluted soil, landfill is not always available and incineration can be costly, thus fungi provide an alternative by breaking down soil-polluting compounds (Aalto University 2014). Contaminated soil is a problem for many plant species and crops whose growth and development are affected by the toxicity of soil pollutants. Only a small number of plants are able to remove soil contaminants and then grow normally. As conventional decontamination methods are costly, alternatives have been studied and one was found in two genes of baker's yeast that increase the resistance of plants, which are then able to grow in contaminated soil (IGC 2017). Whilst further research is needed, the results of these findings offer opportunities for agriculture faced with the challenge of growing crops on polluted lands. Conventional methods are not only costly, but they often require the removal of soil to be treated off-site, and if soil washing is used, polluted water poses an additional problem. Researchers at the American Institute of Physics (2017) found a cheaper and more efficient solution in the use of lasers, which can zap contaminated soil on site and break the chemical bonds of pollutants through heat. Examples of companies offering soil remediation services include Jiangsu DDBS Environment Remediation Co. in Nanjing, China, Nelson Environmental Remediation from Canada, and Ballast Nedam Environmental Engineering in Belgium (see www.environmental-expert.com for more).

Abatement and control of air pollution comes in various forms; traditionally filters and scrubbers have been used, but recent developments have seen more innovate solutions offered. Overall, the air pollution control equipment market is a growing market, which was worth an estimated $14 billion in 2016 and which is expected to increase to over $20 billion by 2021 (BCC Research 2017). Such pollution control equipment is available to industrial, commercial and residential users, with industrial end users the largest segment. The type of products offered include dust collectors and vacuums, mist eliminators, fire and emergency exhausts, fume and smoke collectors, vehicle exhausts and others. The technologies used to control air pollution include HEPA (High Efficiency Particulate Air), ultraviolet light air purifiers, activated carbon, ionic filters and others (BCC Research 2017). Leaders in this market include well-known companies such as Honeywell, 3M, Whirlpool, and Hyundai Wacortec (Globe Newswire 2017). Smaller companies in this market include AirScience Technologies from Canada, Air Clean from Italy and Abatement Technologies from the US (see www.environmental-expert.com for more).

Air cleaning technologies offered to consumer households include for example air purifiers and filters as well as air sterilisers and ionisers. The Healthy House, to name one online retailer (www.healthy-house.co.uk), offers different types of air purifiers from a range of manufacturers. Air purifiers either remove or destroy pollutants, depending on the technology used and whether they are mechanical air filters or electronic air cleaners EPA (2017a). Another technology that can be used comes in the form of air quality apps that do not directly prevent or control pollution but make pollution visible to individuals who can change their behaviour accordingly (Delmas and Kohli 2019).

Other solutions to indoor air quality include the substitution of appliances used for cooking and lighting. Solar cookers avoid the use of open fires and solid fuel stoves. Whilst many solar cookers require moving to optimise sunlight,[8] South African engineer Fritz developed the Water&Solar100 which automatically tracks the sun and adjusts its position. In addition to cooking, the device can also sterilise water and medical equipment as well as generate electricity and charge batteries, thus meeting needs in rural communities off the grid in developing countries (Hodal 2017). Other innovations in this area include the non-electric portable slow cooker Wonderbag, which only requires a limited time of stove cooking in a pot which is then transferred to the Wonderbag, using the principle of retained-heat cooking. It cuts down on air pollution by requiring less fuel (see www.wonderbagworld.com for more information). In terms of lighting, solar lights solve both air and light problems, replacing toxic kerosene lamps. Other inventions that are increasingly replacing kerosene lights and thus reducing air pollution in developing countries include for example GravityLight, which combines kinetic and potential energy to provide instant lighting without batteries, and plastic bottles filled with water and some drops of bleach which are fixed to roofs and then refract sunlight (see https://deciwatt.global/gravitylight and http://literoflight.org/).

Conventionally, ionic liquids and organic solvents are used to reduce the levels of volatile organic compounds both indoors and outdoors. New research has identified an eco-friendlier solution to absorbing VOCs in a mixture made from urea and choline salt. Such deep eutectic solvents are more biodegradable than ionic liquids, which are potentially toxic, and cheaper to produce (Springer 2017).

Other innovative approaches to tackling air pollution include for example smog-eating cement which, in the presence of light, uses a natural process called photocatalysis to break down air pollutants such as CO, NO_x and benzene (Weybrecht 2014). It was initially developed as a self-cleaning material by a chemist working for Italian cement manufacturer Italcementi and was found not only to clean itself but also the surrounding air, neutralising pollutants (Italcementi 2014). It can be used in road construction and for building facades. Other types of facades that reduce air pollution include de-polluting facades constructed from superfine titanium dioxide coating (TiO_2) and microalgae bio facades which are fed by carbon (Fedele 2015). More natural solutions of fighting pollution through buildings can be found in the so-called vertical forest ("bosco verticale"), a high-rise building designed by Italian architect Stefano Boeri, which has seen structural engineers and botanists collaborate to address not only air pollution, but also climate change and housing shortages. The trees on the building filter out fine dust particles and muffle the noise pollution from traffic below. The concept is being replicated across the world (e.g. in Nanjing, China, as well as in Toronto, Canada and Bogota, Colombia) (Xie 2017).

Pollution prevention technologies in terms of products offered to buyers include biodegradable detergents, mercury-free batteries, unbleached recycled paper, and water-based paints (Weybrecht 2014) as well as electric cars which were covered also in the previous chapter and address both climate change and air pollution, provided renewable electricity is used.[9] Companies that offer biodegradable detergents as well as other cleaning products include for example Ecover, Green Works (owned by Clorox), Seventh Generation, Method (owned by Ecover) and Terra Activ (owned by Henkel, producer of Persil).

Biodegradable detergents substitute the materials comprising detergents. Another approach to addressing the environmental impact of detergents is the use of concentrated formulas, which mean less water is used and shipped, and less plastic is used in the production process whilst cardboard use for transport is cut as well. This also entails a reduction in transport and thus air pollution. Once the concentrated detergents reach retailers, less shelf space is used and consumers have less weight to carry. Walmart introduced this in partnership with suppliers ten years ago and eventually offered consumers only the (edited) choice of concentrated formulas (Shelton 2017; Walmart 2007).

In the commercial cleaning sector, the move is not only towards using greener chemicals but also towards eliminating harmful chemicals altogether. Traditional industrial cleaners are being replaced with devices using split stream technology. Ozonisation, electrolysation or oxygenation processes convert tap water with the help of ozone or salt and then electricity into an effective cleaning solution requiring no further detergents. This can be used for example in scrubber-dryers, buckets and trigger sprays. Examples include Orbio® 5000-Sc machines developed by Tennant or the Lotus Pro manufactured by Tursano (McCullough 2014).

Although it could be argued that prevention of water pollution is easier than controlling water pollution due to the difficulties in treating and reaching groundwater, agricultural runoff and dispersed pollutants at sea, it is not impossible. Christianson (2016) highlights how agricultural subsurface drainage systems, also known as tile drainage networks, are used by farmers to channel excess water out of the ground but are also major contributors to water pollution from nitrogen and fertilisers. Research has shown that natural processes that use bacteria can be employed to clean up polluted subsurface drainage water from farms. Denitrifying bioreactors made from woodchips are effective tools to treat water pollution and can be installed and used with little disruption and maintenance. In addition to treating agricultural runoff, bioreactors can also be used to clean up aquaculture wastewater (Christianson 2016).

SUMMARY

Pollution of air, water and soil presents significant environmental, health, social, financial and regulatory challenges that companies – alongside civil society and policymakers – are facing.

In this chapter, we have covered pollution across various aspects, what the challenges are and how we can understand and classify them – by source, type, content, location, behaviour, density and range – what the concerns are in terms of human, animal and plant health, and how it links to other challenges such as waste and climate change. We then looked at regulatory responses across local, national and global levels before covering business responses to it. Corporate responses include the development and use of technologies that remedy, abate, control and prevent pollution internally as well as grasping external opportunities to offer profitable pollution solutions to other companies and consumers. Whilst zero pollution might not always be possible, prevention of pollution in the first place is preferable over remediation and control of pollution, which are, in turn, better than doing nothing. Business as a major cause of pollution has a significant role to play in addressing this challenge, developing technologies, products and processes for internal and external use, thus not only reducing its own polluting impact but also minimising how it is affected by pollution itself and helping others to reduce their contribution to the problem.

In part three, where corporate functions and their importance for managing environmental sustainability are covered, pollution-relevant material is included as well. For example, in Chapter 9 on production and operations, design for environment is covered, where pollution can be designed out of manufacturing processes, whilst in Chapter 11 environmental management systems are discussed which can help companies monitor and control pollution, and Chapter 12 on accounting relates the measuring and reporting of pollution with its management.

The concluding part of the book includes integrative case studies where pollution as well as its management and prevention can be discussed within a broader context.

SUGGESTED QUESTIONS AND ACTIVITIES

- Visit the EPA's list of hazardous air pollutants – do you recognise any? www.epa.gov/haps/initial-list-hazardous-air-pollutants-modifications
- Use the European Environment Agency's Glossary to look up any pollution-related terms: www.eea.europa.eu/help/glossary/eea-glossary
- Use The Natural Step (TNS) framework introduced in Chapter 4 to sketch out a company's pollution prevention approach.
- Choose one of the other frameworks, models or strategies from Chapter 4 and use it to plan a company's approach to control pollution.
- Visit www.environmental-expert.com and investigate which companies offer what kind of equipment and solutions to remedy pollution.
- Visit the webpages of the International Dark-Sky Association and find a Dark Sky Places designation near you: www.darksky.org/
- Have some creative fun and build your own solar oven from a pizza box or Pringles tube – check out YouTube for any advice.
- Debate the following motions, preparing for both agreement and disagreement on the following statements:
  - Zero pollution is neither feasible nor desirable.
  - The polluter pays principle should only apply to companies.
  - The polluter pays principle should apply to companies and consumers.
- Visit the homepage of one of the incubators and hubs or conferences mentioned in the previous chapter and see what the latest innovations are in cleantech regarding pollution, e.g.: http://laincubator.org/, www.impacthub.net/, http://ufl.nyc/, www.cleantechinnovate.com/.

NOTES

1 Nemetz (2013) furthermore differentiates between uniformly mixed pollutants that are assimilative or accumulative. Whilst assimilative pollutants act globally and thus neither source nor damage location are important, for accumulative pollutants the location of discharge is not important but the location of the impact matters.

2 pH measures the acidity and alkalinity, i.e. how acid or basic a water-based solution is based on the concentration of hydrogen ions (Robertson 2017).

3 Chapter 6 in Robertson (2017) is a good starting point, while www.who.int/topics/water/en/ offers information from a global health perspective, and www.unep.org/ecosystems/freshwater/ from an environmental perspective.

4 There are seven industries where air pollution damages exceed their value added – solid waste combustion, sewage treatment, stone quarrying, marinas, and oil and coal-fired power plants – though the results of this study (Muller et al. 2011) need to be interpreted with caution (Nemetz 2013) as possible offsetting benefits from pollution control were not taken into account.

5 Under EU law, limit values are legally binding as soon as the law enters into force, while target values should be adhered to as far as possible by the date set – target values are therefore less strict than limit values (http://ec.europa.eu/environment/air/quality/standards.htm).

6 For example, it does not even appear in the wide-ranging glossary of the European Environment Agency (EEA 2017).

7 If the same standard of decontamination were to be applied to all of China's polluted soil, the cost would – in theory – be $1,000 trillion, which, according to The Economist (2017) would be more than all the wealth in the world.

8 One example is Solari (www.jamesdysonaward.org/2014/project/solari-portable-solar-cooker/).

9 However, electric cars still pollute the air as using brakes and tyres result in iron particles and plastic fragments being emitted.

SOURCES AND FURTHER READING

Aalto University (2014) Fungi Clean Oil-Polluted Soil, Study Shows, *ScienceDaily*, 21.05.2014, available at www.sciencedaily.com/releases/2014/05/140521094743.htm (Retrieved, 30.08.2017).

American Institute of Physics (2017) Lasers Zap Decontaminates from Soil, *ScienceDaily*, 29.08.2017, available at www.sciencedaily.com/releases/2017/08/170829113850.htm (Retrieved, 30.08.2017).

Anderson, D. A. (2014) *Environmental Economics and Natural Resource Management*, London: Routledge, 4th ed.

Batmanian, G. (2017) Exchanging Experiences to Fight Air Pollution, *China Daily*, 07.12.2016, available at www.chinadaily.com.cn/opinion/2016-12/07/content_27594507.htm (Retrieved, 24.08.2017).

Bauer, S. E., Tsigaridis, K. and Miller, R. (2016) Significant Atmospheric Aerosol Pollution Caused by World Food Cultivation, in: *Geophysical Research Letters*, 43(10): 5394–5400.

BBC (2017) Fipronil Egg Scandal: What We Know, 11.08.2017, available at www.bbc.co.uk/news/world-europe-40878381 (Retrieved, 22.08.2017).

BCC Research (2017) Air Pollution Control Equipment: Technologies and Global Markets, Report ENV021B, available at www.bccresearch.com/market-research/environment/air-pollution-control-equipment-markets-report-env021b.html (Retrieved, 31.08.2017).

Berinato, S. (2017) Air Pollution Brings Down the Stock Market, in: *Harvard Business Review*, 95(2): 38–39.

Bigg (2011) Pollution Prevention and Control, in: Brady, J., Ebbage, A. and Lunn, R. (eds.) *Environmental Management in Organizations, The IEMA Handbook*, London: Earthscan/Routledge, 2nd ed., pp. 213–238.

Birchby, D., Cooke, S., Stedman, J., Hurley, F., Searl, A. and Hunt, A. (2014) Valuing the Impacts of Air Quality on Productivity, Report for the Department for Environment, Food and Rural Affairs, Ricardo-AEA/R/3417, available at https://uk-air.defra.gov.uk/assets/documents/reports/cat19/1511251135_140610_Valuing_the_impacts_of_air_quality_on_productivity_Final_Report_3_0.pdf (Retrieved, 22.08.2017).

Carrington, D. (2017) Plastic Fibers Found in Tap Water Around the World, Study Reveals, *The Guardian*, 06.09.2017, available at www.theguardian.com/environment/2017/sep/06/plastic-fibres-found-tap-water-around-world-study-reveals (Retrieved, 08.09.2017).

Cavagnaro, E. and Curiel, G. (2012) *The Three Levels of Sustainability*, Sheffield: Greenleaf Publishing.

Chang, T., Graff Zivin, J., Gross, T. and Neidell, M. (2016a) Particulate Pollution and the Productivity of Pear Packers, in: *American Economic Journal: Economic Policy*, 8(3): 141–169.

Chang, T., Graff Zivin, J., Gross, T. and Neidell, M. (2016b) The Effect of Pollution on Worker Productivity: Evidence from Call-Center Workers in China, NBER Working Paper No. 22328, National Bureau of Economic Research, available at www.nber.org/papers/w22328 (Retrieved, 22.08.2017).

Christianson, L. (2016) Reducing Water Pollution with Microbes and Wood Chips, *The Conversation*, 06.07.2016, available at https://theconversation.com/reducing-water-pollution-with-microbes-and-wood-chips-58852 (Retrieved, 04.09.2017).

Clancy, H. (2017) Investors Urge Cargill, Smithfield to Address Water Pollution, *GreenBiz*, 21.11.2016, available at www.greenbiz.com/article/investors-urge-cargill-smithfield-address-water-pollution (Retrieved, 30.08.2017).

Clean Air Asia (2017) Clean Air Asia webpages, available at http://cleanairasia.org/ (Retrieved, 24.08.2017).

Copeland, B. R. and Taylor, M. S. (2004) Trade, Growth, and the Environment, in: *Journal of Economic Literature*, 42(1): 7–71.

Costanza, R., Cumberland, J., Daly, H., Goodland, R. and Norgaard, R. (1997) *An Introduction to Ecological Economics*, Boca Raton, FL: St Lucie Press.

Date, W. (2017) PM2.5 Targeted in China Emissions Crackdown, 24.08.2017, available at www.airqualitynews.com/2017/08/24/pm2-5-targeted-china-emissions-crackdown/ (Retrieved, 04.09.2017).

Davies, J. (2017) Business Case for Soil, in: *Nature*, 543(7645): 309–311.

Delmas, M. A. and Kohli, A. (2019) Can Apps Make Air Pollution Visible? Learning About Health Impacts Through Engagement with Air Quality Information, in: *Journal of Business Ethics*, available at https://doi.org/10.1007/s10551-019-04215-7 (Retrieved, 03.12.2019).

Eacott, C. (2011) Chemicals, in: Brady, J., Ebbage, A. and Lunn, R. (eds.) *Environmental Management in Organizations, The IEMA Handbook*, London: Earthscan/Routledge, 2nd ed., pp. 153–168.

EEA (2016) The European Environment – State and Outlook 2015 Report, European Environment Agency, available at www.eea.europa.eu/soer (Retrieved, 22.08.2017).

EEA (2017) Glossary of Environmental Terms, European Environment Agency, available at www.eea.europa.eu/help/glossary/eea-glossary (Retrieved, 14.08.2017).

Economist (2017) The Bad Earth: The Most Neglected Threat to Public Health in China Is Toxic Soil, 08.06.2017, available at www.economist.com/news/briefing/21723128-and-fixing-it-will-be-hard-and-costly-most-neglected-threat-public-health-china (Retrieved, 24.08.2017).

Economy, E. C. (2004) *The River Runs Black: The Environmental Challenge to China's Future*, Ithaca, NY: Cornell University Press.

Elkington, J. (2012) *The Zeronauts: Breaking the Sustainability Barrier*, London: Routledge.

EPA (2017a) Air Topics, United States Environmental Protection Agency, available at www.epa.gov/environmental-topics/air-topics (Retrieved, 14.08.2017).

EPA (2017b) Pollution Prevention (P2), United States Environmental Protection Agency, available at www.epa.gov/p2 (Retrieved, 30.08.2017).

Esty, D. C. and Winston, A. S. (2009) *Green to Gold: How Smart Companies Use Environmental Strategy to Innovate, Create Value and Build Competitive Advantage*, Hoboken, NJ: Wiley.

European Commission (2016a) Cleaner Air for All, available at http://ec.europa.eu/environment/air/cleaner_air/index.html (Retrieved, 14.08.2017).

European Commission (2016b) EU Water Policies webpages, available at http://ec.europa.eu/environment/waste (Retrieved, 23.08.2017).

European Commission (2016c) EU Soil webpages, available at http://ec.europa.eu/environment/soil/index_en.htm (Retrieved, 23.08.2017).

European Commission (2016d) EU Noise webpages, available at http://ec.europa.eu/environment/noise/index_en.htm (Retrieved, 23.08.2017).

European Commission (2017) Commission to Review Permits of Large Combustion Plants, press release, 31.07.2017, available at http://ec.europa.eu/environment/pdf/31_07_2017_news_en.pdf (Retrieved, 24.08.2017).

FAO (1996) Control of Water Pollution from Agriculture – FAO Irrigation and Drainage Paper 55, Food and Agriculture Organization of the United Nations, available at www.fao.org/docrep/w2598e/w2598e00.htm (Retrieved, 17.08.2017).

FAO and ITPS (2015) Status of the World's Soil Resources (SWSR) – Technical Summary. Food and Agriculture Organization of the United Nations and Intergovernmental Technical Panel on Soils, Rome, Italy, available at www.fao.org/3/a-i5126e.pdf (Retrieved, 23.08.2017).

Fedele, A. (2015) Five Pollution Reducing Facades, *Architecture News*, 27.02.2015, available at https://sourceable.net/six-pollution-reducing-facades/ (Retrieved, 31.08.2017).

Friedman, F. B. (2011) *Practical Guide to Environmental Management*, Washington, DC: Environmental Law Institute, 11th ed.

Globe Newswire (2017) Air Pollution Control Equipment Market to Reach US$21B by 2021, 27.07.2017 available at https://globenewswire.com/news-release/2017/07/27/1063298/0/en/Air-Pollution-Control-Equipment-Market-To-Reach-US-21B-by-2021.html (Retrieved, 31.08.2017).

Graff Zivin, J. and Neidell, M. (2012) The Impact of Pollution on Worker Productivity, in: *American Economic Review*, 102(7): 3652–3673.

Green, T. (2013) *The Dictionary of Global Sustainability. With Case Studies on the Environmental, Economic, Technological, and Social Aspects of Sustainability*, New York: McGraw-Hill.

Hackett, C. (2011) *Environmental and Natural Resource Economics*, London: Sharpe, 4th ed.

Hawken, P., Lovins, A. B. and Lovins, L. H. (1999) *Natural Capitalism: The Next Industrial Revolution*, London: Earthscan.

Heyes, A., Neidell, M. and Saberian, S. (2016) The Effect of Air Pollution on Investor Behavior: Evidence from the S&P 500, NBER Working Paper No. 22753, National Bureau of Economic Research, available at www.nber.org/papers/w22753 (Retrieved, 22.08.2017).

Hodal, K. (2017) The Solar Cooker that Seeks Its Own Place in the Sun, *The Guardian*, 05.01.2017, available at www.theguardian.com/global-development/2017/jan/05/solar-cooker-place-in-the-sun-water-and-solar-100-wilfred-fritz (Retrieved, 31.08.2017).

Hughes, J. (2011) Food and Agriculture, in: Brady, J., Ebbage, A. and Lunn, R. (eds.) *Environmental Management in Organizations, The IEMA Handbook*, London: Earthscan/Routledge, 2nd ed., pp. 199–212.

IGC (2017) Baker's Yeast Can Help Plants Cope with Soil Contamination, Instituto Gulbenkian de Ciencia, Portugal, *ScienceDaily*, 03.07.2017, available at www.sciencedaily.com/releases/2017/07/170703083248.htm (Retrieved, 30.08.2017).

Italcementi (2014) Inventor of the Self-Cleaning City: Luigi Cassar's "Smog-Eating" Concrete Garners Nomination for the European Inventor Award,

Press Release, 29.04.2014, available at www.italcementigroup.com/NR/rdonlyres/A7D06606-FBED-4FEA-8295-BBD59B49484B/0/2_PressreleaseEIA.pdf (Retrieved, 31.08.2017).

Leahy, K. (2017) 'There Are Hundreds of Sick Crew': Is Toxic Air on Planes Making Frequent Flyers Ill? *The Guardian*, 19.08.2017, available at www.theguardian.com/science/2017/aug/19/sick-crew-toxic-air-planes-frequent-flyers-ill (Retrieved, 04.09.2017).

McCullough, D. G. (2014) 'Chemical-Free' Cleaning Is Trending in the Commercial Sector, *The Guardian*, 25.06.2014, available at www.theguardian.com/sustainable-business/2014/jun/25/chemical-free-eco-cleaning-commercial-sector-households (Retrieved, 30.08.2017).

McIntyre, A, Crowcroft, P. and Chisholm, A. (2011) Emissions and Contamination Standards, in: Brady, J., Ebbage, A. and Lunn, R. (eds.) *Environmental Management in Organizations, The IEMA Handbook*, London: Earthscan/Routledge, 2nd ed., pp. 357–378.

Morgan-Taylor, M. (2015) Global Approaches to Legislation for Light Pollution, 1st International Conference on Sustainable Lighting and Light Pollution, Soul 2014, Proceedings, pp. 80–89.

Muller, N. Z., Mendelsohn, R. and Nordhaus, W. (2011) Environmental Accounting for Pollution in the United States Economy, in: *The American Economic Review*, 101(5): 1649–1675.

Nemetz, P. N. (2013) *Business and the Sustainability Challenge: An Integrated Perspective*, London: Routledge.

Phillips, R. (2017) Air Pollution: The Business Concern, available at www.eco-business.com/opinion/air-pollution-the-business-concern/ (Retrieved, 22.08.2017).

Porter, M. E. and van der Linde, C. (1995) Toward a New Conception of the Environment-Competitiveness Relationship, in: *Journal of Economic Perspectives*, 9(4): 97–118.

Post, J. E. (2012) Business, Society, and the Environment, in: Bansal, P. and Hoffman, A. J. (eds.) *The Oxford Handbook of Business and the Natural Environment*, Oxford: Oxford University Press, pp. 537–555.

Pullen, J. (2011) Measurement and Monitoring, in: Brady, J., Ebbage, A. and Lunn, R. (eds.) *Environmental Management in Organizations, The IEMA Handbook*, London: Earthscan/Routledge, 2nd ed., pp. 379–398.

Puttock, A., Graham, H. A., Cunliffe, A. M., Elliott, M. and Brazier, R. E. (2017) Eurasian Beaver Activity Increases Water Storage, Attenuates Flow and Mitigates Diffuse Pollution from Intensively-Managed Grasslands, in: *Science of The Total Environment*, 576: 430–443.

Reuters (2017) China Launches 8,000 Water Clean-Up Projects Worth $100 billion in First Half of 2017, 24.08.2017, available at www.reuters.com/article/us-china-pollution-water-idUSKCN1B4090 (Retrieved, 24.08.2017).

Robertson, M. (2017) *Sustainability Principles and Practice*, London: Routledge, 2nd ed.

SCU (2013) Soil Contamination: Impacts on Human Health, Science for Environment Policy In-depth Report Produced for the European Commission DG Environment by Science Communication Unit, University of the West of England, Bristol available at http://ec.europa.eu/environment/integration/research/newsalert/pdf/IR5_en.pdf (Retrieved, 17.08.2017).

Shelton, S. (2017) Speaking Sustainably: What Laundry Detergent Can Teach the Electric Industry, *Green Biz*, 03.08.2017, available www.greenbiz.com/article/what-laundry-detergent-can-teach-electric-industry (Retrieved, 30.08.2017).

Slawson, N. (2017) Primark Recalls Men's Flip-Flops after Carcinogenic Chemical Discovery, *The Guardian*, 17.07.2017, available at www.theguardian.com/business/2017/jul/17/primark-recalls-mens-flip-flops-after-carcinogenic-chemical-discovery (Retrieved, 22.08.2017).

Stringer, L. (2014) Chemicals Are Everywhere. Can a New Business Model Make Their Use Greener? *The Guardian*, 09.12.2014, available at www.theguardian.com/sustainable-business/2014/dec/09/chemical-leasing-ecolab-coke-ikea-gm-un-cleaning-environment (Retrieved, 29.08.2017).

Springer, L. (2017) New Green Solvent Could Help Clean Our Air: Eco-Friendly Mixture Made from Choline Salt and Urea Could Absorb Harmful Gaseous Pollutants, *ScienceDaily*, 24.08.2017, available at www.sciencedaily.com/releases/2017/08/170824101812.htm (Retrieved, 31.08.2017).

Tabuchi, H. (2017) As Beijing Joins Climate Fight, Chinese Companies Build Coal Plants, *New York Times*, 01.07.2017, available at www.nytimes.com/2017/07/01/climate/china-energy-companies-coal-plants-climate-change.html (Retrieved, 23.08.2017).

Tingting, D. (2017) In China, the Water You Drink is as Dangerous as the Air You Breathe, *The Guardian*, 02.06.2017, available at www.theguardian.com/global-development-professionals-network/2017/jun/02/china-water-dangerous-pollution-greenpeace (Retrieved, 23.08.2017).

Turner, C. (2015) What Is Aerotoxic Syndrome? *Telegraph*, 21.02.2015, available at www.telegraph.co.uk/news/aviation/11427400/What-is-aerotoxic-syndrome.html (Retrieved, 04.09.2017).

UNEP (2017) United Nations Environment Programme webpages, available at www.unep.org (Retrieved, 24.08.2017).

Walmart (2007) Wal-Mart To Sell Only Concentrated Products in Liquid Laundry Detergent Category by May 2008, News Archive, available at http://corporate.walmart.com/_news_/news-archive/2007/09/26/wal-mart-to-sell-only-concentrated-products-in-liquid-laundry-detergent-category-by-may-2008 (Retrieved, 30.08.2017).

Ward, V. (2017) B&Q Customers Compensated after Complaining That Paint Smells Like Cat Urine, *The Telegraph*, 20.07.2017, available at www.telegraph.co.uk/news/2017/07/20/bq-customers-compensated-complaining-paint-smells-like-cat-urine/ (Retrieved, 29.08.2017).

Weybrecht, G. (2014) *The Sustainable MBA: A Business Guide to Sustainability*, Chichester: Wiley, 2nd ed.

WHO (2016) Household Air Pollution and Health, World Health Organization, Fact sheet N°292, available at www.who.int/mediacentre/factsheets/fs292/en/ (Retrieved, 14.08.2017).

Witherington, P. (2011) Contaminated Land, in: Brady, J., Ebbage, A. and Lunn, R. (eds.) *Environmental Management in Organizations, The IEMA Handbook*, London: Earthscan/Routledge, 2nd ed., pp. 325–334.

World Bank (2016) The Cost of Air Pollution: Strengthening the Economic Case for Action, Report by World Bank and Institute for Health Metrics and Evaluation, available at https://openknowledge.worldbank.org/handle/10986/25013 (Retrieved, 22.08.2017).

Xie, J. (2017) High-Rise Forests in Italy Are Fighting Air Pollution, *The Verge*, 09.08.2017, available at www.theverge.com/2017/8/9/16112758/milan-vertical-forest-stefano-boeri-video (Retrieved, 31.08.2017).

Yang, H. (2016) China's Soil Plan Needs Strong Support, in: *Nature*, 536(7617): 375.

CHAPTER 7

Waste

INTRODUCTION

Waste is natural. As Cavagnaro and Curel (2012: 42) put it, "every form of activity, including human activity, generates waste" and therefore we all produce it as part of the circle of life. As such, it should not be a problem or challenge, but the rise in human (and in particular industrial) activity over the last few decades has seen the creation of such amounts of waste that the capacity of nature to treat and absorb it is exceeded. What was once considered normal practice for business – the disposal and discharge of effluents and waste to water, air or land – is no longer safe or acceptable, neither environmentally, nor socially or economically (Cavagnaro and Curiel 2012) and, like pollution, can no longer be assumed to be part of the industrialisation process as under the "take, make, waste" paradigm (Hart 2010) of cradle-to-grave, which Raworth (2017: 212) calls "take, make, use, lose".

Also underlying this challenge, as with climate change and pollution, is the use and depletion of resources. All economic activity uses up resources and materials, which have to end up somewhere, whether that is on landfill, in the atmosphere or disposed of in water (Pearce et al. 1999), or dealt with in other ways, as we will cover. In order for waste disposal to be sustainable, the waste generated from projects should not exceed the capacity of the environment to assimilate it (Costanza et al. 1997).

We reiterate here that there are overlaps between the various environmental challenges. For example, pollution prevention, as covered in Chapter 6 (or source reduction) prevents waste from being generated in the first place and is therefore different from waste reduction (Friedman 2011). Both pollution prevention and waste minimisation are needed to reduce companies' environmental impact, and often one particular focus of pollution prevention is on reducing waste (Hart 2010). Similarly, reducing the amount of waste that is sent to landfill also leads to fewer methane emissions emitted from landfill sites, thus addressing climate change as well. One example that demonstrates the overlap between the challenges is the water and soil contamination from the Love Canal landfill site in New York, which led to the permanent evacuation of over 900 families and to legislation regulating landfills and nearby development (Anderson 2014).

In this chapter, we cover the final environmental challenge of waste across multiple levels – what is it and why is it a challenge, in general and then more particularly for and by business; how it can be broken down as an issue in terms of waste streams and types, how it can be treated along the waste hierarchy (the 3Rs of reduce, reuse and recycle) and what kind of regulations address it. In the final part, internal and external corporate responses to waste are discussed, including *zero waste* and *waste as a business model* where the waste hierarchy is extended beyond the so-called 3Rs to the principle of re-thinking how waste is managed.

WASTE – WHAT IS IT AND WHY IS IT A CONCERN?

> Waste is defined by where something is, not what it is. It's about context, not content.
>
> (Leonard 2010: 235)

The Dictionary of Global Sustainability (Green 2013: 138) refers to waste as "unwanted materials left over from a manufacturing process" and "refuse from place of human or animal habitation", whilst a legal definition of waste comprises "any substance or object . . . which the producer or the person in possession of it discards or intends or is required to discard" (Woolston 2011: 316). The concept of "discarding" is key to defining waste as the classification of materials as waste provides the basis for regulations in how to handle or treat it (Woolston 2011).

Taiichi Ohno, the originator of the lean manufacturing system at Toyota (the Toyota Production System) who opposed every form of waste, defined it as "any human activity which absorbs resources but creates no value" or, in other words "all elements of production that only increase costs without adding value" (Hawken et al. 1999: 125; Ohno 1988: 54). Carpet company Interface similarly identifies waste as anything or any measurable cost that does not add value for their customers (Anderson 2009). Waste is thus a broader concept, extending it beyond the physical. Although we will restrict our discussion to physical waste, this conceptualisation of waste is useful for businesses. Waste thus constitutes "resources in the wrong place" (Leonard 2010: 235).

For Hawken (1993) waste is any by-product that harms rather than furthers life and which is therefore uneconomical – a sustainable economic system does not create waste, he argues. Similarly, Smith (2007) argues that waste is a major product line of business for which there is no market or use and for which business is potentially liable, and which business has ignored. In the 1990s it was estimated that 99% of original materials used in the production of, or contained within, the goods made in the US become waste within six weeks of sale (Weizsäcker et al. 1998).

In the following we cover the major sources and types of waste, what the concerns are in terms of the environmental impact and how waste can be treated along the waste management hierarchy. The movement of waste and the paths it follows from source to disposal are often referred to as waste stream. It includes the total flow of waste from homes, businesses, institutions and manufacturing plants to be recycled, burned or disposed of in landfills (Green 2013).

Sources of waste

Waste can be categorised in a variety of ways, which leads to some discrepancy in how it is measured and reported. It can be differentiated by its form (e.g. solid versus liquid), by its source, by industry sector or composition (Robertson 2017). In the EU, the major waste streams by source are construction (34%), mining and quarrying (27%), manufacturing (11%), households (9%), energy supply (3%) and other sources (16%) (EEA 2016a). Figures for the US vary, with some estimates placing municipal solid waste (MSW) at 3% in contrast to 97% waste from industrial, construction and demolition, and hazardous (or special) waste (Robertson 2017), though these figures need to be viewed with care as they include self-reporting and imprecise classifications of non-household solid waste in the US (Liboiron 2016). Nevertheless, the distributions indicate the significance of industry and business in waste generation. MSW in the US context is also called trash or garbage and comprises wastes such as durable goods (e.g. furniture, tyres), nondurable goods (such as plastic cups and newspapers), containers and packaging, and other wastes (e.g. food and garden waste) (CSS 2017). MSW as a category includes common household waste as well as office and retail wastes but excludes construction, industrial and hazardous wastes (ibid.). The OECD (2015) in its definition of municipal waste furthermore includes waste from institutions such as schools and government buildings. The term "municipal" here refers to the municipalities that collect the waste generated by households, small commercial activities and institutions. Household or municipal waste is probably the type of waste that comes to mind for most people when waste is mentioned. It is therefore useful to cover it briefly before turning our attention to other types of waste.

Across the OECD, the average municipal waste per person per year generated is 520 kg, which is higher than the 1990 average but lower than the 2000 average. The national average ranges from 300 kg per capita in Poland and Estonia as well as the Czech and Slovak Republic to 500 kg in Finland, Norway and the UK to over 700 kg in Switzerland, Denmark and the United States (OECD 2015). Due to the different definitions, data collection practices and time spans covered, the share of household waste in these municipal waste statistics varies. These so-called waste generation intensities are useful first approximations of the environmental pressures created by waste, though they need to be complemented by other figures. Within the context of this chapter, we need to keep in mind that whilst the management and treatment of municipal waste often constitutes a third of public sector financial efforts to prevent and control pollution, such waste only represents about 10% of the total waste generated (OECD 2015).

Types of waste

As stressed previously, there are different classifications and categorisations of waste, which may lead to overlaps, making comparability more difficult – e.g. hazardous waste from households in the form of batteries or electronic equipment could be counted as household waste, as so-called e-waste or hazardous waste. In the following we cover what we consider significant types of waste in the context of business and environmental sustainability.

Industrial waste

This broad category of waste makes up the majority of waste, possibly accounting for 90% or more of all waste generated (Robertson 2017). It is unwanted materials, leftover and discarded from industrial operations, which includes manufacturing, extraction, power generation and other large-scale commercial organisations. Industrial waste may be liquid, sludge, solid or hazardous (Green 2013).

Construction and demolition waste consist of material that is left over from erecting or taking down buildings and structures (e.g. roads) as well as from renovation and reconstruction. It includes soil, rubble (debris) and building materials, ranging from pieces of wood, chunks of concrete, glass, bricks, wires to metal hardware (EEA 2018; Robertson 2017). Such waste may be hazardous (e.g. in the form of asbestos-containing materials, lead paint or mercury-containing thermostats).

Mining waste results from extracting raw materials from the earth and includes materials that need to be removed to gain access to the resource to be mined (e.g. topsoil and waste rock) as well as tailings that remain after minerals have been extracted from the ore (European Commission 2018a). Some mining wastes are inert and are not likely to post significant environmental threats, whilst others may contain dangerous substances which necessitates careful management to avoid long-term environmental and socio-economic consequences.

E-waste

Electronic waste (e-waste) or waste of electrical and electronic equipment (WEEE) is one of the fastest growing waste streams. It contains a complex mix of over 1,000 different types of embedded materials, some of which are toxic, making this a particularly challenging waste to treat (Robertson 2017). In the EU, the share of both electronic and electrical waste is broadly similar. Electrical waste contains household appliances such as washing machines, vacuum cleaners, refrigerators and toasters, whilst electronic waste includes for example IT and audio equipment as well as consumer electronics and mobile phones. Hazardous materials contained in e-waste include heavy metals, such as mercury, lead and cadmium.

Hazardous waste

Hazardous or toxic wastes (sometimes also called special wastes) are those considered "capable of causing the greatest environmental damage" or "dangerous to human life if not properly handled" (Woolston 2011: 317). They possess at least one of these four characteristics: ignitability (flammable or explosive), corrosivity, reactivity or toxicity (Green 2013). Hazardous wastes are heavily regulated with regard to labelling, movement and treatment, and most jurisdictions publish lists of hazardous wastes based on their classification.

Nuclear waste

One particular form of hazardous or toxic waste is nuclear or radioactive waste, which is any material that contains or is contaminated by radio-nuclides and for which no further use is intended (EEA 2018). We can differentiate between high-level radioactive wastes generated in the fuel cells of nuclear reactors and low-level radioactive wastes, which contain lower quantities of radionuclide and are used in industry, the medical field, in research and in private or government laboratories.

The nuclear arms race during the Cold War has left a "deadly heritage" from weapons production and testing, resulting in nuclear waste that will require costly monitoring for millennia (Costanza et al. 1997: 239). Whilst for some proponents of sustainability and sustainable development, the intergenerational equity associated with it[1] goes as far as the following generation, radioactive waste represents a spill over of current activities for many generations to come (Pearce et al. 1999; Stead and Stead 2014). Once it is produced, it cannot be destroyed, and whichever storage option is taken, none are without risk. Similar issues arise for nuclear energy where so far no acceptable solution for the long-term storage of spent fuel has been found (Russell 2017), although some industry voices argue otherwise (WNA 2017).

Plastic and packaging waste

Plastic pollution and waste has garnered much media attention in the last couple of years. Whether it is the news that both bottled beer and water have been contaminated with microplastic fibres (see previous chapter) or the 2017 BBC series *Blue Planet II* showing footage of how marine life is affected by plastic waste or the plans of EU policymakers to make all plastics packaging recyclable by 2030, this is an issue that has risen to the top of the waste, as well as news, agenda. Businesses and consumers have become more aware of the issue as a result.

Plastic is one type of packaging; other materials include glass, paper and cardboard, metal and wood (Woolston 2011). Of these, plastic is the most problematic one, as there are many types of it, not all of which are recyclable – according to the European Commission (2018b), less than 30% of plastic waste is collected for recycling. Global production of plastic has increased twentyfold since the 1960s, reaching over 320 million tonnes in 2015. In the EU, packaging makes up 40% of plastic production but 60% of plastic waste (European Commission 2018b). Microplastics are one of the most widespread contaminants globally as they turn up in the depths of the oceans as well as in the stomachs of seabirds and whales, far from where they originate (Briggs 2019).

Packaging is often only used once before immediately being discarded (Robertson 2017) – such as for example single-use plastics in the form of small packaging, bags, disposable cups, lids and take-away containers, which in the EU alone 2,025 million are used every year, most of which are not recycled (Gallego-Schmid et al. 2019). The rise of e-commerce and online shopping has particularly contributed to increased packaging waste as products need to be protected during transport to the end consumer, who – rather than buying in-store, with limited packaging – now buys and receives purchases that are doubly and triply packaged. This phenomenon of over-packaging is however not limited to online purchases. Supermarkets have been under scrutiny for excess and unnecessary packaging of fruit and vegetables – in particular, social media coverage has put companies under pressure to address the issue.

Food waste

One reason why food retailers use packaging more than possibly necessary is to avoid food going to waste (e.g. as a result of being damaged in transit), thus posing a potentially difficult trade-off between two types of waste.

The multi-stakeholder partnership Food Loss & Waste Protocol (http://flwprotocol.org) estimates that one-third of all produced food is lost or wasted, intentionally or accidentally – throughout the supply chain, at harvest stage, in transit or storage, or thrown away by the consumer. The effect is economic and social as well as environmental in terms of economic losses (e.g. unsold produce) and food (in)security. The environmental impact goes beyond the issue of waste – the food lost and wasted represents about a quarter of all water used in agriculture and farm land, the size equivalent to China and about 8% of global GHG emissions (FLW Protocol 2016). The impact may be felt across the whole chain from farm to fork: farmers, manufacturers, distributors and retailers as well as consumers.

Two particular issues with regard to food waste concern the desired look and presentation of fresh fruit and vegetables, and the issue of expiry dates. The first often leads to less appealing produce thrown away, often already at harvest stage, whilst the second leads to food (processed or otherwise) being discarded prematurely. Although food may be perfectly edible, misunderstood or rigid date marking rules contribute to confused consumers overcautiously discarding food whilst retailers use conservative expiry dates to avoid legal action and loss of customers.

Other types of waste

The terms and definitions of waste often differ from jurisdiction to jurisdiction, making its management particularly challenging for companies operating internationally. Although the major forms, sources and types have been referred to earlier, some more terms are briefly introduced here.

Chemical waste may fall under the category of e-waste (e.g. batteries or electronic equipment), hazardous waste (e.g. solvent wastes) or medical waste. Medical waste is also known as red-bag waste. Although it is solid waste, it is different from MSW as it may contain pathogens (Robertson 2017). Related wastes are clinical, pharmaceutical, biomedical or biological wastes, some of which are biohazardous.

Biological waste is different from bio-waste, which is biodegradable (or organic) and thus capable of decomposing under the right conditions (anaerobic or aerobic, EEA 2018). It is generated from food or garden waste. Commercial sources of bio-waste include animal waste and manure, and other agricultural waste as well as sewage sludge (human waste).[2] The challenge here is when more waste is produced than can be disposed of safely as in the example of the Netherlands, where 80% of dairy farms produce more manure than they can use legally on their land, leading to groundwater contamination and ammonia emissions to air from illegal cow dung dumping (Levitt 2018).

Environmental impact of waste

As with pollution and climate change, the planet has a natural assimilative (or sink) capacity for recycling waste materials, one of the ecosystem services that natural capital provides (Costanza et al. 1997). Sinks are therefore the places where the output of our activities interact with natural systems. This might go beyond interaction and result in interference and deregulation of the natural system (Cavagnaro and Curiel 2012). It is these systems that provide natural capital for the materials cycle of the economic and productive system in the form of e.g. wood, metal, minerals, oil and gas. These materials are necessary to maintain industrial production and household consumption but are often returned to nature as waste at the end of their useful life. Unlike natural waste (which is not really waste, we could argue), such waste does not naturally recycle, however, and therefore accumulates (Hawken et al. 1999). It has no value to other species or organisms and may even be fatal to them if it is toxic (Hawken 1993).

The different types of waste all have environmental consequences, some more serious than others, and often depend on the treatment of waste (see later discussion). The consequences often come in the form of pollution. Waste that ends up in landfill needs to be transported there, which produces noise and air pollution, is odorous, and airborne dust and litter often result; it attracts vermin (as may food waste) and if not properly separated, might contain toxic substances. Leachate forms, as water that trickles through landfills collects contaminants, resulting in a liquid that can pollute soil and groundwater (Green 2013). Methane emitted from organic sources of waste discarded on landfill contributes to climate change. Waste that is incinerated emits dusts and toxics in the form of dioxins and heavy metals as well as acid gases, exacerbated if the burned waste contains hazardous materials (e.g. from chemicals, electronic equipment). Large quantities of plastic waste end up on land and at sea – it is estimated that 1.5 to 4% (5 to 13 million tonnes) of annual global plastics production end up in the oceans, accounting for 80% of marine debris, causing damage to marine life, as well as economic life (fisheries, tourism, shipping) with the health impact not fully understood (European Commission 2018b; Briggs 2019). Box 7.1 provides an overview of the so-called Great Pacific garbage patch, where plastic waste can be found at all sea levels.

BOX 7.1 GREAT PACIFIC GARBAGE PATCH

Millions of tons of waste end up in the ocean every year, most of it from land, as ocean dumping is prohibited in many countries. Rubbish ends up in drains and rivers before reaching the sea. In particular, plastic waste is not biodegradable and is therefore not dissolved in water, where it can stay for years. Currents transport it to the world's revolving water systems, called gyres, where it becomes trapped. The so-called Great Pacific garbage patch – also called Pacific trash vortex – is the most known one, located in the North Pacific Ocean between Japan and California. Despite what the name suggests, it is not a singular gyre but comprises both the Western and Eastern Pacific garbage patches, which are collectively known as the Great Pacific garbage patch. It has also been called the world's largest rubbish dump or landfill site. It is not visible by satellite as the constant movement of wind and currents break larger plastic pieces into smaller pieces, called microplastics. These present dangers to sea life and birds mistaking the pieces of plastic debris for food, and could, in turn, through the food chain also present a hazard for humans. As it is more a plastic smog in the ocean rather than an island of rubbish, estimates of its size vary, though the latest investigations put it at twice the size of France.

Clearing the ocean from marine debris is challenging since the patches are not stationary, microplastics are found at the surface as well as all the way to the ocean floor, their size is often tiny and disposal could be very costly. Nevertheless, initiatives such as Waste Free Oceans and Upcycle the Gyres are starting to address the issue, which involves using special nets to collect the plastic debris and recycling it into new products such as yarn for carpets (Interface is an example here), swimwear (e.g. Zoggs' Ecolast™ range), car bumpers or sunglasses. Surfers against Sewage is a UK-based campaigning and educational organisation addressing plastic pollution as the new sewage in UK seas. Ocean Cleanup is a young start-up company founded by Dutch student Boyan Slat in 2014. It uses U-shaped floating barriers attached to the ocean floor to act as artificial coastlines for collecting the waste.

- Activity: Find out if your nearest lake, sea or ocean has experienced plastic waste and pollution. What is being done locally to address it, by governments, companies and communities or by public-private partnerships and multi-stakeholder initiatives?

Sources: Yeung 2018; NOAA 2018; Robertson 2017; BBC Newsround 2017; Silverman 2007, Surfers against Sewage (www.sas.org.uk), The Ocean Cleanup (www.theoceancleanup.com/).

Treatment and management of waste

Treatment of waste can be either physical, chemical or biological as well as thermal. It changes the characteristics of waste in terms of volume and hazardousness, facilitates handling or enhances recovery (Robertson 2017; EEA 2018). Physical treatment immobilises waste (e.g. sealing or encapsulating it); chemical treatment changes the chemical makeup through reactions, whilst biological treatment uses bacteria or plant life to break down wastes or immobilise them (anaerobic digestion). Thermal treatment is in the form of combustion.

Waste hierarchy

The waste management hierarchy provides a framework for deciding on waste treatment and for minimising waste streams (Robertson 2017). In its simplest form, it is familiar to many – reduce, reuse, recycle – the so-called 3Rs. This is also known as the eco-efficiency triad (Cavagnaro and

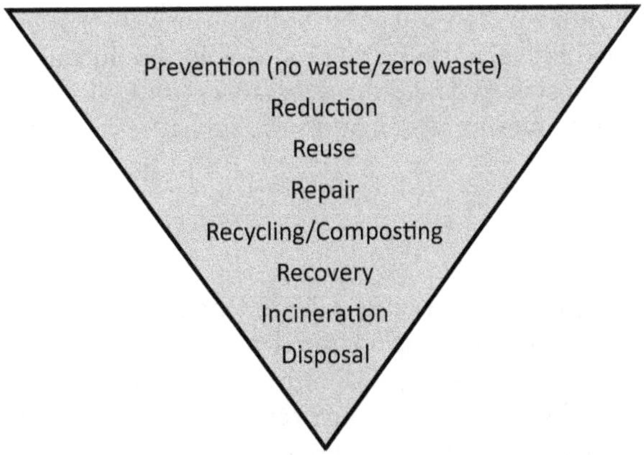

FIGURE 7.1
Waste hierarchy

Curiel 2012), to which we could add that waste minimisation is another term for process efficiency (Wolston 2011).

The waste hierarchy provides a descending order of priority for waste, with prevention or reduction being the highest priority, then reuse, recycling, treatment and then disposal in landfill as a last resort (Robertson 2017). There is a range of terms that are used in how waste can be treated or avoided and prevented in the first place. Repair, reuse, upgrading, remanufacturing and recycling are the main ways (Hawken et al. 1999). Depictions of the hierarchy as a reversed pyramid differ, though we can place composting alongside recycling. Both are sometimes also called recovery, which may also include energy recovery; this differs from incineration without recovery of heat or energy. Figure 7.1 provides a comprehensive depiction of the waste hierarchy as a reversed pyramid.

Whilst waste management has traditionally followed a cradle-to-grave approach, the aim of cradle-to-cradle perspectives is to break the linear model and move towards a cyclical or circular model that closes the loop (Cavagnaro and Curiel 2012). Life cycle analysis (LCA) is a vital input into this approach, which moves the 3Rs of reduce, reuse, recycle to the 4Rs of reduce, reuse, recycle, re-think towards zero waste (McDonough and Braungart 2002). For Esty and Winston (2009) the traditional priorities of the 3Rs are complemented by the new priorities of redesign and reimagine as innovation is key to sustainable competitive advantage in the 21st century.

In the following, we briefly cover the main steps in the waste management hierarchy.

Prevention and reduction

Avoiding the production of waste in the first place is the most effective environmental solution to waste. Prevention of waste entails taking action before materials, products or substances become waste, which can take place throughout the supply or value chain e.g. by avoiding producing faulty products, improving yields or reducing scrap, and focuses on prevention and reduction at the source (Wolston 2011). We can differentiate between quantitative and qualitative waste prevention where measures for quantitative prevention include decreasing the overall volume of inputs used for products and improving the efficiency with which materials are used (EEA

2016b). It also includes limiting unnecessary consumption as well as designing and using products that create less waste. Qualitative waste prevention involves reducing the hazardous content of waste in order to cut the exposure to dangerous substances (ibid.). Reduction of waste also entails reducing (resource) consumption on the demand side, such as e.g. offering unpackaged goods to consumers.

Recovery and Recycling

One definition of recycling is taking used goods and making them into new ones (Anderson 2014). Waste recovery is defined as using material not necessarily in its original form, which can be classed into re-use, direct recycling and indirect recycling, in descending order of priority (Buckingham and Turner 2008). Recovery and recycling involve collecting and separating materials from waste and processing them into marketable products, whether it is paper and cardboard, metal cans, glass bottles, compost, fuel or energy (heat or electricity). These new products may or may not resemble the materials from which they were recycled – e.g. recycled paper and glass versus fleeces made from recycled plastic bottles.

Whilst many materials can be recycled and thus reinjected into the economic system, which reduces demands for virgin materials, there are limits to it – Pearce and colleagues (1999) highlight the example of lead, which can be recycled from batteries but not from petrol exhausts (one reason why it was banned). Similarly, energy cannot be recycled either, as it dissipates once used (ibid.). Here, other means to reduce waste need to be found, such as product redesign, covered also in the next part of the book (Chapter 9), as well as conservation.

Recycling is often the first environmental issue consumers encounter and with which they are most familiar. At the same time, recycling can also result in the rebound effect encouraging even greater consumption as consumers may assume that products displaying a recycling symbol are without environmental cost and thus their purchase, use and disposal are environmentally responsible (Robertson 2017).

Although there are barriers to recycling (e.g. metals lost to corrosion), together with efficiency improvements, it can – in theory – compensate for consumption and consumption's growth (Lomborg 2001). Nevertheless, materials can never be recycled 100% as machinery and energy inputs are required and losses occur at each stage as e.g. the quality of material decreases (Robertson 2017). This often results in downcycling through reprocessing – for example paper cannot be recycled endlessly as the fibres break each time it is recycled.

Recycling offers a number of benefits beyond decreasing the demand for virgin materials. It reduces the amount of non-recyclable product waste, reduces the dependence on fossil fuel-based raw materials and energy and creates cost savings (Anderson 2009). In the case of metals, recycling also reduces the amount of toxic chemicals released in the process of mining the metals. Recycling aluminium cans for example saves about 95% of the energy used to make them from virgin materials. Plants treating wastewater are now so effective that they can produce output that is clean enough to drink, though it is often used for irrigation, cooling and groundwater recharge (Anderson 2014). The latest discovery in recycling is a mutant enzyme that can break down the PET plastic used in bottles. The enzyme has mutated from a bacterium that breaks down the protective wax-like coating on plants. Rather than taking hundreds of years to break down in the environment, the enzyme takes days. Although the development is a long way from widespread scale and deployment, it is expected to result in plastic being truly recycled and reused as the enzyme returns PET into its plastic building blocks, thus reducing the need to produce new plastic from petroleum (Halton 2018).

The boundary between some of the options for treatment within the waste hierarchy is blurry so that recovery may include energy recovery, also called waste to energy. This converts non-recyclable waste into usable heat, electricity or other fuel. Processes used to recover energy include combustion, gasification, anaerobic digestion and landfill gas recovery (EPA 2018).

Disposal – incineration and landfill

Disposal includes incineration without energy recovery and landfill and is the least attractive option for waste treatment or management. Landfills do not actually constitute waste treatment; they simply provide long-term storage.

Landfill sites are increasingly difficult to find, which has led to waste being shipped thousands of miles from industrialised to developing countries in search of unfilled sinks (Costanza et al. 1997). In many countries, not only the number of disposal sites is decreasing, but also the volume of waste is increasing, creating challenges of transport and cost (Anderson 2014). In addition, leaks from landfill can contaminate water, and if the waste is incinerated, the resulting ash might be toxic (Anderson 2014), thus creating an overlap with the challenge of pollution. It is almost impossible to prevent leaks from landfills, and it is expected that all sites do eventually leak (Robertson 2017).

Whilst incineration reduces the volume of MSW by up to 90% and its weight by 75%, it is estimated that over two-thirds of the world's dioxin emissions are produced by incinerators (Robertson 2017). These and other emissions (e.g. NOx, particulate matter) carry health risks. Beyond environmental and health impacts, incinerators often tend to be built in low-income communities which are generally less informed and organised than high-income communities which manage to campaign against and prevent the construction of new plants.

In the US, about half of all MSW goes to landfill, whilst about one-third is recycled and the remainder (ca. 13%) is combusted with energy recovery (EPA 2018). In the EU, broadly similar shares of municipal waste go to landfill (24%), incineration (27%), and recycling (29%), whilst 16% is composted (Eurostat 2018). Of the total waste treated in the EU, almost half goes to landfill, over a third to recycling, 1.5% is incinerated, about 5% goes to energy recovery and 10% is backfilled,[3] though the figures vary vastly between member states (Eurostat 2017).

BUSINESS IMPACT

Cause of waste – business as part of the problem

From the earlier discussion the role of business in waste is already evident, particularly considering that municipal (solid) waste from households, small businesses and public institutions makes up only about 10% of all waste generated. Regardless of what type or source of waste we look at, construction or demolition, retailing, mining or manufacturing, waste generated from companies in their operations is therefore at a vastly different scale compared to consumer waste.

According to Stead and Stead (2014), the so-called *throwaway society* we live in was a deliberate attempt to generate more employment and stimulate economic growth in the US after the Second World War by producing more disposable goods in an expectation that if more products are made and disposed of, more jobs will result to replace them, leading to higher economic growth. Yet, planned obsolescence goes back even further when in the 1920s and 1930s lightbulb manufacturers colluded to agree to limit the lifespan of lightbulbs from 2,500 or more hours to 1,000 hours as the documentary *The Light Bulb Conspiracy* details. Concerns over such built-in limits continue to this day, exemplified in the controversies over batteries in Apple's iPod in the 2000s and suspicions that the company was purposefully slowing down older iPhone models to force users to keep buying new phones in the 2010s. Upgrading as a result of (perceived) obsolescence is a way of life for most consumers, and businesses have played a significant role in this, whether deliberately or inadvertently.

Consumer-facing businesses are of course only one part of the problem – the bulk of waste generated by industry is often produced in the supply chain before it even reaches consumers. Waste is, however, not only a problem caused by business, it also represents a challenge for it.

Effect – waste as a problem for business

Environmental risks and operational challenges are part of the impact of waste on business; above all, waste equals cost. Waste management involves the treating, handling, transportation and disposal of waste, all of which represent significant costs to businesses, particularly if the waste is hazardous (Weybrecht 2014). Furthermore, waste represents an inefficient use of resources, in particular raw materials. Thus, the cost of materials and resources is added to the cost of disposal. Despite waste and its disposal being a cost, Elkington (2012) emphasises that it is the underpricing of resources that make it cheap (and easy) for companies to waste.

Unlike households who do not have to exercise a duty of care for their own household waste as this is dealt with by local authorities, businesses as holders of waste have the duty and responsibility to contain waste securely and transfer it only to authorised carriers or managers, whilst keeping appropriate records (Woolston 2011). Companies that fail to discharge this duty are held liable and are subject to financial or other penalties.

Beyond costs, other problems for companies linked to waste include a lack of waste management responsibilities, performance measurement (waste has to be measured to be managed), inaccurate forecasting, poor information sharing, inappropriate packaging, inappropriate quality management and insufficient awareness and training (Bernon et al. 2012).

REGULATORY RESPONSES TO WASTE: TARGETS, BANS AND EXTENDED PRODUCER RESPONSIBILITY

Without legislation regulating how waste is dealt with or treated, many companies view the proper disposal of waste as an unnecessary cost which is externalised and then picked up by society as a whole (Cavagnaro and Curiel 2012). Taxing goods to cover the costs of their disposal has long been an instrument that encourages goods to be made more recyclable or biodegradable (Daly and Cobb 1989). The practice removes incentives to continue producing waste. Other instruments regulators can use are bans, permits and setting waste reduction targets along the waste hierarchy (reduce, reuse, recycle and so on) e.g. for each type of waste treatment. Similar to the other challenges covered, the policy framework for business in the area of waste covers both regulations and voluntary efforts with phase-in periods possible between the two. As a reminder from Chapter 3, regulations can consist either of command and control legislation and standards, or of market-based policies using incentives.

In the case of waste, regulations are guided by the polluter pays principle, which internalises the costs of waste (e.g. through pay-as-you-throw programmes, Anderson 2014), by the prevention principle and by the proximity principle, which in the case of waste stipulates its treatment as close to its generation as possible, discouraging long-distance shipping of waste and encouraging local responsibility for local waste (Everard 2011).

In terms of the broad environmental policy options of mitigating, adapting, avoiding and restoring (EEA 2016c), the priority for waste regulation is avoidance and prevention. Measures taken include phasing out and banning certain types of waste treatment, such as landfill, setting increasingly stricter targets for recycling (e.g. from 40% to 50% to 75% of all waste or 65% of packaging waste), economic incentives to reduce and discourage landfilling, measures to promote re-use and industrial symbiosis, and incentivising manufacturers to put more sustainable products on the market and to support recovery and recycling (European Commission 2018a).

Here, one important element of waste legislation is the principle of extended producer responsibility (EPR) or product stewardship. The rise in (extended) producer responsibility regulation legally obliges manufacturers to recycle their goods at the end of their useful product life through so-called take-back programmes, examples of which include the EU's Waste of Electrical and Electronic Equipment (WEEE) directive and its end-of-life vehicles (ELV) directive. An early pioneer of extended producer responsibility was Germany, applying the concept to a range of products. In the German car industry, BMW piloted a disassembly plant to recycle its older cars (Hawken 1993).

The EPR concept allows consumers to return regulated products and appliances free of charge to the producers, who are then responsible for the recycling, reprocessing and disposal of the discarded products. It moves manufacturers from a cradle-to-grave approach to cradle-to-cradle design. They can no longer ignore what happens to products once they leave the factory gates. This forces them from saying "this product can be recycled" to "*we* recycle our products and packaging" (Benyus 1997: 256). For example, the EU's WEEE directive provided manufacturers with incentives not only to deal with the end-of-life waste of such equipment but to take waste management aspects fully into account by designing equipment in a more environmentally efficient and friendly way from the start (Visser 2012).

Such end-of-life legislation is often accompanied by regulation of input (Esty and Winston 2009). For example, the WEEE directive has a sister directive on restriction of hazardous substances (RoHS). This provides requirements beyond recycling and the end of life, and includes the ban of heavy metals (lead, mercury, cadmium and hexavalent chromium) in electrical and electronic equipment, and requires the substitution of brominated flame-retardants.

Although there are concerns that the so-called *throwaway society* is running out of space for landfill sites, Lomborg (2001) stresses that this is a political problem rather than a lack of physical space – communities oppose the presence of landfill sites or incinerators due to the environmental problems associated with it – noise, odour, and pollution.

Global and regional responses to waste

In the previous chapters we have seen that responses to climate change and pollution have a global dimension due to their global impact, which may cross borders. As the impact of waste tends to be more localised, global responses to waste focus on the transboundary movement of it, which may violate the proximity principle.

Examples of national and international movements of waste include the so-called garbage barges where potentially hazardous waste originating from the US travelled thousands of miles, often only to end up in the place where it departed as no takers could be found to deal with the rubbish (Anderson 2014). As waste disposal in richer countries became increasingly costly as a result of tightening regulations in the 1970s and 1980s, exporting hazardous waste to low-income countries seemed like a low-cost disposal option. The *Basel Convention on the Control of Transboundary Movements of Hazardous Wastes and Their Disposal*, signed in 1989, addressed this environmental injustice by banning all exports of such waste from rich industrialised countries to lower-income countries (Hackett 2011).

Although recyclable materials can legally be traded internationally to help provide raw materials input for manufacturing products in India and China, illegal waste exports are nevertheless a problem as companies attempt to circumvent environmental regulations that often imply higher costs of legal disposal (Hackett 2011). The transboundary movement of e-waste, for example, raises questions about whether its treatment in third countries is equivalent to regulatory requirements in the country or region of origin, such as for example the EU's WEEE directive.

In a surprising move in 2018, China banned the import of several types of solid waste such as paper and plastics, having been the main importer for more than half of global exports of plastic waste e.g. from the US and Australia and for 15–20% of the EU's paper and plastic waste (Hancock 2018; Tamma 2018). Environmental and health concerns led to the ban alongside dealing with the increase in China's own waste. The country aims to become a world leader in recycling, which would see it move alongside the EU, one of the leading environmental policymaking bodies worldwide, whose regulations are often mirrored in other parts of the world, including China and India.

Box 7.2 (including Figure 7.2) provides an overview of how the EU regulates waste regionally with implications for national and local responses to waste.

BOX 7.2 WASTE LEGISLATION IN THE EU

In the context of growing international economic integration and trade, regulatory standards for waste set in the EU can have significant implications for global production and commerce as companies, especially non-EU ones, operating in multiple markets prefer to manufacture their goods to as few different standards as possible. This often means following the highest standard, such as for example of the EU, which can be seen to take the role of global regulator, also in the area of environmental regulations (Selin and VanDeVeer 2006; Young 2015).

Over the years, since the first regulations dealing with waste were introduced in the 1970s, a whole framework of legislation dealing with all aspects of waste has built up. The following figure displays how waste legislation in the EU was structured in the mid-2000s. Since then, the legislation for waste treatment (or management) operations also includes port reception facilities and industrial emissions, whilst waste streams also comprise POPs (persistent organic pollutants), urban wastewater treatment and ship recycling. Furthermore, end-of-waste directives have been introduced to deal with waste which has ceased to be waste after it has undergone a recovery (including recycling) operation and can be used or sold lawfully for a specific purpose, which includes various metals and glass cullet (European Commission 2018a).

EU waste legislation sets targets for collection, re-use, recovery and recycling of various waste streams, limits incineration to non-recyclable waste, aims to phase out landfill of non-recoverable or non-recyclable waste, and provides a common terminology through its classification of hazardous wastes as a basis for shipment, treatment, permits and data collection for statistical purposes. The latest development on waste in the EU is the circular economy package which aims to close the loop through covering the whole cycle of production, consumption and waste management, with particular measures to transform the way plastic is produced, designed, used and recycled, as well as actions to promote the use of waste as a resource e.g. through encouraging markets for secondary raw materials (European Commission 2018a).

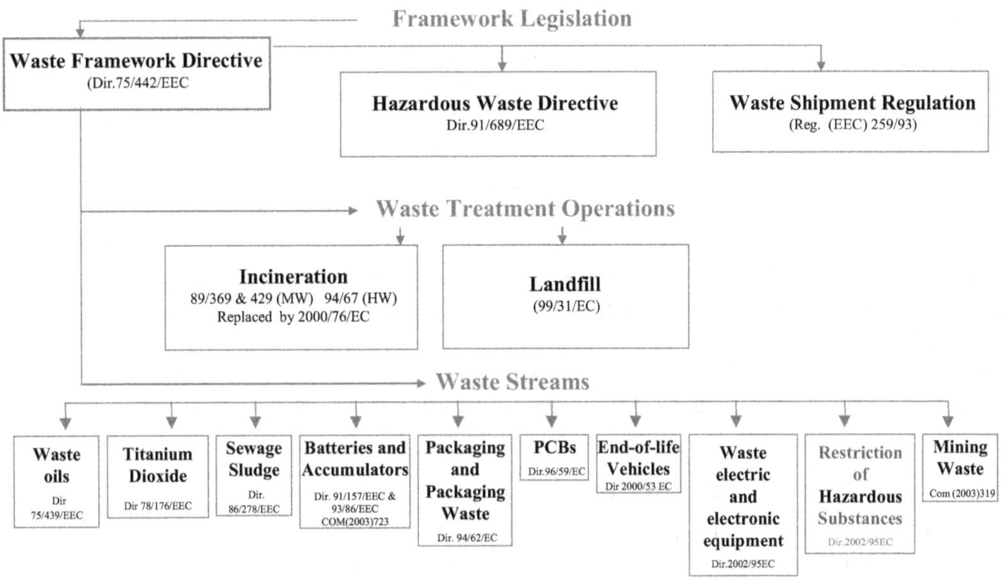

FIGURE 7.2
EU waste legislation

Source: Used with permission.

National responses to waste

National responses to waste are often coordinated through national waste management strategies and action plans, which provide a road map of targets and timelines for each type of waste and an overall framework. Waste management plans are key to national, regional and local policies for managing waste as they include not only objectives and strategies but also detail their implementation (European Commission 2018a). One policy instrument in waste management is in the form of environmental permits – legal documents that can be purchased by commercial organisations to administer their waste. Concrete examples of national responses to waste include the following:

Plastic bags – bans or charges: China banned the use of free ultra-thin plastic bags in 2008, and although compliance is inconsistent, plastic bag use has reduced by more than two-thirds. In 2015, the UK government made retailers charge 5p for single use plastics bags previously available for free, though the proceeds go to charity rather than to the management of waste – in the year before the charge, major retailers and supermarkets gave over 7 billion bags to customers, amounting to 140 bags per person on average. In the first six months of introducing the charge, plastic bag use dropped by 85%. In what is arguably the strictest regulatory change, Kenya banned the production, sale and use of plastic bags in 2017 with breaches risking imprisonment and hefty financial fines.[4] Benefits from such legislative responses include savings in clean-up costs, carbon savings, charitable donations and incentives for alternatives, though for some businesses higher costs are involved in using more expensive alternatives, such as e.g. biodegradable bags.

Deposit schemes: such schemes or so-called bottle bills require a deposit on (recyclable) cans and bottles which consumers pay on purchase but receive back on return of the containers (Anderson 2014). Although such schemes mostly apply to recyclable drinks containers, Germany has applied it to single-use containers since 2003, after the share of reusable bottles fell below the legally required rate of 72%, triggering the introduction of deposits for single-use containers, such as PET bottles. While the share of cans has fallen and littering is no longer such a big issue with 95–98% of cans and bottles being recycled, the scheme has controversially not led to an increase in reusable containers – with the notable exception of glass beer bottles. With the help of reverse vending machines, Norway's deposit-based system achieves a recycling rate of 97%. The planned deposit scheme in the UK also extends to single-use containers for drinks – the country's recycling rate for plastic bottles stands at under 60%. Setting up deposit schemes involves costs for producers and retailers, though the example of Lithuania, which is said to have achieved a return rate of 93% (up from 35%) within three years of launching its scheme, points to the success of such schemes.

Food waste: in 2016, France banned supermarkets of a certain size from throwing away or spoiling unsold food, which they are now required to donate to charities and foodbanks.

Regional and local responses to waste

In line with the proximity principle, waste collection is often local. Local authorities can provide waste minimisation incentives to citizens and businesses in various ways e.g. through the cost of collection, frequency of collections, bins with locks, and educational material, and by making recycling easy.[5] Some local authorities in Germany (e.g. the district of Böblingen in Southern Germany) provide a limited number of free collections as part of the basic fee or council tax, beyond which households have to pay per additional collection. This is administered through a chip on waste bins, and locks are available for purchase to prevent abuse. Other districts in Germany charge per vehicle load or vehicle weight for bigger household items destined for disposal. So-called pay-as-you-throw programmes where people pay per bag or container of waste collected and sent to landfill provide incentives for responsible consumption and recycling, as opposed to fixed-rate fees that might promote excessive waste generation (Anderson 2014). California charges a fee on the retail sale of video display devices which is used to fund the recycling of the goods, often containing hazardous materials (Hackett 2011).

In the Brazilian town of Curitiba, the sprawling slums of the favelas with their narrow or non-existing streets posed problems for rubbish collections. The solution was found in a cyclical, waste-equals-food system where people were paid with public transport vouchers or food tokens for placing and separating their waste in recycling bins around the favelas (Hawken 1993). The result has been a high recycling rate and less waste dumped in nature whilst also saving on the infrastructure of waste collection (Robertson 2017).

Similar practices are found in Cape Town, South Africa, where aluminium cans are collected and returned mostly by poor people for a small fee leading to very high recycling and recovery rates. The Collect-A-Can scheme established in 1993 has created a culture of recycling with environmental, economic and social benefits (Visser 2012).

Various towns and cities in the US, most notably New York City, have banned (or have tried to ban) single-use containers made from expanded polystyrene (EPS), better known as Styrofoam, used for coffee cups and food containers, as it does not biodegrade and is difficult to recycle. In the case of New York City, businesses such as restaurants and plastics manufacturers challenged the ban, arguing that the material is recyclable and should therefore be included in kerbside recycling programmes. The concern for businesses here is the cost of alternatives in an industry that is marked by low margins, where the use of Styrofoam offers cost-effective yet wasteful takeout solutions (Alexander 2017). This illustrates the challenge business is confronted with when it comes to waste – regulatory responses versus consumer convenience, wastefulness versus cost savings.

BUSINESS OPPORTUNITIES – BUSINESS AS PART OF THE SOLUTION TO WASTE

Taking into account the environmental impacts and costs of waste, regulatory frameworks and convenience for industry and consumers, how can businesses address the challenge of waste, both internally and externally? In this final section of the chapter, we explore the opportunities for businesses to address waste, bearing in mind that we cover a selection rather than an exhaustive list of possibilities.

It is no longer enough for companies to assume that their responsibility for their waste ends when a waste contractor removes it from its site. Waste presents firms with opportunities to lower costs and minimise risks by developing competences in pollution prevention and eco-efficiency. Less waste means that inputs are better utilised, which in turn leads to lower costs for raw materials and for the disposal of waste. Waste often means paying twice for products – purchasing them and then paying for their disposal, making waste reduction a significant business imperative (Hart 2010; Weybrecht 2014).

Recycling, for example, not only addresses waste but also helps companies address other environmental issues such as climate change and pollution through reduced energy use, reduced use of virgin materials, reduced land use as well as reduced toxic runoff from e.g. mining and landfill (Esty and Winston 2009).

Internal opportunities – examples of how to minimise impact within business

The importance of waste for business is highlighted by Ray Anderson (2009: 44) who calls waste the "engine that will pull the whole train" in the drive to achieve real sustainability in contrast to conventional perceptions of waste being at the tail end of things. Addressing waste is therefore a natural starting place for any business, particularly as doing so makes not only environmental sense but also economic sense, as demonstrated by the definition Anderson (2009: 46) and his company Interface used:

> Waste is any measurable cost that goes into our product that doesn't add value for our customer.

It is with a definition of waste that Anderson recommends companies start when tackling this challenge, followed by accurate measurements of the waste produced. Only from such a baseline can goals be established and progress be charted and communicated. Anderson (2009) stresses the need to measure waste relative to output and indexing waste costs per unit to a historical baseline, displaying the results transparently.

Measurement and subsequent management as an approach to address waste is similar to climate change (see Chapter 5). Identifying and measuring waste across its operations and across streams enable a company to calculate how much waste it generates and the associated costs. This enables it then to prioritise where waste can be prevented in the first place, how it can be reduced, what can be recycled and reused, and how disposal can be improved and monitored – steps along the waste hierarchy.

Von Weizsäcker and colleagues (1998) also underline the business case of waste reduction efforts for companies, with typical payback periods of 18 months. One example is Dow Chemical company, where both local and company-wide initiatives have led to profitable waste reductions.

In its Louisiana Division, engineer Ken Nelson initiated internal contests among staff to suggest projects that would save energy or reduce waste with a payback period of a year for costs under $200,000. Over the years the projects went on to generate triple-digit returns of investment and millions of cost savings for the company, whose achievements in this area, alongside other firms such as 3M, Xerox and DuPont, have become almost legendary (ibid.). The benefits of reducing waste from current operations represent significant opportunities to lower cost and risk as well as to develop skills and capabilities – Dow Chemical's programme is aptly named Waste Reduction Always Pays (WRAP) (Hart 2010).

Around the turn of the millennium, Xerox was on track to achieve more than 90% waste reduction in its operations (von Weizsäcker et al. 1998) and was close to saving $500 million per year through its Asset Recycle Management programme which set the target of "producing waste-free products from waste-free factories" (Hart 2010: 28). Since 1991 the company has diverted more than 2 billion pounds (over 900,000 metric tons) of waste from landfill – 41,500 metric tons in 2016 alone – and its goal for 2020 of 100% remanufacture, reuse, recycle, energy from waste for equipment and parts, stood at over 99% in 2017 (Xerox 2018). Its aim for waste-free workplaces goes back to the 1980s, introducing the concept of the circular economy before the term was coined.

For 3M, everything leaving a plant is either product, by-product (to be reused or sold) or waste. Asking why there should be any waste at all as it is a sign of inefficiency, 3M's management's answer was its 3P programme – pollution prevention pays, illustrating the closely related nature of pollution and waste (Esty and Winston 2009, see also Chapter 6). Since its inception over 40 years ago, the programme has saved the company over $2.1 billion in financial terms, and over 2 million metric tons of pollution, mostly in the form of waste (the other forms being air and water) (3M 2018). The company is re-imagining waste as a resource, resulting for example in products such as a standing desk which contains 45% postconsumer recycled content, thus reducing the amount of plastic waste that would otherwise end up in landfill. Some of the tools and principles 3M uses in its drive towards sustainability include adopting a product stewardship approach, conducting life cycle assessments and using so-called Lean Six Sigma (see also Chapter 9).

HP (Hewlett Packard) set up its own compliance scheme with waste regulation, designing products with recycling in mind (also called *reverse logistics*[6]), enabling it to retrieve its products at the end of life at relatively low cost (Bernon et al. 2012). Such a product stewardship programme recognises that much of the problem of waste can be dealt with at the design stage by *designing out* waste (Benn et al. 2014).

Interface, under the leadership of Ray Anderson, reduced its total waste in its global operations by 40% during the first years of its efforts, saving the company $67 million in the process (Anderson 1998). When it also added fossil fuels to its definition of waste that should be eliminated, this was revolutionary at the time, but enabled the company to transform its operations towards zero waste (see later discussion).

Developments that have led to waste reduction in the electronics industry include display technologies for computer monitors and televisions moving from cathode ray tubes (CRT) to flat screens (e.g. LED or LCD), reducing the amount and weight of (hazardous) materials needed for manufacturing.

Other advancements in manufacturing more generally, partly in response to take-back programmes, include design for disassembly, moving from a linear production system to a closed-loop system where little or no waste escapes. Instead of glued-together joints, assembly of household appliances or motor vehicles for example may use easy-open snaps, and fewer materials are used for better recyclability (Benyus 1997). For instance, Nike's *Considered Design* integrates sustainability into its product design by making shoes recyclable into new ones at the end of their useful lives (Blowfield 2013).

Looking at other industries, PepsiCo reduced its use of plastic by 75 million pounds annually by redesigning its water bottles (Anderson 2014), whilst Waitrose, a supermarket chain in the UK announced in 2018 that its programme of offering free teas and coffees to loyalty card holders would cease to come in take-away cups and would only be available to card holders bringing their own reusable cups. This has a dual impact in terms of cost savings for the company (procurement of cups and lids), but also in terms of nudging consumers away from disposable cups – depending on the uptake, this could also mean a reduced demand for coffee and therefore purchasing needs. Take-away coffee cups are difficult to recycle as they are made of a mix of paper and plastic – in the UK only 1 in 400 such cups are recycled, prompting the country's largest coffee chain, Costa Coffee, to launch an in-store cup recycling scheme with waste partner Veolia (Smithers 2016). Addressing the difficulties of recycling such containers, vegware™ manufactures compostable foodservice packaging made from plants, whilst Notpla offers edible and biodegradable food and drinks containers.

Both the phasing out of disposable cups from loyalty schemes and the in-store recycling provide an internal opportunity to address waste and an external opportunity for companies producing and selling reusable or compostable coffee cups as well as cost savings for consumers – many coffee chains offer discounts to customers bringing their own cups. The distinction made between external and internal waste solutions to guide the reader is therefore arguably arbitrary.

In order to solve their packaging waste problems, some online retailers have tried delivering without additional packaging, particularly for items already packaged, or with reusable or returnable packaging – one example is LimeLoop which closes the loop by allowing customers to return packaging in the post. The start-up company achieves this by replacing traditional cardboard shipping methods with sturdy reusable shippers, which are modular and expandable (for an illustration see the company's website: www.thelimeloop.com).

Waste as a resource

Beyond reducing the overall amount of waste, waste is increasingly seen as a resource that can be reused and recycled, as demonstrated by some of the earlier examples. McDonough and Braungart (2002) refer to this as "waste equals food" – akin to natural food chains where there is no waste, as organisms at one level provide nutrition for organisms at another level (Anderson 1998). They propose to replace the downcycling often found in recycling with upcycling, i.e. converting waste products into higher value products rather than degrading them (Nemetz 2013).

The use of waste as a resource is central to so-called industrial ecology – industrial ecology transforms linear production processes along the take-make-waste model into ones that mimic the circular processes found in natural ecosystems where hardly anything that is produced or wasted by one organism is not of use for another (Hackett 2011). In mimicking natural processes, industrial ecology is akin to biomimicry briefly introduced in Chapter 4 and further explored in Chapter 9. The term *industrial ecology* was first introduced by Frosch and Gallopoulos (1989) as a strategy for manufacturing that would go beyond reducing waste and would use waste so that it would no longer be waste at all (Hawken 1993). By converting waste into a resource, companies can sell it profitably as raw material whilst preventing it from entering the waste stream, thus contributing to both the financial and the environmental bottom line. Industrial ecology is also known as industrial symbiosis or by-product synergies (Weybrecht 2014), whilst some industries use the term residue valorisation (e.g. metals).

So-called eco-industrial parks use ecological design to encourage partnerships and collaborations between companies in how they manage environmental and energy issues. Examples can be found across the world, though the most elaborate prototype of an ecopark can be found in the Danish town of Kalundborg, described by various authors (e.g. Benyus 1997; Robertson 2017) and displayed on a dedicated webpage (www.symbiosis.dk/en/), detailing the over 20 streams of energy, water and materials connecting organisations as diverse as oil refineries, fish farms, power plants and chemical companies. For example, the power plant recycles its waste by passing on steam to the refinery and the pharmaceutical company, and heat to a fish farm while fly ash is sent to a cement company. In turn, the oil refinery sends its waste cooling water to the power plant.

More generally, waste can generate energy, which can take various forms – for example through incineration (energy recovery) or through harvesting methane or biogas, as piloted in the Netherlands with cow manure. In agriculture, most organic wastes can be recovered or converted e.g. turning inedible vegetable oils into liquid fuels (Hawken et al. 1999). In the automotive industry, end-of-life tyres, which are estimated to number 1 billion per year, can be reincarnated – instead of being burned – into a specialty material called micronised rubber powder (MRP). This is the result of cooling tyres with liquid nitrogen to such a temperature that they can be pulverised into a fine composite. MRP can be used for tyre production and for other applications. It reduces energy demand and the need for virgin material in the industry, moving it towards closed-loop production cycles (Clancy 2018).

It is not necessary for industries to be geographically located close to each other to benefit from industrial symbiosis (though this helps in reducing the transport of waste and associated impacts); it is important that they are connected by information and by a joint desire to make use of waste (Benyus 1997). For example, Resyntex is a project for the textile and chemicals industry with 19 partners including businesses, industry associations and researchers from ten countries in the EU using industrial symbiosis in a bid to create a circular economy making secondary raw materials from unwearable textile waste (see www.resyntex.eu/ for more details). We will return to the concept of industrial ecology and circular economy in Chapters 8 and 9 on supply chains and operations.

Waste as a resource also links to waste as a business model – what companies can offer to consumers and buyers – which we cover further later. The distinction between internal and external opportunities can be blurry – as we have seen, waste as a resource can be used internally as well as offered externally, and effectively addressing waste internally often involves close collaborations with suppliers in terms of selecting materials that minimise waste or facilitate reuse and recycling (Woolston 2011). Amazon for example works both on its own, with suppliers, with charities (e.g. Give Back Box) and with broader initiatives, such as The Recycling Partnership, to reduce packaging waste and increase recycling – all of which the company lists in its Sustainability Question Bank online.

Zero waste

Zero waste is a system of production that aims to eliminate the volume and toxicity of waste and materials through conservation or recovery of all resources (Green 2013). As a concept, zero waste takes a wider look at the context of manufacturing, using and discarding products, and involves for example intelligent product design and phasing out of materials which are not recyclable or easily disposable. Similar to industrial ecology, it aims for a cradle-to-cradle approach in closing the loop. The following two chapters covering supply chain management as well as production and operations will shed further light on some of these aspects of zero waste.

According to the Zero Waste International Alliance (http://zwia.org), zero waste is a visionary goal to guide people and companies in changing lifestyles and practices to emulate sustainable natural cycles. It requires a shift in thinking for producers and consumers, and changes in policies and laws for its wider implementation (Robertson 2017). This shift is embodied in the fourth R of the waste hierarchy – re-thinking beyond the traditional priorities of recycling, reusing, reducing, which requires new priorities, reimagining and redesigning (Esty and Winston 2009). Zero waste begins with design and moves the question from "how can we recycle more?" to "how can we stop producing so much waste in the first place?" (Robertson 2017). Although zero waste firmly puts producer responsibility on the agenda, consumers are affected, too – companies here have a role in changing or nudging consumer behaviour, as for example illustrated in the case of UK supermarket Waitrose phasing out single-use cups for its free coffee offer to loyalty card holders.

Zero waste does not mean literally zero waste – what is often meant by it is that no waste is going to landfill, a critical difference. Despite differences in defining zero waste, aiming for it can focus companies in their efforts to reduce their waste and drive continuous improvement (Elkington 2012). This is also acknowledged by the Zero Waste International Alliance, which stipulates the achievement of 90% of waste diverted from landfill or incineration as a minimum requirement for a successful zero waste business.

Interface is one example of a company targeting zero waste, aiming for a dematerialised business (Cavagnaro and Curiel 2012). The shift in thinking required for zero waste – the first face of mount sustainability (see Box 4.6) – is detailed in the company founder and chairman Ray Anderson's books *Mid-Course Correction* and *Confessions of a Radical Industrialist* (Anderson 1998, 2009). For zero waste to be achievable, Interface measures waste on a macro basis but manages it on the micro level, making sure that while every part of the business is consistently and fairly measured, each facility is only compared with itself, with results widely available (Anderson 2009). To date, Interface has reduced the amount of waste sent to landfill by 91% since starting its *Mission Zero* journey (Interface 2018).

Toyota is another example of a corporation aiming for zero waste. Its British subsidiary Toyota Manufacturing UK was the first car plant in the country to achieve zero waste to landfill generated per vehicle (8.5 kg in 2001, zero in 2005) and the first UK car manufacturer to achieve zero waste to incineration in 2008. The carmaker strives for closed-loop car-to-car recycling where resources from end-of-life vehicles are used to produce new cars. This challenge of creating recycling-based systems sits alongside the company's 2050 strategy of zero CO_2 emissions for new vehicles, the life cycle of cars and for plants as part of Toyota's long-term goal of achieving zero environmental impact (Toyota 2018).

Beyond manufacturing, zero waste is also increasingly attempted in hospitality and catering, early examples of which include experimental Sandwich Me In in Chicago (Bellware 2014), which ran from 2012 to 2015, and the Silo Café in Melbourne, Australia (Snedden 2014), which inspired one of the world's leading restaurants, Noma, in Copenhagen, to install a closed-loop compost system (Closed Loop 2018).

External opportunities – what can be offered to consumers and other customers

Many products have been designed with planned or built-in obsolescence, which means they only last for a period of time before requiring replacement (McDonough and Braungart 2002). Particularly electronic goods, such as for example mobile phones, are known to be built with planned obsolescence. Moving away from this is an important step in addressing the wider issue of waste, not just internally. Similar to industrial ecology introduced earlier, collaboration between public and private organisations provides opportunities to address waste beyond company boundaries.

Landfill requires space and is a major source of emissions from the greenhouse gas methane. In a pioneering project, one of Interface's carpet tile factories teamed up with its host town LaGrange, Georgia, to invest in a unique win-win initiative. Methane is captured from the municipal landfill site and pumped to Interface where it is burned for process heat in the manufacturing process. Any excess gas is sold by the town for a financial return. Benefits from the project include the elimination of landfill odours and of methane emissions to the atmosphere, revenues for the town as well as an extended life span for the landfill site as a result of the extracted gas, reduced fossil fuel use and costs for the company as well as climate neutrality (Anderson 2009).

Another example of a public-private partnership is RecycleBank in the US, which partners up with cities and brands to provide recycling services that reward residents for their waste reduction and recycling efforts, thus helping to make communities into greener and cleaner places (Anderson 2009, see also www.recyclebank.com/).

In the UK, the Waste and Resources Action Programme (WRAP) was set up in 2000 to promote sustainable waste management, working together with international and national governments, businesses, industry associations and charities. In an effort to reduce waste in the food chain, WRAP created the *Love Food Hate Waste* campaign, working together with supermarkets, charities and restaurants as well as consumers to address the costs associated with food waste and packaging waste. On the consumer side, the campaign works with retailers to inform and educate consumers regarding "best before" and "use by" dates as well as food labelling more generally, and to provide advice on how to avoid food waste and what to do with leftover foods. Collaborations with charities (e.g. FareShare) and in some cases restaurants help retailers to address the wastage from unsold surplus food. Supermarkets used to reject so-called wonky vegetables and fruit which did not meet their shape, size and look parameters, leading to unnecessary food waste on the retail side. Companies such as for example UK retailers Asda and Morrison's are now widely offering "wonky veg" boxes at discounted prices (compared to conventional fruit and vegetables) following successful smaller-scale trials. In addition to established supermarkets, smaller and newer companies such as The Wonky Food Company, Waste Not and Rubies in the Rubble use imperfect and surplus fruit and vegetables to make for example juices, sauces and chutneys, whilst Oddbox provides a box subscription service for produce that would have otherwise been discarded, dug back into soil or used as animal feed. In Denmark, WeFood is the first supermarket chain focusing exclusively on surplus food that has either expired or not met conventional quality standards. Although its prices are considerably lower than normal supermarkets (30 to 50%) and attractive for shoppers on a low income, it also attracts consumers concerned about food waste. A similar concept was launched by the former president of Trader Joe's with Daily Table, which sells food other stores plan to throw out, though unlike WeFood, it does not sell food past its expiry date.

In 2018, some of the biggest supermarkets, food manufacturers and processors, including Lidl, Unilever, Nestle and PepsiCo, joined forces in the UK Plastics Pact to launch an industry-wide promise to phase out unnecessary single-use plastic by 2025 and to make all remaining plastic packaging either reusable, recyclable or compostable. Earlier that year, a supermarket in the Netherlands unveiled the first plastic-free supermarket aisle, while in early 2019, Loop teamed up with brands and supermarkets to offer household products in reusable packaging on a circular shopping platform, initially in the US but with plans to expand internationally.[7]

In the textile industry, WRAP launched the *Love Your Clothes* campaign in 2014 with industry organisations to address the amount of used clothing that ends up in landfill. The programme covers all consumer aspects of clothing, from the buying of garments, care and repair, refashion and upcycling to what to do with unwanted clothes (selling, swapping or sharing, giving to charity and recycling rather than letting it go to landfill). As part of the wider Sustainable Clothing Action Plan (SCAP), brands, clothing retailers, suppliers, local authorities and recyclers, trade and public sector bodies as well as charities and academics work together to address the vast environmental impacts associated with textiles – carbon, water and waste.[8] As emphasised in the introduction to part two of the book, these are not isolated challenges but linked together.

Textiles can also be used to address waste in the area of personal hygiene where for example wet wipes have contributed to over 90% of sewer blockages and so-called fatbergs in the case of the UK (Watts and Smithers 2018). Products here include nappies and facial wipes that are washable and thus reusable, whilst female hygiene is addressed with innovations such as absorbent underwear, reusable tampon applicators and menstrual cups for a so-called plastic-free period. Established companies have been somewhat slow in the uptake, prioritising the more profitable disposable products, thus making room for charities and NGOs engaging in social enterprise and so-called period activism against period poverty, and for start-ups financed by crowdfunding campaigns (Radnor 2017; Cooper 2018). Soap and shampoo bars provide an alternative to personal cleansing products sold and packaged in liquid form in plastic bottles or containers, whilst natural detergents such as soapnut fruits are a compostable and waste-free substitute for conventional laundry detergents, which are also mostly packaged in plastic.

Other developments that address waste in the area of textiles and garments include compostable clothes that are made from 100% natural fibres as opposed to the blend of synthetic and natural ones most textiles are made of. In addition to producing compostable as well as recycled and recyclable gear, Houdini Sportswear from Sweden also lets customers rent outdoor wear and offers their products for second-hand purchasing in store in an effort to extend the useful life of their products. The latter has been attempted by British designer Tom Cridland, who followed up his 30-Year sweatshirt, launched with the help of crowdfunding, with the 30-Year T-shirt – both products come with a 30-year guarantee. Online retailer BuyMeOnce addresses the planned obsolescence of many consumer products by only offering products that are made to last, based on a number of criteria covering raw materials, manufacturing, design and after-sales service as well as customer feedback. Addressing both social and environmental aspects of sustainability, The Shoe That Grows provides footwear that expands five sizes and can be used for years – making sure children in poverty have protected feet while at the same time reducing the waste from poorly made shoes that are quickly outgrown (BBC 2017).

In lean manufacturing (see also Chapter 9), the elimination of all forms of waste is emphasised. Combined with cradle-to-cradle production, which moves the linear take-make-waste production system to a circular or closed-loop one, the principle of service and flow or servicisation (also servitization or product-service system) emphasises the provision of a service instead of selling a product (Hawken et al. 1999). This is one of the central principles of natural capitalism introduced in Chapter 4. The elevator company Schindler for example provides the services of a "vertical transportation system" instead of selling elevators as a product, maintaining responsibility for service and repairs (Weybrecht 2014). Contrasting this industrial example of servicisation is the case of baby clothes which Danish company VIGGA is offering as part of a monthly subscription service on a lease basis, which allows consumers to return outgrown clothes and receive garments in the next size of their offspring.

Waste as a business model

Materials that in the past were only considered waste have found new uses or markets (Friedman 2011). Treating waste as a business model is thus one particular route to address waste – not just considering it a resource to be used internally, but also one to sell externally at a profit. There are similarities to industrial ecology, though industrial symbiosis tends to take place between companies rather than between companies and consumers. Where waste is a business model, companies create new products from the waste of old products.

In the following we provide examples of companies that have made waste their business model, offering consumers and other businesses products ranging from food to textiles, accessories and beyond. As more and more enterprises can be found in this industry, it is important to keep in mind that the list is a growing one and we only cover a limited selection.

TerraCycle's motto is "Eliminate the idea of waste", which the company does by collecting, reusing, recycling and upcycling non-recyclable or difficult-to-recycle waste that is not normally recycled by local authorities. The waste collected by the company is wide-ranging and includes arts and craft materials, cigarette waste, sporting goods and used gum, to mention a few examples. One product TerraCycle provides are so-called zero waste boxes, which provide solutions for those wastes not included in municipal recycling or the company's free recycling services. The buyer chooses which box to purchase (e.g. office supplies, contact lenses, flip-flops and so on, see www.terracycle.com for more) and once received, fills the box and returns it for TerraCycle to recycle the collected material. Products made by the company from recycled materials include bags, phone cases and watering cans. It also offers industrial recycling services.

So-called urban mining focuses on reclaiming and recycling valuable industrial materials, such as metals from discarded electronics, buildings and other spent products, and is key for a resource-efficient circular economy. Such mining has advantages over primary mining in terms of reducing the need for long-range transport as materials are already where they are needed for commercial and industrial production, whilst also addressing the overexploitation of natural resources and thus minimising the environmental impact (Jäger 2018). SungEel HiTech is one of 150 urban mining SMEs in South Korea and the country's largest battery recycler. It uses urban mining to recover scarce and precious metals from electronic waste, which it then sells to battery makers such as Samsung and LG (Chung and Park 2018). Umicore is a Belgian urban mining company.

The previously mentioned fatbergs from solidified fat, oil and grease (FOG) block urban sewers when combined with flushable material such as toilet roll but particularly non-flushable items such as wet wipes, nappies and even tights amongst other things. Whilst local authorities and water companies in cities such as New York or London run awareness campaigns to change the way consumers and businesses treat waste, as the clearing up is costly for them, other utility companies have started to turn fatbergs into fuel by reprocessing them and turning them into biofuels, for example used for vehicles in the form of biodiesel or to generate renewable power and heat. 2OC's plant in East London is the world's first industrial-scale power plant to run on waste, including FOG, whilst Argent Energy powers buses in places such as Manchester, London and Glasgow with biofuels made from fatbergs from those cities (Coates 2017). In Scotland, Celtic Renewables turns the by-product of whisky production – nitrogen-rich pot ale – into the biofuel biobutanol.

Gumdrop provides a similar service to TerraCycle's zero waste boxes, focusing on recycling used chewing gum. It offers Gumdrop memberships where subscribers (e.g. companies) receive bins designed for disposing of gum, which are collected once full. The bins themselves are made from recycled gum, a material called Gum-tec®, which is a sustainable compound that can be used by manufacturers in the rubber and plastics industry. Gumdrop itself manufactures products as diverse as lunch boxes, frisbees, sports cones and pencils from Gum-tec® (see http://gumdropltd.com for more examples).

Rubber from used firehoses, tyres and bouncy castles are upcycled to create hand bags, wallets, key rings and other accessories such as belts and jewellery. Companies in this sphere include for example Oxgut Hose Co., Wyatt and Jack, Paguro, and Elvis & Kresse. The latter also reclaims raw materials such as parachute silk, leather, shoe boxes as well as tea and coffee sacks.

Textiles provide a similar area where waste can profitably be recycled, reused or upcycled. Salvage uses 60% pre-consumer recycled organically grown cotton and 40% post-consumer recycled polyester to provide 100% recycled products that are certified under the Global Recycle Standard and Organic Content Standard. Worn Again started off in upcycling – e.g. reworking Virgin Atlantic's old airline seat covers into a range of bags and accessories – before moving into circular recycling technology, working together with big companies and brands such as H&M and Puma.

Used plastic bottles can provide input for various new products. Nube9 started life out as a youth sports uniform supplier, making sportswear from post-consumer plastic bottles before expanding its product collection and using closed-loop recycling with used uniforms. The plastic bottles it sources are chipped or flaked, melted and then extruded into poly fibre by supplier Repreve, which not only works with Nube9 but with a range of companies, including for example Quiksilver, Patagonia and Ford, which uses the material for seats.

Cleaning products company Method makes new plastic bottles by using plastic waste recovered from the sea (Weybrecht 2014). Its PET bottles are made from 100% recycled plastic, and the company was one of the first Cradle to Cradle® endorsed companies, providing over 50 certified products (certification and labelling will be covered in more detail in Chapters 8 and 10). Ecover's cleaning products also come in recycled bottles, made in a zero waste certified factory. A 2019 limited edition washing-up liquid used waste from the beer brewing process, making up a quarter of ingredients. Oceansaver is eliminating single-use plastics by providing refill pods that can simply be dropped in empty bottles that are then filled with water and shaken to make household cleaners which are made up to 90% of water.

We have covered food and drinks earlier in terms of how companies address food waste – the distinction between companies using waste as a business model and helping to address waste more generally and between external and internal waste solutions are of course somewhat ambiguous. In order to avoid repetition, we refer to the other examples covered already but want to draw attention to two companies using waste from food and drink to offer new products.

Toast Ale is an award-winning social enterprise that uses surplus fresh bread no longer suitable for human consumption to brew beer, donating all its profits to charities that address food waste. It works with local suppliers and brewers and has started licensing its brand globally. So far, it is available in the UK, the US, Ireland, Iceland, Brazil and South Africa, though it has inspired other projects elsewhere (see www.toastale.com/international/ for examples).

GroCycle has developed a business model that reduces the amount of coffee grounds going to landfill. As a social enterprise, the not-for-profit community interest company is growing oyster and shiitake mushrooms from waste coffee grounds, selling them to high-end restaurants. In an effort to inspire sustainable food growing, it also started offering mushroom growing kits for consumers as well as online courses to teach small-scale mushroom farming (see https://grocycle.com/).

In Greece, seeing dried up seaweed on beaches inspired Stavros Tsompanidis to start up PHEE, a company that processes dead seaweed into natural surfaces and recyclable products such as sunglasses, smartphone covers and gift boxes. The start-up works with local authorities who spend large amounts of money cleaning up seaweed from the coast and burying it in landfill with economic and environmental consequences (Reuters 2018; The Straits Times 2016). As the country is still reeling from the effects of the financial crisis with record youth unemployment, the start-up is therefore not only addressing the economic and environmental bottom line but also addressing the social side of sustainability.

Whilst some of the mentioned companies here sell their goods directly, these developments have also given rise to specialist retailers offering such products (e.g. Protect the Planet or the aforementioned BuyMeOnce). Increasingly mainstream retailers are also moving into this growing market space – for example Toast Ale and Method products are available in most UK supermarkets.

SUMMARY

Waste and its management is a major challenge facing businesses and the societies within which they operate locally, nationally and globally. In this chapter we have discussed what waste is, which forms it takes in terms of streams, sources and types, and how it and its environmental impact can be treated within the framework of the waste hierarchy (reduce, reuse, recycle). As industry is responsible for up to 90% of waste generated, it is a crucial issue as in turn businesses are faced with the costs and responsibilities of managing waste. Beyond bans and targets, regulatory responses therefore often include extended producer responsibilities to ensure companies take into account what happens to their products once they leave the factory gate and come to the end of their useful lives – designing products with the end of life in mind. Internal recycling programmes and external take-back schemes, whether voluntary or obligatory are characteristic of product stewardship and extended producer responsibility. In order to not only minimise but also to prevent waste, companies need to avoid over-production of components and finished goods, and move away from planned obsolescence, which has distinguished much of the so-called *throwaway society* of the 20th and early 21st century in which "consumption begets waste" (Anderson 2014: 221).

Although zero waste might not always be feasible, it is a useful objective to guide companies in minimising and preventing the waste they generate and to push them to re-think, redesign and reimagine. It encourages industry to see waste as a resource that is costly to dispose of as well as profitable to reuse, providing both internal and external opportunities, where waste can even become a business model for companies, as illustrated with a number of examples.

Whilst technology, research and development, such as the discovery of plastic-eating enzymes provide some solutions, companies cannot solely rely on technology to manage their waste. They also need to work on preventing and avoiding waste in the first place – parallel to the simultaneous nature of adaptation and mitigation for addressing climate change.

In the following part three, corporate functions and their role in managing environmental sustainability are covered, several of which are relevant for this challenge. For example in Chapters 8 and 9 we will cover supply chain management, and production and operations where working with suppliers in the value chain and the design of products and services with the end of life in mind are discussed, whilst in Chapter 10 we look at the role of marketing and consumption. In Chapter 11 we will cover environmental management systems, which can also help companies to measure and monitor their waste and to set stringent internal targets, whilst Chapter 12 on accounting links the measuring and reporting of waste and other environmental challenges with their management. The final chapter in part three covers management of human resources who are key to implementing corporate sustainability initiatives.

The integrative case studies in the concluding part of the book make space for thinking through and planning waste management within a broader context.

SUGGESTED QUESTIONS AND ACTIVITIES

- Explore the materials, energy and water flowing between the various public and private companies in the industrial ecosystem at Kalundborg in Denmark: www.symbiosis.dk/en/ – what other examples of industrial ecoparks can you find?
- Use The Natural Step (TNS) framework introduced in Chapter 4 to sketch out a company's waste management approach.
- Choose one of the other frameworks, models or strategies from Chapter 4 and use it to plan a company's approach to manage its waste.
- Watch the documentary *The Lightbulb Conspiracy* and then research other products that have been designed with planned obsolescence. If you were a sustainability consultant, what actions would you recommend to companies engaging in this practice that are concerned about their profitability?
- Find out about the Chartered Institution of Wastes Management (or the equivalent in your country) – how does it help companies and managers to tackle the environmental challenge of waste?
- Watch the 20-minute short film *Story of stuff* (https://storyofstuff.org/movies/story-of-stuff/) and take the change maker personality quiz (http://action.storyofstuff.org/survey/changemaker-quiz/) – what role could you play in addressing waste?
- Look up the export of toxic waste to Cote d'Ivoire by the Trafigura company.
- Look up Amazon's Sustainability Question Bank online and evaluate how useful and detailed the information is that the company provides to the public. What other material would you like to see?
- Debate the following motions, preparing for both agreement and disagreement on the following statements:
  - Zero waste is neither feasible nor desirable.
  - Scientific and technological advances in waste management mean that we can safely use nuclear energy.
  - Waste transport across national borders should be banned completely.
  - A plastic-free society is not possible in our day and age.
- "Waste is defined by where something is, not what it is. It's about context, not content". (Leonard 2010: 235) – critically discuss this statement.
- Check out http://givebackbox.com/ in the US – are there similar schemes outside the US?
- With regard to waste management, compare the merits of the polluter pays principle with the precautionary principle – generally, or taking into account the external regulatory framework and/or companies' internal approach to waste.
- Explore in-depth the connection between consumption and waste by focusing on one particular consumer product of your choice.
- Go plastic-free for one day, one week or one month. What do you notice? How easy or difficult is it? Why? Did you manage it? If not, why? What recommendations would you give to companies based on your experience?
- Visit your nearest Repair Café (https://repaircafe.org/en/visit/) and find out how it contributes to reducing waste.
- Check out the websites of any of the companies mentioned in the chapter that use waste as a business model. Can you find any more examples of companies that have a similar business model?

NOTES

1 We refer the reader to the definition of sustainable development introduced in Chapter 1, based on the Brundtland report from the World Commission on Environment and Development, which states that sustainable development "seeks to meet the needs and aspirations of the present

without compromising the ability to meet those of the future" (WCED 1987: 40).

2 It should be noted here that the European Commission (2018a) excludes such sources in its definition of bio-waste.

3 Backfilling uses waste in excavated areas for slope reclamation, safety or landscaping (Eurostat 2017).

4 One particular issue observed in Kenya that re-enforced the need for action was that for example abattoirs would find plastic in the guts of three out of ten animals taken to slaughter. Following the ban this has reduced to one out of ten (Watts 2018).

5 In some instances, local authorities can also make it more difficult to recycle for consumers e.g. by not allowing pedestrians to use recycling facilities, making the use of a car necessary, or by restricting the size of items picked up in kerbside collections.

6 Reverse logistics could be seen as "dedistributing" products back from customers for remanufacture (Hawken et al. 1999). See also Chapter 8 on supply chain management for reverse logistics.

7 Loop is a TerraCycle company which is introduced in the following section.

8 It is beyond the scope of the chapter to cover textile in more detail – the industry faces environmental as well as social challenges. The environmental impact of clothing and in particular fast fashion is widely documented in terms of its carbon and water footprint, microfibers of synthetic textiles being washed into rivers and oceans, and the amount of clothes going to landfill instead of being recycled.

SOURCES AND FURTHER READING

3M (2018) 2018 Sustainability Report, available at www.3m.com/3M/en_US/sustainability-report/ (Retrieved, 04.06.2018).

Alexander, P. (2017) New York City Tries to Ban Styrofoam: It's Déjà Vu All Over Again, *Huffington Post*, 27.10.2017, available at www.huffingtonpost.com/entry/new-york-city-tries-to-ban-styrofoam-its-déjavu_us_59f08cefe4b02ace788ca8ea (Retrieved, 24.05.2018).

Anderson, D. A. (2014) *Environmental Economics and Natural Resource Management*, London: Routledge, 4th ed.

Anderson, R. C. (1998) *Mid-Course Correction. Toward a Sustainable Enterprise: The Interface Model*, White River Junction, VT: Chelsea Green.

Anderson, R. C. and White, R. (2009) *Confessions of a Radical Industrialist: How Interface Proved That You Can Build a Successful Business without Destroying the Planet*, London: Random House.

BBC (2018) Growing Feet? Try Expandable Shoes, *BBC World Hacks*, 26.12.2017, www.bbc.co.uk/programmes/p05rkndh (Retrieved, 11.06.2018).

BBC Newsround (2017) How Do You Clean up the Ocean's Biggest Rubbish Dump? *BBC Newsround* 15.05.2017, available at www.bbc.co.uk/newsround/39921749 (Retrieved, 14.06.2018).

Bellware (2014) This Restaurant Hasn't Taken Out the Trash in Nearly Two Years, *Huffington Post*, 28.04.2014, available at www.huffingtonpost.co.uk/entry/zero-waste-restaurant_n_5215019 (Retrieved, 11.06.2018).

Benn, S., Dunphy, D. and Griffiths, A. (2014) *Organizational Change for Corporate Sustainability*, London: Routledge, 3rd ed.

Benyus, J. M. (1997) *Biomimicry: Innovation Inspired by Nature*, New York, NY: William Morrow.

Bernon, M., Baker, P., Mena, C., Palmer, A., Skipworth, H., Smart, A. and Templar, S. (2012) Towards More Sustainable Supply-Chain Management, in: Grayson, D. and Exter, N. (eds.) *Cranfield on Corporate Sustainability*, Sheffield: Greenleaf Publishing, pp. 149–166.

Blowfield, M. (2013) *Business and Sustainability*, Oxford: Oxford University Press.

Briggs, H. (2019) Microplastics: Seeking the 'Plastic Score' of the Food on our Plates, *BBC News*, 13.10.2019, available at www.bbc.co.uk/news/science-environment-49798057 (Retrieved 04.12.2019).

Buckingham, S. and Turner, I. (2008) Waste, in: Buckingham, S. and Turner, M. (eds.) *Understanding Environmental Issues*, London: Sage, pp. 150–174.

Cavagnaro, E. and Curiel, G. (2012) *The Three Levels of Sustainability*, Sheffield: Greenleaf Publishing.

Chung, J. and Park, J. (2018) 'Urban Mining' in South Korea Pulls Rare Battery Materials from Recycled Tech, *Reuters Business News*, 12.04.2018, available at www.reuters.com/article/

us-southkorea-mining/urban-mining-in-south-korea-pulls-rare-battery-materials-from-recycled-tech-idUSKBN1HJ14T (Retrieved, 14.06.2018).

Clancy, H. (2018) How Michelin is Finding a Second Life for Tires, *GreenBiz*, 29.05.2018, available at www.greenbiz.com/article/how-michelin-finding-second-life-tires (Retrieved, 04.06.2018).

Closed Loop (2018) World's Number One Restaurant Noma Adopts Australian-Designed Closed Loop Zero-Waste System, available at www.closedloop.com.au/news-and-events/worlds-number-one-restaurant-noma-adopts-australian-designed-closed-loop-zero-waste (Retrieved, 11.06.2018).

Coates, A. (2017) The Fat-Powered Buses That Are Coming to a Street Near You, *The Independent*, 16.08.2017, available at www.independent.co.uk/news/long_reads/fat-powered-buses-are-coming-to-a-street-near-you-a7882171.html (Retrieved, 14.06.2018).

Cooper, K.-L. (2018) The People Fighting Pollution with Plastic-Free Periods, *BBC News*, 01.05.2018, available at www.bbc.co.uk/news/world-43879789 (Retrieved, 06.06.2018).

Costanza, R., Cumberland, J., Daly, H., Goodland, R. and Norgaard, R. (1997) *An Introduction to Ecological Economics*, Boca Raton, FL: St Lucie Press.

CSS (2017) Municipal Solid Waste Factsheet, Center for Sustainable Systems, University of Michigan, Pub. No. CSS04–15, available at http://css.umich.edu/factsheets/municipal-solid-waste-factsheet (Retrieved, 19.04.2018).

Daly, H. E. and Cobb, J. B. (1989) *For the Common Good: Redirecting the Economy towards Community, the Environment, and a Sustainable Future*, Boston: Beacon Press.

EEA (2016a) Europe's Waste Streams, European Environment Agency, available at www.eea.europa.eu/media/infographics/europe2019s-waste-streams-1/view#tab-based-on-data (Retrieved, 19.04.2018).

EEA (2016b) Prevention of Hazardous Waste in Europe – The Status in 2015, European Environment Agency, available at www.eea.europa.eu/publications/waste-prevention-in-europe (Retrieved, 01.05.2018).

EEA (2016c) The European Environment – State and Outlook 2015 Report, European Environment Agency, available at www.eea.europa.eu/soer (Retrieved, 02.05.2018).

EEA (2018) Glossary of Environmental Terms, European Environment Agency, available at www.eea.europa.eu/help/glossary/eea-glossary (Retrieved, 19.04.2018).

Elkington, J. (2012) *The Zeronauts: Breaking the Sustainability Barrier*, London: Routledge.

EPA (2018) Land, Waste, and Cleanup Topics, United States Environmental Protection Agency, available at www.epa.gov/environmental-topics/air-topics (Retrieved, 02.05.2018).

Esty, D. C. and Winston, A. S. (2009) *Green to Gold: How Smart Companies Use Environmental Strategy to Innovate, Create Value and Build Competitive Advantage*, Hoboken, NJ: Wiley.

European Commission (2018a) EU Waste Policies webpages, available at https://ec.europa.eu/environment/waste/index.htm (Retrieved, 23.04.2018).

European Commission (2018b) A European Strategy for Plastics in a Circular Economy, Communication from the Commission to the European Parliament, the Council, the European Economic and Social Committee and the Committee of the Regions, Brussels, COM (2018) 28 final, 16.01.2018.

European Commission (undated) EU Waste Policy – The Story behind the Strategy, available at www.ec.europa.eu/environment/waste/pdf/story_book.pdf (Retrieved, 03.05.2018).

Eurostat (2017) Waste Statistics, European Commission, available at http://ec.europa.eu/eurostat/statistics-explained/index.php?title=Waste_statistics (Retrieved, 02.05.2018).

Eurostat (2018) Municipal Waste Statistics, European Commission, available at http://ec.europa.eu/eurostat/statistics-explained/index.php?title=Municipal_waste_statistics (Retrieved, 02.05.2018).

Everard, M. (2011) Sustainability and Sustainable Development, in: Brady, J., Ebbage, A. and Lunn, R. (eds.) *Environmental Management in Organizations, The IEMA Handbook*, London: Earthscan/Routledge, 2nd ed., pp. 39–61.

FLW Protocol (2016) Food Loss and Waste Accounting and Reporting Standard, Executive Summary, Food & Waste Loss Protocol, available at http://flwprotocol.org/wp-content/uploads/2017/05/FLW_Standard_Exec_Summary_final_2016.pdf (Retrieved, 30.04.2018).

Friedman, F. B. (2011) *Practical Guide to Environmental Management*, Washington, DC: Environmental Law Institute, 11th ed.

Frosch, R. A. and Gallopoulos, N. E. (1989) Strategies for Manufacturing, in: *Scientific American*, 261(3): 144–153 (September).

Gallego-Schmid, A., Mendoza, J. M. F. and Azapagic, A. (2019) Environmental Impacts of Takeaway Food Containers, in: *Journal of Cleaner Production*, 211: 417–427.

Green, T. (2013) *The Dictionary of Global Sustainability. With Case Studies on the Environmental, Economic, Technological, and Social Aspects of Sustainability*, New York: McGraw-Hill.

Halton, M. (2018) Recycling Hope for Plastic-Hungry Enzyme, *BBC News*, 16.04.2018, available at www.bbc.co.uk/news/science-environment-43783631 (Retrieved, 06.06.2018).

Hancock, T. (2018) China Recyclers Grind to Halt Amid Crackdown on Imported Waste, *Financial Times*, 15.01.2018, available at www.ft.com/content/63cf220c-f8ee-11e7-9b32-d7d59aace167 (Retrieved, 24.05.2018).

Hart, S. L. (2010) *Capitalism at the Crossroads: Next Generation Business Strategies for a Post-Crisis World*, Upper Saddle River, NJ: Pearson, 3rd ed.

Hawken, P. (1993) *Ecology of Commerce: A Declaration of Sustainability*, New York: Harper Business.

Hawken, P., Lovins, A. B. and Lovins, L. H. (1999) *Natural Capitalism: The Next Industrial Revolution*, London: Earthscan.

Interface (2018) Mission Zero: Measuring Our Progress, available at www.interface.com/US/en-US/campaign/climate-take-back/Sustainability-Progress-en_US (Retrieved, 05.06.2018).

Jäger, K. (2018) Urban Mining: Hidden Riches in Our Cities, *Deutsche Well*, 12.03.2018, available at www.dw.com/en/urban-mining-hidden-riches-in-our-cities/a-42913985 (Retrieved, 14.06.2018).

Leonard, A. (2010) *The Story of Stuff*, London: Constable.

Levitt, T. (2018) Dutch Cows Produce More Waste Than Farms Can Dispose of Safely, *The Guardian*, print edition, 17.02.2018.

Liboiron, M. (2016) Municipal Versus Industrial Waste: Questioning the 3–97 Ratio, Discard Studies: Social studies of waste, pollution & externalities, available at https://discardstudies.com/2016/03/02/municipal-versus-industrial-waste-a-3-97-ratio-or-something-else-entirely/ (Retrieved, 19.04.2018).

Lomborg, B. (2001) *The Skeptical Environmentalist: Measuring the Real State of the World*, Cambridge, UK: Cambridge University Press.

McDonough, W. and Braungart, M. (2002) *Cradle to Cradle: Remaking the Way We Make Things*, New York: North Point Press.

Nemetz, P. N. (2013) *Business and the Sustainability Challenge: An Integrated Perspective*, London: Routledge.

NOAA (2018) NOAA Marine Debris Program, Office of Response and Restoration, webpages, National Oceanic and Atmospheric Administration, U.S. Department of Commerce, available at https://marinedebris.noaa.gov (Retrieved, 14.06.2018).

OECD (2015) *Environment at a Glance 2015: OECD Indicators*, Organization for Economic Cooperation and Development, Paris: OECD Publishing, available at http://dx.doi.org/10.1787/9789264235199-en (Retrieved, 23.04.2018).

Ohno, T. (1988) *The Toyota Production System: Beyond Large Scale Production*, Portland, OR: Productivity Press.

Pearce, D., Markandya, A. and Barbier, E. (1999) *Blueprint for a Green Economy*, London: Earthscan (first published 1989).

Radnor, A. (2017) 'We're Having a Menstrual Liberation': How Periods Got Woke, *The Guardian*, 11.11.2018, available at www.theguardian.com/society/2017/nov/11/periods-menstruation-liberation-women-activists-abigail-radnor (Retrieved, 06.06.2018).

Raworth, K. (2017) *Doughnut Economics: Seven Ways to Think Like a 21st-Century Economist*, London: Random House Business Books.

Reuters (2018) Greece's New Startup Culture: Technology and Seagrass Sunglasses, *Reuters Business News*, 03.05.2018, available at www.reuters.com/article/us-eurozone-greece-startups/greeces-new-startup-culture-technology-and-seagrass-sunglasses-idUSKBN1I417D (Retrieved, 11.06.2018).

Robertson, M. (2017) *Sustainability Principles and Practice*, London: Routledge, 2nd ed.

Russell, R. (2017) *Nuclear Waste: Where to Store it for Eternity? Deutsche Well*, available at www.dw.com/en/nuclear-waste-where-to-store-it-for-eternity/a-40449893 (Retrieved, 23.04.2018).

Selin, H. and Van Deveer, S. D. (2006) Raising Global Standards: Hazardous Substances and E-Waste Management in the European Union, in: *Environment: Science and Policy for Sustainable Development*, 48(10): 6–18.

Silverman, J. (2007) Why Is the World's Biggest Landfill in the Pacific Ocean? *HowStuffWorks.com*, 19.09.2007, available at https://science.howstuffworks.com/environmental/earth/oceanography/great-pacific-garbage-patch.htm (Retrieved, 14.06.2018).

Smith, W. G. B. (2007) Accounting for the Environment: Can Industrial Ecology Pay Double Dividends for Business? in: Nemetz, P. N. (ed.) *Sustainable Resource Management: Reality or Illusion?* Cheltenham: Edward Elgar, pp. 304–341.

Smithers, R. (2016) Costa Coffee Launches in-Store Cup Recycling, *The Guardian*, print edition, 26.11.2016.

Snedden, M. (2014) Green Coffee Shops Cut Rubbish, but at What Cost? *BBC Capital*, available at www.bbc.com/capital/story/20130722-erasing-coffee-shop-waste (Retrieved, 11.06.2018).

The Straits Times (2016) Seaweed Makes Waves, *The Straits Times*, 25.06.2016, available at www.straitstimes.com/world/europe/21-seaweed-makes-waves (Retrieved, 11.06.2018).

Stead, J. G. and Stead, W. E. (2014) *Sustainable Strategic Management*, Sheffield: Greenleaf Publishing, 2nd ed.

Tamma, P. (2018) China's Trash Ban Forces Europe to Confront Its Waste Problem, *Politico*, 27.02.2018, available at www.politico.eu/article/europe-recycling-china-trash-ban-forces-europe-to-confront-its-waste-problem/ (Retrieved, 24.05.2018).

Toyota (2018) Environment Pages, Toyota Motor Corporation, available at www.toyota-global.com/sustainability/environment/index.html (Retrieved, 05.06.2018).

Visser, W. (2012) *The Quest for Sustainable Business: An Epic Journey in Search of Corporate Responsibility*, Sheffield: Greenleaf Publishing.

von Weizsäcker, E., Lovins, A. B. and Lovins, L. H. (1998) *Factor Four: Doubling Wealth, Halving Resource Use*, London: Earthscan.

Watts, J. (2018) Eight Months on, Is the World's Most Drastic Plastic Bag Ban Working? *The Guardian*, 25.04.2018, available at www.theguardian.com/world/2018/apr/25/nairobi-clean-up-highs-lows-kenyas-plastic-bag-ban (Retrieved, 24.05.2018).

Watts, J. and Smithers, R. (2018) Easy Living: How the West Became Obsessed with Wet Wipes, *The Guardian*, print edition, 12.05.2018.

WCED (1987) *Our Common Future*, World Commission on Environment and Development, Oxford: Oxford University Press.

Weybrecht, G. (2014) *The Sustainable MBA: A Business Guide to Sustainability*, Chichester: Wiley, 2nd ed.

WNA (2017) Radioactive Waste Management, World Nuclear Association, available at www.world-nuclear.org/information-library/nuclear-fuel-cycle/nuclear-wastes/radioactive-waste-management.aspx (Retrieved, 23.04.2018).

Woolston, H. (2011) Waste Management and Producer Responsibility, in: Brady, J., Ebbage, A. and Lunn, R. (eds.) *Environmental Management in Organizations, The IEMA Handbook*, London: Earthscan/Routledge, 2nd ed., pp. 311–324.

Xerox (2018) 2017 Global Citizenship Report, available at www.xerox.com/corporate-citizenship/2017/ (Retrieved, 04.06.2018).

Yeung, J. (2018) French Man to Cross Great Pacific Garbage Patch on 6-Month Swim, *CNN*, 05.06.2018, available at https://edition.cnn.com/2018/06/05/asia/cross-pacific-swim-intl/index.html (Retrieved, 04.06.2018).

Young, A. R. (2015) The European Union as a Global Regulator? Context and Comparison, in: *Journal of European Public Policy*, 22(9): 1233–1252.

PART III

Corporate Functions
Managing for Environmental Sustainability

INTRODUCTION

Having provided the foundations in part one by looking at the scientific and economic foundations for sustainability (Chapter 2), by translating the theoretical perspectives into principles and philosophies for corporate environmental sustainability by illustrating the tensions inherent in sustainability and how the legislative framework addresses these to provide a context for business (Chapter 3), and by finally translating this into practice in terms of the strategies and models business can use to guide them towards environmental sustainability (Chapter 4), we then covered three specific challenges in part two of the book by covering climate change (Chapter 5), pollution of air, water and soil (Chapter 6) and waste (Chapter 7). We are therefore now at a point where we can look at how different corporate functions can address environmental sustainability and manage their environmental impact based on a solid underpinning of the context, principles and models of sustainability and on an understanding of the environmental challenges faced by businesses. This also includes how businesses contribute to both the problem and the solution of these challenges through the various functions.

Broadly speaking, management consists of three elements, input, process and output, and can be defined as the achievement of goals (in the form of performance as output) in an effective and efficient way (process) working with people and other resources (input). The management process to attain pre-established goals is achieved through planning, organising, leading and controlling. These four are considered the core or primary functions of management within which all business or corporate functions are reflected. Although classifications vary in the literature (see e.g. Laasch and Conaway 2015; Daft et al. 2010), generally speaking, *planning* includes goal setting, strategy and decision-making, as well as entrepreneurship; *organising* comprises organisational structure, operations and supply chain management and human resources management, though the latter can also come under *leading*, which covers marketing and communications as well as organisational behaviour, whilst *controlling* encompasses accounting and finance, quality management and information technology/systems and may also include operations and supply chains. Other authors, such as for example Michael Porter (1985), classify corporate functions according to whether they are primary or support activities within the value chain.

Whilst sustainability is an interdisciplinary subject, business and management studies are still mostly based around the core functional areas (Weybrecht 2014). Throughout the book, we stress the systemic nature of environmental sustainability for business. In part two we highlighted the interrelationships and interlinkages between the various challenges. Similarly, for part three, there are interdependencies and overlaps between the various corporate functions which need to be coordinated in order to address environmental sustainability convincingly and in a holistic manner. We recognise that a comprehensive coverage of each function is not within the scope of the book, and we refer the reader to books and sources that are focused on singular/single functions for a more in-depth understanding of each. We should also note here that the boundaries between different functions as treated in the literature can be artificial, as in practice, for example, in smaller companies and start-ups, there are no clear distinctions between functions as staff (have to) cover several areas. Regardless of size, Laszlo and Zhexembayeva (2011: 127) emphasise that "management is all about decision making" and in the context of sustainability, important choices are made by managers.

The chapters at hand are a selection of functions that reflect the core management processes of planning, organising, leading and controlling focused on achieving the goal of corporate environmental sustainability effectively and efficiently. In terms of the input in form of resources we have referred to the unsustainable use of resources which underpin the environmental challenges covered in part two of the book. In terms of the process we introduced eco-efficiency and eco-effectiveness in part one which mirror efficiency – the ratio between input of resources and output, and effectiveness – the extent to which a stated goal is achieved. The output as a result should be environmentally sustainable performance.

Before we turn our attention to the functions chosen for part three, some brief notes on functions not explicitly covered in the following are in order.

Economics was covered in Chapter 2, underpinning the business case for sustainability whilst strategy was discussed as part of business models in Chapter 4. Entrepreneurship is not considered as a separate function but as something that can be done within most functions in the form of corporate entrepreneurship in businesses of all sizes and therefore is not limited to small businesses or start-ups as conventionally perceived. A brief discussion on environmental entrepreneurship is included in Chapter 13.

In the first three chapters of part three, we take an overview of the whole supply chain and its management as a starting point in Chapter 8 before focusing on operations and production in Chapter 9 and concluding with marketing as the intersection between companies and customers (whether they are commercial buyers or end consumers) in Chapter 10. These functions broadly reflect primary activities in Porter's (1985) value chain, though procurement is also covered in Chapter 8 and research and development in Chapter 9.

In the second three chapters of part three, the functions that largely represent the support activities in Porter's value chain are covered, starting in Chapter 11 with information technology and management systems with a focus on environmental management systems to support management processes. This leads to how such data and information are used in sustainability reporting – covered in Chapter 12 on accounting and finance. Finally, Chapter 13 on human resources looks at organisational structure, governance and the role of leadership and entrepreneurship in initiating and achieving environmental sustainability.

We should stress here that there is some degree of overlap, particularly in the areas of supply chain management (Chapter 8) and operations and production (Chapter 9) and information systems (Chapter 11). For example, life cycle analysis can be used across the whole value chain or in a very focused manner in design, and it consists of systematic information for which benchmarks of the International Standards Organisation (ISO) may be used. Similarly, location decisions may fall under supply chain management but can also come under production and operations decisions, underpinned by data obtained through information systems.

In order to provide consistency – in line with the previous part – the structure of the chapters in this part all follow a broadly similar pattern. Each chapter starts with an introduction to the conventional nature of the function before the environmental impact of the function is explored – how does it contribute to the state of the natural environment as illustrated in the environmental challenges of part two? We will then consider the impact of the environment on the function; how is it affected by the environment in terms of risks and challenges? This illustrates the business case for why it is important for the function to consider environmental sustainability. Subsequently we introduce the sustainable corporation function – how can this function contribute to the solution? What opportunities are there? We then proceed by covering various tools and methodologies and by providing examples from corporate practice.

As managing for environmental sustainability goes beyond company and country borders, some chapters also highlight the international nature of the function covered (e.g. international supply chain management), and examples are provided from a variety of countries.

SUGGESTED QUESTIONS AND ACTIVITIES

- Which corporate function do you consider the most important one when it comes to addressing environmental sustainability? Why?
- Using an industry or company of your choice, discuss which corporate function is the most important one for it in addressing environmental sustainability. Have there been changes over time?
- Do you agree with the choice of corporate functions covered in part three? Why? Why not?

SOURCES AND FURTHER READING

Daft, R. L., Kendrick, M. and Vershinina, N. (2010) *Management: International Edition*, Andover: Cengage Learning EMEA.

Laasch, O. and Conaway, R. (2015) *Principles of Responsible Management: Glocal Sustainability, Responsibility and Ethics*, Stamford, CT: Cengage Learning.

Laszlo, C. and Zhexembayeva, N. (2011) *Embedded Sustainability: The Next Big Competitive Advantage*, Sheffield: Greenleaf Publishing.

Porter, M. E. (1985) *Competitive Advantage: Creating and Sustaining Superior Performance*, New York: The Free Press (Republished with a new introduction, 1998).

Weybrecht, G. (2014) *The Sustainable MBA: A Business Guide to Sustainability*, Chichester: Wiley, 2nd ed.

CHAPTER 8

Supply chain management for sustainability

INTRODUCTION

The three basic issues that business faces are "what it takes, what it makes, and what it wastes", according to Hawken (1993: 12), which is, in short, what supply chain management encompasses. In the following, we will define supply chain management in more detail before covering its environmental impact and the business case for sustainability in the supply chain. In their review of sustainable supply chain management, Carter and Easton (2011) emphasise that sustainability is not just a management fad or buzzword but has become the license to do business in the 21st century of which supply chain management is an integral part. We then provide an overview of sustainable supply chain management before covering the various elements and tools as well as exemplary companies, trends and practices that contribute to sustainability in the supply chain.

Supply chain management is a complex topic that covers a wide range of activities. As mentioned in the introduction to this part of the book, the delineation of management functions varies, and some would argue that operations management would be a more suitable starting point, i.e. within the boundaries of one firm. However, organisations have increasingly recognised that managing operations across the whole chain has benefits for customer satisfaction, efficiency and effectiveness (Slack et al. 2001). When taking into account the integrated and holistic nature of corporate sustainability as emphasised throughout the book, looking at the whole supply chain in terms of what business takes, makes and wastes, or of what it takes, makes, uses and loses (Raworth 2017), is a more natural departure for covering corporate functions. There are, of course, some overlaps; hence, cross-references will be made, and some tools are covered in this chapter but not in others and vice versa. Srivastava (2007) also highlights the overlaps for example between green manufacturing, reverse logistics and waste management, which we have covered in the previous chapter, whilst we look at design, production and operations in the following chapter.

SUPPLY CHAIN MANAGEMENT

Overview

A definition of supply chains and their management that is widely used (e.g. Seuring and Müller 2008; Sarkis and Dou 2018) comes from Handfield and Nichols (1999: 2):

> The supply chain encompasses all activities associated with the flow and transformation of goods from raw materials stage (extraction), through to the end user, as well as the associated information flows. Material and information flow both up and down the supply chain. Supply chain management (SCM) is the integration of these activities through improved supply chain relationships to achieve a sustainable competitive advantage.

Within this context, both upstream and downstream activities need to be considered, the network of suppliers as well as the channels of distribution. A range of functions can be included under this understanding of supply chain management (SCM), including information systems, procurement and sourcing, production, processing of orders and managing inventories and warehousing, customer service and after-market disposal of materials and packaging (Handfield and Nichols 1999) as well as facilities management, finance, marketing and distribution (Oxborrow and Lund-Thomsen 2017).

The systemic and integrated nature of supply chain management is also highlighted by Carter and Rogers (2008) as it involves the coordination of traditional business functions and integration of processes within and across organisations. This is in line with how the Council of Supply Chain Management Professionals defines the function, highlighting its role in the integration of supply and demand management, whilst also acknowledging ambiguities in defining the supply chain.

Although there are debates on whether the value chain and the supply chain are the same thing, the value chain is typically more concerned with firm-internal activities that add value or contribute to value creation (Porter 1985) and thus seems to have a narrower focus than the supply chain, whose scope goes beyond corporate boundaries. Nevertheless, the definition of supply chain management provided by Slack and colleagues (2001: 43) explicitly includes value:

> Supply chain management is the management of the interconnection of organizations which relate to each other through upstream and downstream linkages between the different processes that produce value in the form of products and services to the ultimate customer. It is a holistic approach to managing across company boundaries.

In making the value chain more responsible and sustainable, Laasch and Conaway (2015) include broader concerns of supply chain management, green logistics and design for the environment in their depiction of the value chain, thus demonstrating the somewhat blurry boundaries between the two chains. This is reflected in Fogel's (2016) assertion that the supply chain's objective is the maximisation of the overall value a firm generates.

Primary performance criteria for supply chain management include cost, quality and delivery, which managers of the supply chain have to address for products and processes on both sides of the supply chain – upstream and downstream (Vachon and Klassen 2006), with constant flows of information, resources, products as well as money.

The supply chain of a product is also known as the life cycle of the product (Seuring and Müller 2008), which conventionally has been seen as cradle to grave, whereas integration of sustainability concerns incorporates the whole circle of cradle to cradle, where products at the end of their useful life become input again. This is echoed by Bernon and colleagues (2012) who add design, consumer use and end-of-life management to the five main processes of supply chain management: plan, source, make, deliver and return.

A simplified depiction of supply chain management is provided in Figure 8.1. Arrows indicate flows of goods, information and money. Managing the supply chain includes management of transformation of materials, information flows and supply chain relationships.

The increase in globalisation since the 1970s has had a significant effect on logistics and SCM over recent decades and has seen a rapid growth in outsourcing, off-shoring and global sourcing, resulting in the development of global supply chains and production networks (Oxborrow and Lund-Thomsen 2017).

Environmental impact of supply chain management

The complex, holistic and systemic nature of the supply chain, as illustrated in the overview, already hints at the wide-ranging environmental impact of the function – how do the activities within the supply chain contribute to the state of the natural environment as illustrated in the environmental challenges of part two? This short section serves the purpose of highlighting (rather than detailing) the environmental impact of upstream and downstream supply chain activities. For many industries, the greatest environmental risks lies in the supply chain, particularly upstream (Esty and Winston 2009).

The environmental impact of supply chain activities comprises greenhouse gas emissions, use of natural resources and energy, production of harmful substances, pollution of air, water and soil as well as waste created in the process of resource extraction, manufacturing and packaging.

In particular, trends such as outsourcing where third-party suppliers are (sub-)contracted to carry out specific activities, off-shoring, where such activities are carried out in a different country, and global sourcing where supply chains stretch across different countries (Oxborrow and Lund-Thomsen 2017) have increased the geographical length of supply chains.[1] Such trends have been based particularly on the development and adoption of standard shipping containers, trade

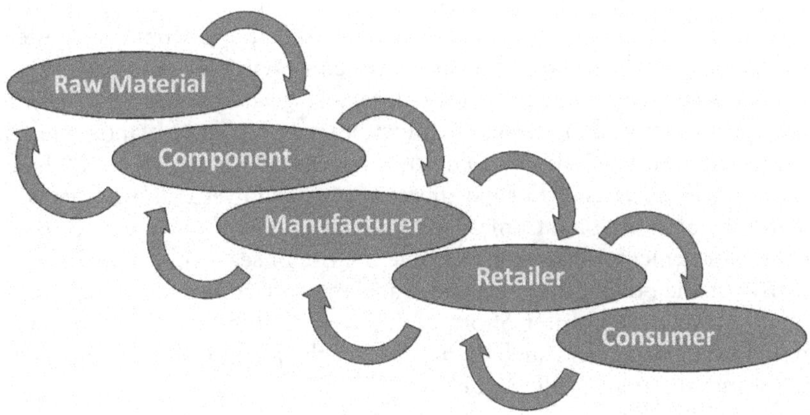

FIGURE 8.1
Supply chain management

liberalisation, as well as improved and expanded transport infrastructures with consequences for cost differentials between developed and developing country locations (Grant et al. 2017).

In brief, the various elements of the supply chain affect the natural environment as follows (based on e.g. Slack et al. 2001; Grant et al. 2017):

- *Network design*: the choice of location and of buildings and their layout affect the energy efficiency companies can achieve as well as the distance products need to be shipped; local-global considerations of production.
- *Logistics, distribution and transportation*: transport-related energy (particularly in the case of empty miles) and greenhouse gas emissions, choice of fuel and of transport mode, packaging waste, energy needed to operate warehouses (especially if temperature-controlled), noise pollution, air pollution, impact of congestion, risk of transporting hazardous material.
- *Procurement, sourcing, and purchasing*: environmental footprint of raw materials, regulatory non-compliance of suppliers, less stringent environmental legislation in sourcing locations, packaging needs and waste, standards can verify origin and security of supply, lack of traceability of each stage of the supply chain may exacerbate environmental impact.
- *Collaboration and partnerships with suppliers*: poor supplier commitment to sustainability (e.g. environmental practices of suppliers and providers not in line with the focal firm) and non-compliance with legislation.

According to Abukhader and Jönson (2004), there are two questions when it comes to SCM and the environment – what is the impact of logistics and supply chains on the environment and what is the impact of the environment on logistics and supply chains, to which we now turn, asking how the function is affected by the environment in terms of risks and challenges.

Impact of environment on supply chain management

Environmental factors that can affect supply chain management include for example floods and droughts. Extreme weather events as a result of climate change have for example affected Coca-Cola in India, where the company lost its operating licence as a result of water shortages, whilst in 2008 factories in Thailand supplying Nike had to be temporarily shut down due to flooding (Dharmasasmita et al. 2017). In 2011, floods in Thailand forced disruptions in the supply chain of the global electronics and car parts industries, which heavily rely on close integration of suppliers and producers as well as just-in-time manufacturing and thus limited inventory (Reuters 2011). In 2018, historic rainfalls halted car production at a number of plants in Japan, and floods in Mexico affected the American supply chain of Japanese car manufacturer Honda. In the same year, droughts across large parts of Europe affected crops and livestock, leading to an increase in feeding costs and threatening some farmers with bankruptcy. Other impacts of the environment on supply chain management include for example the rise of sea levels, which may put buildings and plants at risk, or the pollution of soil and water may affect organisations' ability to provide uncontaminated food and clean water. Shortage and depletion of natural resources affect the supply of raw materials for firms, and such disruptions in the form of storms or flooding can affect supply chains for months at a time (Grant et al. 2017).

Such risks are often primarily operational, but also legal and financial, and can potentially turn into reputational risks if suppliers are not complying with regulations leading to firms serving end consumers facing potential boycotts (Vachon and Klassen 2006). Financial risks of environmental events include for example higher commodity and insurance costs, and lost work capacity and reduced labour productivity as heatwaves, floods and droughts affect the health of the workforce through infectious diseases, weather-related mortality and malnutrition. The risks may be short-term or long-term, both of which force companies to consider their sustainability not just economically but also environmentally and socially.

Business case for greening the supply chain

The two preceding sections underpin the business case for why it is important for supply chain management to consider environmental sustainability due to the interplay between the two. In addition there are internal and external drivers for SCM to incorporate sustainability concerns (Walker et al. 2008).

Beyond the conventional performance criteria for supply chain management, such as cost, quality, flexibility, time and delivery, firms are faced with scrutiny from stakeholders regarding sustainability (Vachon and Klaasen 2006). Customers and governments in the form of regulations, and society more generally, have been driving forces for companies to recognise the importance of greening the supply chain (Seuring and Müller 2008; Walker et al. 2008).

Pressure from major customers in the supply chain has been a powerful driver for businesses to improve their environmental performance – buyers can use their purchasing power to encourage and put pressure on firms to address sustainability concerns, which can extend beyond the focal firm and first-tier suppliers deeper into the supply chain (Brady 2011).

Other external drivers include investor concerns, NGO campaigns and peer pressure within industry (Brady 2011). There are thus pressures as well as incentives that drive sustainability in supply chains (Seuring and Müller 2008). Competitive advantage through first-mover advantages for corporate environmental leaders, as covered in Chapter 3 (based on Porter and van der Linde 1995a, 1995b), as well as competitors play a role in motivating companies to green their supply chains (Walker et al. 2008). Furthermore, supply chain transparency and subsequent reporting is increasingly demanded in order to manage reputational risks for firms.

Internal drivers include the commitment of employees as well as of investors, quality improvement and the desire to reduce costs (Walker et al. 2008). Some firms are actively values-driven by the degradation of the environment in the form of resource depletion, waste and pollution levels, as well as by good business sense and profit margins as green SCM is seen as a business value driver, not a cost centre. Green SCM can reduce the environmental impact of industrial activity whilst also maintaining quality, cost, reliability, performance and efficiency (Srivastava 2007). It can also help the resilience of the supply chain by addressing the various risks inherent in the function (Grant et al. 2017).

SUSTAINABLE SUPPLY CHAIN MANAGEMENT

Sustainable SCM (SSCM) emerged partly as a response to the rapid globalisation of supply chains of the 1980s and 1990s which together with a growth in consumption contributed to more and more complex supply chain activity and subsequent greater environmental impact (Oxborrow and Lund-Thomsen 2017).

A broad definition of SSCM is provided by Seuring and Müller (2008: 1700):

> the management of material, information and capital flows as well as cooperation among companies along the supply chain while taking goals from all three dimensions of sustainable development, i.e. economic, environmental and social, into account which are derived from customer and stakeholder requirements.

Similarly, Carter and Rogers (2008: 368) state that SSCM is characterised by

> the strategic, transparent integration and achievement of an organization's social, environmental, and economic goals in the systemic coordination of key interorganizational business processes for improving the long-term economic performance of the individual company and its supply chains.

Carter and Easton (2011) point out that in the SCM literature sustainability and environment have often been used interchangeably, though increasingly sustainability is applied to the triple bottom line comprising also social aspects as well as environmental and economic ones, as demonstrated by the definitions presented here.

We focus here on environmentally sustainable supply chain management, also called green supply chain management, which is broadly defined as incorporating environmental thinking into supply chain management (Srivastava 2007). Nieuwenhuis and Touboulic (2017) make the case for so-called ecological supply chain management as a sub-field of SSCM that focuses on resilience and is based on ecological principles, mimicking natural systems.[2] For Srivastava (2007) supply chain management here includes design, sourcing and selection of material, manufacturing processes, distribution of the final product to the consumers as well as end-of-life management of the product after its useful life (also called reverse logistics).

Touboulic and Walker (2015) stress that there are a range of definitions for SSCM, while Fahiminia and colleagues (2015) emphasise that there is no consensus definition for green and sustainable supply chains, particularly as for some authors logistics, reverse logistics, procurement and purchasing may be seen as synonymous for supply chain management. Generally, green supply chains include manufacturing/production, operations, purchasing, sourcing, performance measurement, product development and design, according to Fahiminia and colleagues (2015), to which logistics (reverse and forward) and closed-loop systems may be added.

As already emphasised in the introduction to this part of the book, the boundaries are not always clear. In our case, this means that we cover design, production and operations as well as the role of technology in the next chapter, whilst management systems are covered in Chapter 11 and reporting and measuring in Chapter 12. Product stewardship and extended producer responsibility, which are crucial in greening operations and supply chains, have already been covered to some extent in Chapter 7 on waste, and we will return to these in the next chapter. Nevertheless, cross-references will be made to emphasise the integrated and systemic nature of sustainability.

In the following we provide a brief overview of environmental (or green) supply chain management before detailing the elements and tools that help companies incorporate environmental thinking into SCM. We will use the terms green and sustainable SCM interchangeably, though acknowledging that we imply environmentally sustainable supply chain management since SSCM includes the social aspects of the triple bottom line as outlined earlier.

Overview

The focus of the remainder of the chapter is on how the supply chain management function can contribute to the solution for environmentally sustainability business, what opportunities are there for businesses to incorporate sustainability into their supply chains and what tools and concepts can they apply to make use of those opportunities.

Handfield and Nichols (1999) centre their approach to greening the supply chain around four areas: supplier selection and evaluation, surplus and scrap disposition, carrier selection and transportation of hazardous materials, and lastly, product design, packaging and labelling. Within these environmentally friendly supply chain practices they differentiate between proactive and reactive approaches, which we encountered in the approaches of reactive compliance and proactive prevention in Chapter 4. Green SCM signifies a paradigm shift where companies go from compliance and end-of-pipe control to minimising and preventing environmental damage whilst being overall economically profitable (Srivastava 2007).

Abukhader and Jönson (2004) highlight three main themes in SSCM – reverse logistics, assessment of emissions, and the greening of activities in the supply chain, which Grant et al. (2017) draw on, emphasising the increased use of life cycle assessment (LCA). Grant and colleagues (2017) cover warehousing, product design, cleaner production and packaging, purchasing and procurement as well as freight transport, reverse logistics and recycling in their coverage of principles and practices for sustainable operations whilst Zhu and Sarkis (2004) point to five major green SCM practices of eco-design, green purchasing, internal environmental management, customer cooperation with environmental concerns and investment recovery practices. Sarkis and Dou (2018) cover the following functions or dimensions for green supply chains: eco-design and customer-supplier relationships, procurement and purchasing, supplier collaboration and development, logistics and transportation, reverse logistics and circular economy, international relationships and multi-tier supplier management.

Seuring and Müller (2008) identify two strategies for making supply chains more sustainable: managing suppliers for sustainability in terms of risk and performance, and managing the supply chain for sustainable products. The first strategy uses supplier evaluation on the basis of set criteria and standards to avoid risks and disruption whilst striving for win-win situations rather than trade-offs between economic and environmental performance goals. The second strategy uses life cycle assessment and environmental criteria to manage impact and requirements along a product's life cycle. The two strategies are complementary and heavily emphasise communication and cooperation with suppliers (Seuring and Müller 2008).

In the previous chapter we encountered several "R's" of waste management, and SSCM can draw on these as well. Srivastava (2007: 54) lists reducing, re-using, reworking, refurbishing, reclaiming, recycling, remanufacturing and reverse logistics as some of the more proactive green SCM practices. To these we can add recovering, revalorising, and repairing from the previous chapter, and regulations as part of the overall legislative framework within which businesses operate and which have at times either forced or encouraged companies to incorporate environmental concerns into their operations. On a broader spectrum, the R's of rethinking, redesigning and reimagining remind us that a shift in thinking is often required to put companies on a path to environmental sustainability.

Bearing in mind that the functions covered by SCM vary, we now turn to major elements of supply chain management and how they can be made more environmentally sustainable.

Elements of sustainable supply chains

The full product life cycle from cradle-to-cradle includes not just the five main SCM processes of plan, source, make, deliver and return but also design, consumer use and end-of-life management (Bernon et al. 2012). Design and end-of-life management have already been referred to in the previous chapter on waste in the context of extended producer responsibility and product stewardship, and we will return to design in the next chapter on production and operations where we will cover the making of products. Nevertheless, design is crucial in greening the supply chain as three-quarters of life cycle costs of a product are determined at the design stage. In this chapter we consider the design of networks and processes in terms of logistics and purchasing. Consumer use was referred to in our discussion of waste and will be addressed again in Chapter 10 on marketing where consumer behaviour is included as well.

Network design

As supply chains are increasingly geographically dispersed and often global in nature, complex networks are the result, characterised by outsourcing and off-shoring. Integrating sustainability into supply chains therefore requires careful planning and design. Network design is concerned with determining the location choices of key facilities and how goods are moved between them (Oxborrow and Lund-Thomsen 2017).

Supply chains almost always come in the form of supply networks with multiple tiers of buyers and suppliers, upstream and downstream. These networks go beyond the traditional simple dyadic buyer-supplier relationship of the focal firm and are characterised by complex interdependence (Laasch and Conaway 2015; Sarkis and Dou 2018). Such networks of inter-dependent relationships encourage integration and collaboration along the supply chain (Vachon and Klassen 2006). The change in conceptualisation and organisation of the supply chain into supply webs and networks is an important one, as networks are critical to sustainability (Greening of Industry Network 2017).

Important design decisions as a result of a network view of the supply chain comprise location choices, firm boundaries in terms of internal and external activities, location of upstream and downstream activities as well as size considerations. This leads for example to distributed and networked factories, using microfactories and micromanufacturing as well as additive manufacturing, all recent developments in greening the supply chain (Greening of Industry Network 2017). One aspect of location design is the co-location of manufacturing, where companies in different or related industries make use of by-product synergies in eco-industrial parks – the concept is also known as industrial ecology, which was introduced in Chapters 3 and 7.

Network design for supply chains covers decisions as varied as for example quantity, location, size and layout of warehouses and manufacturing sites, distance of sites to customers, which products to produce and distribute from which site and when, whether to use a hub-and-spoke model, or whether to outsource or produce in-house, supplier choice, trade-offs between facilities and costs as well as service level, capacity and flexibility (Watson et al. 2013). How these decisions are made and optimised, for example with the help of mathematical modelling using quantitative data or by considering non-quantitative features, affects the cost structure, efficiency and effectiveness of any firm, and in each aspect opportunities for reducing the environmental impact and increasing overall resilience can be found, whether it is energy efficiency of sites, distances travelled and transport energy used or level of pollution as a result of these decisions.

Logistics, distribution and transportation

Although logistics and supply chain management are often used interchangeably, logistics comprises the transport, storage and handling of goods from raw material source through production and operations to the point of sale or consumption (McKinnon 2015). Activities associated with logistics management include for example inbound and outbound transportation, warehousing and inventory management, handling of materials, fulfilment of orders, planning of supply and demand, and manging third-party logistics (3PL) service providers (Grant et al. 2017). Green logistics involves studying the environmental effects of all activities involved in transporting, storing and handling physical goods as they move in forward and reverse directions along the supply chain (McKinnon 2015).

Location choices for manufacture, warehousing and distribution made in the first step of network design and supply chain configuration have an impact on logistics in terms of the planning, controlling and implementation of moving goods and information exchange through the supply chain (Oxborrow and Lund-Thomsen 2017). It is the freight transport of goods that is the most visible supply chain activity and environmentally the most damaging with regards to climate change and air quality primarily, with other impacts being noise pollution, traffic congestion and road accidents. Transport uses about one-quarter of the world's energy and is responsible for 20 to 25% of global GHG emissions; thus decisions made in this area have the potential to either increase a firm's environmental burden or to significantly reduce it.

In addition to achieving the aim of minimising their costs in the supply chain whilst providing the agreed level of service for customers by managing the costs of transport, warehousing and inventory, companies with logistics operations are increasingly under pressure from stakeholders to mitigate their transport impact (Bernon et al. 2012). In particular, in light of the aforementioned trends of global production networks, outsourcing and off-shoring, there are trade-offs between costs and distance travelled as part of the local-global argument as to whether supply should be sourced off-shore or locally – also called nearshoring (Grant et al. 2017; Oxborrow and Lund-Thomsen 2017).

There is a general consensus that transporting goods less often and in larger volumes is more cost effective and less environmentally damaging, which has led to greater consolidation. In the case of global shipping, this trend of consolidation has given rise to super-ships, which also cut the cost per good shipped. However, contemporary supply chain models such as just-in-time processes and fulfilment of online orders focus on shipping smaller amounts and moving storage from physical warehouses onto the road where consolidation might not be appropriate (Oxborrow and Lund-Thomsen 2017). Consolidation often leads to centralised hub-and-spoke logistical networks with often negative impacts at the centre of these networks. Similar to the trade-offs between local and global arguments, there are inherent paradoxes of green logistics, which in addition to hub-and-spoke models include warehousing and just-in-time logistics, minimising costs while meeting the requirements for speed, flexibility and reliability, as well as e-commerce which requires small individual shipments which affect the efficiency of logistics in terms of more packaging and customised transportation (Laasch and Conaway 2015).

The main transport modes are road, rail, air and sea, which are chosen on the basis of operational and consignment factors – such as availability, accessibility and quality of infrastructure, taxation and legislation – as well as product characteristics, cost and service requirements such as for example speed, value and weight density (Grant et al. 2017). Strategies to reduce the environmental impact include changing the mode of transport, reducing the need for transport, and reducing the (GHG and pollutant) emissions of modes used by considering alternative fuels, operational options, making use of information technology, and infrastructure and transport design.

We briefly cover each mode highlighting the environmental impact and what can be done to reduce this (based on Grant et al. 2017; Robertson 2017; Oxborrow and Lund-Thomsen 2017; Weybrecht 2014; Bernon et al. 2012):

- *Road*: the most used form of transport in logistics and therefore the one with the most emissions contributing to climate change and air pollution as well as noise pollution; highly accessible and flexible. Switching to alternative fuels such as LNG (liquefied natural gas) and biofuels[3] as well as to more fuel-efficient vehicles, such as hybrid vehicles and electric cars, as well as to differently designed trucks that are more aerodynamic and thus more fuel-efficient, can reduce impact, whilst changes to current practices may include voluntarily restricting vehicle speeds, regulating tyre pressure, optimising capacity and load configuration, avoiding empty trucks (known as deadhead trips or empty miles) and therefore underutilisation, use of telematics systems to optimise routes and to reduce the number of miles or kilometres driven, voluntary (or automatic) idle engine turn-off, and driver training programmes.

- *Rail*: slower mode of land transport, often used for heavy loads; based on fixed and expensive infrastructure, but less susceptible to disruptions. Environmental impact depends on how trains are powered (electricity or conventional fuels), but generally considered a greener form of transport – on average second lowest GHG emissions per distance travelled, though carbon intensity could be reduced by greater electrification.[4] Increased use of rail transport would need greater investment into infrastructure which would decrease environmental impact from other modes.
- *Air*: fastest mode of transport and also the most carbon-intensive as well as the most expensive, fast-growing sector. Advances in aircraft technology and design can improve fuel efficiency, whilst the use of composite materials reduces the weight and fuel use.
- *Sea*: slow and cheap mode for transporting low-value, high-volume items, maritime shipping underpins global trade, responsible for 90% of international trade. Lower environmental impact than air or road, but bunker fuel used by most ships particularly pollutant. Ways to improve environmental sustainability of this mode include provision of electricity whilst at port, use of wind energy through towing kite systems to assist propulsion and thus reduce fuel needs, switching to LNG fuel, so-called slow steaming which saves fuel by running ships at slower speeds, and altered design of shape and surface that mimic nature to improve fuel efficiency.

In addition to the mentioned modes, pipelines are used to move large quantities of gas or liquids but require high investment, high volume and predictable demand.

In practice, the environmental impact of each journey varies and depends not just on mode of transport, but on type of fuel used, size of vessel, type of vehicle, engine efficiency and distance travelled, making comparisons not a straightforward exercise.

This is exemplified in the concept of *food miles,* which uses a single characteristic of miles travelled to gauge the environmental impact, in particular CO_2 emissions, of transporting food over large distances. It does not take into account the mode of transport used, and, more significantly, does not consider the environmental impact of production, seasonality, storage (e.g. temperature-controlled storage to reduce food waste) and transport efficiency – i.e. the whole life cycle of the product. Whilst there is some correlation between journey distance and emissions, there are also many exceptions, rendering this seemingly intuitive concept more complex and relative than at first glance[5] (Bernon et al. 2012; Garnett 2015).

In many cases, switching transport mode entirely to other forms is not realistic, particularly as sea and rail very often require road transport for the final part of the journey. Multi-modal transport across different modes is therefore frequently the reality, which has been made easier with the spread and use of intermodal solutions such as standardised pallets and containers or roll-on roll-off (RORO) ferries or trains (Grant et al. 2017).

One particular challenge for this part of the supply chain is last mile logistics, which is the final step of the distribution process, delivering the product to the end user, though the actual distance can extend to dozens of miles. With the increase in e-commerce, the importance of last mile logistics has increased significantly in terms of costs, particularly in the context of free and fast shipping options, and inefficiencies in the cases of failed delivery and repeated delivery attempts for order fulfilment. It is often the most time-consuming and expensive part of the shipping process (Dolan 2018). Environmentally, it increases congestion and air pollution, which has led to calls for banning internet shopping delivery vehicles from city centres, such as for example London.

Last mile logistics have increasingly been outsourced to so-called third-party logistics (3PL) providers with over three-quarters of warehousing and transportation activities performed by specialist firms (Grant et al. 2017). These can make use of data and technology such as telematics and tracking systems to optimise routing and reduce environmental impact. Other developments in reducing environmental impact of last mile logistics include urban consolidation centres, click-and-collect models, piloting delivery by drone and increased use of electric vehicles.

Two further aspects of logistics, distribution and transportation where environmental impacts can be reduced include warehousing and packaging. The conventional warehouse function is the receiving, storing and dispatching of goods, though cross-docking does not include storing, but simply receiving, processing and dispatching (Grant et al. 2017). Despite just-in-time initiatives and other processes to reduce inventory, warehouses are still found in almost all supply chain stages. Although transport makes up most of the GHG emissions generated in supply chains, warehousing constitutes about 13% of emissions in the supply chain, and in the case of the UK, is responsible for 3% of the country's total GHG emissions[6] (Grant et al. 2017). The overall environmental impact is during the construction phase and the operations stage, whilst their location determines the amount and mode of transport needed in logistics. The three main ways to reduce the environmental impact of warehouses is through improving energy efficiency, using renewable energy, and designing and constructing buildings with sustainability in mind (Baker and Marchant 2015).

In the construction phase, schemes such as LEED and BREEAM can be used – these are voluntary sustainable building award schemes and set best practice standards for sustainable design. BREEAM (Building Research Establishment Environmental Assessment Method) was established in the UK, whilst LEED (Leadership in Energy and Environment Design) is a US-based scheme, with other countries developing similar programmes. During the operational phase, using green energy can either be in the form of sourcing from energy suppliers or through generating electricity on site through small-scale windfarms, solar panels or other forms of renewable energy (Grant et al. 2017). Regardless of what kind of energy is used, energy efficiencies can be achieved by making changes to warehouse lighting, handling equipment and temperature control (Baker and Marchant 2015; Grant et al. 2017).

We discussed packaging in the previous chapter on waste and how this can be addressed but briefly return to it here as it is required for practically all goods ranging from drinks, food, consumer products such as electronics or cosmetics, healthcare, cleaning products and clothing to larger items such as white goods and furniture. It serves to protect goods and allows transport as well as marketing to consumers and is thus crucial to SCM and in particular logistics (Grant et al. 2017). Materials used for packaging include glass, paper, cardboard, plastics, wood and metal, which differ in their environmental impact in terms of raw materials and energy used in production, fuel used for transport based on their weight and the waste at the end of the packaging's useful life. For logistics providers, weight of packaging is an important consideration, where companies along the supply chain, from retailers to producers to logistics providers can all contribute to reducing the environmental impact of packaging. For example in response to Walmart's requirement to reduce packaging by 5%, detergent manufacturers removed water from their laundry detergent formula to sell concentrated products in smaller bottles requiring less plastic, water and cardboard, leading to more efficient loading of trucks and therefore fewer miles on the road (Robertson 2017).

The discussion so far pertains to what companies can do to reduce their environmental impact in logistics. Before moving on, we want to emphasise that regulatory changes and the design of infrastructure – for which public bodies are responsible – can also contribute to making logistics more environmentally sustainable. Infrastructure planning and design can make railways a more viable mode of transport whilst congestion charging or environmental zones, taxation of fuels and subsidies for alternative fuels can encourage changes in road transport. Various pieces of national and international legislation have led and can lead to improvements in vehicle design such as through emissions regulations, in energy performance of buildings and in packaging and its waste.

In this section we have mostly covered the one-way flow of forward logistics from raw materials and producers to consumers. Reverse logistics are a growing area in SCM and concern the opposite direction of typical supply chains. As it is almost synonymous with waste management and material recycling, which we have partly covered in the previous chapter on waste, and as it is a broader concept within supply chain optimisation overall, we will return briefly to it later in the chapter when we cover general tools and concepts for greening the supply chain.

Procurement, sourcing and purchasing

Procurement and purchasing are often used interchangeably and even synonymously with SCM – similarly to logistics – as they could be seen as precursors to SCM which has evolved from procurement and purchasing (Browne 2011; Fahiminia et al. 2015; Touboulic and Walker 2015; Grant et al. 2017; Sarkis and Dou 2018). As a general rule, procurement is a more strategic operation including all activities of acquiring goods or services, whereas purchasing is more narrowly focused on the buying process and thus on operational activities (Grant et al. 2017). Traditionally, purchasing has been very cost-driven without much regard for the environment (or society), but recent developments have moved the function more towards procurement and a more strategic supply chain approach to risk and value management which has also included sustainability (Browne 2011). Other trends in this area include the increase in outsourcing and off-shoring, which means that there is greater complexity and a much higher proportion of third-party spending for products and services bought from suppliers, often over 50% of operating costs. Such bought-in elements can generate the highest environmental impact of organisations. This is not just an environmental risk for buying organisations, but also a commercial risk regarding supply, demand, security and operations, and, more significantly reputation and credibility, which all drive sustainable procurement (Grant et al. 2017; Walker et al. 2008).

Procurement focuses on the upstream supply chain and includes not just services and goods, but also raw materials, contracts for services (e.g. facilities management) and utilities as well as contracts for works such as construction (Browne 2011). Sustainable procurement minimises the environmental impact by including sustainability considerations into procurement strategy and decisions, such as for example sourcing or selection of suppliers – as Grant et al. (2017: 32) phrase it – "sustainability is also about ensuring the source of product".

The environmental impact of procurement varies, depending on type of good, service or contract sourced and their environmental footprint, logistics and packaging used (see earlier discussion), waste generated, environmental legislation, compliance (or lack thereof) and supplier practices, to name just a few. In order to prioritise which areas to focus sustainable procurement

TABLE 8.1

Matrix for sustainable procurement strategies

	Low spending value	High spending value
High sustainability impact	*High risk, low spend – Bottleneck* – Buying power possibly low, push for industry-wide standards or regulation – Procurers need effective relationship and risk management – Choose suppliers based on sustainability criteria	*High risk, high spend – Strategic* – Cost and relationship management key for procurers – Manage cost and sustainability – Choose suppliers based on sustainability support and cooperative culture – Sustainability as performance criterion through contract and close supplier engagement – New product development and innovation at focal firm and supplier
Low sustainability impact	*Low risk, low spend – Non-critical* – Generic and simple sustainability criteria in contract – Use certifications to select suppliers	*Low risk, high spend – Leverage* – Best practice sharing with suppliers for improvements – Simple and basic sustainability criteria in contract

Source: adapted from Browne (2011), Grant et al. (2017), Krajlic (1983).

efforts on beyond quick wins and fast deliverables such as for example eliminating paper and printing costs by moving purchasing processes online, companies can draw on the matrix framework developed by Kraljic (1983) and adapted for sustainability purposes. The original framework categorises items to be supplied by profit impact and supply risk (low versus high), which for a sustainable procurement strategy is modified to position expenditures according to sustainability impact and value of spend (Browne 2011; Grant et al. 2017). Some possible strategies within the matrix are provided in Table 8.1.

Once sustainability has become an essential criterion in the procurement decision-making process, it is important to understand how the environmental footprint of the purchased and sourced goods, services, works or contracts can be assessed, which in turn allows comparison between suppliers during selection (Grant et al. 2017). Certification and life cycle assessment enable informed decision-making for sustainable procurement.

Sustainability labels and certifications play an important role in selecting suppliers. As they can be used more broadly in greening the supply chain (and other corporate functions), we cover them in the next part of the chapter on general tools and concepts for SSCM. Certifications play an important role upstream, whereas labels are more significant for end consumers, which implies that marketing and purchasing functions need to work together in order to make supply chains more sustainable (Oxborrow and Lund-Thomsen 2017). Within procurement, the use of certifications takes the burden of auditing and monitoring off the buying side and simplifies sustainability, as standards are set externally (Grant et al. 2017).

Life cycle assessment (LCA) was introduced in Chapter 3 as a tool to measure the impact of a good or service on the environment from cradle to grave, i.e. its entire life cycle. It is the most comprehensive way of determining impact and also the most accepted method. LCAs vary in resource requirement, level of detail and sophistication, resulting in either detailed LCAs, which are the most accurate and more quantitative in nature, or in more simplified LCAs, such as streamlined LCAs based on average or generic data, and conceptual LCAs which are more qualitatively focused (Grant et al. 2017). As with certification, LCA can be used more widely in making SCM (and other functions) more sustainable, and will be covered in the next part of the chapter.

One variation of LCA that focuses on greenhouse gas emissions is the Greenhouse Gas Protocol (GGP) introduced in Chapter 5 on climate change. The GGP demonstrates the importance of suppliers and therefore of procurement and purchasing, as identifying impact according to Scope 1, 2 and 3 often reveals that most of the carbon impact of firms is found in Scope 3 emissions, which are outside the control of the focal firm.[7] Supplier selection, management and development is therefore becoming increasingly crucial for companies in managing their environmental impact – an element of SCM we cover in the next section.

Looking specifically at the purchasing function within procurement, green purchasing or environmental preferable purchasing (EPP) provides companies with a systematic process for selecting goods and services not based just on price but on full lifetime costs taking into account environmental impacts throughout life cycles using e.g. LCAs (Robertson 2017; Fogel 2016). Buying green products can occur in any supply chain stage, and green purchasing policies can be developed to guide procurement in the form of specific selection criteria, contract specifications, collaboration with suppliers and communication with end users (Robertson 2017). Common impact categories in assessing environmental performance for EPP consider the maximisation of biodegradability, recyclability and recycled content, durability, reusability and overall product performance as well as the minimisation of product safety hazards, noise, odour, radiation, waste, land, water and energy use as part of wider ecological consequences (Fogel 2016).

Taking into account risk assessment and management, supplier selection and development, the marketing of sustainable procurement, and integration of environmental considerations into the procurement process, and maintaining excellence throughout, one other element of a sustainable procurement strategy is staff training and awareness (Browne 2011). Staff development is also emphasised by Seuring and Müller (2008) and will be covered in Chapter 13 on human resource management.

As we have seen, supplier selection and broader sourcing and purchasing decisions play a significant role in controlling and reducing environmental impact in the supply chain and depend to a great extent on collaborating and interacting with suppliers (Grant et al. 2017), which extends the chain or responsibility in line with extended producer responsibility (EPR) and product stewardship.

Collaborations and partnerships with suppliers

Compared to conventional supply chain management, sustainable SCM needs greater cooperation among partnering companies (Seuring and Müller 2008), also called supply chain integration. This section is therefore equally relevant and applicable to the previous sections on network design, logistics and procurement, especially as the complexity involved in managing environmental sustainability means that embedding and integrating it requires great commitment and solid understanding of the issues at stake across supply chain partners (Bernon et al. 2012). Furthermore, close supplier relationships help to mitigate the risks to brand, image and reputation that can arise when such relationships fail, for example when companies are unknowingly exposed to negative environmental impacts through supplier practices (Weybrecht 2014). Companies are increasingly aware that "responsibility does not stop at the factory gate" and that they are now expected to use their leverage to raise the environmental performance of all stakeholders, in particular suppliers (Blowfield 2013: 122). Essential to such product stewardship across the whole supply chain is effective communication and partnering with suppliers.

The relationship with suppliers is especially important as most of a focal firm's environmental impact can be in the supply chain, as highlighted previously. The Greenhouse Gas Protocol (GGP) as an accepted standard for measuring carbon emissions helps to identify risks and opportunities for addressing a company's environmental impact in the supply chain, in particular when including the indirect emissions in Scope 3 from processes outside the control of an organisation. For some sectors, measuring Scope 1, 2 and 3 GHG emissions shows that most of the carbon footprint is in Scope 3. For example, for the British construction company Willmott Dixon 99% of emissions are from Scope 3, whilst Scope 1 and 2 amount to 1% of the firm's GHG emissions, whereas for Telefonica Scope 3 accounts for 90% (UK Forum for Supply Chain Sustainability 2017). Similarly for pharmaceutical and healthcare firm GSK, 20% of the carbon footprint is from internal operations and 80% from the rest of the value chain (Paddison 2013). It is an area that companies in the past have not paid much attention to, particularly in the context of distant supply chains characterised by outsourcing and off-shoring. Building resilience into the supply chain is vital for ensuring the economic and environmental sustainability of companies in the light of climate change and other environmental risks.

According to Oxborrow and Lund-Thomsen (2017), greening the supply chain can happen in two ways, either as a result of imposition or as a result of collaboration. Imposition may be the result of regulations or codes of conducts and strategic objectives by more powerful companies further down the supply chain, such as for example brands and retailers. Collaboration is however more likely to yield positive results – as suppliers are a vital factor, careful relationship management and development are key. If suppliers are motivated to improve and green their own practices, this can reach further up and down the supply chain (Browne 2011). Sharing of information and of goals are key elements (Vachon and Klassen 2006) as are traceability and transparency (Oxborrow and Lund-Thomsen 2017).

Weybrecht (2014) provides the following pointers for building strategic supplier relationships:

- Policy to help suppliers improve environmental sustainability performance.
- Collaboration with suppliers to assist them in developing own sustainability strategies.
- Perform baseline assessment of current sustainability performance of suppliers.
- Consider second- and third-tier suppliers – suppliers of suppliers.
- Consistent communication.
- Work with and support local suppliers.
- Use a step-by-step approach with tiers or levels in recognition of different stages of sustainability development within supplying companies.

Seuring and Müller (2008) also confirm the cooperation beyond first-tier suppliers (often the focus of conventional SCM) and the use of extended criteria for sustainability to complement conventional economic ones to evaluate suppliers.

Companies can use environmental management systems to audit suppliers on regulatory compliance, product and service specifications (e.g. phasing out of harmful materials), and environmental practices or management systems, which can help suppliers improve their environmental performance (Marsden 2011). Such quality management systems also entail total quality management (TQM) approaches that are built on continuous improvement (Laasch and Conaway 2015). Similar to TQM, total quality environmental management (TQEM) is a pervasive programme that can be used within and between organisations, and that should include customers and suppliers (Sarkis 2003). Other tools include the balanced scorecard, which can be used internally as well as externally (Blowfield 2013). We cover environmental management systems in more detail in Chapter 11.

Before we move to more general tools and concepts used in greening supply chains, we should emphasise that with global supply chains, focal firms are often faced with suppliers in countries with lower environmental standards, often chosen on the basis of cost and flexibility. Whilst this represents significant challenges, it also provides focal firms with an opportunity (and responsibility) to share good practice and help suppliers improve their environmental performance. For example, Walmart's commitment to environmental sustainability and low cost requires its global suppliers to move towards sustainable practices if they want to retain Walmart's business (Stead and Stead 2014).

Conditions that support what Sharfman and colleagues (2009) call cooperative supply-chain environmental management include interfirm trust, uncertainty surrounding environmental issues, strong regulation and pro-active and shared commitment to addressing sustainability, which also affects the economic bottom line.

The relationship with suppliers may also include collaboration from the design phase in order to embed sustainability from the start – so-called design for the environment (DfE, Grant et al. 2017). As sustainability covers all aspects of a product's life cycle, a cyclical understanding is key for suppliers and buyers, and thus the relationship spans the entirety of the supply chain and a product's life from cradle to grave. We will focus on design in the next chapter, though it is important to stress the role of suppliers here.

Tools and concepts

Some of the following tools have been introduced in previous chapters, illustrating the overlap of concepts and the holistic and systemic nature of environmental sustainability, regardless of which function or area or industry we consider. Here we briefly return to them, focusing specifically on their contribution to greening the supply chain.

Crane and Matten (2016) highlight three key levels of sustainable supply chains: firstly, sustaining the supply chain to ensure continuity and effective sustainability; secondly, turning supply chains into supply loops within the broader concept of a circular economy which uses end-of-life strategies to eliminate waste and reduce material inputs; and thirdly, industrial ecosystems which move companies beyond single product supply, recapture loops and encourage a more system-oriented view. Some of what we have covered thus far has focused on sustaining the supply chain and ensuring effectiveness, whilst touching on circular and systemic thinking. The following tools and concepts contribute to all three levels of SSCM, with particular emphasis on closing the loop and industrial ecosystems.

Circular economy – closing the loop

The circular economy is a concept that has gained increasing traction in the context of sustainability within industry, research and policymaking (Geissdoerfer et al. 2017). We have referred to it in previous chapters (e.g. in Chapters 3 and 4, and in Chapter 7 on waste), where it was introduced as an alternative to the traditional take-make-waste (or dispose) linear economy (Bocken et al. 2017).

The Ellen MacArthur Foundation (EMF 2013: 7) has played an important role in promoting the transition to a circular economy, which it defines as an "industrial system that is restorative or regenerative by intention and design". Geissdoerfer and colleagues (2017: 766) provide a more detailed definition of the concept[8]:

> A regenerative system in which resource input and waste, emission, and energy leakage are minimised by slowing, closing, and narrowing material and energy loops. This can be achieved through long-lasting design, maintenance, repair, reuse, remanufacturing, refurbishing, and recycling.

A circular economy is often seen as a condition for sustainability, though other interpretations involve a trade-off, with more emphasis on the economic bottom line and less focus on long-term solutions (Geissdoerfer et al. 2017). The circular economy concept has been influenced by cradle-to-cradle and industrial ecology – to which we turn later – and by biomimicry which we cover in the following chapter. Other influential schools of thought that have refined and developed the circular economy concept include the performance economy, natural capitalism, blue economy and regenerative design (EMF 2017). Central to all is the idea of closed loops (EMF 2013).

In the context of this chapter, this involves turning supply chains into supply loops. Whilst in a circular economy one company's waste can become another company's raw material, closed-loop supply chains often recover materials for reuse by the same organisation, also called a cradle-to-cradle approach to manufacturing (Laasch and Conaway 2015). This means that the traditional forward (or downstream) flow of supply chains is complemented by a reverse supply chain moving back upstream to enable revalorisation, which reintegrates products or services back into the supply chain by techniques such as repair, refurbishment, remanufacturing, recycling, upcycling or downcycling (Laasch and Conaway 2015).

A closed-loop supply chain that is well-designed and implemented can underpin a business system that is both economically and environmentally sound (Abbey and Guide 2012). This should comprise design, operation and control of a system for maximising value creation across the entire life cycle of a product with recovery of value from different volumes and types of return over time (Guide and Van Wassenhove 2009). Product recovery can be broken down into reduce, reuse and recycle, which can be achieved through various means, one of which is servicising or product-service systems, which build reuse into consumption (Abbey and Guide 2012). Servicisation (or servitization) is a crucial concept to sustainability, as encountered before and one we will return to in the next two chapters.

Closed loops are critical to eco-effectiveness, which uses cradle-to-cradle design to create products that are not only less bad (eco-efficient) but also net positive (Blowfield 2013). Like cradle-to-cradle design – as developed by McDonough and Braungart (2002) – the circular economy differentiates between biological and technical nutrients or ingredients. Whilst biological nutrients are considered consumable components that are non-toxic and can return to the biosphere, technical nutrients are durable and form part of closed-loop technical cycles that circulate within industry, designed from the start for reuse (EMF 2013; Robertson 2017).

The transition to a circular economy will take time, which first requires "bending" the linear take-make-waste model by addressing the two ends of raw materials and end products through e.g. recycling before a full circular model can be achieved that is characterised by total recovery. In some industries, progress has already been made – for example almost three-quarters of glass bottles in Europe are collected and recycled, and scrap steel accounts for up to 50% of new steel products (Accenture 2017). The circular economy approach helps to prioritise the management of reverse logistics and recycling by providing a hierarchical framework similar to the waste hierarchy introduced in the previous chapter (Grant et al. 2017).

One example of a circular business model is Danish company Vigga, which offers a subscription model for renting baby clothes. This involves reverse logistics – returning used baby clothes, which are then repaired and cleaned, recycled and reused. Its founder Vigga Svensson has launched The Continued™ Platform to help other companies in the fashion industry transition towards circular models by remodelling fashion as a service rather than product, allowing brands to sell more but buy less as well as substantially reduce their negative environmental impact.

The increasing importance of a circular economy is reflected in regulation as well; for example the EU has launched a Circular Economy Action Plan which includes proposals on plastics and its recycling as well as addressing the interface between chemical, product and waste legislation, whilst China has implemented the Circular Economy Promotion Law which aims at facilitating a circular economy, raising resource utilisation, environmental protection and sustainable development.

Cradle-to-cradle

Cradle-to-cradle design takes sustainability into account across the whole lifetime of a product, and is an example of a circular economy approach. Whilst we have introduced the concept in Chapters 3 and 4 and return to it in the following chapter on design and productions, we highlight its relevance for the supply chain here. Differentiating between biodegradable biological nutrients and recyclable technical ones allows firms to source materials and design products that are easy to disassemble and to recover (EMF 2013). Restoring technical nutrients is a core objective of reverse logistics (Grant et al. 2017).

Firms can apply for Cradle to Cradle (C2C) certification for products which are assessed according to their material health, material reutilisation, use of renewable energy, water stewardship and social responsibility (Robertson 2017). As emphasised before, use of certification and accreditation is an important method to green the supply chain.

Industrial ecology

Industrial ecology facilitates the transition from the linear take-make-waste production processes to a more circular economy (Saavedra et al. 2018). The concept was briefly referred to in Chapter 3 and then introduced in Chapter 7 – a central aspect of it is the use of waste as a resource. It is aimed at understanding how material and energy circulate and flow, which starts first with understanding the industrial ecosystem and then the application of tools such as industrial symbiosis and eco-industrial parks (Saavedra et al. 2018). A key concept within industrial ecology is industrial metabolism, which links this concept to the cradle-to-cradle approach of nutrient cycles.

Industrial ecology broadens the supply chain as it is not confined to a single industry. Industrial ecosystems go beyond single product supply and recapture loops, and can be seen as a network where the number of supply chain links are growing. Some of these networks have developed without careful planning but have emerged as a result of bi-lateral arrangements. Industrial ecosystems are characterised by symbiosis and often by co-location, leading to the emergence of so-called eco-industrial parks (Klassen and Vachon 2012).

Ehrenfeld (1997) draws a comparison with The Natural Step, which has much in common with industrial ecology, although not in name. We introduced The Natural Step framework in

Chapters 3 and 4 – it stipulates that extraction and production of substances as well as the physical degradation of nature should be eliminated. This covers for example fossil fuels and minerals, toxins and resource extraction.

Industrial ecology can provide a practical framework that along with material and energy flows encourages product life cycle planning, assessment and design, extended producer responsibility (or product stewardship), co-location of industrial facilities and eco-effectiveness (Laasch and Conaway 2015). It is characterised by controlling metabolic pathways and toxics use reduction, loop-closing practices, dematerialisation and systematising energy usage (Ehrenfeld 1997). In the context of eco-effectiveness, we will return to industrial ecology in Chapter 9.

Life cycle analysis

The life cycle of products is a core concept in industrial ecology (Robertson 2017). We introduced life cycle analysis/assessment (LCA) in Chapter 3 on tools that guide companies and regulators. As a reminder, it is a process to assess and evaluate the environmental and resource-related consequences of a product in all its life phases, which of course reflect the supply chain – extraction and processing of raw material, production, logistics, use, remanufacturing, recycling and final disposal (Srivastava 2007).

LCA can be applied either to the whole product, to only one aspect of the system or to one particular material decision. It covers design, sourcing material, production and manufacturing, packaging and transportation, use as well as disposal and end of life (Weybrecht 2014). This makes it particularly suitable for use in SSCM as it is a total systems approach for products and processes along the supply chain. Grant et al. (2017) outline the four steps involved in LCA:

- Definition of scope and goals – choice of function, system or product, and boundaries.
- Data collection and inventory analysis: recording and assessing of input.
- Impact assessment: evaluation of emissions.
- Interpretation of inventory analysis and impact assessment.

LCA is increasingly used in logistics and SCM to consider the environmental impacts beyond cost-benefit analysis. Frameworks that companies can draw on include the ISO standard 14040 which covers the principles and framework for LCA, and ISO 14044 on the requirements and guidelines for LCA (Weybrecht 2014).

In the context of procurement, life cycle thinking can be used in contracts for products as well as in service contracts, though the issues considered might differ (Browne 2011). In wider supply chain management, LCA shows where there is scope for better environmental performance within the focal firm as well as regarding supplier selection (McLaren 2011).

Reverse logistics and reverse supply chains

Reverse supply chains are a growing phenomenon as a result of regulation, cost pressures and shortage of materials. They form a significant element of closed-loop supply chains and cradle-to-cradle design within the broader context of transforming the economy towards a circular economy (Grant et al. 2017; Oxborrow and Lund-Thomsen 2017). Reverse logistics is concerned with moving products from the point of consumption or end user towards the producer or point of origin and may include not just end-of-life products but also product returns and recalls, amongst others (Grant et al. 2017). The purpose of this reversal of the supply chain is to recapture or create value, or to dispose of products properly. Reverse logistics can apply to products or to packaging and includes returns and secondary markets, recycling, remanufacturing and reusable packaging (Rogers and Tibben-Lembke 2001). The management of the process is also called product recovery management (PRM, Grant et al. 2017).

Reverse logistics, product recovery and remanufacturing fall under the concept of product stewardship (Vachon and Klassen 2006), which we covered in the previous chapter and will return to in the following when we look at product design. Designing products with the end of life and recycling in mind allows companies to address both customer returns as well as producer responsibility legislation such as take-back schemes.

Examples of companies incorporating reverse logistics into their supply chain and production processes include Kodak, Bosch, IBM, BMW, Xerox and Hewlett Packard, amongst others.

Certifications and labels – using standards to green the supply chain

Within all stages of the supply chain and for all tools and concepts covered up to now, certification can help companies in making their supply chain more sustainable. In general, we can differentiate between labels which are used for marketing and consumers and thus downstream, whereas accreditation, standards and certification are more often used upstream. Whilst labels such as the Marine Stewardship Council (MSC), Rainforest Alliance, Fairtrade or the Forest Stewardship Council (FSC) are used to encourage more sustainable practices throughout the supply chain, they are also more focused on consumer-facing environments, and so we will return to them in Chapter 10 on marketing.

Within the supply chain, certification and labels help with traceability and transparency, which are increasingly expected by stakeholders, particularly customers and consumers (Oxborrow and Lund-Thomsen 2017). They are therefore often used as a generic supplier selection criterion.

One drawback of certification is that it often focuses on particular aspects of sustainability, which can make comparison difficult and sourcing decisions more complex (Grant et al. 2017). The use of life cycle assessment can address this limitation, though LCAs are more complex and require more resources and data. Nevertheless, certification provides a level of benchmarking against which suppliers can be measured as well as compared.

The International Standards Organisation developed the ISO 14000 series, which is a suite of standards for establishing and improving environmental management, based on continuous improvement (Grant et al. 2017). ISO 14020 specifies the different types of labels and certifications – they can either be verified independently by a third party based on certain criteria (type I), they can be self-declarations (type II) or third party verified labels based on a full LCA of the product (type III) (Robertson 2017). Third party verification provides the most meaningful and most credible set of standards.

Standards that address the sustainability of buildings include LEED (Leadership in Energy and Environmental Design), which is used in the US, BREEAM (Building Research Establishment Environmental Assessment Method) in the UK, DGNB in Germany and Green Star in Australia (Fogel 2016). These can be used to incorporate sustainability in warehousing and inventory for new and existing buildings and refurbishments (Grant et al. 2017). Other examples of certifications include the aforementioned C2C certification, Carbon Trust, EMAS (Eco-Management and Audit Scheme), as well as the Forest and Marine Stewardship Councils.

Overall, certification and standardisation in the supply chain can signal the compliance of suppliers, enabling more informed supplier selection based on sustainability criteria, whilst also signalling attractiveness to potential customers (Laasch and Conaway 2015).

Spotlight on . . . examples and practices of sustainable supply chain management

In the following, we provide examples of companies and practices in sustainable supply chain management. Although some have been mentioned in previous sections, this part draws examples and practices together as these often overlap across the different elements, tools and concepts used to green the supply chain.

Examples of network design for more sustainable supply chains include the use of co-creation and microfactories by GE Appliances which works with additive manufacturing such as 3D printing and open-source technologies (Economist 2016), e-waste microfactories that reclaim and recycle smartphones into valuable materials (Nogrady 2016), and the co-location of processes within and across companies, such as for example the Green Zone in Sweden which involves Ford, McDonald's and Statoil (see https://greenzone.se/ for details). Chemical company BASF's integrated system Verbund provides cost advantages for logistics, energy and infrastructure as well as reduces environmental impact such as pollutant emissions and waste. In 2018 BASF announced it would invest in a Verbund chemical production site in Southern China (Junwei 2018), thus spreading the use of sustainable supply chain practices through network design.

Sourcemap is a technology platform that enables supply chain transparency by allowing companies to trace products to their origins and to collect data from suppliers which is then visually presented in interactive maps. The tool allows companies to make more informed sustainability decisions, particularly with regards to sourcing and purchasing, and to communicate more transparently. Chocolate manufacturer Hershey used Sourcemap to plot its production sites and origins of key ingredients. Hershey also led the development of SmartLabel™ QR codes to provide consumers with detailed information about products, ingredients and origins.

Regarding distribution and transportation, the German logistics provider and national mail company Deutsche Post invested in its own electric vehicles when taking over the start-up Streetscooter after major car companies had failed to offer suitable solutions. The electric van has proven so successful that in 2018 a second plant was opened, and the vehicle is not only distributed and used internally but also offered to third parties, pre-empting planned bans of diesel vehicles in major city centres (Tagesschau 2018). In Stockholm, the *älskadestad* (beloved city) project combines last-mile distribution and delivery of parcels with waste collection, transported by electric vehicles.

Across different modes of transport, FedEx replaced 25% of its van fleet with hybrids which are over 40% more fuel efficient than its conventional part of the truck fleet, and invested in Boeing 777s to improve carrying capacity by 20% which also cut daily fuel use by over 30%, underpinned by new software programmes for optimising aircraft scheduling (Stead and Stead 2014). IKEA has saved up to 15% on fuel per item by packing trucks and trains much tighter, in addition to using The Natural Step framework (see e.g. Box 4.7).

Finland-based RePack has developed reusable packaging in adjustable sizes to help realise a circular economy in e-commerce where it partners with retailers to offer a RePack choice for consumers when buying goods online. The packaging reduces CO_2 emissions by up to 80% and can be (re)used at least 20 times. Consumers who choose this delivery option and return the packaging are rewarded with vouchers. The approach brings the circular economy concept to the consumer level and relies on working with other companies.

As highlighted in the chapter, collaboration and partnerships with suppliers is crucial to green the supply chain. The Supply Chain Sustainability School is an industry-wide collaboration of clients, contractors and suppliers providing British companies in the construction, homes, infrastructure and facilities management sector access to free industry-relevant training in sustainability. Over 70 partnering companies pay for the school, whose resources are used by over 2,500 companies and over 4,500 individual members for free. It enables self-assessment followed by an action plan that breaks down the complexity of sustainability in the supply chain into manageable action (Greening of Industry Network 2017).

Examples of retail companies building relationships and collaborations for sustainability in the supply chain include Sainsbury's sustainability scorecard and supplier training, Tesco's Knowledge Hub for suppliers to share their carbon efficiency boosts, and Marks & Spencer's Supplier Exchange programme, where carbon-cutting expertise can be shared.

Dutch company Philips uses LCA as an eco-innovation tool and has applied circular economy principles through its EcoDesign process since 1994. This has been developed to include energy, packaging, substances, weight and materials, circularity and lifetime as six key so-called Green Focal Areas. When the company's products outperform competitor products, existing products or minimum regulatory requirements by at least 10% in one or more of these areas, they receive the Philips Green Award with a Green tick mark, verified by an external auditor. In addition to making greener products, as part of its move to circularity Philips is making the shift from selling products to providing products-as-a-service, such as for example its lighting-as-a-service where customers pay per use or per service through renting or leasing rather than outright purchasing (Phipps 2018).

In recognition that collaboration enables circular economy, Philips offers LCA and Ecodesign workshops for industry professionals through its Innovation Services. Furthermore, Philips is a global partner of The Ellen MacArthur Foundation (EMF), which works towards accelerating the transition to a circular economy in cooperation with business, government and academia. The aim is to establish a framework for an economy restorative and regenerative by design (Saavedra et al. 2018; EMF 2013).

Reverse supply chain initiatives also require cooperation within and across supply chains. Examples of organisations that reclaim, recycle and reuse materials include Adidas, which launched a shoe range made from plastic debris reclaimed from the ocean, and McDonald's, which collects used cooking oil to convert it into biodiesel (Oxborrow and Lund-Thomsen 2017).

One company that uses eco-effectiveness and cradle-to-cradle principles is Herman Miller, which produces office furniture. An early mover in the area, it formed an Environmental Quality Action Team in 1989 and recognised early on that using eco-design principles requires working not just with designers and engineers but also with supply chain managers (Blowfield 2013). It is now committed to achieve zero waste by 2023 and 100 level 3 certified products (Herman Miller 2018). LEVEL is a comprehensive evaluation and certification system for environmentally sustainable and socially responsible commercial furniture – similar to what LEED is for buildings – and is only awarded after an approved and independent third party has verified certification (LEVEL 2018).

Certifications and standards can be used throughout the supply chain to green it and are a useful tool to work and collaborate with suppliers on sustainability. ISO 14040 and 14044 provide standards on LCAs for example which can be used internally but also externally to assess the environmental impact of a product across its whole life cycle. Furthermore, companies can specify their own codes of conducts to develop suppliers (Grant et al. 2017). Walmart requires its suppliers to conduct environmental impact assessments of their goods which generates eco-ratings for the company's Sustainability Index, which was developed in collaboration with The Sustainability Consortium. The index is used to build a more resilient supply chain and to offer more sustainable products. Similar to other examples mentioned earlier, Walmart provides a Sustainability Hub for suppliers with tools and resources to improve the sustainability of goods supplied.

BOX 8.1 SPOTLIGHT ON INTERNATIONAL SUPPLY CHAIN MANAGEMENT AND SUSTAINABILITY

As emphasised earlier in the chapter, the rise in globalisation witnessed since the 1970s has shaped logistics and supply chain management immensely, for example in the form of off-shoring of production, global sourcing of raw or other materials, and outsourcing of activities to more cost-effective locations for example in developing countries, leading to increased interdependence.

These developments have increased the environmental impact of companies and their SCM for example through increased distances that need to be covered, leading to increased air pollution, higher emissions of greenhouse gases and demand for more (secure) packaging.

Global production networks have made it possible for companies to offer their products cost-effectively worldwide, but often with little regard for the environment, particularly when environmental standards vary between countries (e.g. developed versus developing countries).

Geographically extensive and complex supply chains and webs involving first-, second- and third-tier suppliers make it more difficult for firms to know exactly who is providing what and from where, as illustrated in this quote by Mary Teagarden on Chinese businesses (Stern 2007):

> "[Retailer X] squeezes [Toymaker Y who] squeezes its supplier, that supplier squeezes its supplier, and at the end of the chain you have a remote business far out in the countryside that takes a different approach. They don't put lead in paint because they are wicked, it's just what works for them. China is so large, and industrialisation has been so rapid, that maintaining any control over multiple sites is extremely difficult".

Collaboration and partnerships with suppliers in global supply chains are understandably challenging across borders, languages and time zones, which makes it all the more important for companies to engage. Global frameworks such as the Global Reporting Initiative or the GGP and internationally accepted standards such as the ISO 14000 series, which are available in a range of languages, can help companies to green their international supply chains. International regulatory frameworks and protocols also provide leverage for companies to green their supply chains. With major economies such as the EU and China implementing circular economy packages in legislation affecting SCM, cases as mentioned in the quote presented earlier will become increasingly rare. Although imposition of sustainable practices through regulations can work, collaboration tends to lead to better results and greater commitment of suppliers.

Diverging environmental regulations can lead to challenges for companies operating globally, needing to comply with various legislative frameworks. Some international manufacturers have taken the approach of producing according to (or even beyond) the strictest environmental standards, amongst them Honda, 3M and Xerox. Such a strategy ensures regulatory compliance worldwide as well as environmental leadership which can be further used to develop supplier networks and to green the supply chain.

Another development contributing to sustainability in international supply chains is blockchain, which is a digital ledger technology that offers opportunities to improve traceability in the supply chain through a digital trail e.g. by verifying that standards are followed and confirming certification (Clancy 2017, 2019).

SUMMARY

We opened the chapter reminding ourselves that the three basic issues that business faces are "what it takes, what it makes, and what it wastes" (Hawken 1993: 12), which in short encapsulates supply chain management. Supply chain management is an encompassing and complex corporate function which has company-wide implications for how sustainability is achieved. Whilst there are many opportunities to green the supply chain across network design, logistics, procurement and supplier selection, there are also barriers to implementing sustainable supply chains. These include the costs, effort and complexity of coordination, and inadequate or lacking communication in the supply chain (Seuring and Müller 2008). Factors supporting sustainable SCM include communication across company boundaries, management systems, measures to monitor, evaluate, report and sanction, training of employees and suppliers, and integration into corporate policy and strategy (Seuring and Müller 2008). These factors illustrate the importance of functions not operating in isolation but working together to support sustainability efforts, which is reflected in the book content – communication and marketing are included in Chapter 10, management systems in Chapter 11 and reporting in Chapter 12, whilst strategy is covered in Chapter 4.

Broader challenges in greening supply chains pertain to economies more generally. For example, in economies that are growing, sole reliance on postconsumer or recycled materials in a circular economy is not possible, as infrastructure and thus material stock is still building up (Bocken et al. 2017), which thus serves as an inherent limit to circularity as primary resource extraction cannot be altogether avoided. Furthermore, there is the challenge of the rebound (or Jevons) effect, whereby increased efficiency enables cheaper consumption of goods, which can lead to increased consumption (Bocken et al. 2017) – this may influence the ability to close loops in a circular economy. A question that consumers face more broadly is also a key challenge for sustainable procurement – do we know what we buy and what its environmental impacts are? – which arguably makes procurement one of the most significant issues facing society today (Browne 2011). We will look at the consumer aspect in Chapter 10.

Despite these challenges and barriers, the chapter has illustrated what changes can be made to make supply chains and their management more environmentally sustainable, whether it is in location choices and micromanufacturing, in network design, reduction of freight transport and avoidance of empty miles alongside changes to transport modes in logistics, more local sourcing and careful supplier assessment and selection for procurement using sustainability criteria and standards, and overall in the collaboration and partnering with suppliers including training and development. Various tools and concepts were introduced that can be used to green the supply chain both upstream and downstream, whilst also being more widely applicable, such as for example product-service systems (Bernon et al. 2012), which have been introduced as service and flow economy in Chapter 4 and are also known as servicisation or servitization where customers pay per service or per use, and companies deliver functionality rather than ownership.

As emphasised in the introduction to part three and to this chapter, the functions and activities that SCM comprises vary, and all stages impact on the environment and all can thus contribute to environmental sustainability. Product stewardship and the development of goods and services influences all stages of logistics and SCM (Grant et al. 2017). Having provided an overview of the whole supply chain in this chapter, we now therefore turn to this crucial part of it, that of design, production and operations, in the next chapter.

SUGGESTED QUESTIONS AND ACTIVITIES

- Debate: what is the most important element in the supply chain when considering sustainability?
- Where from? Examine the labels of your clothes, your snacks or the contents of the bag you're using the most. Identify any aspects of sustainability related to the items' supply chain (transport, packaging, materials, production processes, waste/recycling).
- Take one of the items from the previous exercise and identify which data you would need for a life cycle analysis of the product in question.
- In the fashion industry https://continued.fashion/ offers a platform for circular fashion. Choose another industry and explore the progress towards a circular economy.
- Watch the film *The true cost* on the phenomenon of fast fashion and find out what is behind our clothes, i.e. trace the supply chain: https://truecostmovie.com/
- Look up some of the corporate examples and practices mentioned in the chapter. Here are some starting points: www.sourcemap.com/, www.originalrepack.com, www.supplychainschool.co.uk, www.innovationservices.philips.com/looking-expertise/environment-health-safety/life-cycle-analysis-lca-services/, www.ellenmacarthurfoundation.org, www.hermanmiller.com/, https://corporate.walmart.com/global-responsibility/environment-sustainability/sustainability-index-leaders-shop.
- For more in-depth material on sustainable/green SCM, check out the following:
  - Grant, D. B., Trautrims, A. and Wong, C. Y. (2017) *Sustainable Logistics and Supply Chain Management: Principles and Practices for Sustainable Operations and Management*, Kogan: London, 2nd ed.
  - McKinnon, A., Browne, M., Piecyk, M. and Whiteing, A. (2015) *Green Logistics: Improving the Environmental Sustainability of Logistics*, Kogan: London, 3rd ed.
  - Sarkis, J. and Dou, Y. (2018) *Green Supply Chain Management: A Concise Introduction*, New York: Routledge.
  - Articles from the Global Environmental Management Initiative http://gemi.org/solutions/.

NOTES

1 In recognition Goldstein and Newell (2019) recommend studying the supply chains of individual corporations to fully appreciate the environmental (and social) impact in geographically specific areas, which is in contrast to the majority of research looking at products, industries or sectors.

2 Nieuwenhuis and Touboulic (2017) differentiate between sustainability and resilience as follows – they see sustainability as a conservation model that tries to control and limit change, whereas resilience focuses on the ability to cope and to manage impact, which are inherent parts of dynamic natural systems that are always changing.

3 Whilst the switch to biofuels is, in principle, generally environmentally desirable, socially and politically, their promotion is contentious if agricultural products and land are diverted from food production to biofuel development. Emphasis should therefore be placed on using waste and uneatable crops as well as by-products from food production (Grant et al. 2017).

4 The source of electricity needs to be taken into account here, whether it is generated from fossil fuels such as coal or gas, or from nuclear power or renewable sources.

5 Bernon et al. (2012) provide the example of tomatoes bought by UK consumers – due to the artificial heat required to grow them in Britain, Spanish tomatoes have a lower CO_2 footprint even when taking into account the distance travelled.

6 Grant et al. (2017) point out that in simplified terms there is a trade-off between transport costs (longer distances travelled) and the costs of holding inventory, which affects the number of warehouses.

7 As a reminder, Scope 1 emissions measure direct emissions over which the company has full control, whereas Scope 2 measures indirect emissions, where there is limited control, and Scope 3

includes indirect emissions outside an organisation's control.

8 For context, the authors define sustainability as "the balanced integration of economic performance, social inclusiveness, and environmental resilience, to the benefit of current and future generations" (Geissdoerfer et al. 2017: 766).

SOURCES AND FURTHER READING

Abbey, J. D. and Guide, V. D. R. (2012) Closed-Loop Supply Chains, in: Bansal, P. and Hoffman, A. J. (eds.) *The Oxford Handbook of Business and the Natural Environment*, Oxford: Oxford University Press, pp. 290–309.

Abukhader, S. M. and Jönson, G. (2004) Logistics and the Environment: Is it an Established Subject? in: *International Journal of Logistics: Research and Applications*, 7(2): 137–149.

Accenture (2017) Taking the European Chemical Industry into the Circular Economy, Executive Summary, available at www.accenture.com/gb-en/insight-circular-economy-european-chemical-industry (Retrieved, 31.07.2018).

Baker, P. and Marchant, C. (2015) Reducing the Environmental Impact of Warehousing, in: McKinnon, A., Browne, M., Piecyk, M. and Whiteing, A. (eds.) *Green Logistics: Improving the Environmental Sustainability of Logistics*, Kogan: London, 3rd ed., pp. 194–226.

Bernon, M., Baker, P., Mena, C., Palmer, A., Skipworth, H., Smart, A. and Templar, S. (2012) Towards more Sustainable Supply-Chain Management, in: Grayson, D. and Exter, N. (eds.) *Cranfield on Corporate Sustainability*, Sheffield: Greenleaf Publishing, pp. 149–166.

Blowfield, M. (2013) *Business and Sustainability*, Oxford: Oxford University Press.

Bocken, N. M. P., Olivetti, E. A., Cullen, J. M., Potting, J. and Lifset, R. (2017) Taking the Circularity to the Next Level: A Special Issue on the Circular Economy, in: *Journal of Industrial Ecology*, 21(3): 476–482.

Brady, J. (2011) The Response of Organisations, in: Brady, J., Ebbage, A. and Lunn, R. (eds.) *Environmental Management in Organizations, The IEMA Handbook*, London: Earthscan/Routledge, 2nd ed., pp. 251–260.

Browne, C. (2011) Sustainable Procurement, in: Brady, J., Ebbage, A. and Lunn, R. (eds.) *Environmental Management in Organizations, The IEMA Handbook*, London: Earthscan/Routledge, 2nd ed., pp. 275–294.

Carter, C. R. and Easton, P. L. (2011) Sustainable Supply Chain Management: Evolution and Future Directions, in: *International Journal of Physical Distribution & Logistics Management*, 41(1): 46–62.

Carter, C. R. and Rogers, D. S. (2008) A Framework of Sustainable Supply Chain Management: Moving Toward New Theory, in: *International Journal of Physical Distribution & Logistics Management*, 38(5): 360–387.

Clancy, H. (2017) The Blockchain's Emerging Role in Sustainability, *GreenBiz*, 06.02.2017, available at www.greenbiz.com/article/blockchains-emerging-role-sustainability (Retrieved, 10.12.2019).

Clancy, H. (2019) Blockchain, Meet Supply Chain, *GreenBiz*, 27.03.2019, available at www.greenbiz.com/article/blockchain-meet-supply-chain (Retrieved, 10.12.2019).

Crane, A. and Matten, D. (2016) *Business Ethics*, Oxford: Oxford University Press, 4th ed.

Daft, R. L., Kendrick, M. and Vershinina, N. (2010) *Management: International Edition*, Andover: Cengage Learning EMEA.

Dharmasasmita, A., Kennedy, E., Puntha, H. and Holmes, R. (2017) Climate Change and Greenhouse Gas Management, in: Molthan-Hill, P. (ed.) The *Business Student's Guide to Sustainable Management*, Sheffield: Greenleaf Publishing, 2nd ed., pp. 412–462.

Dolan, S. (2018) The Challenges of Last Mile Logistics & Delivery Technology Solutions, *Business Insider*, 10.05.2018, available at http://uk.businessinsider.com/last-mile-delivery-shipping-explained (Retrieved, 26.07.2018).

Economist (2016) Co-creation and Microfactories: Exploring the Future of Global Manufacturing, available at http://producersoftomorrow.economist.com/co-creation-and-microfactories (Retrieved, 29.08.2018).

Ehrenfeld, J. R. (1997) Industrial Ecology: A Framework for Product and Process Design, in: *Journal of Cleaner Production*, 5(1–2): 87–95.

EMF (2013) *Towards the Circular Economy: Economic and Business Rationale for an Accelerated Transition*, Ellen MacArthur Foundation, Vol. 1, Isle of Wight.

EMF (2017) Ellen MacArthur Foundation, webpages, available at www.ellenmacarthurfoundation.org (Retrieved, 29.08.2018).

Esty, D. C. and Winston, A. S. (2009) *Green to Gold: How Smart Companies Use Environmental Strategy to Innovate, Create Value and Build Competitive Advantage*, Hoboken, NJ: Wiley.

Fahiminia, B., Sarkis, J. and Davarzani, H. (2015) Green Supply Chain Management: A Review and Bibliometric Analysis, in: *International Journal of Production Economics*, 162: 101–114.

Fogel, D. S. (2016) *Strategic Sustainability: A Natural Environmental Lens on Organizations and Management*, London: Routledge.

Garnett, T. (2015) The Food Miles Debate: Is Shorter Better? in: McKinnon, A., Browne, M., Piecyk, M. and Whiteing, A. (eds.) *Green Logistics: Improving the Environmental Sustainability of Logistics*, Kogan: London, 3rd ed., pp. 358–371.

Geissdoerfer, M., Savaget, P., Bocken, N. M. P. and Hultink, E. J. (2017) The Circular Economy – A New Sustainability Paradigm? in: *Journal of Cleaner Production*, 143(1): 757–768.

Goldstein, B. and Newell, J. P. (2019) Why Academics Should Study the Supply Chains of Individual Corporations, in: *Journal of Industrial Ecology*, https://doi.org/10.1111/jiec.12932.

Grant, D. B., Trautrims, A. and Wong, C. Y. (2017) *Sustainable Logistics and Supply Chain Management: Principles and Practices for Sustainable Operations and Management*, London: Kogan, 2nd ed.

Greening of Industry Network (2017) SDGs & Sustainable Supply Chains in the Post-Global Economy, International Symposium, 27.05.2017, organised by Royal Holloway, University of London, The Greening of Industry Network and University of Twente.

Guide, V. D. R. and Van Wassenhove, L. N. (2009) The Evolution of Closed-Loop Supply Chain Research, in: *Operations Research*, 57(1): 10–18.

Handfield, R. B. and Nichols, E. L. (1999) *Introduction to Supply Chain Management*, New Jersey: Prentice-Hall.

Hawken, P. (1993) *Ecology of Commerce: A Declaration of Sustainability*, New York: HarperBusiness.

Herman Miller (2018) Herman Miller, webpages, available at www.hermanmiller.com (Retrieved, 29.08.2018).

Junwei, W. (2018) BASF to Establish $10b Site in Guangdong, *China Daily*, 10.07.2018, available at www.chinadaily.com.cn/a/201807/10/WS5b446a63a3103349141e1e59.html (Retrieved, 29.08.2018).

Klassen, R. D. and Vachon, S. (2012) Greener Supply Chain Management, in: Bansal, P. and Hoffman, A. J. (eds.) *The Oxford Handbook of Business and the Natural Environment*, Oxford: Oxford University Press, pp. 269–289.

Kraljic, P. (1983) Purchasing must Become Supply Management, in: *Harvard Business Review*, 61(5): 109–117.

Laasch, O. and Conaway, R. (2015) *Principles of Responsible Management: Glocal Sustainability, Responsibility and Ethics*, Stamford, CT: Cengage Learning.

LEVEL (2018) LEVEL Certified, webpages, available at www.levelcertified.org (Retrieved, 29.08.2018).

Marsden, J. (2011) Auditing, in Brady, J., Ebbage, A. and Lunn, R. (eds.) *Environmental Management in Organizations, The IEMA Handbook*, London: Earthscan/Routledge, 2nd ed., pp. 399–417.

McDonough, W. and Braungart, M. (2002) *Cradle to Cradle: Remaking the Way we Make Things*, New York: North Point Press.

McKinnon, A. (2015) Environmental Sustainability: A New Priority for Logistics Managers, in: McKinnon, A., Browne, M., Piecyk, M. and Whiteing, A. (eds.) *Green Logistics: Improving the Environmental Sustainability of Logistics*, Kogan: London, 3rd ed., pp. 3–31.

McLaren, S. (2011) Life Cycle Assessment, in: Brady, J., Ebbage, A. and Lunn, R. (eds.) *Environmental Management in Organizations, The IEMA Handbook*, London: Earthscan/Routledge, 2nd ed., pp. 511–522.

Nieuwenhuis, P. and Touboulic, H. (2017) Is SSCM Sustainable? – The Case for 'Ecological Supply Chain Management', Presentation at International Symposium, Greening of Industry Network, 27.05.2017.

Nogrady, B. (2016) The Rise of Miniature 'Microfactories', *BBC*, 18.11.2016, available at www.bbc.com/future/story/20161117-the-rise-of-miniature-microfactories (Retrieved, 29.08.2018).

Oxborrow, L. and Lund-Thomsen, P. (2017) Sustainable Supply Chain Management, in: Molthan-Hill, P. (ed.) The *Business Student's Guide to Sustainable Management*, Sheffield: Greenleaf Publishing, 2nd ed., pp. 380–410.

Paddison, L. (2013) Climate Change: Are Companies Prepared for the Impact? *The Guardian*, 11.10.2013, available at www.theguardian.com/sustainable-business/climate-change-companies-supply-chain-impact (Retrieved, 30.07.2018).

Phipps, L. (2018) How Philips became a Pioneer of Circularity-as-a-service, *GreenBiz*, 22.08.2018, available at www.greenbiz.com/article/how-philips-became-pioneer-circularity-service (Retrieved, 29.08.2018).

Porter, M. E. (1985) *Competitive Advantage: Creating and Sustaining Superior Performance*, New York: The Free Press (Republished with a new introduction, 1998).

Porter, M. E. and van der Linde, C. (1995a) Toward a New Conception of the Environment-Competitiveness Relationship, in: *Journal of Economic Perspectives*, 9(4): 97–118.

Porter, M. E. and van der Linde, C. (1995b) Green and Competitive: Ending the Stalemate, in: *Harvard Business Review*, 73(5): 120–134.

Raworth, K. (2017) *Doughnut Economics: Seven Ways to Think Like a 21st-Century Economist*, London: Random House Business Books.

Reuters (2011) Thai Floods Batter Global Electronics, Auto Supply Chains, *Reuters World News*, 28.10.2011, available at www.reuters.com/article/us-thai-floods/thai-floods-batter-global-electronics-auto-supply-chains-idUSTRE79R0QR20111028 (Retrieved, 23.07.2018).

Robertson, M. (2017) *Sustainability Principles and Practice*, London: Routledge, 2nd ed.

Rogers, D. S. and Tibben-Lembke, R. (2001) An Examination of Reverse Logistics Practices, in: *Journal of Business Logistics*, 22(2): 129–148.

Saavedra, Y. M. B., Iritani, D,. R., Pavan, A. L. R. and Ometteo, A. R. (2018) Theoretical Contribution of Industrial Ecology to Circular Economy, in: *Journal of Cleaner Production*, 170(1): 1514–1522.

Sarkis, J. (2003) A Strategic Decision Framework for Green Supply Chain Management, in: *Journal of Cleaner Production*, 11(4): 397–409.

Sarkis, J. and Dou, Y. (2018) *Green Supply Chain Management: A Concise Introduction*, New York: Routledge.

Seuring, S. and Müller, M. (2008) From a Literature Review to a Conceptual Framework for Sustainable Supply Chain Management, in: *Journal of Cleaner Production*, 16(15): 1699–1710.

Sharfman, M. P., Shaft, T. M. and Anex Jr, R. P. (2009) The Road to Cooperative Supply-Chain Environmental Management: Trust and Uncertainty Among Pro-Active Firms, in: *Business Strategy and the Environment*, 18(1): 1–13.

Slack, N. Chambers, S. and Johnston, R. (2001) *Operations Management*, Harlow: Pearson Education, 3rd ed.

Srivastava, S. K. (2007) Green Supply-chain Management: A State-of-the-art Literature Review, in: *International Journal of Management Reviews*, 9(1): 53–80.

Stead, J. G. and Stead, W. E. (2014) *Sustainable Strategic Management*, Sheffield: Greenleaf Publishing, 2nd ed.

Stern, S. (2007) West Must Take Some Blame for Tainted Chinese Goods, *Financial Times*, 03.09.2007, available at www.ft.com/content/bd5c5754-5a2a-11dc-9bcd-0000779fd2ac (Retrieved, 29.08.2018).

Tagesschau (2018) Streetscooter: Wenn die Milch mit dem Postwagen kommt, *Tageschau*, 06.06.2018, available at www.tagesschau.de/wirtschaft/boerse/post-streetscooter-101.html (Retrieved, 26.07.2018).

Touboulic, A. and Walker, H. (2015) Theories in Sustainable Supply Chain Management: A Structured Literature Review, in: *International Journal of Physical Distribution & Logistics Management*, 45(1/2): 16–42.

UK Forum for Supply Chain Sustainability (2017) Sustainability as a Source of Innovation and Competitive Advantage, 20.09.2017, organised by The Carbon Trust and Loughborough University.

Vachon, S. and Klassen, R. D. (2006) Extending Green Practices Across the Supply Chain: The Impact of Upstream and Downstream Integration, in: *International Journal of Operations & Production Management*, 26(7): 795–821.

Walker, H., Di Sisto, L. and McBain, D. (2008) Drivers and Barriers to Environmental Supply Chain Management Practices: Lessons from the Public and Private Sectors, in: *Journal of Purchasing & Supply Management*, 14(1):69–85.

Watson, M., Lewis, S., Cacioppi, P. and Jayaraman, J. (2013) *Supply Chain Network Design:*

Applying Optimization and Analytics to the Global Supply Chain, Upper Saddle River, NJ: Pearson Education.

Weybrecht, G. (2014) *The Sustainable MBA: A Business Guide to Sustainability*, Chichester: Wiley, 2nd ed.

Zhu, Q. and Sarkis, J. (2004) Relationships Between Operational Practices and Performance Among Early Adopters of Green Supply Chain Management Practices in Chinese Manufacturing Enterprises, in: *Journal of Operations Management*, 22(3): 265–289.

CHAPTER 9

Design, production and operations
Greener and (c)leaner

INTRODUCTION

Following the wide-ranging approach in covering the whole supply chain, we now focus on a central part of it, that of design, production and operations. Similarly to supply chain management, key issues for companies here are what they take, make and waste (Hawken 1993). Crucially, in the design stage, important decisions can be made to reduce what they take, to design out waste, and to devise ways of making (things) that produce less impact on the environment.

We start the chapter with a brief overview of operations management in general before covering the environmental impact and the business case for sustainable production and operations, which we then define. Various elements and tools of sustainable operations as well as examples of companies and practices follow. A comprehensive coverage of sustainable operations is beyond the remit of this textbook, so the following represents an overview and some snapshots rather than a detailed treatment, and readers are referred to other sources for more in-depth material.

OPERATIONS MANAGEMENT

Overview

Broadly, operations management concerns the way companies and other organisations produce goods and services, which can be modelled as a process that transforms inputs into outputs (Slack et al. 2016). Input comes in the form of resources to be transformed, such as materials, information or customers, and in the form of resources that are doing the transforming, such as facilities and staff, which are core building blocks of operations management. Key elements of operations management include design, planning and control, and improvement (Slack et al. 2001), which can be rephrased as design, deliver and develop with direct added as the link to the organisation's overall strategy (Slack et al. 2016). Daft and colleagues (2010) stress that production and operations apply to all organisations, including services, and thus not just to manufacturing, despite what the name suggests.

Environmental impact of production and operations

Linear cradle-to-grave production systems that take, make and waste focus on products that are made and delivered to the customer in the fastest and cheapest possible way (Doppelt 2003). Such industrial processes often produce waste as a major output, with estimates that for every kilogram of finished goods, the waste generated during production amounts to about 20 kilograms, and the purchases often ending up as waste within six months (Benn et al. 2014). Production processes require heat and pressure, intense use of petrochemicals and materials as well as chemicals that may be toxic and hazardous. The result is often a ratio of waste to durable products of close to 100 to one (Lovins et al. 1999). In addition to waste, the environmental impact extends to climate and pollution of air, water and soil, as covered in part two of the book.

McDonough and Braungart (2002) take a broader, historical perspective and highlight the environmental degradation of the cradle-to-grave model that dominates much of modern manufacturing as a fundamental flaw of the industrial revolution's design. The negative consequences such as toxic materials released to water, air and soil, and enormous amounts of waste were never intended but took shape gradually as designers, engineers and industrialists worked on specific projects that were limited to the practical, efficient, profitable and linear without considering their designs as part of a greater system (McDonough and Braungart 2002). The environmental problems of past industrial practices are also emphasised by Fiksel (2012a) who points to mineral depletion, pollution, global warming, sea level rise, scarcity of freshwater, land degradation and biodiversity loss.

Impact of environment on production and operations

The environmental impact both of and on production and operations mirror what was covered in the previous chapter on supply chain management. Extreme weather events such as floods or droughts as a result of climate change for example can halt manufacturing operations entirely and jeopardise just-in-time manufacturing which relies on limited inventory. Pollution can affect the availability of clean air and water as production inputs as well as the health of employees, both of which may put constraints on a company's operations and resources. Natural resources that have been depleted may affect an organisation's ability to source raw materials. Whilst these impacts are mainly operational, financial costs are also a major factor as a result of higher commodity and insurance costs as well as of reduced labour productivity.

Business case for greener production and operations

We can see from these two brief sections why making productions and operations more sustainable is needed. The way resources such as energy, water and materials are currently managed is not sustainable, and the environmental problems created pose a threat to the natural capital that underpins economies and societies (Fiksel 2012a). Growing societal concerns put pressure on businesses to clean up their operations in conjunction with concerns over cost and profitability (Kleindorfer et al. 2005). Cleaner production processes that prevent pollution include good housekeeping practices that ensure an efficient use of resources (where material losses are avoided), substitution or elimination of toxic materials, changes to manufacturing processes, and recovery of resources where waste materials are reused as input (Fiksel 1996). Other internal drivers pushing companies towards greener operations include employee commitment and quality improvement as well as a desire and need to reduce costs (Walker et al. 2008). Examples of corporate initiatives such as WRAP (Waste Reduction Always Pays, Dow Chemical), PPP (Pollution Prevention Pays, 3M), SMART (Save Money and Reduce Toxics, Chevron) underline the business case for more environmentally sustainable operations where preventing pollution by redesigning products or processes can save money, reduce risk and lead to product improvements (Hart 2010).

SUSTAINABLE OPERATIONS MANAGEMENT

Overview

Efficient and successful operations management, as the activity of managing resources that produce and deliver products and services, goes hand-in-hand with recognising the environmental impact of a firm. Sustainable operations management is not just about the process of transforming raw materials through to delivery but also about design, development, by-products and disposal of waste (Zeng 2014). In some ways, it could be argued that sustainability is an enhanced version of conventional lean manufacturing ideas and therefore efficient operations management naturally goes together with recognising the environmental impact of the firm (Zeng 2014).

For Kleindorfer and colleagues (2005: 489), sustainable operations management constitutes the

> set of skills and concepts that allow a company to structure and manage its business processes to obtain competitive returns on its capital assets without sacrificing the legitimate needs of internal and external stakeholders and with due regard for the impact of its operations on people and the environment.

This encapsulates a broad definition of sustainability, which includes people, profit and planet. Focusing on the environmental aspect, we also find the term green operations or green manufacturing. Green manufacturing aims to cut the environmental impact through the use of suitable material and technologies, whilst green operations concerns all aspects to do with the manufacturing of goods, their remanufacturing, usage, handling, logistics and management of waste following the finalisation of design (Srivastava 2007).

Since operations management concerns the process of transforming inputs into outputs, productivity as the ratio of output per unit of input is a central concern. In order to increase energy and material productivity, companies can use design, new technologies, controls, corporate culture, new processes and saving materials (Lovins et al. 1999). In the following, we focus in particular on design, processes and technologies. Controls in the form of e.g. information technologies are covered in Chapter 11, and in Chapter 12 on accounting, whilst corporate culture is part of Chapter 13 on human resource management. Saving materials is linked to reducing waste, which we covered in Chapter 7, and is a theme that runs through various parts of the book.

Cleaner production is core to sustainable operations management. It traces its routes to pollution prevention, focusing initially on processes and reducing the use of toxic materials before integrating life cycle-wide issues (Lifset and Boons 2012). The terms pollution prevention and cleaner production are often used synonymously (Grant et al. 2017), and both are closely interlinked with industrial ecology which takes a systems perspective encompassing the whole life cycle of products from cradle to grave. The goal is to optimise resource efficiency, which echoes eco-efficiency, and to close material loops in order to pursue sustainable production (Lifset and Boons 2012). These concepts were introduced earlier in the book (see e.g. Chapters 2 and 3), underlining the systemic nature of environmental sustainability – here they are specifically applied to design, production and operations.

Elements of sustainable production and operations

Extended producer responsibility and product stewardship

Extended producer responsibility (EPR) takes a broader view of manufacturers' duties, extending them beyond the factory gates and placing cradle-to-grave responsibility for products on producers. EPR shifts the downstream costs and responsibilities of disassembly, recycling or disposal upstream onto producers, often through legislation (covered in Chapter 7), as these costs are often not reflected in the pricing of goods.[1] In the past, companies have often responded by developing collection and treatment mechanisms for post-consumer waste rather than incorporating life cycle thinking into their operations (Lifset and Boons 2012). EPR provides incentives for manufacturers to design products and their packaging with less waste, for easier and less costly reuse, disassembly and recycling. It is also known as product stewardship. Product stewardship strategies emphasise reducing a firm's footprint by making products that are safer, use less materials and energy, and are for example more recyclable, biodegradable, and durable as well as less wasteful (Stead and Stead 2014). It is a step further than eco-efficiency as it goes beyond reducing material inputs and enables a firm to differentiate its products in the marketplace. It requires a firm to evaluate a product's environmental impact throughout the value chain and thus its life cycle, covered in the next section.

In short, core objectives of extended producer responsibility are the following:

- Cost internalisation: waste-related costs are transferred to producers (primarily) and consumers of products, shifting the cost of waste collection away from local authorities.
- Waste diversion: increased reuse and recycling of products.
- Design for environment: product design that minimises or eliminates resource consumption and hazardous/non-hazardous waste generation associated with the product.

EPR requires a shift in thinking for manufacturers as responsibility for a product at the end of its life no longer rests with the consumer – who might have little influence beyond the choice of buying or not buying a product (Robertson 2017) – but with the designers and producers who are in a position to consider the whole life cycle of a product and its cost. It is to design we now turn as the starting point for any product or service.

Eco-design: design for environment

> Design for Environment is the systematic consideration of design performance with respect to environmental, health, safety and sustainability objectives over the full product and process life cycle.
>
> (Fiksel 2012b: 6)

Making decisions is a core management task. In order to decide what is the more sustainable option, choices need to exist in the first place, which is the remit of design (Laszlo and Zhexembayeva 2011). While bolted-on sustainability is often based on changing what already exists, embedding sustainability requires new thinking and the generation of new choices based on clear intent. With a different design mentality, methods and technologies of design can enable resource productivity to improve not only by factor 4 or 10, but by factor 100 or more (Visser 2012, drawing on Lovins et al. 1999 and von Weizsäcker et al. 1998). Design is crucial as much of the environmental impact of a product is established in the design phase, whether it is water, chemicals, waste or energy (Esty and Winston 2009). It is estimated that 75% of the life cycle costs of a product are determined by its design (Bernon et al. 2012).

Design for Environment (DfE for short) is a systematic approach that integrates environmental thinking in the design of products and processes from the start (Esty and Winston 2009). DfE is also known as eco-design, life cycle design, sustainable design or design for eco-efficiency (Fiksel 2012b). Whichever term is used, eco-design integrates cost, performance, quality and environmental concerns and can be effective in advancing eco-efficiency in organisations (Saavedra et al. 2018).

DfE moves business beyond environmental compliance and environmental remediation and opens the opportunity to avoid costly and detrimental environmental mistakes from the start (Lifset and Boons 2012). It encourages a cradle-to-cradle approach, taking design cues from nature, also called biomimetics or biomimicry (Whitfield and McNett 2014). Biomimicry[2] was first introduced in Chapter 4 and referred to again in Chapter 7. Biomimicry uses nature as a model, measure and mentor where design imitates or is inspired by nature's processes, is measured by what works in nature, and learns from it (Benyus 1997). Hey (2000: 323) emphasises that engineering was one of the earliest professions to realise that natural processes "do engineering work" which would therefore make it wise to design alongside rather than against such processes. Biomimicry is particularly useful for designing certain production processes but can also be used to shape the structure and function of the economy (Lovins et al. 1999), thus underlining the systematic nature of sustainability. Box 9.1 provides further details on biomimicry.

BOX 9.1 BIOMIMICRY – INNOVATION INSPIRED BY NATURE

On a systemic level, biomimicry encourages the redesign of industrial systems to copy nature in order to reduce the wasteful throughput of materials. Such so-called biomimetic production is centred around closed-loop manufacturing focused on preventing waste. Biomimicry forms the basis for industrial ecology where waste equals food. Biomimicry is one of the four principal strategies in natural capitalism alongside resource productivity, service and flow economy, and investment in natural capital. It also provides general guidance for greening industry and how to conduct business based on organisms in mature ecosystems.

In eco-design and cradle-to-cradle design, biomimicry is the basis that signifies a shift from built-in obsolescence to copying the effectiveness of nature and its processes. It firstly uses nature as a *model* to be studied and then imitated in order to solve human problems. Secondly, it employs nature as a *measure* to judge whether innovations inspired by nature work, as after billions and millions of years of evolution natural processes have learned what functions. Thirdly, biomimicry takes nature as a *mentor* from which we can discover new ways of viewing and valuing natural processes. The emphasis is on learning rather than extraction from nature. The learning ranges from how food is grown, how materials are made and energy is harvested to how information is stored and business is conducted. The fields in which biomimicry can be applied thus include for example agriculture, materials science, medicine, computing and commerce.

Biomimicry is based on a number of design principles derived from nature, one of which is

BOX 9.1 (CONTINUED)

collaboration. Reflecting the holistic and integrated nature of sustainability, Benyus (1997) encourages the cooperation of biologists and engineers as well as of designers and marketers in bringing green products to the market. Other principles include the use of solar energy, using only the energy needed, fitting form to function, recycling, diversity, local expertise, curbing excesses from within and tapping the power of limits. Overall, biomimicry offers an integrated framework that is flexible and durable, and ultimately sustainable.

Products and services inspired by nature and designed with biomimicry include some of the following examples:

- Food: natural prairies can be used as a model for food production that mimics stable natural ecosystems that are resilient, self-enriching, productive and sustainable.
- Energy: termite mounds have inspired energy efficient buildings; making aerodynamic improvements to transport modes based on skin surfaces of various animals.
- Manufacturing: nature's low-pressure, life-temperature and solar-powered assembly techniques are mimicked; products include for example non-toxic adhesives inspired by the gecko and self-cleaning surfaces.
- Medicine: resistance-free antibiotics modelled after red seaweed; needles modelled on mosquito mouths to minimise pain.
- Computing and Communication: carbon-based rather than silicon-based processing; underwater data transmission modelled on dolphins' frequency-modulating acoustics.

Sources: Benyus 1997; Hawken et al. 1999; Matten 2010; Collins 2014, https://biomimicry.org/biomimicry-examples/.

Three guiding principles underpin a cradle-to-cradle approach to design, which addresses the three basic issues business faces: what it takes, makes and wastes (Hawken 1993):

1 Waste equals food
2 Use current solar income
3 Celebrate diversity.

These tenets are derived from the so-called Hannover Principles on Design for Sustainability, developed by McDonough and Braungart (1992) in preparation for Expo 2000, The World Fair hosted by the city. The principles provided the foundations for a new design philosophy and a coherent framework to help rethink design through the lens of sustainability. This was further developed by McDonough and Braungart in their work on cradle-to-cradle design (2002) and upcycling (2013).

As DfE (or green or environmentally conscious design) is a broad term, various other "design for" terms have been used to provide further detail, such as design for biomimicry, design for energy efficiency, design for waste minimisation, design for disassembly, end-of-life product design, design for remanufacturing, design for recycling, design for reuse. In his seminal book, *Design for Environment: A Guide to Sustainable Product Development*, Joseph Fiksel (2012b) provides some structure to these terms by providing four key DfE strategies:

1 Design for dematerialisation
2 Design for detoxification
3 Design for revalorisation
4 Design for capital protection and renewal.

These are further detailed by the following guidelines in Table 9.1, which are suggestive rather than prescriptive, to account for differences between firms and industries. Nevertheless, they provide some consistency and continuity to lead to design processes that are systematic rather than subject to individual designers' biases or idiosyncrasies (Fiksel 2012b).

There are interrelationships between these practices and overlaps with other design disciplines (also called Design for X or DfX), such as design for manufacturability or for reliability, and the guidelines variously pertain to the stages of the product life cycle, from procurement to assembly, from distribution to use and support, and recovery (Fiksel 2012b).

The guidelines for DfE are underpinned by the following seven basic principles in table 9.2.

TABLE 9.1

DfE guidelines

Principal strategy	Guidelines
1. Design for dematerialisation	Design for energy and material conservation Design for source reduction Design for servicisation
2. Design for detoxification	Design for release reduction Design for hazard reduction Design for benign waste disposition
3. Design for revalorisation	Design for product recovery Design for product disassembly Design for recyclability
4. Design for capital protection and renewal	Design for human capital Design for natural capital Design for economic capital

Source: Fiksel 2012b.

TABLE 9.2

DfE principles

Headline	DfE principle
Life cycle thinking	1. Embed life cycle thinking into the product development process
System perspective	2. Evaluate the resource efficiency and effectiveness of the overall system
Indicators and metrics	3. Select appropriate metrics to represent product life cycle performance
Design strategies	4. Maintain and apply a portfolio of systematic design strategies
Analysis methods	5. Use analysis methods to evaluate design performance and trade-offs
Information technology	6. Provide software capabilities to facilitate the application of DfE practices
Learning from nature	7. Seek inspiration from nature for the design of products and systems

Source: Fiksel 2012b.

To summarise Fiksel's (2012b) work, DfE practice consists of three main elements – indicators and metrics for performance, rules and guidelines for design, and analysis methods, and serves the objective of creating eco-efficient and eco-effective products within cost, quality and time constraints.

As a reminder, eco-efficiency is focused on reducing the environmental impact and the resource intensity of a product throughout its life cycle, whilst eco-effectiveness goes beyond doing less harm and focuses on doing good by seeking restorative environmental impacts. Eco-effectiveness is core to cradle-to-cradle design and is based on the principles introduced in Chapter 4 (McDonough and Braungart 2002):

1 Get "free of" known culprits
2 Follow informed personal preferences
3 Create a "passive positive" list
4 Activate the positive list
5 Reinvent.

In steering such design, the guiding principle of *waste equals food* differentiates between biological and technical nutrients. McDonough and Braungart (2013) consider differentiating materials one of the most important concepts in cradle-to-cradle design, where biological nutrients return to nature and technical nutrients become inputs for another product rather than ending up in landfill.

Both the guidelines and the principles of DfE as well as cradle-to-cradle design reflect the broader principles of natural capitalism (radical resource productivity, biomimicry, service and flow economy, investing in natural capital, Lovins et al. 1999) and of The Natural Step framework (eliminate extraction and production of substances as well as physical degradation, meet basic needs, Robèrt 2002), as introduced in Chapters 3 and 4. Whilst focusing on environmental sustainability, the broader context of social and economic sustainability is evident.

Hoffmann (2012) emphasises the importance of integrating users in sustainable product development as such sustainable innovations often necessitate changes in user behaviour. In contrast to conventional product development, sustainable product development often starts with a need rather than a want and considers the wider societal and environmental context. It can support servicisation, where products are substituted by services, and can help realise sufficiency-oriented production strategies (Hoffman 2012).

For Laszlo and Zhexembayeva (2011), participation is particularly relevant in the process of embedding sustainability in design thinking. They point to the examples of employees helping Walmart develop its sustainability commitment and GE's Ecomagination Challenge as an innovation initiative open to anyone wanting to contribute to product development. Similarly, Sarkis and Dou (2018) underline the importance of cooperation with customers for eco-design. This can extend to so-called co-creation (Weybrecht 2014).

On a related note, Bernon and colleagues (2012) emphasise that working together with different corporate functions during the development phase is more efficient in getting new products to market than the conventional sequential approach. This is called concurrent (or simultaneous) engineering which can also reduce the cycle time for new product development as different organisational disciplines work in parallel to address life cycle requirements (Fiksel 2012b). Concurrent engineering is core to integrated product development which encompasses both the business life cycle and the physical life cycle, with which we are mostly concerned in the context of sustainability. Concurrency helps avoid some of the pitfalls of a traditional "pass over the wall to the next function" approach, such as for example unnecessary transportation, unresponsive supply chains and failing to meet customer needs (Bernon et al. 2012). Embedding DfE within a wider cross-functional process (including e.g. marketing, Chapter 10) enables a product life cycle management that makes enterprises more resilient and sustainable (Fiksel 2012b). In addition, developing eco-design capability has an impact on performance – products that are designed with the natural environment as a central aspect have a higher potential for success in the market (Dangelico et al. 2016).

Cleaner production: lean and green manufacturing

Cleaner production is key to making operations management more sustainable. Cleaner production goes beyond compliance and initially involved managing waste and emissions before focusing on pollution prevention and efficiency in the growing context of extended producer responsibility (Fiksel 2012b). It signifies a shift from mere compliance, pollution control or end-of-pipe solutions (Grant et al. 2017). It is an integrated preventive strategy which can be applied to processes and products as well as services (Weybrecht 2014). Pollution is not inevitable and often a sign of inferior design and inefficiency (Robertson 2017). In the context of operations, production processes and technologies need modifications to reduce the amount of waste and pollution generated.

Many companies are using lean and Six Sigma techniques in their environmental management to make their manufacturing processes and systems leaner and more efficient (Fiksel 2012b). Resource productivity and eco-efficiency are closely linked, allowing companies to do more with less. This covers inputs such as material, natural resources and energy. Options for companies to address their environmental impact here include reduction of material variety and of material weight as well as exploring new materials (Weybrecht 2014). Alongside eco-efficiency and pollution prevention, other concepts used in cleaner production include waste minimisation and zero waste, green productivity, which, similar to resource efficiency, aims to improve productivity and environmental performance, as well as industrial symbiosis, aimed at closing loops of industrial production (Grant et al. 2017).

Similarly to eco-efficiency, lean manufacturing involves using less of everything in producing goods compared to mass production (Green 2013). It is based on the methods used in the Toyota production system (TPS) and focuses on eliminating waste while simultaneously improving quality, delivering value to customers and, specific to our context, achieving environmental performance goals as well as cost efficiencies (Weybrecht 2014; Grant et al. 2017). Simplification is key to lean thinking (Lovins et al. 1999). The concept of waste as a sign of inefficient production goes beyond traditional understanding in lean manufacturing, which targets several kinds of waste ("muda"): overproduction, inventory, transportation, motion, defects, over-processing and waiting (Weybrecht 2014). These can be broadly categorised as waste from irregular flow, inexact supply, inflexible response and variability (Slack et al. 2016).

Being lean is the first step towards becoming green, i.e. environmentally sustainable, as waste reduction leads to lower resource consumption in the form of energy, raw materials and solid waste (Souza 2018). Kleindorfer et al. (2005) emphasise the synergies between lean and green operations management, which share common values and goals (Zeng and Stratton 2017). Closely linked and almost synonymous with lean is just-in-time (JIT), stockless production, and lean synchronisation which highlight the focus on flow, continuity and "pull" control in lean operations. Lean is not just a method of planning and controlling, it is also a philosophy of running operations as well as a set of tools for improving the performance of operations (Slack et al. 2016).

In operations management, we find various approaches to improvement (Slack et al. 2016): Lean Operations, Total Quality Management (TQM), Business Process Re-engineering (BPR), Six Sigma; others include enterprise resource planning (ERP), total preventive maintenance (TPM) and Lean Sigma, which is a combination of lean and Six Sigma. It is beyond the scope of this chapter to provide further detail – standard operations textbooks cover these in greater depth. Relevant for our purposes is total quality environmental management (TQEM), which, like TQM and lean operations, builds on continuous improvement and combines TQM principles with environmental management. TQM encourages the search for and fixing of the root causes of problems (Benyus 1997). Beyond continuous improvement, TQEM involves three more basic elements: understanding customer requirements, doing the job right the first time, and a systems approach to work (Friedman 2011). TQEM can be used within and across companies, underlining the integrated approach that is needed for businesses and their environmental sustainability. We will go deeper into environmental management systems (of which TQEM is an example) in Chapter 11.

Continuous improvement is based on the so-called Deming cycle or wheel, named after a leading quality management thinker (Laasch and Conaway 2015). The cycle is also abbreviated as PDSA or PDCA for the stages plan, do, study (or check) and act and is evident in other approaches to continuous improvement, such as for example the ISO 14000 series of environmental standards (Blowfield 2013).

Six Sigma goes beyond incremental improvement and aims at breakthrough improvement (Laasch and Conaway 2015). It was initially developed by Motorola before being adopted and popularised by GE, 3M and other firms. It is centred around the DMAIC cycle (pronounced "de-make"), a systematic approach that defines (the problem), measures, analyses, improves and controls (Slack et al. 2016). Lean Sigma (also known as Lean Six Sigma or Six Sigma Lean) combines lean methods and Six Sigma concepts.

One further tool for companies for improving their operations is benchmarking, which can stimulate creativity in improvement practice by learning from others, whether externally or internally (Slack et al. 2016). One form of benchmarking, so-called best practices benchmarking, can help organisations identify practices and processes that lead to superior performance[3] (Laasch and Conaway 2015).

Just-in-time and agile manufacturing are, one could say, two sides of the same coin, both developed as a response to a greater need for operational efficiency and reduced logistics and supply chain times (Grant et al. 2017). Lean operations emphasise efficiency, whilst agile operations focus on responsiveness and flexibility, both of which address crucial performance objectives in operations, such as quality, speed, dependability, flexibility and cost (Slack et al. 2016).

In just-in-time manufacturing and lean operations, inventory is something to be avoided, as it is considered a form of waste. We briefly covered inventory and warehousing in the previous chapter on supply chains – the environmental impact of warehouses occurs both during construction and operations, which can be addressed at the design and building stage as well as in the use phase through addressing energy needs. Whilst warehousing makes up 13% of GHG emissions in the supply chain, there is – in simplified terms – a trade-off between the cost of transport (and distance travelled) and the cost of holding inventory (Grant et al. 2017), which means that from an environmental perspective a case for holding inventory may be made.[4]

Taken together, it is very much a question of capacity planning and inventory management which, alongside layout and flow, need careful consideration. Business process reengineering (BPR) as part of process management can (through e.g. JIT and TQM) help with the redesign of such processes (Kleindorfer et al. 2005).

Other aspects of cleaner production include repairable inventory and remanufacturing (Srivastava 2007). Repairable inventory concerns systems for items that are cheaper to repair than to replace (Guide and Srivastava 1997), while remanufacturing similarly restores end-of-life products to like-new condition, also known as recycling-integrated manufacturing (Srivastava 2007). Both are closely related to reverse logistics and supply chains, covered in the previous chapter.

Various other new developments in manufacturing more generally can all contribute to making production cleaner, such as distributed and networked manufacturing, microfactories and micromanufacturing, additive manufacturing, smart materials, miniaturisation, dematerialisation as well as new technologies, such as e.g. nanotechnologies. Servicisation, the move from manufacturing a product to providing a service represents a shift in thinking, allowing companies to take ownership of a product throughout its life cycle and thus truly practice product stewardship and extended producer responsibility. Lean thinking and cradle-to-cradle production are combined in a service and flow economy as included in natural capitalism.

One such new development, additive manufacturing, deserves some consideration here. Although three-dimensional (3D) printing is often used as a synonym for additive manufacturing, there are, in fact, a number of different processes involved in layer manufacturing, subject to which material and which technology is used. It involves digital design and producing objects layer by layer. Unlike conventional production and operations, which involves removing material through machining, carving and shaping (also called subtractive manufacturing), and often produces waste as a result, additive manufacturing carefully adds material when creating objects and is considered more sustainable as a result (Kellens et al. 2017). Its environmental merits are still under investigation, though it is already clear that it leads to less waste, lighter products and reduced transport needs as production can be more localised, though the energy requirements of additive manufacturing exceed that of conventional production.

As conventional subtractive manufacturing often produces waste as a by-product, waste management has been an important part of making production and operations more sustainable, to which we now turn.

The following table summarises the general integrated and preventive principles of cleaner production for the whole production cycle.

TABLE 9.3

Principles of cleaner production
Reduce use and consumption of raw materials and energy in manufacturing
More efficient use of raw materials, energy and water to increase productivity
Reduction at source of emissions and waste to improve environmental performance
Design environmentally friendly but cost-effective products to reduce environmental impact of products throughout lifecycle
Reduce quantity and toxicity of emissions and waste at source
Eliminate use of toxic and dangerous materials

Source: Grant et al. 2017.

Waste management

As we covered the environmental challenge of waste in a dedicated chapter, and looked at waste management in the previous chapter on supply chain management, we keep this section short, briefly highlighting production and operations' role in waste management.

Waste management is crucial to production and operations (Weybrecht 2014) and takes different forms. Starting points include prevention of waste in the first place, reducing, recycling and reusing it, and if it cannot be prevented or reused, disposal and improved monitoring (Weybrecht 2014). Broader strategies encompass extended producer responsibility which encourages and forces manufacturers to internalise the full costs of waste, servicisation, making use of by-product synergies (industrial symbiosis or ecology), quality management and lean manufacturing, making new products from old ones (also called upcycling), and urban mining, which involves tapping into landfill sites for resources.

Waste is very much the product of the linear economy that relies on taking, making and disposing of materials. By using circular economy principles, zero waste is made more achievable, which is supported by the introduction of extended producer responsibility and the push towards reverse logistics and recycling (Grant et al. 2017). Moving towards zero waste can involve reuse, refurbishing or remanufacturing, sharing, upcycling, recycling and using energy-from-waste as well as dematerialisation more generally.

Technology and Innovation

Whether it is a move to zero waste in a circular economy, making production cleaner, greener and leaner, or designing products for the environment with the end of life in mind, technology and innovation are vital ingredients in making operations management and production more sustainable. One important principle of operations management is the exploitation of technological developments (Slack et al. 2016) and thus is key in sustainable operations management.

Whilst early environmentalists might have often been anti-technology (Elkington 1997), it is technology that can be considered key in cleaning up operations and, in particular, production, as aptly illustrated by Ray Anderson. Anderson (1998, 2009) uses the IPAT equation (introduced in Chapter 3[5]) to distinguish between the old industrial paradigm where technology (T_1) is part of the problem (extractive, linear, wasteful) and a new industrial paradigm where technology (T_2) becomes part of the solution (renewable, circular, resource productive).

$$I = P \times A \times T_1 \text{ versus } I = \frac{P \times A}{T_2}$$

By shifting technology from the numerator to the denominator, technology is used to reduce the environmental impact of population and affluence. Although new technologies may have unintended environmental consequences (Huesemann and Huesemann 2011, see also Box 3.4 in Chapter 3), technological innovation is key to dealing with environmental challenges.

Technology can be characterised in various ways. Weybrecht (2014) differentiates between *remediation technologies*, that address environmental problems after the fact and are therefore rather expensive; *abatement technologies*, which treat or capture pollutants before they are released into the environment, a capital-, resource- and energy-intensive method, and; *pollution prevention technologies*, that avoid pollution from the start, using alternative processes or products. The ultimate aim is *sustainable technologies*, which focus on resource efficiency and low or no environmental impact. Blowfield (2013) also lists *efficiency technologies* in his categorisation of technologies, where in particular eco-efficiency has long been a focus. Other technologies he classifies include *substitution technologies*, where alternatives for damaging resources are developed, *effectiveness technologies*, which go beyond efficiency and seek to close loops and to be restorative (add value in sustainability terms), and *mitigation technologies*, which aim to counteract climate change risks in particular.

Furthermore, innovation in existing technologies can be harnessed. Innovation is, however, not synonymous with technology as e.g. behavioural innovation can be used to encourage more sustainable consumption, and economic innovation can be used to put a value on the different dimensions of sustainability, for example through incentives or sanctions. Similarly, technology is not just found in hardware, but also in softer factors (e.g. thought technology or organisational technology).

Green innovation, or eco-innovation, is key for the development of new products and processes (Klassen and Vachon 2012). This also underlines the importance of research and development (R&D) leading to cleaner production based on design for the environment. When underpinned by a proactive environmental (or sustainability) strategy, eco-innovation helps firms push beyond regulatory requirements to satisfy customer needs (Tsai and Liao 2017). Technology innovation is a path to cleaner production (Grant et al. 2017). Examples of innovations leading to cleaner production include smart materials, miniaturisation, dematerialisation, nanotechnologies and servicisation. More examples are provided in the final section of the chapter.

Clean technologies (or cleantech) as introduced in Chapter 5 (see Table 5.6) are not confined to one particular technology and cover a wide range of technological innovations, pertaining to e.g. buildings, energy, and transport, and advanced green materials using principles such as resource efficiency, performance improvement and waste minimisation. Other terms that are used in the context of (clean) technology are leapfrog technology, restorative technology and the umbrella term sustainable technology (Hart 2010) as well as life cycle technology which shifts focus from product to function, supporting servicisation, based on life cycle design.

To summarise, the role of technology is often overestimated in the short term, but underestimated in the long run. Technology is therefore key for companies on the path to sustainability in the long term.

Tools for sustainable production and operations

Many of the tools and concepts covered in the previous chapter on supply chain management apply to production and operations as well, demonstrating the fluid boundary between the two corporate functions as well as the holistic and systemic nature of environmental sustainability. Here we focus on the contribution of these tools and concepts to production and operations particularly and refer the reader to previous chapters for more general introductions.

Circular economy

The circular economy is the broader context and underlying condition in which sustainable operations management is made possible. It moves the economy away from a linear take-make-waste approach to an industrial system which is regenerative or restorative by design and intent (EMF 2013). Important influences on the circular economy have been industrial ecology and cradle-to-cradle principles as well as biomimicry as a key design concept. One of the aims of the circular economy is to create closed-loop systems for production.

Closed-loop systems and industrial ecology

Industrial ecology typically focuses on resource efficiency not just within a firm, but beyond the boundary of individual organisations into the supply chain across whole product life cycles and industrial clusters (Lifset and Boons 2012). It helps us situate the context of sustainable operations within the wider environment and emphasises the need to collaborate, as discussed in the previous chapter on supply chain management.

Industrial symbiosis, as it is also called, can help companies close the cycle of industrial production as it provides a practical framework of material and energy exchange that fosters product life cycle planning, EPR and co-location of industrial operations where the by-products of one industry or company can be used by another, whether it is waste, heat or water (Grant et al. 2017). Carried out on a large scale, this can help the emergence of so-called eco-industrial parks.

Cradle-to-cradle

EPR shifts manufacturers' attitudes and approaches from cradle-to-grave (or cradle-to-gate) to cradle-to-cradle, from design to production, use and disposal. Design is central here and represents a move from built-in obsolescence to biomimicry which imitates the effectiveness of nature. Materials are conceptualised as nutrients (biological and technical) which recycle continuously. The principles listed previously enable companies to remove harmful materials from production processes and then to introduce new materials that are safer and healthier to use (McDonough and Braungart 2002). Such design for eco-effectiveness that aims to add good on top of subtracting bad is a form of transformative innovation that often takes shape in industrial ecology (Fiksel 2012b). Nevertheless, McDonough and Braungart (2013) acknowledge that such transformation is not easy and emphasise that starting is important as no company can become 100% sustainable overnight. Their Cradle to Cradle Certified[CM] programme recognises this and starts by certifying intentionality which provides a baseline from which optimisation is possible. An important question for manufacturers to ask in their design process is, what's next? as it forces companies to take into account what happens to products and their components in the next period. If companies start to conceive of every design component as borrowed which needs to be returned in good condition, products are designed not just with the next reuse period in mind, but with the ones after that as well (McDonough and Braungart 2013). The intention is that materials can be reused endlessly in a closed-loop system.[6] In order to move from downcycling (which downgrades material use) to upcycling (which upgrades material reuse), companies need to know which of the materials they use are harmful and what their impact is. A vital tool in cradle-to-cradle approaches is therefore life cycle analysis which helps organisations shift production from cradle-to-grave or cradle-to-gate, i.e. beyond the end of life and final manufacturing to the whole product cycle (Fiksel 2012b).

Life cycle analysis

We introduced life cycle analysis (or assessment, LCA) as an important tool to measure environmental impact more generally in Chapter 3 and outlined its use for supply chain management in Chapter 8. It helps companies identify areas of operations and production where improvements may be made. Although much of the impact of a product is during its use phase and thus outside factory gates, LCA can support design for environment decisions that address this phase of the life cycle. Within the context of operations management, LCA is a natural bridge from more conventional topics such as lean manufacturing and quality to EPR and product stewardship (Klassen and Vachon 2012).

LCA has been standardised and documented through the ISO 14040 series. It is a systematic method to capture all input and output of energy, materials, resources and wastes for the whole life cycle of goods and services: resource extraction and processing, design, production and retail, distribution, use, waste collection, end-of-life processes (reuse, recycling, energy recovery, disposal) (Zeng and Stratton 2017). We can differentiate between conceptual, detailed, simplified, comprehensive, conventional, streamlined or hybrid LCAs with some trade-offs between specificity, detail and effort involved. There are various limitations to LCAs, such as the time and expense involved, defining system boundaries and the difficulty of interpreting results (Fiksel 2012b). Nevertheless, it is the most commonly accepted form of environmental impact assessment (Grant et al. 2017).

There are four stages in LCA of initiation (definition of scope and goals), inventory (data collection), impact analysis (interpretation) and improvement. Concrete analysis steps are as follows (Stead and Stead 2004: 149):

- Analysis of type and amount of raw materials and energy (inputs into products and processes) and ecological impacts of acquiring them.
- Assessment of ecological impact of manufacturing process (including examining materials and energy used in component manufacturing processes, final product manufacturing processes, and product assembly processes).
- Analysis of transportation and distribution systems (related to delivering the product to market, analysed in terms of distribution modes, distances, fuel consumption and so on).
- Analysis of environmental consequences of product use (including product durability, energy requirements, polluting potential and the like (this may include social consequences, such as impacts on health and safety of consumers).
- Analysis of potential for reuse or recyclability of product.
- Examination of product's ultimate disposal in terms of its toxicity, volume, biodegradability and so on.

Results of LCAs can be grouped e.g. by media (land, air, water) or can be aggregated for a total score to be compared with past results, competitor products or services, or with external benchmarks or standards (Stead and Stead 2004).

Similar tools include input-output analysis, footprint analysis, material-energy-toxicity matrix, material intensity, materials flow analysis and circularity measurements. These are all designed to help with life cycle management and to support product development decisions that improve environmental performance and reduce resource intensity of products as well as extend a product's life. The key point in time for LCA results to be considered is at the design stage of new products to enable life cycle design with all its DfE subcomponents.

Spotlight on . . . examples and practices of sustainable design, production and operations

As in the previous chapter, in the following we will highlight examples of practices and of products in sustainable production and operations, loosely structured around the elements and tools covered in the preceding sections.

As an example of extended producer responsibility, Germany initiated an ambitious take-back programme in 1991 with the German Packaging Ordinance on avoiding and recycling packaging waste. It forced distributors and manufacturers to take comprehensive responsibility for disposing of packaging waste. The regulation obliged companies to take part in an authorised system for waste disposal and recycling or to join an industry solution where a third party handles how packaging is collected and recycled – also called a dual system as it is parallel to public-sector waste disposal services. Germany's so-called The Green Dot (*Der Grüne Punkt*) was one of the first systems worldwide of a dual take-back system.

HP Inc., the manufacturer of electronics such as printers and laptops, provides an example of a company exercising product stewardship through following principles of design for environment, cleaner production, waste management and innovation, underpinned by circular economy values. The company has started to design recovered materials into its products lines, using post-consumer recycled content, including recycled plastic from electronics, thus exercising closed-loop recycling. Widening its impact, it has also started to collect ocean-bound plastic waste in order to upcycle it into new ink cartridges. The company's sustainability efforts have contributed significantly to new revenues through sales or contracts (Clancy 2018).

Visually, one of the most powerful examples of eco-design, and in particular design for dematerialisation, are flat screens that have almost completely replaced CRT (cathode ray tube) screens for computers and televisions. The introduction of LED and LCD (light-emitting diode; liquid-crystal display) screens has led to reduction of weight, volume and materials for electronics manufacturers. Similarly, desktop computers have benefited from miniaturisation – the conventional heavy computing towers have often been replaced by slim or mini versions.

Another example where redesign has led to reductions in weight, volume and materials in manufacturing and therefore also to greater efficiencies in transportation and logistics can be found in washing machines. These appliances are normally fitted with 25kg of concrete which stop the machines from moving during a spin cycle. The redesign involves replacing the block of concrete with plastic containers that are filled with water once the machine is installed. This not only makes the machines lighter and easier to transport, it also limits the production of concrete, which is carbon-intensive (Harrabin 2017).

So-called green chemistry is one approach to make production cleaner. It is a key area in sustainability as chemistry in the form of chemicals and chemical processes provides over 95% of the products society uses (Weybrecht 2014). Green chemistry (also called sustainable chemistry) is a design philosophy based on 12 principles[7] that encourage innovation as well as developments in chemical technology to eliminate (or at least reduce) the production or use of toxic and hazardous substances (Archer et al. 2008). Examples of companies that have design and manufactured products or materials based on green chemistry principles include chemicals producers Croda, BASF and Dow as well as pharmaceutical companies Merck and Lilly.

An example where a company uses both waste management and remanufacturing following circular economy principles is Boldking™ which offers a subscription and mailing service for shaving products. The packaging is made of recyclable cardboard, and it offers consumers recycling envelopes to send back old razor blades which are then disassembled and remanufactured into new metal and plastic.

Huadu Worldwide Transmission is a Chinese remanufacturing company based in Guangzhou that has partnered with a retail franchise (4S), enabling the return of used car parts which are then remanufactured and sold as certified as-new spare parts. Over 30 major car companies – Chinese and international ones – have authorised the partnership's system for collecting and distributing used and remanufactured transmission boxes, which is an example of closing the loop (EMF 2019).

Cradle-to-cradle approaches address the built-in obsolescence that dominates cradle-to-grave design and manufacturing (Nemetz 2013). Its core principle of eco-effectiveness replaces eco-efficiency and thus moves from reducing negative impacts to producing positive ones, and changes downcycling of materials to upcycling. The steps are outlined in Chapter 4 and summarised earlier. Swiss company Rohner Textil attempted to find environmentally benign dyes – a challenging task as only one major chemical company was prepared to cooperate, leading to 16 environmentally benign chemicals identified out of a total of 8,000. This reengineering process resulted in the elimination of toxic wastewater as well as cost reduction and the new product Climatex Lifecycle™ contributing to Rohner's bottom line (Nemetz 2013, see also company website www.climatex.com). Climatex® is a certified cradle-to-cradle fabric.

Other examples of products certified to cradle-to-cradle standards (ranging from basic, bronze, silver to gold and platinum) include cleaning products from Method, personal care products from Garnier (owned by L'Oréal), inks by green4print, carpets from Takrett and Interface (see also Chapter 4) as well as building supplies, basic materials, packaging and paper (for more examples, see www.c2ccertified.org/products/registry).

Although the products from geoorganics™ are not C2C certified, the company follows a similar philosophy of avoiding synthetic ingredients and using natural ingredients in its production of oral care products. On its website, the company lists all the (natural) ingredients it uses as well as all the ingredients it avoids using. Similar to Boldking, the company offers consumers the possibility of sending back any products or parts that cannot be recycled at home.

In an example of how challenging sustainability is when engaging in international or global production networks and operations, Mattel had to recall almost 10 million toys in 2007 after a supplier in China had used lead paint in manufacturing some of them (Esty and Winston 2009). In 2018, only a third of countries have legally binding controls on the manufacturing, import, sale and use of paints containing lead, according to the World Health Organization (WHO 2018).

Life cycle analysis (or assessment) and similar tools are used by companies across a range of industries and can be used to underpin any of the elements of operations management, but in particular design for environment. Many of the pollution prevention programmes start with LCAs, such as for example 3M's 3P (Pollution Prevention Pays) programme and Dow Chemical's WRAP (Waste Reduction Always Pays) programme. Other examples include Unilever, Procter & Gamble, Tropicana and Levi Strauss & Co. which was one of the first apparel companies to use LCA to study the impact of their products' life cycles (Zeng and Stratton 2017). As the cultivation of cotton and consumer use were found to have the biggest impact, the company has since focused its sustainability efforts on the supply chain and on consumer education (see company website for detailed results).

Checklists of criteria (e.g. material selection or product or process design) provide a qualitative assessment tool. Although they have various limitations and are not very detailed, they are nevertheless a good starting point for companies to consider environmental issues and take action. For example, DuPoint developed a comprehensive product stewardship checklist that the company has shared freely (Fiksel 2012b).

Examples of servicisation (or servitization) innovations include manufacturing companies such as Rolls-Royce, Caterpillar, Alstom, MAN Truck and Xerox which offer service packages to customers, often with the help of remote monitoring (Baines et al. 2017). Benefits include better responses to customer needs and increased loyalty, improved product innovation and financial benefits in terms of revenue and profit.

Examples of smart materials, nanotechnologies and additive manufacturing include smart building materials that can e.g. respond to changes in heat, nanofiltration for cleaning toxic gases from air, 3D printable batteries powered by vegetable oil, amongst others – all part of technological advancements that can help companies towards a more sustainable future, the full potential of which is not yet fully explored (see e.g. Jackson 2019).

SUMMARY

In this chapter on design, production and operations we have further explored central aspects of the supply chain and the challenges that businesses face in terms of what they take, make and waste. As much of this is decided in the product development stage, we heavily emphasised design. This is where important decisions are made regarding the environmental impact of a product, from how toxic production is in terms of materials used, how much energy is needed for its manufacture and for its use, and what is left in terms of useful material or waste at the end of the production process and the end of a product's useful life. In cleaner production, preventative and integrated approaches are emphasised. In particular, design, innovation and technology are closely interlinked.

Design for Environment is driven by life cycle principles which are supported by tools such as LCA and cradle-to-cradle design protocols covered in the chapter (Klassen and Vachon 2012). Industrial ecology and its core concept of LCA help identify interdependencies for producers in terms of which waste equals food and thus set priorities for how to make manufacturing (c) leaner and greener. Waste minimisation is core to both green and lean operations. Lean thinking in particular challenges the need for goods and services, and raises questions about functionality of, demand for and value proposition of products, thus encouraging a move to servicisation (Elkington 2012).

Other concepts that are key to making productions more sustainable include eco-efficiency and resource productivity. Operations as well as logistics are significant functions when it comes to creating the cost advantages gained through eco-efficiency (Stead and Stead 2014). Although there may be trade-offs between cost and other performance objectives such as quality or speed, price is not always the order-winning criterion (Zeng and Stratton 2017), which leaves room for investment into new technology and for taking a wider systems view encompassing the whole life cycle.

In summary, various shifts are necessary for business to make production and operations, including design, more sustainable:

- The costs of disassembly, recycling, or disposal need to be shifted from "downstream" buyers and consumers to "upstream" producers and designers through extended producer responsibility and design for environment
- Companies need to shift their thinking to provision of service and functionality rather than manufacturing of products and provision of ownership
- Design shifts from cradle-to-grave to cradle-to-cradle and thus from short-term thinking to long-term perspectives
- Moving from eco-efficiency to eco-effectiveness, shifting from less bad to more good
- Shifting from linear models to closed loops and circular systems
- Technology shifts from part of the problem to part of the solution.

In the following chapter, we move to the final stage of the supply chain, that of the end consumer, and thus cover the other side of production's coin, that of consumption and the role of marketing.

SUGGESTED QUESTIONS AND ACTIVITIES

- Eco-design and biomimicry in action: Think of a kind of product you might like to design, for example a piece of furniture, a support structure, a propeller, a fastener, a waterproof coating, an air filter and so forth. Investigate how you could use biomimicry to help with the design. Biomimicry is an approach to design and problem-solving which uses strategies developed by living organisms. AskNature is the Biomimicry Institute's online database of solutions from nature, organised by type of challenge. You can find the database at www.asknature.org/. First, identify the function you want to design for, rather than the thing you want to design. (For example, you might want "a place to sit" rather than "a chair"). Then go to the strategies database and identify which categories you think could address your design goal. List some of the ways other organisms perform the function you have in mind. Include a brief description of how each strategy works.
- Debate: what is the most important element of production and operations when considering sustainability?
- Consider the LCA implications of the following products:
  - Coffee cups: ceramic versus Styrofoam versus paper
  - Light bulbs: CFL (compact fluorescent lamp) versus LED (light-emitting diodes) versus incandescent
  - Carpets: nylon versus woollen
  - Grocery bags: paper versus plastic
  - Diapers: disposable versus cloth.
- Find out what the twelve principles of Green Chemistry are.
- Explore the international dimension of production (and its environmental impact) by watching some of the Story of Stuff films made on e.g. electronics or cosmetics: https://storyofstuff.org/movies/.
- Look up some of the corporate examples and practices mentioned in the chapter, and research them in more detail. Here are some starting points: www.gruener-punkt.de/en/company/der-gruene-punkt.html, http://www8.hp.com/us/en/hp-information/global-citizenship/index.html, www.boldking.com/, www.levistrauss.com, https://georganics.co.uk
- For more in-depth material on sustainable and green design, production and operations, check out the following:
  - Fiksel, J. (2012) *Design for Environment: A Guide to Sustainable Product Development*, New York: McGrawHill, 2nd ed.
  - Grant, D. B., Trautrims, A. and Wong, C. Y. (2017) *Sustainable Logistics and Supply Chain Management: Principles and Practices for Sustainable Operations and Management*, Kogan: London, 2nd ed.
  - Articles in the following journals: *Environmental Quality Management*, *Journal of Cleaner Production*, *Journal of Industrial Ecology*, *International Journal of Life Cycle Assessment*.
  - Articles from the Global Environmental Management Initiative: http://gemi.org/solutions/.

NOTES

1 See also the concept of externalities as discussed in Chapter 2.

2 The word comes from the Greek word for life, *bios*, and for imitation, *mimesis*.

3 Other types of benchmarking include internal, external, non-competitive, competitive, performance and practice benchmarking (Slack et al. 2016) to which strategies and parameter benchmarking could be added (Laasch and Conaway 2015).

4 See Chapter 8 on supply chain management for more material on warehousing.

5 I equals the environmental impact, P population, A affluence and T technology, summarising the drivers of environmental degradation (Robertson 2017).

6 Following their Cradle To Cradle publication in 2002, McDonough and Braungart (2013) saw the term closed loops and "waste equals food" often misinterpreted to include the design and production of toxic products that were then configured into other toxic products. As this does not positively define materials based on cradle-to-cradle design, the authors suggest the

term "nutrient management" instead of closed loops.

7 The principles were developed by Anastas and Warner (1998) and include some general, familiar ones (such as design for energy efficiency and waste prevention) as well as chemistry-specific ones (such as atom economy and safer solvents and auxiliaries). For a complete list see The American Chemical Society: www.acs.org/content/acs/en/greenchemistry/principles/12-principles-of-green-chemistry.html.

SOURCES AND FURTHER READING

Anastas, P. T. and Warner, J. C. (1998) *Green Chemistry: Theory and Practice*, New York: Oxford University Press.

Anderson, R. C. (1998) *Mid-Course Correction. Toward a Sustainable Enterprise: The Interface Model*, White River Junction, VT: Chelsea Green.

Anderson, R. C. and White, R. (2009) *Confessions of a Radical Industrialist: How Interface Proved that you can Build a Successful Business Without Destroying the Planet*, London: Random House.

Archer, G., Larson, A., White, M. and York, J. G. (2008) Green Chemistry and EVA: A Framework for Incorporating Environmental Action into Financial Analysis, in: Wankel, C. and Stoner, J. A. F. (2008) *Innovative Approaches to Global Sustainability*, New York: Palgrave MacMillan, pp. 83–102.

Baines, T., Bigdeli, A. Z., Bustinza, O. F., Shi, V. G., Baldwin, J. and Ridgway, K. (2017) Servitization: Revisiting the State-of-the-art and Research Priorities, in: *International Journal of Operations & Production Management*, 37(2): 256–278.

Benn, S., Dunphy, D. and Griffiths, A. (2014) *Organizational Change for Corporate Sustainability*, London: Routledge, 3rd ed.

Benyus, J. M. (1997) *Biomimicry: Innovation Inspired by Nature*, New York: William Morrow.

Bernon, M., Baker, P., Mena, C., Palmer, A., Skipworth, H., Smart, A. and Templar, S. (2012) Towards more Sustainable Supply-chain Management, in: Grayson, D. and Exter, N. (eds.) *Cranfield on Corporate Sustainability*, Sheffield: Greenleaf Publishing, pp. 149–166.

Blowfield, M. (2013) *Business and Sustainability*, Oxford: Oxford University Press.

Clancy, H. (2018) HP Links $700 Million in Revenue to Sustainability Strategy, *GreenBiz*, 09.07.2018, available at www.greenbiz.com/article/hp-links-700-million-revenue-sustainability-strategy (Retrieved, 14.01.2018).

Collins, K. (2014) *The Nature of Investing: Resilient Investment Strategies through Biomimicry*, Brookline, MA: Bibliomotion.

Daft, R. L., Kendrick, M. and Vershinina, N. (2010) *Management: International Edition*, Andover: Cengage Learning EMEA.

Dangelico, R. M., Pujari, D. and Pontrandolfo, P. (2016) Green Product Innovation in Manufacturing Firms: A Sustainability-Oriented Dynamic Capability Perspective, in: *Business Strategy and the Environment*, 26(4): 490–506.

Doppelt, B. (2003) Overcoming the Seven Sustainability Blunders, in: *The Systems Thinker*, 14(5), available at https://thesystemsthinker.com/overcoming-the-seven-sustainability-blunders/ (Retrieved, 21.11.2017).

Elkington, J. (1997) *Cannibals with Forks: The Triple Bottom Line of 21st-Century Business*, Oxford: Capstone.

Elkington, J. (2012) *The Zeronauts: Breaking the Sustainability Barrier*, London: Routledge.

EMF (2013) *Towards the Circular Economy: Economic and Business Rationale for an Accelerated Transition*, Ellen MacArthur Foundation, Vol. 1, Isle of Wight.

EMF (2019) *Scaling up Remanufacturing – Huadu Worldwide Transmission, Case Study*, Ellen MacArthur Foundation, available at www.ellenmacarthurfoundation.org/case-studies/remanufacturing-at-scale (Retrieved, 21.01.2019).

Esty, D. C. and Winston, A. S. (2009) *Green to Gold: How Smart Companies use Environmental Strategy to Innovate, Create Value and Build Competitive Advantage*, Hoboken, NJ: Wiley.

Fiksel, J. (1996) Achieving Eco-Efficiency through Design for Environment, in: *Total Quality Environmental Management*, 5(4): 47–54.

Fiksel, J. (2012a) A Systems View of Sustainability: The Triple Value Model, in: *Environmental Development*, 2(1): 138–141.

Fiksel, J. (2012b) *Design for Environment: A Guide to Sustainable Product Development*, New York: McGrawHill, 2nd ed.

Friedman, F. B. (2011) *Practical Guide to Environmental Management*, Washington, DC: Environmental Law Institute, 11th ed.

Grant, D. B., Trautrims, A. and Wong, C. Y. (2017) *Sustainable Logistics and supply Chain Management: Principles and Practices for Sustainable Operations and Management*, London: Kogan, 2nd ed.

Green, T. (2013) *The Dictionary of Global Sustainability. With Case Studies on the Environmental, Economic, Technological, and Social Aspects of Sustainability*, New York: McGrawHill.

Guide, V. D. R. and Srivastava, R. (1997) Repairable Inventory Theory: Models and Applications, in: *European Journal of Operational Research*, 102(1): 1–20.

Harrabin, R. (2017) Device could Make Washing Machines Lighter and Greener, *BBC*, 04.08.2017, available at www.bbc.co.uk/news/uk-40821915 (Retrieved, 14.01.2018).

Hart, S. L. (2010) *Capitalism at the Crossroads: Next Generation Business Strategies for a Post-Crisis World*, Upper Saddle River, NJ: Pearson, 3rd ed.

Hawken, P. (1993) *Ecology of Commerce: A Declaration of Sustainability*, New York: HarperBusiness.

Hawken, P., Lovins, A. B. and Lovins, L. H. (1999) *Natural Capitalism: The Next Industrial Revolution*, London: Earthscan.

Hey, R. (2000) River Processes and Management, in: O'Riordan, T. (ed.) *Environmental Science for Environmental Management*, Harlow: Pearson, 2nd ed., pp. 323–347.

Hoffmann, E. (2012) *User Integration in Sustainable Product Development: Organisational Learning through Boundary-spanning Processes*, Sheffield: Greenleaf Publishing.

Huesemann, M. H. and Huesemann, J. A. (2011) *Techno-Fix: Why Technology won't Save us or the Environment*, Gabriola Island: New Society Publishers.

Jackson, F. (2019) Top Ten Technologies To Save The World, *Forbes*, 17.01.2019, available at www.forbes.com/sites/feliciajackson/2019/01/17/top-ten-technologies-to-save-the-world/#68af3a578742 (Retrieved, 21.01.2019).

Kellens, K., Baumers, M., Gutowski, T. G., Flanagan, W., Lifset, R. and Duflou, J. R. (2017) Environmental Dimensions of Additive Manufacturing: Mapping Application Domains and their Environmental Implications, in: *Journal of Industrial Ecology*, 21(S1): 49–68.

Klassen, R. D. and Vachon, S. (2012) Greener Supply Chain management, in: Bansal, P. and Hoffman, A. J. (eds.) *The Oxford Handbook of Business and the Natural Environment*, Oxford: Oxford University Press, pp. 269–289.

Kleindorfer, P. R., Singhal, K. and Van Wassenhove, L. N. (2005) Sustainable Operations Management, in: *Production and Operations Management*, 14(4): 482–492.

Laasch, O. and Conaway, R. (2015) *Principles of Responsible Management: Glocal Sustainability, Responsibility and Ethics*, Stamford, CT: Cengage Learning.

Laszlo, C. and Zhexembayeva, N. (2011) *Embedded Sustainability: The Next Big Competitive Advantage*, Sheffield: Greenleaf Publishing.

Lifset, R. and Boons, F. (2012) Industrial Ecology: Business Management in a Material World, in: Bansal, P. and Hoffman, A. J. (eds.) *The Oxford Handbook of Business and the Natural Environment*, Oxford: Oxford University Press, pp. 310–326.

Lovins, A. B., Lovins, L. H. and Hawken, P. (1999) A Road Map for Natural Capitalism, in: *Harvard Business Review*, 77(3): 145–158.

Matten, D. (2010) Biomimicry, in: Visser, W., Matten, D., Pohl, M et al. (eds.) *The A to Z of Corporate Social Responsibility*, Hoboken, NJ: Wiley, 2nd ed., available at http://ezproxy01.rhul.ac.uk/login?url=https://search.credoreference.com/content/entry/wileyazcsr/biomimicry/0?institutionId=8498 (Retrieved, 24.12.2018).

McDonough, W. and Braungart, M. (1992) The Hannover Principles: Design for Sustainability, available at www.mcdonough.com/wp-content/uploads/2013/03/Hannover-Principles-1992.pdf (Retrieved, 22.11.2018).

McDonough, W. and Braungart, M. (2002) *Cradle to Cradle: Remaking the Way we Make Things*, New York: North Point Press.

McDonough, W. and Braungart, M. (2013) *The Upcycle: Beyond Sustainability – Designing for Abundance*, New York: North Point Press.

Nemetz, P. N. (2013) *Business and the Sustainability Challenge: An Integrated Perspective*, London: Routledge.

Robèrt, K.-H. (2002) *The Natural Step Story: Seeding a Quiet Revolution*, Gabriola Island: New Society Publishers.

Robertson, M. (2017) *Sustainability Principles and Practice*, London: Routledge, 2nd ed.

Saavedra, Y. M. B., Iritani, D,. R., Pavan, A. L. R. and Ometteo, A. R. (2018) Theoretical Contribution of Industrial Ecology to Circular Economy, in: *Journal of Cleaner Production*, 170(1): 1514–1522.

Sarkis, J. and Dou, Y. (2018) *Green Supply Chain Management: A Concise Introduction*, New York: Routledge.

Slack, N., Brandon-Jones, A. and Johnston, R. (2016) *Operations Management*, Harlow: Pearson Education, 8th ed.

Slack, N., Chambers, S. and Johnston, R. (2001) *Operations Management*, Harlow: Pearson Education, 3rd ed.

Souza, G. C. (2018) *Sustainable Operations and Closed Loop Supply Chains*, New York: Business Expert Press, 2nd ed.

Srivastava, S. K. (2007) Green Supply-chain Management: A State-of-the-art Literature Review, in: *International Journal of Management Reviews*, 9(1): 53–80.

Stead, J. G. and Stead, W. E. (2014) *Sustainable Strategic Management*, Sheffield: Greenleaf Publishing, 2nd ed.

Stead, J. G., Stead, W. E. and Starik, M. (2004) *Sustainable Strategic Management*, Armonk, NY: M.E. Sharpe.

Tsai, K.-H. and Liao, Y.-Ch. (2017) Sustainability Strategy and Eco-Innovation: A Moderation Model, in: *Business Strategy and the Environment*, 26(4): 426–437.

Visser, W. (2012) *The Quest for Sustainable Business: An Epic Journey in Search of Corporate Responsibility*, Sheffield: Greenleaf Publishing.

von Weizsäcker, E., Lovins, A. B. and Lovins, L. H. (1998) *Factor Four: Doubling Wealth, Halving Resource Use*, London: Earthscan.

Walker, H., Di Sisto, L. and McBain, D. (2008) Drivers and Barriers to Environmental Supply Chain Management Practices: Lessons from the Public and Private Sectors, in: *Journal of Purchasing & Supply Management*, 14(1): 69–85.

Weybrecht, G. (2014) *The Sustainable MBA: A Business Guide to Sustainability*, Chichester: Wiley, 2nd ed.

Whitfield, R. and McNett, J. (2014) *A Primer on Sustainability: In the Business Environment*, New York: Business Expert Press.

WHO (2018) Regulations and Controls on Lead Paint, World Health Organization, website, available at www.who.int/gho/phe/chemical_safety/lead_paint_regulations/en/ (Retrieved, 21.010.2019).

Zeng, M. (2014) Sustainable Operations Management, in: Molthan-Hill, P. (ed.) The *Business Student's Guide to Sustainable Management*, Sheffield: Greenleaf Publishing, pp. 198–217.

Zeng, M. and Stratton, R. (2017) Sustainable Operations Management, in: Molthan-Hill, P. (ed.) The *Business Student's Guide to Sustainable Management*, Sheffield: Greenleaf Publishing, 2nd ed., pp. 357–379.

CHAPTER 10

Marketing, communication and sustainable consumption

INTRODUCTION

Having covered sustainable supply chains and cleaner production, we now move to the other side of the coin, that of consumption and the role marketing plays in companies' efforts to move towards environmental sustainability. As marketing is a (possibly *the*) key interface between companies and consumers, its role in moving businesses towards sustainability is crucial (Goworek 2017). Consumer behaviour and consumption are closely related to marketing, and thus this role of consumers needs to be acknowledged as well. In this context, consumers entail commercial buyers as well as end consumers.

By taking into account the role of buyers and consumers, the broader context of sustainable consumption becomes part of the picture, which also highlights the potential limits to corporate sustainability efforts as a number of natural tensions become apparent. Firstly, on the one hand consumers' willingness to go green and on the other, their willingness to pay. Secondly, on the one hand the encouragement of consumption – a key goal of marketing – and on the other, the reduction of resource consumption – a key requirement for sustainability (see e.g. Goworek 2017). Finally, the tension between credible green product claims and so-called greenwashing, where companies use environmentally sustainable language to sell products that do not live up to such claims or standards. This also highlights the role of information and education.

The chapter at hand therefore deals with a multifaceted topic, which we cannot treat exhaustively in the confines of a textbook that covers a broad range of corporate functions. Interested readers are therefore referred to other material for a more in-depth coverage of the topics touched on here.

In particular, we will focus on two aspects – the marketing of sustainable products and services (which might involve convincing consumers to switch from conventional to greener products) and sustainable marketing, i.e. marketing sustainably. In this, we will also touch upon the encouragement of sustainable (or reduced) consumption and the role of information and education. We discussed the roles of certification and labelling in helping companies to green their supply chain (e.g. in procurement in Chapter 8). Here we focus on their role in marketing green products to end consumers and in promoting sustainable consumption.

MARKETING

Overview

In short, marketing is about meeting needs profitably, which involves human and social needs that are first identified and then met (Kotler and Keller 2016). For more detail, we can turn to the American Marketing Association's formal definition of marketing (AMA 2019):

> Marketing is the activity, set of institutions, and processes for creating, communicating, delivering, and exchanging offerings that have value for customers, clients, partners, and society at large.

Marketing is both an art and a science, and its management involves choosing target markets and "getting, keeping, and growing customers" through a process that creates, delivers and communicates superior customer value (Kotler and Keller 2016: 27).

The traditional marketing mix of the so-called four Ps classifies the various marketing activities of product (e.g. quality, packaging), price (e.g. list price, discounts), promotion (advertising, sales force) and place (e.g. distribution and communication channels, locations, transport). New realities and market forces such as technology, globalisation and social responsibility have seen marketing move towards holistic marketing, which comprises relationship marketing, integrated marketing, internal marketing and performance marketing (Kotler and Keller 2016).

In developing marketing strategies, companies use the formula of segmentation, targeting and positioning (STP) for choosing and creating value before providing value through the product, place and price, and communicating the value through promotion, i.e. the marketing mix. The value proposition companies put forward to meet customer needs is strengthened if the value offering is from a known source, i.e. a brand (Kotler and Keller 2016).

Environmental impact of marketing

Broadly speaking, marketing contributes to environmental degradation by encouraging consumption and wasteful lifestyles. The environmental impact is across the challenges covered in part two and can be illustrated using the four Ps of the marketing mix.

- Product: The most obvious impact is the waste of product packaging and the waste at the end of a product's useful life. Furthermore, embedded in each product are the resources used to make it, such as energy and raw materials, which contribute to climate change and pollution. Depending on the quality and design, the rate of repurchasing for replacing broken or obsolete products results in further environmental impacts, particularly if goods end up as waste within a short space of time after purchasing.
- Price: Low prices encourage consumption, in turn leading to increased emissions of pollutants and greenhouse gases in production and to more waste. Negative externalities such as polluted air and water are not reflected in the price of products (Nemetz 2013).
- Promotion: Printed marketing campaigns such as traditional mailshots of promotional materials require natural resources for their production, which emits air pollutants and greenhouse gases. Leafleting and other promotional materials such as free give-aways easily end up as waste and litter the environment. Although online advertising is seemingly less tangible and physical, it requires energy, which (more often than not) involves the burning of fossil fuels and the emission of greenhouse gases.

- Place: the choice of distribution channels, location, warehousing and transport all have environmental implications, which were covered in Chapter 8 on logistics and supply chain management. In the context of marketing, the rise in e-commerce and online shopping has led to an increase in individual shipments to homes (as opposed to bulk shipments to shops) requiring e.g. more packaging and transport.

Impact of environment on marketing

Unlike the previous two chapters, it is less clear or obvious how environmental factors can affect marketing. Extreme weather events as a result from climate change and air pollution may influence a company's ability to distribute its products. Greater environmental awareness among consumers puts greater pressure on companies to justify how and where they make, distribute and price their products and to provide more environmentally friendly products.

Business case for greener marketing

Marketing is focused on creating customers – a main purpose for business (Drucker 1974) – and thus plays a key role in the commercial success of businesses, as without customers value cannot be realised (Schaltegger et al. 2003). As consumers are increasingly using environmental considerations in their buying decisions, companies need to respond through offering and marketing eco-friendly products, which can increase their market share and revenues (Esty and Winston 2009). Offering greener products can open new markets and can be a factor of differentiation, which may command a price premium.

In the broader context of making companies more sustainable, marketing has the potential to contribute greatly to greening a company with beneficial consequences for the company beyond marketing e.g. through lower costs as a result of reduced material consumption and increased eco-efficiency. Although much can be done upstream in the supply chain, much of the environmental impact of products is downstream during their use phase; marketing thus plays a crucial role as it has a great impact on buyers as a key interface to consumers. Effective communication and marketing of environmental performance has the potential to "make or break" a business. In particular, accusations of greenwashing where a company does not "walk the talk" can greatly damage corporate reputations and brand values (Laasch and Conaway 2015).

Engaging in sustainable marketing – in marketing greener products and in marketing more sustainably – therefore makes business sense regarding risk and cost management as well as new markets.

SUSTAINABLE MARKETING

Overview

There are a number of terms that are used in the context of marketing and environmental sustainability, with different meanings and nuances. Sustainable marketing is comprehensively defined by Martin and Schouten (2014: 6) as

> The process of creating, communicating, and delivering value to customers in such a way that both natural and human capital are preserved or enhanced throughout. Marketing must be done in a sustainable manner so that all the marketing processes are environmentally and

> socially benign while helping to bring about a society in which striving for sustainability is the norm.

Similarly, Stead and Stead (2014) emphasise that sustainability marketing places the commitment to product stewardship at the core of a company's marketing efforts in the broader context of moving towards a sustainable society. Product stewardship comprises the whole life cycle of goods, including marketing. The concept has also influenced marketing perspectives of people not just as customers but more holistically as people (Ottman 2011). Sustainable marketing acknowledges the limitations of marketing philosophy as inherently unsustainable and recognises the need for (self-) restraint in markets (van Dam and Apeldoorn 1996). Blowfield (2013) similarly emphasises that marketing has become more complex than simply selling more products and that a sustainability-based approach to it can take into account the carrying capacity of the planet.

In the wider context of sustainability encompassing environmental and social elements, we can furthermore introduce the overlapping practice of social marketing, which is a broader approach to changing consumer behaviour to the benefit of society, whilst societal marketing takes the long-term impact of goods and services into account (Goworek 2017). This is different from cause-related marketing where products are promoted by linking their sales with the contribution to a good cause, which may be environmental (Laasch and Conaway 2015). For example, 1% of the planet works with companies that contribute 1% of total sales to preserve and restore the natural environment.

Sustainable marketing recognises the interconnections of the various sustainability dimensions of the triple bottom line and could be categorised as a macro-marketing concept as it aims at changing the behaviour of consumers as well as producers across the value chain (Blowfield 2013). This is embodied in the three criteria sustainable marketing should satisfy: meeting customer meets, attaining organisational goals and making processes compatible with ecosystems (Stead et al. 2004, based on Fuller 1999). In contrast, green marketing pays more attention to individual problems.

Although green marketing is a term that is widely used, definitions vary as do the terms used, from environmental and ecological marketing to eco-marketing (Fogel 2016; Fuller 1999). As a sub-section of sustainable marketing, green marketing focuses on environmental sustainability (Goworek 2017). Dahlstrom (2011: 5) defines green marketing as

> the study of all efforts to consume, produce, distribute, promote, package, and reclaim products in a manner that is sensitive or responsive to ecological concerns.

This reflects a product stewardship approach across the whole life cycle of products, and consequently green marketing management is the process of planning and implementing the marketing mix to facilitate the various aspects from consumption to reclamation in an ecologically responsible manner (Dahlstrom 2011).

Similarly, eco-marketing goes beyond the marketing of more environmentally friendly products and integrates consumer education on consumption's environmental impact in an attempt to make consumer choice easier (Schaltegger et al. 2003). Eco-marketing provides consumers with information about product quality, use, maintenance, repair and disposal to facilitate informed purchasing decisions. Sustainability marketing goes beyond the sticker price for the customer and incorporates other social and environmental costs that buyers incur when purchasing, using and disposing of goods (Blowfield 2013).

For Zinkhan and Carlson (1995) green advertising is a corporate response to environmental concerns of consumers and consists of promotional messages appealing to the desires and needs of such consumers. Drawing on Banerjee and colleagues (1995), Fogel (2016) highlights that green advertising addresses one or more of three criteria:

1 It talks to the relationship between a good or service and the natural environment,
2 It promotes a green lifestyle with or without emphasising a good or service, and
3 It presents an environmentally responsible image of a business.

This provides a broader picture of what is involved in sustainable marketing and requires marketing to move from consumer demands that have to be immediately met to considering longer-term environmental impacts (Sharp 2013). In the following, we will look at three particular aspects: environmentally sustainable consumption, the marketing of goods and services that are sustainable (green or environmental marketing) which also involves portraying the companies that offer them as environmentally sustainable, and finally, marketing practices that are themselves sustainable.

Sustainable consumption and consumer behaviour

> Consumption begets waste.
>
> (Anderson 2014: 221)

Consumption – which in simple terms means using resources in order to live (Middlemiss 2018) – highlights the inherent tension for marketing and environmental sustainability. The traditional goal of marketing was selling more things to people, which is the antithesis to the reduction in resource consumption that is needed to meet the environmental sustainability challenge (Blowfield 2013). Over-consumption and consumerism are characteristic of contemporary capitalism and the consumer society, which is why sustainability is often portrayed as a consumption problem since the environmental impact of households stems mainly from what people eat, how they run their homes, and get around and holiday (Blowfield 2013). Ultimately, consumption is the reason why products are made in the first place (Crane and Matten 2016). The impact of consumption is further compounded by the inability of consumers to differentiate between wants and needs (real or artificial) and by so-called conspicuous consumption, which is led by hedonistic values and status seeking (Piasecki et al. 1999; Fogel 2016; Crane and Matten 2016). It is estimated that household consumption contributes to a significant share of greenhouse gas emissions globally (over 60%) and more than half of land, material and water use (Scales 2018), which reflects the fact that the main impact of products (80%) is during the use and disposal phase (Weybrecht 2014).

Despite increasing willingness to take into account environmental criteria when making buying decisions, consumer understanding is often lacking as a result of confusion over what green products are and of uncertainty over their benefits. In particular, consumers have in the past perceived trade-offs between product characteristics (e.g. quality and low cost) and environmental benefits (Fogel 2016). In a similar line, Wackernagel and Rees (1996: 153) emphasise that "sustainability-as-sacrifice will not sell".

In contrast, sustainable consumption is behaviour that enhances life quality whilst also minimising or eliminating environmental (and social) impacts throughout the life cycle of a product (Crane and Matten 2016), which mirrors the holistic nature of green marketing as defined previously. A more detailed definition of sustainable consumption that draws explicitly on sustainable development is used by the European Environment Agency based on the 1994 Oslo Symposium (Crane and Matten 2016: 370):

> the use of goods and services that respond to basic needs and bring a better quality of life, while minimising the use of natural resources, toxic materials and emissions of waste and pollutants over the life-cycle, so as not to jeopardise the needs of future generations.

This is similar to ethical consumption as a broad concept where consumers deliberately make decisions based not just on price or quality but also on ethical attributes. Green consumption (or green consumerism) is more focused than sustainable or ethical consumption as it aims to reduce the environmental impact of household consumption by changing how consumers shop, use energy and handle their waste (Scales 2018).

Bringing business into the pictures, the following (based on Crane and Matten 2016; Lorek 2016; Middlemiss 2018) are some steps that companies – supported by consumers, civil society and governments – can take to encourage sustainable consumption, which includes both consuming less and consuming differently.

- *Producing environmentally responsible products*: producing goods that are less harmful and use less material is an obvious approach for companies to respond to sustainable consumption. This starts with new product development where green product development under DfE (Design for Environment, see Chapter 9) can be seen in the context of sustainable marketing.
- *Product recapture*: redesigning products with the end of life in mind involves take-back systems, and as costs for inputs are lower, prices can also be brought down, encouraging consumers to purchase goods made using circular rather than linear flows of resources.
- *Service replacements for products and other systemic approaches such as industrial ecology and circular economy*: moving from offering products to offering services and functionality involves a transformation of consumer demand and a change in business-to-consumer relationships. It also changes the boundaries of production and consumption (Ryals et al. 2012a). Servicisation (or servitization) through product-service systems moves away from ownership and has been most common in industrial contexts (see also previous chapters), whereas in consumer contexts the lack of ownership and of economic incentives are challenging for companies. Similarly, industrial ecology and circular economy are strongly business oriented though there is potential for consumer involvement as co-producers and co-designers as well as in collaborative consumption.
- *Product sharing and other collective approaches*: this is similar to servicisation and reduces consumption as sharing increases the utilisation of existing products. Although not always convenient, consumers benefit from savings in money, space and maintenance. There are various approaches to the sharing economy, both for-profit and not-for-profit, which have attracted various criticisms. Alternative terms include collaborative economy and access economy.
- *Reducing demand*: ultimately, the environmental sustainability challenge can only be fully met if consumers and businesses accept that we have to buy less. This might seem counterintuitive for companies and can only be done with the support of governments and consumers, such as for example implementing the polluter pays principle to create financial incentives for lower consumption or banning the use of single-use plastics (e.g. bags and packaging). Companies can contribute to reducing demand by making products with greater

longevity, which could command a price premium and which the company could service throughout their life cycle, thus engaging in a different (more sustainable) business model.

- *Encouraging sustainable consumption through standards and transparency, certification and labelling*: so-called green consumerism, which, to some extent, allows business as usual with some attention to environmental concerns, can be seen as a form of greenwash as well as a risk for rebound effects. In order to avoid greenwashing and to enable more informed purchasing decisions, companies are increasingly turning to certification and labelling, which we will cover in more detail later in the chapter. Through increased transparency, trust can be built across the supply chain. Incorporating the whole of the value chain is crucial as inattention to supply chain practices carries a reputation risk for companies with the potential to affect their ability to market goods and services.

These steps are reflected in the key assumptions of green consumption that Scales (2018) summarises as firstly, clear information to consumers about the environmental impacts of the production and the consumer use of purchased goods, which then secondly, enables consumers to make informed (and possibly changed) choices of consumption and behaviour, which in turn, thirdly, prompts manufacturers to react and change how they provide goods and services. Pearce and colleagues (1999) add to the need for information the role of government in making green consumerism effective through improved information (in terms of amount and quality).

A big challenge for sustainable consumption lies in resolving the tension between consumers' willingness to go green and their willingness to pay, which is also known as the attitude-behaviour gap or value-action gap between what buyers say and their actual purchasing decision (Blowfield 2013; Middlemiss 2018). Although consumers might hold green values, this might not necessarily show in their behaviour for a variety of reasons (Ryals et al. 2012a). Nevertheless, in a survey conducted in 2019, over two-thirds of consumers said they would feel more positive about companies that can demonstrate their efforts to reduce their products' carbon footprint (Carbon Trust 2019).

Greener products are often perceived as more expensive than conventional consumer choices. This is often as a result of companies internalising negative externalities in production, thereby incorporating the full life cycle costs of products that are not reflected in the prices of other goods (Weybrecht 2014). The resulting price premium is therefore arguably justified alongside the fact that increasingly greener products often work equally or better. This is a change from the past when such products were not necessarily of good quality or high value (Ottman 2011; Esty and Winston 2009). With the exploitation of economies of scale and unsustainable products becoming more expensive, we can expect a levelling of prices in the future (Weybrecht 2014). Still, whilst some consumers are willing to pay more for green options, companies should not expect a price premium. As Paul Hawken (1993: xv) puts it:

> We can't just ask people to pay more to save the planet. They won't do it in some cases – and can't in most.

When talking about sustainable consumption we need to keep in mind that we cannot "shop our way out of our troubles", though empowered and informed consumers can pressure manufacturers and retailers and thus send market signals (Elkington 1997: 57). We could also argue that sustainable products can only be beneficial for the environment if consumers are actually willing to buy them (Fogel 2016). This speaks to the other two tensions inherent in sustainable marketing – the reduction of consumption needed for sustainability versus the encouragement of consumption through marketing, and the challenge for consumers to know whether the green products they are buying are what companies say they are.

In this, consumers face a similar question that companies face in their procurement – do we know what we buy and what its environmental impacts are? In the following, we look at the question of what companies can therefore do to inform consumers and how they can market their green products.

Marketing green products – green marketing

The two objectives green marketing has to meet is improved environmental quality and customer satisfaction. Green or environmental products in turn are used to describe goods that aim to protect or enhance the environment through for example energy and resource conservation, and reduction or elimination of toxic ingredients, pollution and waste (Ottman et al. 2008). They are thus less damaging to the environment than traditional equivalents (Martin and Schouten 2014). For Stead and colleagues (2004), the objectives of environmental marketing go further – the development of products that effectively balance price, performance, convenience and environmental compatibility whilst also projecting an image to customers reflecting these criteria.

Companies wanting to use environmental marketing face a number of challenges (Stead et al. 2004; Ottman et al. 2008; Gershoff and Irwin 2012):

- Defining the term "green" in the context of methodological issues and implied complexities.
- Finding ways to convince consumers to change their lifestyle towards greener behaviours and products, particularly where consumers are sceptical about green outcomes of green purchasing actions.
- Educating consumers about the true nature of environmental issues and how they are connected to the goods they buy and consume.
- Finding legitimate ways that are not misleading in order to communicate environmental benefits and features of products.
- Gaining credibility for the idea that corporate interests are not necessarily in conflict with environmental responsibility.
- Making green products appeal to consumers outside traditional green niches.
- Meeting customers' needs rather than managing products.

Environmental claims do not necessarily narrow the market appeal of a company, but marketers need to be careful to realise the broad appeal of their brand in order to attract mainstream buyers (Sharp 2013). This is particularly important as in the past consumers have perceived sustainable brands and green products to be of lower quality or functionality (Esty and Winston 2009) – something referred to as 'green marketing myopia' where companies either misjudge one or both of the green marketing objectives or overemphasise environmental quality improvements at the expense of customer satisfaction (Ottman et al. 2008). In line with past expectations, companies cannot necessarily expect a price premium for green products (Esty and Winston 2009), though Ottman (2011) highlights that some green products work better and longer, and thus justify a premium price.

In order to meet the challenges identified earlier, Stead et al. (2004) suggest seven components for sustainable marketing strategies:

1. A foundation of clear understanding of how environmental issues affect the organisation.
2. Strong organisational commitment to sustainability at all levels.

3 Nature is used as a framework for product and packaging design – biomimicry and source reduction – to cultivate innovative ways for meeting consumer needs in a sustainable way.
4 Honest and nondeceptive promotional strategies where companies under-promise and over-deliver to avoid being accused of greenwashing.
5 Close consultation with external stakeholders.
6 Pricing should reflect and include true ecological costs in the prices of goods and services.
7 Marketing channels used should minimise environmental impacts.

These components suggest a comprehensive approach that goes beyond marketing, encompassing other functions and parts of the organisation as well. We will cover promotion and pricing as well as marketing channels later in the chapter and return now to other ways companies can address some of the green marketing challenges listed previously.

In order to appeal to customers who are outside traditional green niches, companies need to highlight the desirable benefits of green products for desirable benefits for non-green consumers (Ottman et al. 2008). Marketing a product on environmental credentials alone is not sufficient, as other concerns such as quality and price remain key for customers (Esty and Winston 2009). There is naturally a greater acceptance of eco-marketing when the consumption of environmentally beneficial or preferred products are combined with some personal benefit (Schaltegger et al. 2003). This can be done through building necessary "fit-for-purpose" associations around functional and other attributes (Ryals et al. 2012a).

Ottman and colleagues (2008) suggest five desirable benefits that companies can use to position their green products for non-green consumer value:

- *Efficiency and cost effectiveness*: fuel-efficient or energy-saving appliances and laundry detergents are more resource efficient and thus provide consumers with cost savings.
- *Health and safety*: as most green products aim to eliminate or at least minimise toxic ingredients, they appeal to health-conscious consumers who are prepared to pay a price premium for example for organic foods, household cleaning products that are safe to use for families or green buildings that reduce indoor air pollution.
- *Performance*: the negative perceptions of green products in the past have changed and many such products now perform better than conventional ones, thus justifying a higher price. Examples include low emissivity windows that reduce a building's energy use, composite building materials without toxic chemicals, and preservatives or products with life-time guarantees.
- *Symbolism and status*: green products may be chosen and used by consumer to signal their status, particularly if others can observe their behaviour (Gershoff and Irwin 2012). For example, the first petrol-electric hybrid cars had a green chic appeal that attracted the so-called LOHAS market (Lifestyles of Health and Sustainability) – an example of conspicuous green consumption (Middlemiss 2018; Weybrecht 2014).
- *Convenience*: many green products are more convenient as a result of needing fewer replacements or fewer refuelling stops because of their energy-efficiency. Other benefits include for example free parking for electric or hybrid vehicles.

Implied in these propositions is the need to align the consumer value of environmental products to the relevant segments of the consumer market – for example saving money for cost-conscious buyers. Underpinning this is the importance of consumer knowledge and credible product claims. We will look at these concepts now in turn.

Marketing strategy builds on segmentation, targeting and positioning (STP) (Kotler and Keller 2016). Market segmentation helps a company divide a market into separate groups of consumers with different characteristics, needs and behaviours that might necessitate offering distinct products or marketing programmes (Kotler et al. 2013). Companies can then target selected segments of the market by developing a marketing mix that places the organisation, brand or product in a particular position in the mind of the consumer (Goworek 2017).

Often a cost-leadership strategy is associated with low market segmentation, whereas a differentiation strategy is linked with high market segmentation (Stead and Stead 2014). The increase in the number of consumers demanding environmentally friendly products provides green companies with segmentation opportunities (Laszlo and Zhexembayeva 2011). In the UK, a market segmentation exercise for the government based on pro-environmental behaviours resulted in the following categories (see Figure 10.1), which all differ according to their ability (or potential) and willingness to act (Goworek 2017, based on Defra 2008):

1. Positive greens – prepared to actively limit their environmental impact, knowledgeable, early adopters of green products.
2. Waste watchers – more concerned with avoiding waste, summarised by "waste not, want not", committed recyclers, concerned with local issues.
3. Concerned consumers – broadly pro-environmental beliefs, less committed and knowledgeable than 1 and 2.
4. Sideline supporters – acknowledge need to act, but barriers to action result in gap between attitudes and behaviours (attitude-action-gap).
5. Cautious participants – willing to act, but perhaps pessimistic about impact of own actions.
6. Stalled starters – limited knowledge on environmental issues, other life priorities acting as barriers to ability to act.
7. Honestly disengaged – opposite to positive greens, negative or indifferent attitudes to environmental issues.

Dahlstrom (2011) highlights the following market segments based on psychographic segmentation: true blues, greenback greens, sprouts, grousers, and apathetics ranging from consumers with strong environmental values to consumers with little or no such values. The Natural Marketing Institute differentiates four "shades of green consumers" as demonstrated in values, activities and purchases: LOHAS consumers (also called deep greens), naturalites, drifters and conventionals, with the non-green segment labelled unconcerned (Ottman 2011). Within the green segment, consumers can be further divided according to their priorities: resource conservers, health fanatics, outdoor enthusiasts and animal lovers (Ottman 2011).

Devinney (2012) stresses that in reality there are few day-in-day-out green consumers and that like the ethical consumer, the ubiquity of the LOHAS consumer is a myth – whilst many consumers profess to care about the natural environment, this does often not equate to everyday consumer behaviour (see also Devinney et al. 2010). Green consumption therefore requires a solid understanding of non-green aspects of consumption and how consumers trade off functional and social aspects of their consumption (Devinney 2012) – as illustrated earlier in how products are positioned.[1]

Whichever categories companies use (and there are probably more than the ones highlighted previously), market segmentation can then be used to tailor communication accordingly, whether

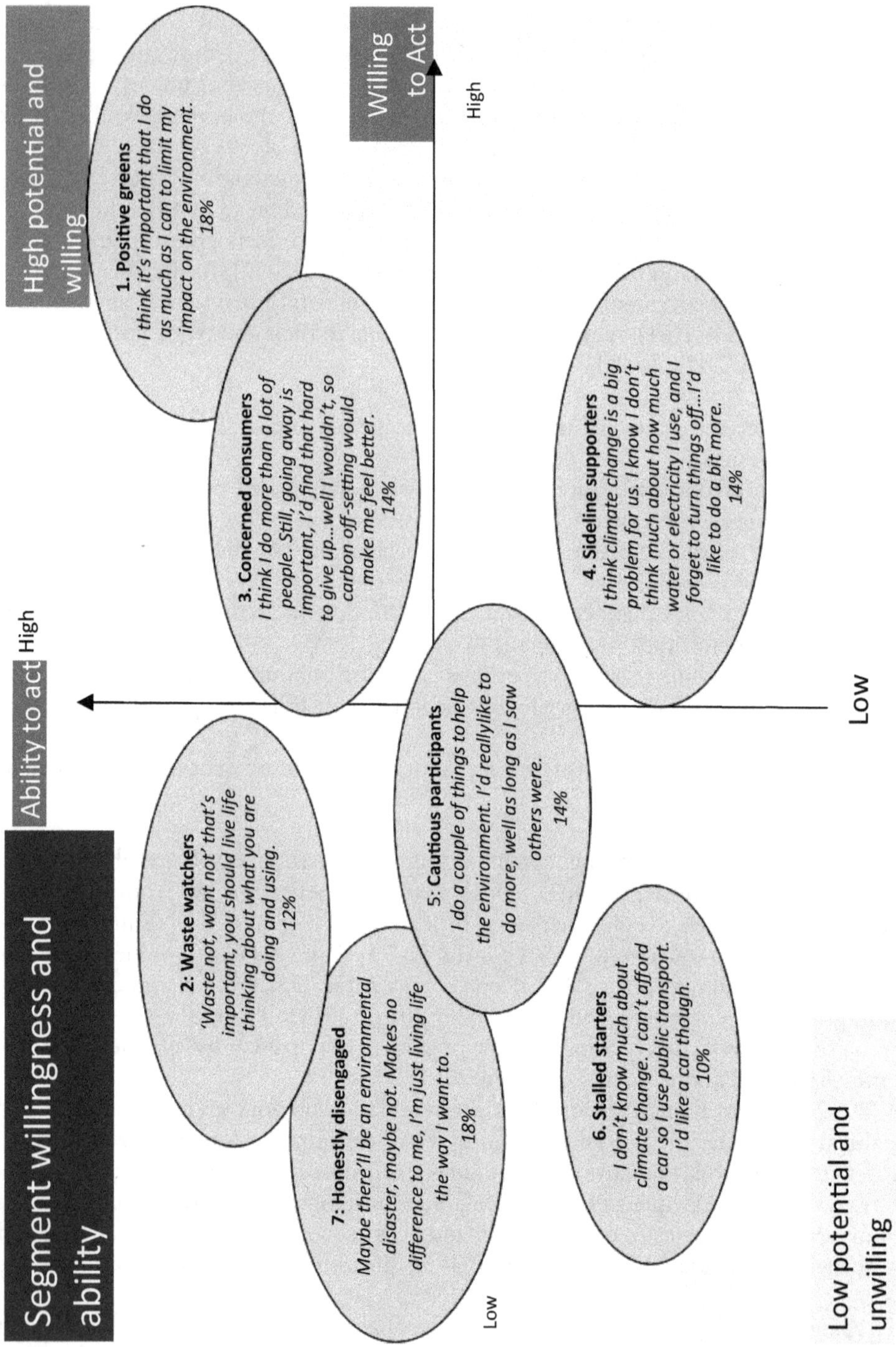

FIGURE 10.1
Market segmentation according to pro-environmental behaviour

Source: Defra, 2008; nationalarchives.gov.uk, used with permission.

to exemplify, encourage, enable or engage consumers[2] with regard to sustainable consumption and behaviour change more broadly, or concerning green products specifically.

In the context of green products, it is useful to include branding as well. As a reminder, a brand is

> a name, term, sign, symbol, or design, or a combination of them, intended to identify the goods or services of one seller or group of sellers and to differentiate them from those of competitors.
>
> (Kotler and Keller 2016: G1)

In turn, a sustainable brand consists of

> economically enduring trademarks and meanings of a product or company; its associations in the minds of customers and other stakeholders rightfully include social justice and ecological sustainability.
>
> (Martin and Schouten 2014: 6)

Some brands are inherently green and sustainable as they are built on environmental or sustainable values, whereas other brands are more "selectively" sustainable. Examples of sustainable brands include Body Shop, Ben & Jerry's, and Howies, whereas companies such as Toyota or Hyundai offer a limited range of e.g. hybrid cars, making them more selectively sustainable[3] (Goworek 2017).

Green products come in many different forms, whether they are made from recycled materials, recyclable, repurposed, or reused (or a combination thereof, Goworek 2017). The emphasis is often on durability and longevity, a marked contrast to conventional products manufactured for disposability and planned (or built-in) obsolescence (see also Chapter 7 on waste).

Marketing professionals tasked with advertising such eco-friendly products often face a dilemma – if they promote the environmental benefits of their products too loudly, they risk being accused of greenwashing. However, if they fail to highlight proven benefits, the opportunity of attracting customers who care and are willing to buy (possibly even at a higher price) is missed (Whitfield and McNett 2014).

This highlights the importance of credible product claims. Providing truthful and reliable information is important not only in order to avoid greenwashing, but also for companies to operate within the law as several countries have legislation in place that punishes companies for providing misleading environmental claims (Visser 2012). One tool to help companies provide such credible product claims is eco-labelling, which we cover separately in the section on tools. As greenwashing has been mentioned a number of times already, we provide some more details in the following box and move on from marketing green products (or green marketing) to making marketing itself more sustainable.

BOX 10.1 GREENWASHING

Greenwashing is something companies need to be aware of regardless of industry or corporate function as it contains a risk to their reputation if they are found guilty, but it is particularly relevant for marketing. It broadly occurs when stakeholders perceive an organisation creating a misleading impression of its environmental, social or ethical performance – which can even happen to

BOX 10.1 (CONTINUED)

well-intentioned businesses (Laasch and Conaway 2015). In the context of marketing specifically, greenwashing refers to using marketing and communication tools to create such a misleading impression of creating stakeholder value (ibid.). Greenwashing concerns the difference between real and ceremonial corporate environmental practices (Lounsbury et al. 2012) and answers the question of whether companies "walk the talk" of corporate environmental practice. Similarly, Esty and Winston (2009) highlight that greenwashing is more a failure of action than a failure of intent.

The term was originally coined by Jay Westerveld in 1986 to highlight that hotels were asking customers to reuse towels to protect the environment whilst at the same time performing poorly in other environmental aspects (Seele and Gatti 2017). Since then, various definitions of greenwashing have been provided, often emphasising strategic intention to mislead and not taking into account unjustified allegations. Whilst the Oxford English Dictionary defines greenwashing as "disinformation disseminated by an organisation so as to present an environmentally responsible public image", Greenpeace explains it as "the act of misleading consumers regarding the environmental practices of a company or the environmental benefits of a product or service" (Seele and Gatti 2017: 241). In contrast, Seele and Gatti (2017) stress that the element of accusation is a crucial element of greenwashing which therefore encompasses falsity and an accusation of misleading. It contains an internal information-related element (communication) and an external distortion-related element (accusation) – both of which need to be present to constitute greenwashing. Bowen (2018) also underlines that it is a challenge to identify when greenwashing is intentional and when it is an unintended consequence of companies managing a range of environmental issues. Furthermore, when a company performs well but does not communicate this, there is misperceived greenwash (Laasch and Conaway 2015).

Companies engage in green marketing and communication in order to obtain legitimacy but can be rightly (or wrongly) accused of greenwashing when there is inconsistency between symbolic communication and substantive action (Seele and Gatti 2017).

Examples of greenwashing include the use of green and social imagery in packaging, general statements ("natural", "environmentally friendly"), making one particular claim to distract from key environmental issues (i.e. missing the point), unsupported labels, claims that are not actionable (e.g. recyclable products but no facilities for recycling), claims that sound good but are not lived up to (e.g. carbon neutral status that is only achieved by offsets rather than through genuine reductions), environmentally friendly products made by companies that are anything but (e.g. lobby against environmental regulations) and overemphasise small improvements (Weybrecht 2014).

Similarly, in a Terrachoice report on greenwashing, seven so-called "sins of greenwashing" were identified: hidden trade-off, no proof, vagueness, irrelevance, lesser of two evils, fibbing and worshipping false labels (Nemetz 2013).

The risk of greenwashing is that well-intentioned consumers are misled into making purchases that do not live up to their environmental claims. This in turn can lead to cynicism and doubt regarding any environmental product claims (Nemetz 2013). In the context of sustainable consumption, green consumerism can be seen as a form of greenwash that might run the risk of a rebound effect (Middlemiss 2018). For individual consumers it is difficult to know the difference between misleading corporate spin and authentic environmental improvements (Bowen 2018).

In response, eco-labelling is used by consumers to make decisions that are more informed and by companies to emphasise their adherence to standards. Other tools to counter greenwashing include honest, creative and consistent communication, transparency and focus based on what is important for consumers (Weybrecht 2014).

Marketing sustainably – sustainable marketing techniques

From promoting and marketing sustainable or green products, we now move to the second aspect of sustainable marketing – promoting goods with the help of sustainable methods.

In the preceding section, minimisation of ecological and social impacts of the marketing channels the firm uses was mentioned as one of the components of sustainable marketing strategies (Stead and Stead 2004). Here, we briefly highlight various aspects such as the role of retailing, channel partner selection, and implementation issues more broadly before covering tools in the following section, which also includes an extended marketing mix, such as for example place and lean channels.

In direct marketing, traditional mailshots are increasingly replaced by more sustainable and economical online communications such as e-mail that save on paper, printing, postage and distribution (Goworek 2017). Whilst online and digital advertising is arguably more sustainable than the use of printing promotional materials, the energy requirements for computers and servers often entail the use of fossil fuel-based energy in contrast to print materials that may be recyclable. Online communications and direct marketing can also support relationship marketing where the focus shifts to the lifetime value of customers rather than single purchasing transactions.

In demanding suppliers to remove water from detergent formulae and only offer concentrated versions, Walmart reduced its environmental impact on transportation and distribution channels, whilst at the same time removing more impactful products from the market and pre-selecting its product range for customers, thus choice editing (Robertson 2017). Other options for retailers to make their marketing practices more sustainable in addition to choice editing and selecting partners according to environmental criteria (see also Chapter 8) include making store design more sustainable, minimising energy use in transport, stores and offices, offering less wasteful and more recyclable packaging, and, above all, educating consumers and enabling sustainable consumption (Goworek 2017).

We return to some of these practices in the following section where we cover further tools to make marketing more sustainable – both in terms of green products and in terms of marketing tools and techniques.

Tools for sustainable marketing and marketing green products

In this section we cover tools and concepts that have been touched upon in the chapter up to here and provide further details and signposting how companies can engage in sustainable marketing, ranging from the marketing mix to eco-labelling, choice editing and communication more generally.

Extending the marketing mix: from 4Ps to 7Ps to 4Cs

In the following, we briefly review how each P in the marketing mix of product, price, place and promotion can address environmental impact before we extend it to 7Ps and present an alternative in the 4Cs. The marketing mix is the set of tools firms can combine to influence demand for its products and services in the target market (Kotler et al. 2013).

Product

Whilst production and supply chain management can address the environmental impact of products in the design stage taking into account the whole life cycle, marketing can support product development by questioning the underlying assumptions, for example around product experience

(Weybrecht 2014). Do consumers need to purchase the physical product, or can it be borrowed, rented, reused or repaired? Environmentally improved product alternatives for consumers can be developed on the basis of principles such as extended producer responsibility, life cycle orientation, regionality and seasonality, durability, consistency, dematerialisation, intensification of use as well as sufficiency (Schaltegger et al. 2003). Sustainability can be built into products through various choices for materials and processes, where marketers (and designers) are key decision-makers (Goworek 2017). Depending on what kind of sustainable product options consumers are looking for, companies can offer them products that present a solution, that are safe as well as money-saving, high quality, not harmful, make them feel and look good, and that make it easy to switch from more harmful products (Weybrecht 2014).

Marketing can also address the branding and packaging of a product. Reduced packaging lowers costs and can position a product in the mind of the consumer as eco-friendly (Laasch and Conaway 2015). Paying attention to packaging is particularly important as it may be the first part of a consumer's experience with a product – this concerns both the physical packaging and the messaging on the package (Weybrecht 2014). Packaging and branding are closely related as names, symbols, labels and layout are used by companies in their branding to differentiate their products from competitors' which may reduce the information needed by customers in making buying decisions (Schaltegger et al. 2003). Trust and credibility are key for eco-branding for which companies can use eco-labels, which we cover in the next section. Having an established brand and brand recognition can allow companies to use their brand equity not only to add value to their product but also to charge a price premium (Laasch and Conaway 2015). As mentioned previously we can differentiate between inherently sustainable brands and selectively sustainable brands (Goworek 2017) – or corporate branding versus product branding for specific products as Schaltegger et al. (2003) call it. Similarly, Fogel (2016) distinguishes between product or service claims and other claims such as process, image or social ones.

Price

Sustainable products are often perceived to be more expensive and, for example, organic products are arguably more expensive as the yield per hectare of land is lower than for conventional products due to avoidance of synthetic pesticides and fertilisers, and thus justify a higher price. There is also the price producers have to pay for certification and the premiums paid to farmers for e.g. fair-trade goods. As a more holistic approach to pricing is taken, which includes social and environmental costs, premium prices could thus be explained and justified (Goworek 2017). Since the environmental impacts of production in the form of negative externalities are not reflected in the price of goods (Nemetz 2013), more information and communication is needed so individual consumers can make informed purchasing decisions. For some products and services, consumers have the opportunity to factor in those externalities by paying an additional price for carbon offsetting for example when booking flights or other modes of transport (Laasch and Conaway 2015).

However, if companies want their sustainable products to appeal to consumers more broadly, beyond dedicated green and ethical consumers, price needs to be competitive as consumers might want to buy greener products but are not necessarily willing (or able) to pay more (Weybrecht 2014). With companies reducing costs through sustainability efforts such as for example waste elimination, recycled content and fuel-saving transport, this is a feasible prospect (Goworek 2017). Schaltegger and colleagues (2003) furthermore propose initial cross-subsidisation of conventional products in favour of environmental products in order to establish these on the market.

Place

Place is where products are made, where and when consumers purchase products and services and also includes from/to where and how products are distributed (Weybrecht 2014; Laasch and Conaway 2015). We covered transport, logistics and distribution in Chapter 8 on supply chain management and highlight here the importance of one particular distribution channel, that of retail, which is in a unique position of connecting suppliers and customers and thus influencing sustainability efforts (Goworek 2017). Retailers can address sustainability through place in a number of ways (Weybrecht 2014), such as site selection and store management, where products are sourced from, elimination of unsustainable products, where products are placed in store (easy to find, easy to buy?) and how they are sold, presentation in-store, retailing differently and consumer education, which has been highlighted before and to which we return again.

Ryals and colleagues (2012b) also emphasise that channel design offers opportunities to cut not only companies' carbon footprint but also their costs by moving e.g. simpler transactions, services or queries online or to call centres, and cutting down on carbon-intensive field salespeople. Laasch and Conaway (2015) include the move from product distribution to servicisation under place.

One specific example where place plays a significant role is food consumption, where locally produced food requires less transport and thus cuts the environmental impact whilst eating seasonally requires less energy consumption compared with food produced out of season and far away (Robertson 2017). Whilst seasonality is particularly relevant to food, regionality and short distances from production to consumption applies to other products as well (see e.g. Schaltegger et al. 2003).

Promotion

Promotion is the final P of the traditional marketing mix and is perhaps what is perhaps commonly perceived as marketing as it involves advertising and communication, which is part of all marketing (see e.g. Schaltegger et al. 2003). This can be either one-way (monologue) or two-way (dialogue) with implications for credibility and legitimacy where one-sided communication can be problematic whilst relational marketing is more customer-oriented and more suitable for customer-driven eco-marketing that places relationship over transactions (ibid.).

In addition to lean transport channels under *place*, lean communication channels can help companies reduce their environmental impact by eliminating printed promotional materials and introducing e-mail and web-based channels (Laasch and Conaway 2015). These can include blogs, tweets, Facebook, Pinterest or Instagram posts, which often contain direct purchasing links for consumers. New social media are of course constantly being developed. Social media provide "word-of-mouth" opportunities, which in an old adage is considered "the best advertising" (Weybrecht 2014: 248). This, however, comes with risks attached as bad word-of-mouth can be damaging to reputation, and consumer attitudes and are often beyond the control of firms (Kotler et al. 2013; Goworek 2017). Nevertheless, such social networks and media provide companies with platforms for relational (or relationship) marketing including co-creation and crowdfunding.

Integrated marketing communications concerns both the promotion of sustainable goods as well as the use of sustainable methods in promoting such products and thus encompasses both corporate communications as well as brand or product communication (Goworek 2017). Integrated marketing communication involves and connects all explicit and implicit activities of communication by an organisation. This includes a range of activities, tools and channels such as media and stakeholder relations, sponsoring and events, co-operation, internal communication, websites, reporting and marketing more generally (Laasch and Conaway 2015). Reporting as a form of communication will be covered in Chapter 12 on accounting and finance.

One particular tool companies can use in their promotions of green products which addresses credibility problems in communication is eco-labelling, to which we turn in the next section.

The marketing mix has been updated and complemented in various ways[4] (Kotler and Keller 2016). In the context of sustainability, the 4Ps have been extended to 7Ps. This extended marketing mix of the 7Ps includes the 4Ps plus people, physical environment and process (Goworek 2017). Alternatively, the additional three Ps are arguably the people, planet, and profit from the triple bottom line, thus firmly embedding the whole concept of sustainability in the marketing mix.

The 4 Cs of sustainability marketing

In addition to the 7Ps, another alternative to the 4Ps especially proposed by Belz and Peattie (2009) for marketing for sustainability is found in the 4Cs of customer solution, customer cost, convenience, communication. The focus here moves from the narrow product-based language of the 4Ps' marketing mix to the solution of customer problems in a relationship of co-creation (Scammon and Mish 2012). As the 4Cs encompass a broader and longer-term perspective on customers and stakeholders, multiple needs, relationships and a full(er) life cycle, the approach is more appropriate for sustainability marketing. In reconfiguring the marketing mix from the 4Ps to the 4Cs, marketing moves from producer orientation to customer orientation, integrates socio-ecological issues from the start and treats consumer behaviour more holistically, which together can better harness the transformational potential of marketing (Peattie and Belz 2010). Briefly, the 4Cs include the following (based on Peattie and Belz 2010; Belz and Peattie 2009, 2012):

- *Customer solutions*: rather than focusing on the product, the emphasis here shifts on what solutions to customer needs and problems can be offered, which can for example be in the form of a service rather than a product (servicisation). A life cycle orientation is taken to include pre-purchase, use, and post-use solutions with an emphasis on significant and continuous improvements that lead to customer satisfaction.
- *Consumer cost*: as price is only one of many elements of costs for consumers at the point of exchange, companies need to consider the total customer cost across a product's lifetime, which also includes use costs and post-use costs.
- *Convenience*: this goes beyond the conventional place and physical distribution and addresses convenience in purchase, use and post-use, and includes online convenience, though the challenge for companies here is a potential trade-off between convenience and environmental aspects.
- *Communication*: communication shifts from the one-way messaging of promotion to two-way communication with customers in order to make consumers aware of sustainability solutions that companies have developed, and to establish relationships. It is part of making marketing communications more integrated, encompassing product as well as corporate communications. The challenge here is to avoid greenwashing by emphasising openness and dialogue with stakeholders and to appeal not just to green niche consumers, thus avoiding green marketing myopia.[5]

One particular tool that can help companies market sustainable products and services, and avoid greenwashing is the use of eco-labels, which provide credibility to product claims.

Eco-labelling: using labels and certification to provide credible marketing claims

We covered labelling in Chapter 8 on supply chain management, where we highlighted the role of standards and accreditation in improving transparency and sustainability throughout the whole value chain with an emphasis on upstream operations. Here we focus on labels in consumer-facing (downstream) environments as these can help make consumption more sustainable, as pointed out in the relevant section earlier. Whilst the terms labels, standards and certifications are often used loosely and in the same context, it is worth highlighting that companies commit to standards which are controlled through certification schemes which then allow companies through a licence to display the relevant label on a product (Middlemiss 2018).

Claims in advertising environmental characteristics of goods or services cover several categories: product or service claims, process claims, image and social claims (Fogel 2016). The use of labels and certification can particularly underpin product and process claims and can provide a means of verifying these. Consumers are faced with confusing labels and complexity when it comes to green product claims which, when simplified, can lead to trade-offs between different environmental benefits (Scales 2018). This provides a challenge for companies when consumers feel they cannot trust or differentiate different claims that are possibly "misleading corporate spin" or authentic environmental improvements (Bowen 2018: 689). Trust is therefore key when establishing markets for green products, which can be supported by certification and control that is trustworthy and effectively communicated (Nuttavuthisit and Thøgersen 2017). While trust can for example be established through branding and name recognition, further credibility can be ascertained through verification of compliance with eco-label criteria (Schaltegger et al. 2003). Such labels in effect relieve consumers of the challenge of verifying green product claims whilst providing a means of product differentiation for companies (Gershoff and Irwin 2012; Weybrecht 2014).

Based on ISO standards for environmental labelling, we can differentiate between three different types of eco-labels (Brady 2011; Weybrecht 2014; Daddi et al. 2015; Fogel 2016):

- *Type I eco-labels* (ISO 14024): voluntary eco-labels based on multiple criteria across the product life cycle that are certified by an independent third party which checks and verifies compliance and then awards a licence, authorising companies the use of its label or logo. Such a seal of approval provides assurance to consumers. Examples include the US Green Seal or the Nordic Swan amongst national labels, whilst sector-or issue-specific programmes include the Rainforest Alliance and the Fair Trade label.
- *Type II self-declared environmental claims* (ISO 14021): green claims made by retailers or manufacturers without requiring independent third-party certification. They are possibly the most widespread form of environmental communication to consumers and include self-declarations such as "100% recycled", "made from 50% recycled material" or "compostable". Company private labels come under this category, such as for example Philips' Green Logo on electronic products meeting certain criteria. Although most statements in this type are accurate reflections of environmental performance, this type is most prone to accusation of greenwash and being perceived as misleading or vague (e.g. "environmentally safe" or "non-polluting") as claims are difficult to substantiate or verify.
- *Type III environmental declarations* (ISO 14025): extensive environmental product information of a quantitative nature based on life cycle analysis or environmental management systems and pre-established parameters. These independently verified labels

allow comparison but are mainly used in a business-to-business context as they are relatively complex and, unlike other labels, leave the task of assessing products to buyers. Examples include Environmental Product Declarations and carbon footprints, amongst others.

Other categories of eco-labels include mandatory versus voluntary labels, single-issue labels versus multiple-criteria labels, and we can also differentiate according to who provides the labels and accompanying standards – a neutral institution (e.g. Marine Stewardship Council), a state or regional institution (Germany's Blue Angel or the EU eco-label), a non-governmental organisation (e.g. WWF or Greenpeace), a private sponsor (such as the International Standards Organisation) or a manufacturing association (Fogel 2016; Brady 2011; Schaltegger et al. 2003). Companies can choose multiple certification – for example tea can be certified Fair Trade as well as organic which allows a company to signal its wider efforts to achieve sustainability according to triple bottom line considerations. An example of the author's campus catering provision of coffee being certified organic, Fair Trade as well as Rainforest Alliance is shown in Figure 10.2.

Whilst life cycle assessment is particularly used for Type III labels, all eco-labels can be supported by LCA, and the relationship between the two is for example illustrated by the American label Safer Choice that meets the US Environmental Protection Agency's safer product standards – it was previously known as the Design for the Environment, or the "DfE", label. As highlighted in the previous chapter on production and operations, LCA is a fundamental tool for DfE, and embedding DfE within a wider cross-functional process including marketing enables product life cycle management that also involves consumers through co-creation and integrated product development (Fiksel 2012). Eco-labelling is a further manifestation of integration in sustainability with regard to the consumer and the use phase as part of the overall product life cycle.

Although eco-labels can help to avoid accusations of greenwashing, companies as well as organisations administering the labels and certifications need to be wary of widening their appeal too broadly to make them attractive as this can lead to a counter-productive watering down of standards, in turn leading to further confusion among consumers (Middlemiss 2018; Fogel 2016). Examples of standards where this risk has been highlighted include the Marine Stewardship Council (MSC), which was criticised for allowing certified products to be caught by the same vessel alongside uncertified ones during a single trip, making the distinction between sustainable and unsustainable fish more difficult to ascertain. The council has since taken the decision to eliminate this practice.

Labels are used by consumers to make informed buying decisions. Similarly, retailers can use labels to inform and underpin their product selection, which can be edited to such an extent that consumers can only buy sustainable products in certain categories. For example, the range of MSC certified fish available in UK supermarkets varies widely, and several UK supermarkets have made the choice of only offering Fair Trade certified bananas or tea, often starting with their own brand ranges. In 2007, UK supermarket Sainsbury's was one of the first retailers to sell only Fair Trade certified bananas. Similarly, in the 1990s, B&Q, the DIY retailer, made a decision to sell only certified timber. Such a practice is also known as choice editing.

Choice editing for sustainable consumption

Choice editing addresses one of the challenges of green marketing: finding ways to convince consumers to change their lifestyle towards greener behaviours and products. The approach can be taken by retailers as well as by governments when setting standards, which may result in the

FIGURE 10.2
Multiple certification

removal of the most environmentally damaging products from the market which consumers are thus prevented from buying (Robertson 2017). It is a form of pre-selection and has proved to effect significant changes in consumer behaviour. Choice editing can be supported by certification – for example when energy efficiency ratings for domestic appliances were mandated by law, retailers increasingly offered only the most efficient models whilst manufacturers developed and innovated further, introducing appliances going beyond regulated standards and phasing out inefficient models (Blowfield 2013). Choice editing can thus be practiced by retailers as well as manufacturers in pre-screening products for buyers (Gershoff and Irwin 2012).

Choice editing can also be practiced by governments and can thus be seen in the wider context of behavioural economics where nudge strategies change the environment in which consumers make choices without necessarily limiting those choices (Middlemiss 2018). For Thaler and Sunstein (2008), who were instrumental in making this approach more widely known, a nudge is part of a wider choice architecture that alters how people behave, not ruling out any options. Nudges are thus not mandates but provide relatively easy and cheap options to make more beneficial choices through tools such as social comparison, defaults, anchoring, prompts and changing the physical environment (Middlemiss 2018). In using choice editing, companies – specifically retailers – move from choice providers to choice architects in offering more sustainable products, but they can also choice edit unsustainable goods, thus making sustainable ones the mainstream option and making every consumer a green consumer for the particular item or category in question (Gunn and Mont 2014).

Nudging and choice editing can thus encourage more sustainable consumption and can be considered as sustainable marketing tools for companies which have used the marketing mix as a form of choice architecture already without necessarily using this term (Ryals et al. 2012a). Social marketing is a particular form of marketing directed at changing behaviour, which companies can use to change behaviour and improve their reputation by communicating commitment to environmental causes for example (Laasch and Conaway 2015).

Communication: advertising, education and beyond

Although communication has been mentioned in various places, such as under promotion in the 4P marketing mix and in the 4Cs, it is worth emphasising briefly the role of communication beyond advertising, for example in educating consumers as an integral part of sustainability marketing. Relationship or relational marketing forms part of sustainable integrated marketing communications and allows companies to build up longer-term relationships with consumers (Goworek 2017), enabling dialogue rather than monologue (Schaltegger et al. 2003). Other external and internal sustainable communication tools include cause-related marketing and social marketing, as well as issues and crisis communication, codes of conduct, and vision and mission statements (Laasch and Conaway 2015). Beyond focusing on advertising particular green products, communication of an organisation's environmental performance can also take the form of annual environmental, CSR or sustainability reports. We will cover this aspect separately in Chapter 12 on accounting and reporting.

By educating consumers on sustainability issues, companies – retailers in particular – can enable consumers to behave more sustainably (Goworek 2017). In many European countries, each local authority may have a different recycling regime, making recycling from a consumer point of view challenging. This has implications for how companies can inform and educate consumers in addition to providing (edited) choices on offer when it comes to the disposal or end-of-life options for products, whether they end up as waste, are recycled or otherwise reused.

Spotlight on . . . examples and practices of sustainable marketing

Some examples have already been provided throughout the chapter; here we briefly highlight further practices of sustainability marketing that use the various tools and elements covered in the preceding sections.

Outdoor clothing and equipment manufacturer Patagonia uses cause-related marketing to demonstrate its commitment to sustainability – 10% of its profits (or 1% of sales, whichever is higher) are donated to environmental charities with the help of the aforementioned 1% for the planet and a grant system. In its garments, the company uses 100% organic cotton that is licenced through accredited third-party certification, and other materials are following similar standards and labels. It has also used promotion to encourage sustainable consumption by running an advertising campaign on the busiest day in US retail, Black Friday, with the tagline "Don't buy this jacket". It further promotes the repair, reuse and resale of used products through its Worn Wear tools underpinned by extensive care advice for each material, thus informing and educating consumers. Patagonia's approach can also be seen as an example of the 4Cs of customer solutions, cost, convenience and communication in its customer orientation across the life cycle.

Italian company WeWOOD, which produces wooden watches, plants a tree for every watch purchased with the goal of planting 1 million by 2020. When household consumer products provider Clorox launched a range of naturally sourced cleaning products, which unlike previous ones did not trade-off performance and environmental impact, it formed a partnership with an environmental organisation, the Sierra Club, which endorsed the products and received a share of the profits in turn (Whitfield and McNett 2014).

As behaviour change is a cornerstone of sustainability, social marketing is an important complement to education and advertising (McKenzie-Mohr et al. 2012). Traditionally used by public bodies, for example in the area of energy efficiency, it can also be a tool for companies, though care needs to be taken if responsibility for behaviour change and sustainability is seen as being shifted away from the manufacturer towards the consumer. As a result of this possible criticism, there are limited examples of companies using social marketing to effect environmental change. The now defunct website thefuntheory.com was a social marketing campaign by Volkswagen to nudge people towards less wasteful behaviours with the help of videos (Ryals et al. 2012a, Laasch and Conaway 2015). An online search for videos with the terms *fun theory* and *Volkswagen* shows up some of the campaign's clips.

On the manufacturing side, Bocken (2017) cites Unilever as an example of a company that uses choice editing to drive changes as part of its Sustainable Living Plan. On the retail side, Walmart, in working together with its suppliers on more compact packaging for laundry detergents, shifted the frame of choice for its consumers without raising costs (Robertson 2017). Similarly, supermarket Sainsbury's decided to offer only Fair Trade cotton in its own-label range of T-shirts (Goworek 2017).

Sweden-based think tank Doconomy launched a credit card called Do Black in 2019 that sets a credit card limit not based on monetary spending but based on annual carbon footprints using country-specific calculation and industry footprints for goods purchased. This recognises the power in consumer wallets as 60% of individuals' carbon footprint is linked to consumption and can help understand and shift consumption patterns (Peters 2019). Similarly, the non-profit start-up Poseidon Foundation is working with retailers to study the impact of consumer purchases and enabling customers to buy carbon offset with every purchase, trialled e.g. by Ben & Jerry's which piloted paying a penny to offset each ice cream scoop's carbon footprint whilst also asking buyers if they would like to donate an additional penny to make the offset go further (Peters 2018).

Both examples demonstrate the significance of consumer education and information, as well as showing the power of direct cause-related marketing – behaviour change can start small, and the barrier to it is thus broken down. Informing and educating consumers is also done by consumer associations or consumer rights organisations (e.g. Which? in the UK) which are increasingly taking sustainability into account when reviewing products and helping consumers make better choices.

When practicing sustainability marketing in an international context, companies have to consider not only issues of linguistics and translation concerning advertising and other communication, but they also have to take into account different cultural and social values and norms with regard to environmental sustainability. This is against the background of developed versus developing countries where traditionally economic development has taken precedence over sustainable development (Laasch and Conaway 2015). For firms this has implications for whether and how they can market sustainable products and encourage sustainable consumption depending on markets in which they are active or to which they want to expand.

SUMMARY

In this chapter on marketing, the final part of the supply chain has been covered in terms of consumption in contrast to production, and in exploring the role of marketing and communication for sustainability. The significance of marketing as the key interface between companies and consumers has major implications for its role in making companies more sustainable and in encouraging sustainable consumption.

There are a number of tensions we have highlighted that companies need to navigate. As one would expect, marketing aims at encouraging consumption whilst sustainability requires reducing it. Whereas consumers might want to go green, they are not necessarily willing to pay the price for it. Moreover, if they want to purchase green products, they are faced with the challenge of identifying which products live up to their green claims and which companies engage in greenwashing.

There are, of course, limits to consumer power and their contribution to sustainable consumption, as they cannot replace legislation, for example, and are, overall, less powerful than companies and governments. Nevertheless, they play an increasing role in pushing political action on green consumption (Crane and Matten 2016; Scales 2018). Their role is furthermore reflected in the growth of the consumption of ethically sourced products. The use of eco-labelling (alongside information and education) has contributed to this, which has overcome one of the highlighted tensions of credibility and greenwashing. Beyond the tension that eco-labelling addresses, companies have further reduced the gap between consumers' willingness to pay and their willingness to go green by making their sustainable brands and products appeal to a wider audience beyond green market segments through a number of tools and strategies covered in the chapter. These include the marketing mix of the 4Ps as well as the 4Cs, choice editing, labelling, and communication for advertising, information and education. Companies can use these tools to market their green products and services and to encourage sustainable consumption, as often the main environmental impact of products is during the use phase by the end consumer. At the same time, in order to fully integrate sustainability into their operations, companies also need to use these tools in a different way in order to market sustainably, i.e. they need to make marketing itself sustainable. The chapter has thus covered a multitude of aspects to marketing and environmental sustainability. As it is beyond the scope of the chapter to treat all facets in detail, recommendations for more in-depth material are provided in the section for suggested questions and activities presented next.

This completes the first half of part three on corporate functions, where we broadly covered the key primary activities of Porter's (1985) value chain. In the second half, we will focus on the supporting activities of accounting, human resource management and information systems, to which we turn in the next chapter. As we have stressed before, companies need to work across individual corporate functions in order to achieve environmental sustainability. In particular, communication, reporting and accounting are closely related, providing a natural link between the communication aspect of marketing and the reporting element of accounting, which is covered in Chapter 12. A key to making consumption as well as production more sustainable consists in decoupling resource consumption from economic growth (Blowfield 2013). As a result of their decisions on what to buy (or not to buy[6]), consumers play an important role in making supply chains and production more sustainable based on available information. One function that supports companies in this endeavour deals with systems comprising information as well as environmental management systems to which we now turn in Chapter 11.

SUGGESTED QUESTIONS AND ACTIVITIES

- Choose one of the environmental challenges covered in part two and explore its connection to consumption.
- Look up some of the corporate examples and practices mentioned in the chapter, and research them in more detail.
- Can "green" marketing claims be trusted?
- What eco-labels are you aware of? Visit www.ecolabelindex.com/ to see if you recognise any of the over 400 eco-labels used in over 190 countries.
- Check out the World Resources Institute's Better Buying Lab: www.wri.org/our-work/project/better-buying-lab/
- Visit www.goodguide.com and find out which products score high on environmental variables.
- Explore the ethical consumer website: www.ethicalconsumer.org/. Select one area of consumption and research with the help of other sources how (un)sustainable it is and why. How could this area of consumption be made more environmentally sustainable by companies?
- Search online for the graphic *The Buyerarchy of Needs* as designed by Sarah Lazarovic and discuss how this could be used to help consumers (and companies) encourage sustainable consumption.
- Go to https://buymeonce.com/ and review the range of durable products sold – which other products could you suggest that would appeal to consumers and could be sold under the site's headline "love things that last"?
- For more in-depth material on marketing and sustainable consumption, check out the following:
  - Belz, F.-M. and Peattie, K. (2012) *Sustainability Marketing: A Global Perspective*, Chichester: Wiley, 2nd ed.
  - Daddi, T., Iraldo, F. and Testa F. (2015) *Environmental Certification for Organisations and Products*, London: Routledge.
  - Devinney, T. M., Auger, P. and Eckhardt, G. M. (2010) *The Myth of the Ethical Consumer*, Cambridge, UK: Cambridge University Press.
  - Martin, D. and Schouten, J. (2014) *Sustainable Marketing*, Harlow: Pearson Education Limited.
  - Middlemiss, L. (2018) *Sustainable Consumption: Key Issues*, London: Routledge.
  - Ottman, J. A. (2011) *The New Rules of Green Marketing*, Sheffield: Greenleaf Publishing.
  - Reisch, L. A. and Thøgersen, J. (2017) *Handbook of Research on Sustainable Consumption*, Cheltenham: Edward Elgar.

NOTES

1 See also Luchs and Kumar (2017) for trade-offs between sustainability and other product attributes.

2 The "4Es" behaviour change model of enabling, engaging, encouraging and exemplifying is based on the UK's strategy for sustainable development

and includes government, businesses and citizens as stakeholders (DEFRA 2008).

3 If this term can be applied to the car industry at all (as some would arguably question in the context of e.g. the diesel emissions scandal).

4 Kotler and Keller (2016) for example introduce People, Processes, Programmes and Performance as the Four Ps of modern marketing management, which is more reflective of holistic marketing concepts, and realities of modern marketing. An alternative are the four As of marketing reflecting consumer values – Acceptability, Affordability, Accessibility and Awareness (Kotler and Keller 2016, based on Sheth and Sisodia 2012).

5 Ottman et al. (2008) introduce the 3Cs as principles to avoid green marketing myopia – consumer value positioning, calibration of consumer knowledge and credibility of product claims.

6 As it is beyond the scope of the chapter, we have not covered the topic of boycotts as a particular form of making consumption more sustainable by discouraging consumers from buying particular products or purchasing from particular companies. Weybrecht (2014) highlights the role of so-called Buycotts where consumers deliberately buy from businesses in support of their sustainability efforts.

SOURCES AND FURTHER READING

AMA (2019) Definitions of Marketing, American Marketing Association, available at www.ama.org/the-definition-of-marketing/ (Retrieved, 13.03.2019).

Anderson, D. A. (2014) *Environmental Economics and Natural Resource Management*, London: Routledge, 4th ed.

Banerjee, S., Gulas, C. S. and Iyer, E. (1995) Shades of Green: A Multidimensional Analysis of Environmental Advertising, in: *Journal of Advertising*, 24(2): 21–31.

Belz, F.-M. and Peattie, K. (2009) *Sustainability Marketing: A Global Perspective*, Chichester: Wiley.

Belz, F.-M. and Peattie, K. (2012) *Sustainability Marketing: A Global Perspective*, Chichester: Wiley, 2nd ed.

Blowfield, M. (2013) *Business and Sustainability*, Oxford: Oxford University Press.

Bocken, N. (2017) Business-led Sustainable Consumption Initiatives: Impacts and Lessons Learned, in: *Journal of Management Development*, 36(1): 81–96.

Bowen, F. (2018) Greenwashing, in: Castree, N., Hulme, M. and Proctor, J. D. (eds.) *Companion to Environmental Studies*, London: Routledge, pp. 689–693.

Brady, J. (2011) Environmental Product Development, in: Brady, J., Ebbage, A. and Lunn, R. (eds.) *Environmental Management in Organizations, The IEMA Handbook*, London: Earthscan/Routledge, 2nd ed., pp. 295–309.

Carbon Trust (2019) Research Reveals Consumer Demand for Climate Change Labelling, *Carbon Trust*, 03.04.2019, available at www.carbontrust.com/news/2019/04/footprint-labelling/ (Retrieved, 03.12.2019).

Crane, A. and Matten, D. (2016) *Business Ethics*, Oxford: Oxford University Press, 4th ed.

Daddi, T., Iraldo, F. and Testa, F. (2015) *Environmental Certification for Organisations and Products: Management Approaches and Operational Tools*, London: Routledge.

Dahlstrom, R. (2011) *Green Marketing Management*, Mason, OH: South-Western Cengage Learning.

Defra (2008) A Framework for Pro-environmental Behaviours, Department for Environment, *Food and Rural Affairs*, available at https://assets.publishing.service.gov.uk/government/uploads/system/uploads/attachment_data/file/69277/pb13574-behaviours-report-080110.pdf (Retrieved, 28.03.2019).

Devinney, T. M. (2012) Using Market Segmentation Approaches to Understand the Green Consumer, in: Bansal, P. and Hoffman, A. J. (eds.) *The Oxford Handbook of Business and the Natural Environment*, Oxford: Oxford University Press, pp. 384–402.

Devinney, T. M., Auger, P. and Eckhardt, G. M. (2010) *The Myth of the Ethical Consumer*, Cambridge, UK: Cambridge University Press.

Drucker, P. (1974) *Management: Tasks, Responsibilities, Practices*, London: Heinemann.

Elkington, J. (1997) *Cannibals with Forks: The Triple Bottom Line of 21st-Century Business*, Oxford: Capstone.

Esty, D. C. and Winston, A. S. (2009) *Green to Gold: How Smart Companies use Environmental*

Strategy to Innovate, Create value and Build Competitive Advantage, Hoboken, NJ: Wiley.

Fiksel, J. (2012) *Design for Environment: A Guide to Sustainable Product Development*, New York: McGrawHill, 2nd ed.

Fogel, D. S. (2016) *Strategic Sustainability: A Natural Environmental Lens on Organizations and Management*, London: Routledge.

Fuller, D. A. (1999) *Sustainable Marketing: Managerial-Ecological Issues*, London: Sage.

Gershoff, A. D. and Irwin, J. R. (2012) Why not Choose Green? Consumer Decision Making for Environmentally Friendly Products, in: Bansal, P. and Hoffman, A. J. (eds.) *The Oxford Handbook of Business and the Natural Environment*, Oxford: Oxford University Press, pp. 366–383.

Goworek, H. (2017) Sustainable Marketing, in: Molthan-Hill, P. (ed.) *The Business Student's Guide to Sustainable Management*, Sheffield: Greenleaf Publishing, 2nd ed., pp. 290–319.

Gunn, M. and Mont, O. (2014) Choice Editing as a Retailers' Tool for Sustainable Consumption, in: *International Journal of Retail & Distribution Management*, 42(6): 464–481.

Hawken, P. (1993) *Ecology of Commerce: A Declaration of Sustainability*, New York: Harper Business.

Kotler, P., Armstrong, G., Harris, L. C. and Piercy, N. (2013) *Principles of Marketing*, Harlow: Pearson, 6th ed., European ed.

Kotler, P. and Keller, K. L. (2016) *Marketing Management*, Boston: Pearson Education, 15th ed., Global ed.

Laasch, O. and Conaway, R. (2015) *Principles of Responsible Management: Glocal Sustainability, Responsibility and Ethics*, Stamford, CT: Cengage Learning.

Laszlo, C. and Zhexembayeva, N. (2011) *Embedded Sustainability: The Next Big Competitive Advantage*, Sheffield: Greenleaf Publishing.

Lorek, S. (2016) Sustainable Consumption, in: Brauch, H. G., Spring, Ú. O., Grin, J. and Scheffran, J. (eds.) *Handbook on Sustainability Transition and Sustainable Peace*, New York: Springer, pp. 559–570.

Lounsbury, M., Fairclough, S. and Lee, M.-D. P. (2012) Institutional Approaches to Organizations and the Natural Environment, in: Bansal, P. and Hoffman, A. J. (eds.) *The Oxford Handbook of Business and the Natural Environment*, Oxford: Oxford University Press, pp. 211–228.

Luchs, M. G. and Kumar, M. (2017) 'Yes, but this Other One Looks Better/Works Better': How do Consumers Respond to Trade-offs Between Sustainability and Other Valued Attributes? in: *Journal of Business Ethics*, 140(3): 567–584.

Martin, D. and Schouten, J. (2014) *Sustainable Marketing*, Harlow: Pearson Education Limited.

McKenzie-Mohr, D., Lee, N. R., Wesley Schultz, P. and Kotler, P. (2012) *Social Marketing to Protect the Environment: What Works*, London: Sage.

Middlemiss, L. (2018) *Sustainable Consumption: Key Issues*, London: Routledge.

Nemetz, P. N. (2013) *Business and the Sustainability Challenge: An Integrated Perspective*, London: Routledge.

Nuttavuthisit, K. and Thøgersen, J. (2017) The Importance of Consumer Trust for the Emergence of a Market for Green Products: The Case of Organic Food, in: *Journal of Business Ethics*, 140(2): 323–337.

Ottman, J. A. (2011) *The New Rules of Green Marketing*, Sheffield: Greenleaf Publishing.

Ottman, J. A., Stafford, E. R. and Hartmann, C. L. (2008) Avoiding Green Marketing Myopia: Ways to Improve Consumer Appeal for Environmentally Preferable Products, in: Russo, M. V. (ed.) *Environmental Management: Readings and Cases*, London: Sage, 2nd ed., pp. 296–314.

Pearce, D., Markandya, A. and Barbier, E. (1999) *Blueprint for a Green Economy*, London: Earthscan (first published 1989).

Peattie, K. and Belz, F.-M. (2010) Sustainability Marketing – An Innovative Conception of Marketing, in: *Marketing Review St. Gallen*, 27(5): 8–15.

Peters, A. (2018) This Tech Offsets the Carbon Footprint of Each Item you Buy, *Fast Company*, 24.05.2018, available at www.fastcompany.com/40576812/this-tech-offsets-the-carbon-footprint-of-each-item-you-buy (Retrieved, 16.05.2018).

Peters, A. (2019) This Credit Card Won't let you Buy Anything Else After you've hit your Annual Carbon Limit, *Fast Company*, 06.05.2019 available at www.fastcompany.com/90344150/this-credit-card-wont-let-you-buy-anything-else-after-youve-hit-your-annual-carbon-limit (Retrieved, 16.05.2019).

Piasecki, B. W., Fletcher, K. A. and Mendelson, F. J. (1999) *Environmental Management and Business Strategy: Leadership Skills for the 21st Century*, New York: John Wiley & Sons.

Porter, M. E. (1985) *Competitive Advantage: Creating and Sustaining Superior Performance*, New York: The Free Press (Republished with a new introduction, 1998).

Robertson, M. (2017) *Sustainability Principles and Practice*, London: Routledge, 2nd ed.

Ryals, L., Abdollahi, S., Maklan, S. and Wilson, H. (2012b) Implementing Sustainable Marketing, in: Grayson, D. and Exter, N. (eds.) *Cranfield on Corporate Sustainability*, Sheffield: Greenleaf Publishing, pp. 133–148.

Ryals, L., Baines, P., Dimitriu, R., Knox, S., Macdonald, E. and Marcos-Cuevas, J. (2012a) Issues in Sustainable Marketing, in: Grayson, D. and Exter, N. (eds.) *Cranfield on Corporate Sustainability*, Sheffield: Greenleaf Publishing, pp. 119–132.

Scales, I. R. (2018) Green Consumption, in: Castree, N., Hulme, M. and Proctor, J. D. (eds.) *Companion to Environmental Studies*, London: Routledge, pp. 684–688.

Scammon, D. L. and Mish, J. (2012) From Green Marketing to Marketing for Environmental Sustainability, in: Bansal, P. and Hoffman, A. J. (eds.) *The Oxford Handbook of Business and the Natural Environment*, Oxford: Oxford University Press, pp. 347–365.

Schaltegger, S., Burritt, R. and Peterson, H. (2003) *An Introduction to Corporate Environmental Management*, Sheffield: Greenleaf Publishing.

Seele, P. and Gatti, L. (2017) Greenwashing Revisited: In Search of a Typology and Accusation-Based Definition Incorporating Legitimacy Strategies, in: *Business Strategy and the Environment*, 26(2): 239–252.

Sharp, A. (2013) Sustainable Marketing in Principle and Practice, in: Wells, G. (ed.) *Sustainable Business: Theory and Practice of Business under Sustainability Principles*, Cheltenham: Edward Elgar, pp. 108–199.

Sheth, J. N. and Sisodia, R. S. (2012) *The 4 A's of Marketing: Creating Value for Customers, Companies, and Society*, New York: Routledge.

Stead, J. G. and Stead, W. E. (2014) *Sustainable Strategic Management*, Sheffield: Greenleaf Publishing, 2nd ed.

Stead, W. E., Stead, J. G. and Starik, M. (2004) *Sustainable Strategic Management*, London: M. E. Sharpe.

Thaler, R. H. and Sunstein, C. R. (2008) *Nudge: Improving Decisions about Health, Wealth and Happiness*, London: Penguin Books.

van Dam, Y. K. and Apeldoorn, P. A. C. (1996) Sustainable Marketing, in: *Journal of Macromarketing*, 16(2): 45–56.

Visser, W. (2012) *The Quest for Sustainable Business: An Epic Journey in Search of Corporate Responsibility*, Sheffield: Greenleaf Publishing.

Wackernagel, M. and Rees, W. (1996) *Our Ecological Footprint: Reducing Human Impact on the Earth*, Gabriola Island: New Society Publishers.

Weybrecht, G. (2014) *The Sustainable MBA: A Business Guide to Sustainability*, Chichester: Wiley, 2nd ed.

Whitfield, R. and McNett, J. (2014) *A Primer on Sustainability: In the Business Environment*, New York: Business Expert Press.

Zinkhan, G. M. and Carlson, L. (1995) Green Advertising and the Reluctant Consumer, in: *Journal of Advertising*, 24(2): 1–6.

CHAPTER 11

Information technology and management systems for the environment

INTRODUCTION

Whilst the first three chapters of part three broadly covered primary activities in Porter's Value Chain, the second three chapters concern the secondary activities in the chain that support the primary ones. As stressed before, there are natural connections and overlaps between the various functions, and in this chapter the importance and significance of integrating data and information to support environmental efforts in a systematic way emphasises the holistic and integrated nature of sustainability.

When it comes to environmental performance, companies can easily capitalise on low-hanging fruits by adopting best practice, but in order to go further companies need to implement environmental management systems and determine their continuous improvement in performance (Fiksel 2012). Weybrecht (2014) emphasises the role information technology and systems can play in enabling organisations to move forward with sustainability efforts. This is made possible through information that is more easily collected and analysed, and can then be used to improve decision-making. Information technology (IT) has changed the way organisations work and therefore offers challenges (e.g. disruptions of industries) as well as advantages (Daft et al. 2010), which can be used to the benefit of environmental sustainability by making corporate activity more transparent and bringing environmental issues to prominence (Hoffman and Bansal 2012). IT is particularly pertinent, as data-driven environmental management has become more and more the norm (Esty and Winston 2009).

In the following, we provide a brief outline of information technology and management systems and what their environmental impact is before introducing the business case for making both greener. We continue by overviewing green and sustainable information and management systems before detailing standards for environmental management systems, such as ISO 14000 and EMAS, and briefly highlighting the role of balanced score cards and other tools and applications for environmental management.

INFORMATION TECHNOLOGY AND MANAGEMENT SYSTEMS

Overview

In order to establish a joint understanding of what we mean by management systems, information technology or systems, and other terms, we start with a few clarifying definitions.

According to Fiksel (2012: 381), a management system is "a management approach that enables an organisation to identify, monitor and continuously improve its performance, including financial, environmental, and/or social aspects". In simpler terms, it is the "set of procedures which systemizes how things are done in an organisation" (Robertson 2017a: 88). A management information system (MIS) is a computer-based system providing information and support for more effective decision-making by managers (Daft et al. 2010).

Technology is characterised by applying knowledge to achieve practical outcomes – this includes tools, techniques as well as skills and processes (Meadowcroft 2018). Information technology (IT) is also called information and communication technology (ICT) and refers to the technology and further equipment used for displaying, processing, transmitting or storing information (Weybrecht 2014). It includes software, hardware, database management and any other technologies that support organisational decision-making through information (Daft et al. 2010). IT can be used for knowledge management, enterprise resource planning (ERP) or customer relationship management (CRM), to name a few applications (ibid.).

Information Systems (IS) denote any combination of IT and human activities that provide support to operations, management and decision-making (Weybrecht 2014). They are vital support tools for businesses as they can help manage environmental information across a variety of functions, such as accounting, production or human resource management (Piasecki et al. 1999).

Environmental impact of information technology and management systems

A decade ago, the energy used by modern ICT as well as consumer electronics accounted for 15% of the world's residential electricity use with estimates that it might become the largest energy end-user by 2020 (Nemetz 2013; IEA 2009). More recent figures indicate that by 2014, appliances accounted for 20% of residential energy consumption, though despite energy efficiency gains manufacturing and transport are still more significant energy users than residential ones (IEA 2017a, 2018). Depending on type of fuel, energy generation uses natural resources, produces pollution and contributes to climate change. In 2016, coal, oil and natural gas accounted for over 80% of the worldwide primary energy supply (IEA 2018). Although electrification and use of renewable energy is on the rise, fossil fuels are therefore a significant input for information technology and management systems, which in turn impact on the environment. Whilst digitalisation – the growing application of ICT across the economy – can be used to increase efficiency and energy savings, and to improve productivity for the major energy users of industry, transport and buildings (for example through smart grids, intelligent systems, automated driving, smart lighting and so on), it also contributes significantly to energy use itself (e.g. through data processing in data centres and networks) (IEA 2017b) and potential rebound effects (Melville 2012). It is estimated that the carbon footprint of the IT industry is roughly equal to that of the entire airline industry due to the energy needed to power data centres (Griffin 2019).

Beyond the impact with regard to air pollution and climate change, IT also contributes to the issue of waste in the form of e-waste or WEEE (waste of electrical and electronic equipment), one of the fastest growing waste streams, as highlighted in Chapter 7. Such waste also contains hazardous materials such as heavy metals which may be explosive, flammable or toxic, and which pose substantial risks to health and safety as well as the environment, depending on how they are treated. It is in particular the short life cycles of such electronic equipment – mostly due to the speed of technological change making hardware quickly obsolete – that is problematic as toxic e-waste proliferates (Melville 2012; vom Brocke et al. 2013).

Impact of environment on information technology and management systems

Extreme weather events such as floods or droughts as a result of climate change for example can disrupt the electricity supply that is vital for the smooth functioning of information and management systems. Equipment supplies may also be affected if production of electronics is halted, as happened with floods in Thailand in 2011 (Reuters 2011). Rising temperatures can make it more difficult and expensive to keep for example data centres cool. Other impacts such as buildings and plants put at risk by rising sea levels because of climate change are less specific to IT and management systems and are part of the broader impact of the environment on corporate activities.

Business case for greener information technology and management systems

Whilst the impact of the environment on information and management systems is less tangible, it is clear that there are opportunities for these activities to reduce their environmental impact, which is arguably not only the right thing to do but can also have tangible financial benefits for business and can thus make a big difference. Measures such as reducing energy consumption through efficiency gains result in cost savings, whilst having better metrics, data and information on environmental impacts available allows companies to identify areas for improvement and enables fact-based decision-making (Esty and Winston 2009). For example, Environmental Management Information Systems (EMIS) can improve productivity at factory, business and corporate levels (Friedman 2011). An important internal driver for organisations to adopt environmental management systems is the need to improve environmental management and resource efficiency alongside development of organisational capabilities, whilst external drivers can be seen in ensuring regulatory compliance and improving corporate image and reputation in the eyes of external stakeholders (Daddi et al. 2015). For Esty and Winston (2009), implementing an environmental management system is crucial for companies as environmental issues can be viewed as potential liabilities that companies need to include in financial statements.

Environmental management is however not only a response to external drivers but also a key part of an organisation's strategic vision aiming at better commercial and environmental results (Daddi et al. 2015, see also Piasecki et al. 1999). Friedman (2011) emphasises the importance of avoiding environmental policies and strategy being developed in a vacuum and of purposefully linking them to an organisation's overall strategy. Shifting environmental strategies also creates new or additional information requirements necessitating better environmental information systems (Melville 2012).

There are a number of ways information and management systems can contribute to reducing companies' environmental impact, whether it is directly or indirectly (e.g. by providing data to other corporate functions). In the following, we will explore greener information technology and systems, including environmental management systems.

GREEN AND SUSTAINABLE INFORMATION SYSTEMS, TECHNOLOGY AND ENVIRONMENTAL MANAGEMENT SYSTEMS

Overview

As earlier, we start with clarifying some terms before exploring tools that can be used to make information and management systems more sustainable.

Green information systems refer to designing and implementing information systems that help businesses with sustainable practices (Weybrecht 2014). Information systems have the potential to enable better visualisation of environmental performance (also called eco-visualisation) e.g. by providing real time data and customised dashboard solutions (Melville 2012). In turn, this can help improve environmental performance. Green IS is different from green IT, which is more limited in its focus on energy efficiency and equipment utilisation of IT – green IS applies a broader focus on designing and implementing IS that support sustainable business processes and their implementation (vom Brocke et al. 2013). As both green IS and IT are fairly recent developments, other authors apply a more comprehensive understanding of green IT which encompasses not only new technologies and innovations (such as energy efficiency and cloud computing) but also new organisational practices and strategies (e.g. e-waste management and power management) (Liu 2011).

Green technology is more general and often refers to technology that is "kind to the environment" i.e. avoiding, reducing or remediating the environmental impact of organisations and individuals (Meadowcroft 2018: 694). Related terms include clean technology, sustainable technology and environmental technology.

An environmental management system (EMS) is a formal system for how an organisation develops, implements and maintains environmental policies and procedures (Robertson 2017a). It is a framework that allows it to manage the quality of the environment over which it has influence (Robertson 2017b). EMSs are centred around the iterative Deming cycle of Plan-Do-Check-Act (PDCA) found in quality management, with a focus on continuous improvement (ibid., Daddi et al. 2015). We briefly referred to this cycle in Chapter 9 on production and operations.

Although the adoption of an EMS is voluntary, as a tool it can ensure a company's regulatory compliance (Weybrecht 2014). Standards and certification systems for EMS can be public or private, and relate to an organisation itself or products, whilst Nemetz (2013: 423) categorises three basic types of standards: principle-based standards, performance standards and process standards. Standards are formulated by standard-setting organisations, and companies can audit their performance against these standards (Schaltegger et al. 2003). The most widely used EMS standards are the ISO 14000 series and the Eco-Management and Audit Scheme (EMAS), both of which we cover in more detail in the following sections.

An Environmental Management Information System (EMIS), particularly when computerised, provides organisations with a collection of data-management tools that can help them achieve their environmental objectives by improving regulatory compliance, reducing risks and liabilities and managing environmental costs (Friedman 2011). EMIS is part of broader corporate EMSs, which comprise a number of tools (Nemetz 2013). One component is for example an environmental financial statement, which allocates costs in a way that allows organisations to identify where its costs and opportunities for savings are located (Nemetz 2013). There is thus a close link between information systems and accounting, which we cover in the following chapter.

Broadly, environmental management can be contrasted with conventional economic management. Whilst conventional management tools consist of calculation and costing, accounting, reporting, total quality management and control, environmental management tools can be found in life cycle assessment (LCA), environmental accounting and reporting, environmental auditing, and total quality environmental management (TQEM) with eco-control as the foundation (Schaltegger et al. 2003). These all support the key functions of environmental management, which Schaltegger and colleagues (2003: 317) list as goal-setting, information management, supporting decision-making, organising or planning environmental management programmes, steering, implementation and control, communication as well as internal and external auditing and review.

As we will cover environmental accounting and reporting as well as auditing and controlling in the following chapter, we briefly remind ourselves of LCA and TQEM (both also covered in previous chapters), before highlighting eco-control, and moving on to covering standards for EMS in more detail.

Life cycle assessment (LCA) is a comprehensive method for measuring environmental performance, allowing organisations a cradle-to-grave approach in evaluating activities across the value chain (Grant et al. 2017). Such identification and assessment of the impact an organisation has on the environment and which activities, products or services cause them, is a key part within an EMS (Baxter 2011). In the broader corporate context, LCA is key to strategy and decision-making for environmental sustainability (Nemetz 2013). As data are integral to any LCA, information technology and management systems have an important role to play in enabling environmentally sustainable decisions. For example, information technology and software capabilities can facilitate the application of Design for Environment (DfE) practices (Fiksel 2012). The ISO 14040 series provides standardisation and documentation for LCAs, embedded in the wider ISO 14000 series of environmental management standards.

Applying quality principles to environmental management is also known as Total Quality Environmental Management (TQEM), which is a derivative of Total Quality Management (TQM) (Piasecki et al. 1999). Both were briefly covered in Chapters 8 and 9. TQM is a management approach focused on quality that builds on employee participation and customer satisfaction for long-term success and is particularly suitable for integrating environmental performance as an important aspect of total quality across various functions (Fiksel 2012; Friedman 2011). Piasecki and colleagues (1999) also emphasise the systematic and integrated nature of the total quality philosophy, which comprises the entire organisation. The confluence of quality management and environmental concerns (which Hart calls "a marriage made in heaven", 2010: 25) has been instrumental in the emergence of EMSs and TQEM protocols, in particular the ISO 14000 series. The basic elements of TQEM involve identifying customers (internal or external), continuous improvement, doing the job right the first time and a system approach, as developed by GEMI, the global environmental management initiative (GEMI 1993). Similar to LCA, TQEM uses data analysis and statistical tools for managing information, supporting decisions and communication (Schaltegger et al. 2003).

Schaltegger and colleagues (2003) highlight that a foundation for EMS is eco-control, a concept that aims to guide and co-ordinate the various environmental management activities of an organisation. It is an attempt to manage environmental issues in their totality despite businesses (as well as society) being organised to manage problems in isolation rather than in an integrated manner. Control is a key function of management as well as of external regulation (*command*

and control regulation) and offers a systematic approach for companies to achieve and measure economic and environmental objectives (ibid.). Buhr and Gray (2012) emphasise that environmental management is, in essence, about control. Although the control function is traditionally associated with accounting, control in this context is not just financial control in isolation but more systemic, covering non-financial elements as well. EMSs are a particular form of management control systems, which can be defined as information-based and formal practices that can be used to keep or change the pattern of organisational activities (Buhr and Gray 2012, based on Simons 1995). The information can be of a financial or non-financial nature. Environmental management control systems in turn are defined as a package of formal organisational measures that are specifically concerned with the environmental aspects of organisational performance and are thus based on environmental information and a firm's environmental strategy (Pondeville et al. 2013). An alternative term for management control systems that incorporates environmental concerns is sustainability control systems (Gond et al. 2012) or sustainable management control systems (Lueg and Radlach 2016). As the chapter is not solely on control, and as our focus is on environmental sustainability, we will use the term environmental management system[1] (EMS).

Companies may develop EMSs internally, though it is much more common that they are developed according to established standards, such as EMAS or the ISO 14000 series, to which we now turn.

Environmental Management Systems (EMSs): standards and tools

Grant and colleagues (2017) highlight that when designing, implementing and evaluating EMSs, various factors need to be considered, most importantly that there is not a single approach suitable for all organisations. The following should thus be read in this context.

The benefits of adopting EMSs are not restricted to the environmental performance of organisations and can improve organisational procedures and structures more generally (Daddi et al. 2015; Schaltegger et al. 2003). In addition to regulatory compliance and reduced environmental costs there are also opportunities for lower purchasing costs, and reduced waste and disposal costs, which in turn can lead to competitive advantages through improved efficiencies (ISO 2019). Furthermore, government relations can be improved as well as stakeholder relations, particularly employee engagement, without which environmental improvements cannot be achieved – a topic further explored in Chapter 13 on human resource management. As highlighted in Chapter 8, EMSs can be used to audit suppliers on their regulatory compliance, product and service specifications (e.g. phasing out of harmful or toxic materials), and environmental practices or management systems, which can help improve environmental performance in the supply chain (Marsden 2011).

We focus on the most widely used EMS standards, the ISO 14000 series and the Eco-Management and Audit Scheme (EMAS), thus contrasting a private (industry-based) standard-setting organisation with a public one. Whilst for example the International Standards Organisation ISO does not provide certification, external third parties can provide confirmation that an organisation meets ISO's EMS requirements and thus certifies the standard. This provides recognition for achievements, which may be useful in the eyes of customers and environmental regulators (Baxter 2011).

Broadly speaking, standards provide a foundation for continual improvement in relation to established benchmarks. The principle of continual improvement is at the core of an EMS, which is based on the Deming cycle of Plan-Do-Check-Act (PDCA) of planning what you are going to do, doing what you planned to do, checking to ensure that you did what you planned to do and acting to make improvements. For EMSs, the planning and doing involves environmental policies, planning and then implementation and operation, followed by checking and management review (Baxter 2011). It is the reiterative nature of the Deming cycle that reinforces continuous improvement (Robertson 2017b).

ISO 14000 series

The ISO 14000 series of standards consists of a wide-ranging set of tools for companies to manage their environmental performance. The series was developed in the early 1990s in an effort to harmonise what was happening at a national level (e.g. standards developed by the British Standards Institutions, Deutsche Industrie Norm or Standards Australia).

In particular, ISO 14001 specifies the requirement of an EMS regardless of sector or activity (ISO 2019). It provides a structured framework to identify, evaluate, manage and improve the environmental performance of an organisation (Baxter 2011). The intended outcomes beyond enhancing environmental performance are fulfilling compliance obligations and achieving environmental objectives (ISO 2019).

The different elements of an EMS defined in ISO 14001 are the following:

Environmental policy

This element is central to an EMS (Baxter 2011) and defines the commitment the organisation intends to adopt for its environmental improvements, comprising broad objectives and principles for action (Daddi et al. 2015). It is important to be integrated with other organisational policies and should be endorsed by senior management (Weybrecht 2014).

Planning

This often involves an initial environmental review to identify and evaluate the organisation's environmental impacts across a number of different criteria, whether it is the carbon footprint, pollution or waste. Such an environmental review is a crucial phase in the implementation process, providing details of the environmental problems, impacts and performances of an organisation (Daddi et al. 2015). Particular importance should be attached to identifying the boundaries of the EMS and considering which impacts (direct or indirect) are of particular significance (Baxter 2011). Once the significance of these impacts is assessed, controls, targets and objectives can be set (Weybrecht 2014).

Implementation and operation

An environmental programme translates the targets and objectives set into practical actions to improve environmental performance (Baxter 2011). This also involves assigning responsibilities and creating appropriate structures. Communication is particularly important for an EMS to be effective – this applies both internally and externally. The implementation and operation phases are supported by procedures and documentation so that employees know how the EMS operates and what the requirements are.

Checking

The internal audit following the EMS implementation is a crucial activity of the "check" phase of the Deming cycle (Daddi et al. 2015) and involves monitoring and auditing, followed by review. Data on the environmental performance is key for tracking progress against the set objectives and thus for continuous improvement (Baxter 2011).

Management review

In line with the Deming cycle's core principle of continuous improvement, the management review can help identify areas for improvement. Daddi and colleagues (2015) advise paying particular attention to the review of the environmental policy, objectives and programmes.

Other standards in the ISO 14000 family of international tools and standards include environmental auditing, environmental performance evaluation, vocabulary and communications, environmental labels and declarations, life cycle assessment, product design and development, as well as specific environmental challenges such as climate change in the form of guidelines for greenhouse gas reporting (Robertson 2017b). Life cycle assessment has been covered in various chapters, whilst the previous chapter on marketing explained the three different types of eco-labels according to ISO. Although the ISO certificate may be used for general marketing, it may not be used for product marketing. Other standards developed by ISO that are relevant to sustainability include for example ISO 9000 on quality management, ISO 50001 on energy management, ISO 26000 on social responsibility, and ISO 19001 on auditing the 14000 series (Fiksel 2012; Robertson 2017b). The standards are regularly reviewed and updated – for example ISO 14001 was first developed in 1996, revised in 2004 and last updated in 2015 (ISO 2019).

Whilst the ISO 14000 is not prescriptive in setting detailed environmental management objectives or how they should be obtained, it sets the requirements for a formal system that should be followed – summarised by Robertson (2017b: 269) as "say what you do, then do what you say". As firms self-declare, there is, however, a danger of symbolism and greenwashing associated with EMS, as previously highlighted by Melville (2012). To counter this risk, companies can use external bodies to audit their EMS for accredited certification – for ISO 14001 more than 300,000 certifications to this standard have been issued in over 170 countries (ISO 2019). This sends a signal to stakeholders such as customers, buyers and suppliers that the EMS standard has been implemented.

EMAS (Eco-Management and Audit Scheme)

EMAS is the European Union's Eco-Management and Audit Scheme, which consists of sustainability management and report standards developed by the European Commission for use by organisations in Europe (Robertson 2017b). Unlike ISO 14001, which is a private standard, EMAS is publicly regulated, and, although initially valid only in Europe, it is now also valid at the international level (Daddi et al. 2015). Whilst there is no general register of certified organisations that meet the ISO 14001 standard and no requirement for publishing environmental statements, EMAS provides a publicly accessible list of registered organisations that have to publish their environmental statement (Schaltegger et al. 2003).

Common to both ISO and EMAS is the implementation of an internal environmental management scheme that involves an initial analysis, development and adoption of an environmental policy, definition and application of an environmental programme, a management and organisational structure to pursue the programme and a periodic verification process to evaluate the environmental performance and management scheme's efficiency and effectiveness (Daddi et al. 2015).

Both standards lead to substantial cost savings when adopted, and whilst ISO 14001 has a smaller administrative workload, EMAS goes further, as it incorporates ISO 14001 and adds supplementary requirements such as developing and publishing an environmental declaration to demonstrate the commitments undertaken, the programmes developed and the results obtained (Daddi et al. 2015). The unique features that differentiate EMAS from ISO 14001 is the demonstration of legal compliance, commitment to continuous improvement, employee participation

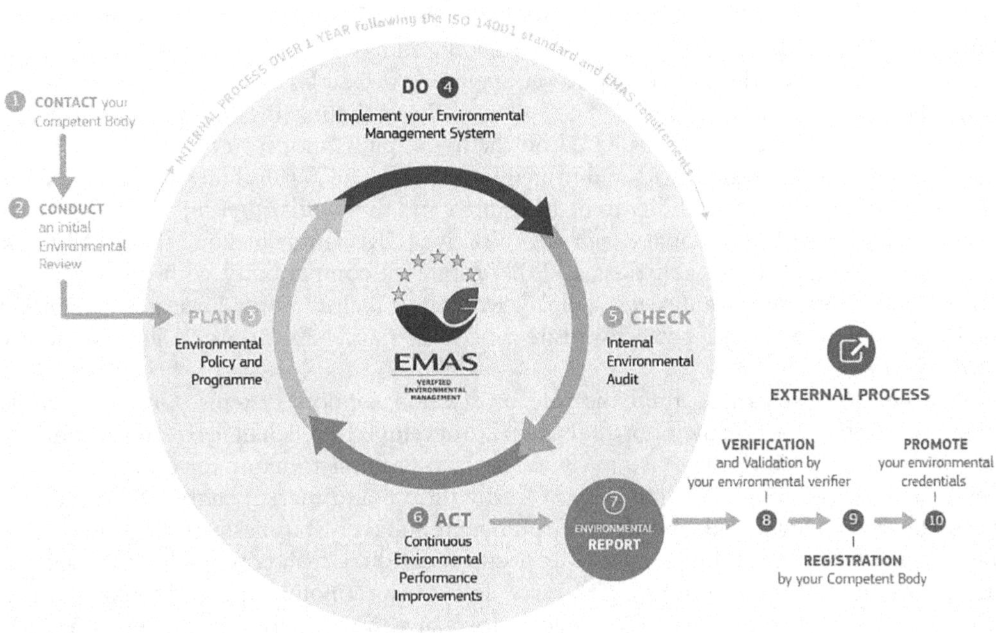

FIGURE 11.1
EMAS process

Source: European Commission (http://ec.europa.eu/environment/emas/images/join_emas/step00.jpg).

and compulsory communication to the public (European Commission 2019). As ISO 14001 is an integral part of EMAS, these added benefits are obtained by implementing a few adaptations. Other existing environmental management systems are recognised by EMAS to avoid multiplication of efforts, though guidelines are provided for companies wishing to step up or move from local or regional labels or non-formal EMS to the more recognised and established EMAS registration.

Beyond cost savings, EMAS leads to better environmental performance, improved risk and opportunity management that ensures legal compliance, enhanced credibility, transparency and reputation through reporting, and finally, EMAS can empower and motivate employees (European Commission 2019). As with ISO, the EMAS logo can be used as a marketing tool.

The process for obtaining EMAS certification follows the Deming cycle of Plan-Do-Check-Act for continuous improvement and is depicted in Figure 11.1.

As of late 2019, the EMAS network consisted of over 3,700 organisations of all sizes and sectors, covering over 12,000 sites (European Commission 2019).

Other tools and standards: from SIGMA sustainability integrated guidelines for management and eco-efficiency portfolio matrix to sustainability balanced scorecard

The SIGMA project, which stands for "Sustainability Integrated Guidelines for Management", was an initiative launched in 1999 by the British Standards Organisation in partnership with

Forum for the Future and AccountAbility, funded by the UK government's Department of Trade and Industry. The guidelines were published in 2003 and based on the Plan-Do-Check-Act cycle of management frameworks and included a set of guiding principles and a toolkit (Sheldon 2011). SIGMA should not be confused with Six Sigma or Lean Sigma, the tools used in operations management (see Chapter 9). It was based on the five capitals model (introduced in Chapter 2) and drew from existing frameworks and principles such as The Natural Step and the UN Global Compact. Its strength was the ability to be integrated and used with other initiatives, such as ISO 14001 or EMAS or with the Global Reporting Initiative (GRI) Guidelines and the AA1000 Assurance Standard (Knight 2006; Leipziger 2003). Despite its compatibility with other framework and its helpful integrationist approach (Leipziger 2003), it seems to only have attracted attention in the UK and, as of 2019, the project website is no longer accessible, thus raising questions over its viability.

Management systems also fulfil the role of decision support systems. One such decision-making tool is the eco-efficiency portfolio matrix, developed by Schaltegger and Sturm (1998) based on the Boston Consulting Group (BCG) matrix, familiar to many for evaluating strategic business units according to the dimensions of growth rate and market share (Daft et al. 2010). The eco-efficiency matrix evaluates the environmental impact of business activities as well as the economic performance, resulting in the four positions of dirty cash cow (high economic performance, low environmental performance), dirty dogs (low economic and environmental performance), green question marks (low economic performance, high environmental performance) and green stars (high economic and environmental performance) (Schaltegger et al. 2003). Depending on the results, companies can use different ways to improve the environmental performance of its products and processes – e.g. by exchanging a particular input, eliminating a product line or by adopting a functional approach to replace a product with a service.

A comprehensive management control system can be found in the balanced scorecard (BSC) that complements traditional financial measures with other critical operational success measures such as internal business processes, customer service and an organisation's capacity for learning and growth (Daft et al. 2010). It was developed by Kaplan and Norton as a strategic management tool with the organisation's vision and strategy at its centre to which four processes contribute – translating the vision into operational objectives and measures, communicating and linking the strategy to all levels of the organisation, business planning for allocation of resources and, finally, feedback and learning to facilitate feedback and review (Kaplan and Norton 1996).

The BSC can be adapted for use in sustainability as it can combine information about environmental impacts with economic success measures and allows consideration of long- and short-term goals (Schaltegger et al. 2003). This helps inform decision-making and reporting results whilst also changing individual behaviours and actions (Nemetz 2013). Adapting the scorecard for sustainability includes for example incorporating sales revenues from "green" products in the financial dimension, percent of workforce trained in sustainability in the learning and growth dimensions, or certification of suppliers in the business process dimension (Nemetz 2013; Epstein 2008). This scope for integration and moving beyond financial measures alone is a significant factor in the use of the BSC for sustainability that can be used internally as well as externally (Dharmasasmita et al. 2017; Blowfield 2013; Cavagnaro and Curiel 2012). Figge and colleagues (2002) outline the development of a sustainability balanced scorecard (SBSC) as a strong tool for integrated sustainability management, which can improve strategic alignment (Crawford and Scaletta 2005). The type of information used for a SBSC can be qualitative or quantitative in terms of measurability, monetary or physical in terms of measure, and environmental, social or economic in terms of the sustainability perspective (Schaltegger et al. 2015).

Apps for environmental management

With the rise of smart technology and the internet of things, it is necessary to briefly highlight their use for corporate environmental sustainability in the context of this chapter. Smart phone applications (so-called apps) have the potential to make information and data more easily digestible and more clearly visible. They can also serve as communication tools, engagement platforms or to support processes and projects.

Whilst there is little information on corporate apps, partially due to commercial sensitivity, some tools and apps have been highlighted publicly for driving green business change and helping companies as well as consumers with sustainability efforts, ranging across food, water, waste, supply chain tracking and design for example (Clancy 2013). Other types of sustainability apps include personal efficiency, information sharing or sharing economy apps such as Tap, joulebug, HappyCow or allgreenup, recycling and food waste prevention apps such as OLIO or NoFoodWasted, product and firm assessment apps such as HowGood, and climate change apps, such as Oroeco (see e.g. Bozhinova 2018; Carrara 2019). Many of these are on the consumer side of things but can also be used by companies. As such apps are in constant development, they often have a short life cycle and are superseded by new additions to the portfolio of apps supporting sustainability. Delmas and Kohli (2019) highlight in their study of a bespoke air quality app – which makes pollution visible – the potential of apps to change stakeholder behaviour, but caution that mobile apps are still limited in effecting long-term individual behavioural change. AirVisual is one such live air quality app that is crowdsourced.

An example of software for companies that also comes as a mobile app is Fleetmatics, a GPS-based fleet tracking software that can lead to substantial fuel savings and thus reduce operating expenses for fleet owners.

Spotlight on . . . examples and practices of green, sustainable and environmental information or management systems and information technology

As in previous chapters, we briefly highlight selected examples and practices of green technology and environmental management systems not already mentioned.

Examples of standard-setting institutions for EMSs include the aforementioned International Standards Organisation (ISO), the British Standards Institution (BSI), Standards Australia (SA) and the German Deutsche Industrie Norm (DIN). Whilst ISO is a private organisation, the others are public institutions.

In the hospitality sector, companies that have adopted ISO standards to improve their sustainability include Hilton, which built on its existing ISO 9001 quality management standard and ISO 14001 EMS to achieve energy management certification according to ISO 50001 (ISO 2019). The company has used its triple ISO certification to continuously reduce its environmental footprint.

Within the same industry, Belgian hotel group Martin's Hotels is certified under EMAS, the first in the country to receive the certification. The first company in the outdoor industry to introduce EMSs in accordance with both EMAS and ISO 14001 is VAUDE from Germany. Other manufacturing companies certified under EMAS include BMW, Bombardier, Hyundai's European operations and MAN Trucks (European Commission 2019).

The Sustainability Integrated Guidelines for Management (SIGMA) was used by the British Aviation Authority at Heathrow in the mid-2000s to inform and guide its efforts towards sustainable development, particularly addressing sustainability aspects that were not covered by its existing environmental management system (Knight 2006).

ABB uses an eco-efficiency portfolio approach to establish targets for its products and services across three areas – energy efficiency, renewable energy and resource efficiency – which the company wants to account for 60% of overall revenues (ABB 2019). One of its products is a digital water management solution, which allows organisations and cities to track water consumption in order to then improve water management.

Companies that have adapted Kaplan and Norton's balanced scorecard for sustainability purposes include Nike and its Nike Considered Index, as well as Starbucks and Herman Miller (Blowfield 2013).

Other examples of green technology practices that comprise and go beyond IT include energy-efficient computing, data centre design, layout and location, responsible disposal and recycling, regulatory compliance, green metrics, assessment tools and methodology, use of renewable energy sources and eco-labelling of IT products, to name a few.

With its Project Natick, Microsoft has started experimenting with submerging data centres in the ocean – first, off the coast of California in 2015, and, in 2018, off the coast of Scotland's Orkney Islands in the North Sea. The environmental benefits include savings of energy needed to cool such data centres and the use of ocean currents to power the centres, which are envisioned to be fully recyclable. Other advantages include reducing the distance data need to travel to coastal customers, thus speeding up access to cloud-based services (Goldman 2016; Cellan-Jones 2018; Roach 2018).

In order to improve its data centre efficiency, Google used artificial intelligence (AI) to reduce the energy needed to cool its data centres by 40%, which provides environmental as well as financial benefits for the company (Riffle 2017). The company is also one of the world's largest purchasers of renewable energy to power its data centres (Griffin 2019).

Artificial intelligence, machine learning, blockchain technology and big data have great potential to help companies with their sustainability efforts, offering products and services even recently unthinkable. The development of these technologies is at an early stage and the opportunities also carry risks with them, neither of which have yet been fully explored (Riffle 2017). In particular, artificial intelligence brings mixed blessings with it as on the one hand it can be effectively used to study and lessen environmental impacts, whilst on the other hand the energy requirements for machine learning contribute to a rise in CO_2 emissions (Schmelzer 2019). To harness the power of these technologies Microsoft launched AI for Earth in 2017 to bring together cloud computing and AI tools for environmental innovation. In the US, climate scientists are using virtual reality programmes to show what communities would look like under the impacts of climate change, helping people and communities to become more engaged with a future that is often hard to imagine and plan for, and thus to enable more timely solutions (NPR 2019). Blockchain technology enables a company to track the entire lifecycle of a product, providing relevant data e.g. of energy use or of the supply chain (Clancy 2017).

SUMMARY

This chapter started the second half of part three where we look at the supporting functions within Porter's value chain that back the primary functions discussed in the first half. The role of information, technology and management systems in companies' sustainability efforts was covered by firstly providing an overview of and considering the environmental impact of information technology and management systems before then looking at the impact of the environment on these functions and making the business case for greening them. A broad overview of greener information and management systems followed, where we also introduced eco-control as a differentiator of conventional management from environmental management, and revisited the role of life cycle assessment and total quality environmental management in the context of data and information systems.

The ISO 14000 series, in particular the ISO 14001 standard, and EMAS were explored in further detail as specific environmental management systems. These can be complemented by other tools, such as the eco-efficiency portfolio matrix and the sustainability balanced scorecard. Newer developments to help make information (technology) and management systems as well as other corporate functions more sustainable include smart applications, artificial intelligence, big data and machine learning as well as blockchain technology.

Despite the potential of information and management systems to improve environmental performance, Melville (2012) cautions that such systems may be used for greenwashing e.g. by creating a green image through adoption of an EMS when this is not fully backed up by action. Similarly, Buhr and Gray (2012) emphasise that EMSs focus on eco-efficiency and slow improvements, thus helping companies to manage their environmental resources but not whether to increase or decrease their use. Other challenges in this context include the costs for small firms and unsuccessful certification of EMSs, which may be due to a lack of understanding by organisations and their failure to address all activities and to recognise the boundaries of their activities (Schaltegger et al. 2003).

Beyond harvesting low-hanging fruits and implementing environmental management systems, a greater challenge is the integration of sustainability thinking into all processes for long-term change (Fiksel 2012). The adoption of an EMS (whether in-house or accredited) signals the corporate commitment to sustainability (Nemetz 2013).

In a similar line, Grant and colleagues (2017) emphasise that EMSs are a key mechanisms for building awareness and providing training to employees. A good environmental management programme (or system) needs continued awareness of current status and trends in environmental technology and management. A good foundation for this consists of a continuing provision of education and training (Friedman 2011), which we will cover in Chapter 13 on human resource management. Robertson (2017b) also emphasises that continuous improvement is not the result of top-down management but of employees working together at all levels.

An important role for information and management systems is the support of corporate decision-making by bringing together existing information and data around sustainability (Weybrecht 2014). EMSs can help companies to measure, monitor and control pollution, waste and their carbon footprint and other environmental challenges to improve their decision-making. With data and information (e.g. from life cycle assessments informing environmental policies), EMSs set a solid foundation for setting targets and reporting against them. A further key element in EMS and in eco-control is environmental accounting that relies on collecting data, which can then be used for analysis, reporting, strategy and target setting. Sustainability reporting and accounting as well as finance are the subject of Chapter 12, to which we now turn – as information and data are central to reporting, there is a clear and natural link between these two chapters.

SUGGESTED QUESTIONS AND ACTIVITIES

- Look up some of the corporate examples and practices mentioned in the chapter, and research them in more detail.
- Select a company of your choice and browse its webpages to see if you can find examples of green technology practices and how they are being used to improve organisational sustainability.
- Debate: what is the most important element of information technology and management systems when considering sustainability?

- Discuss the likely organisational challenges faced in supporting green technology for sustainability.
- Choose one free sustainability app mentioned in the chapter and play with it. How helpful do you think it is to support your own sustainable behaviour? How could you use it to transform corporate behaviour? Can you find other apps that have recently been developed?
- Evaluate and propose ideas for a corporate sustainability app to improve companies' general corporate environmental performance or the environmental performance of a particular organisation of your choice. What steps could be taken to develop the app?
- For more in-depth material on sustainable and green (information) technology, information systems and (environmental) management systems, check out the following:
  - Daddi, T., Iraldo, F. and Testa F. (2015) *Environmental Certification for Organisations and Products: Management Approaches and Operational Tools*, London: Routledge.
  - Linger, H. Fisher, J., Barnden, A. Barry, C., Lang, M. and Schneider, C. (2013) *Building Sustainable Information Systems*, New York: Springer.
  - Mulvaney, D. (2011) (ed.) *Green Technology: An A-to-Z Guide*, Los Angeles: Sage Publications.

NOTE

1 Guenther and colleagues (2016) emphasise that environmental management systems, environmental management control systems and environmental management accounting are related and overlapping concepts but caution against blurring the boundaries between them. Whilst Henri and Journeault (2010) consider environmental management control systems ("eco-control") to be part of environmental management accounting, Buhr and Gray (2012) equate such environmental management control systems with environmental management systems. There is insufficient space to go into detail here, so the interested reader is referred to the review article of Guenther et al. (2016).

SOURCES AND FURTHER READING

ABB (2019) Sustainability Report 2018, *ABB*, available at https://sustainabilityreport2018.abb.com/servicepages/downloads/files/sustainability_performance_2018_abb_csr18.pdf (Retrieved, 17.07.2019).

Baxter, M. (2011) Environmental Management Systems, in: Brady, J., Ebbage, A. and Lunn, R. (eds.) *Environmental Management in Organizations, The IEMA Handbook*, London: Earthscan/Routledge, 2nd ed., pp. 261–273.

Blowfield, M. (2013) *Business and Sustainability*, Oxford: Oxford University Press.

Bozhinova, K. (2018) 16 Apps Helping Companies and Consumers Prevent Food Waste, *GreenBiz*, 12.10.2018, available at www.greenbiz.com/article/16-apps-helping-companies-and-consumers-prevent-food-waste (Retrieved, 15.07.2019).

Buhr, N. and Gray, R. (2012) Environmental Management, Measurement, and Accounting: Information for Decision and Control? in: Bansal, P. and Hoffman, A. J. (eds.) *The Oxford Handbook of Business and the Natural Environment*, Oxford: Oxford University Press, pp. 425–443.

Carrara, J. (2019) The Best Sustainability Apps, *Eco-Age*, 01.01.2019, available at www.eco-age.com/news/best-sustainability-apps (Retrieved, 15.07.2019).

Cavagnaro, E. and Curiel, G. (2012) *The Three Levels of Sustainability*, Sheffield: Greenleaf Publishing.

Cellan-Jones, R. (2018) Microsoft sinks Data Centre off Orkney, *BBC News*, 06.06.2018, available at www.bbc.co.uk/news/technology-44368813 (Retrieved, 17.07.2019).

Clancy, H. (2013) 20 Sustainability Apps that Made Waves in 2013, *GreenBiz*, 18.12.2013, available at www.greenbiz.com/blog/2013/12/18/20-tools-and-apps-driving-green-business-change (Retrieved, 15.07.2019).

Clancy, H. (2017) The Blockchain's Emerging Role in Sustainability, *GreenBiz*, 06.02.2017, available at www.greenbiz.com/article/blockchains-emerging-role-sustainability (Retrieved, 10.12.2019).

Crawford, D. and Scaletta, T. (2005) The Balanced Scorecard and Corporate Social Responsibility: Aligning Values for Profit, *GreenBiz*, 23.10.2005, available at www.greenbiz.com/news/2005/10/23/balanced-scorecard-and-corporate-social-responsibility-aligning-values-profit (Retrieved, 15.07.2019).

Daddi, T., Iraldo, F. and Testa, F. (2015) *Environmental Certification for Organisations and Products*, London: Routledge.

Daft, R. L., Kendrick, M. and Vershinina, N. (2010) *Management: International Edition*, Andover: Cengage Learning EMEA.

Delmas, M. A. and Kohli, A. (2019) Can Apps Make Air Pollution Visible? Learning About Health Impacts Through Engagement with Air Quality Information, in: *Journal of Business Ethics*, available at https://doi.org/10.1007/s10551-019-04215-7 (Retrieved, 15.07.2019).

Dharmasasmita, A., Erlandsson, L., Willats, J. and Brown, S. (2017) Integrating the Three Pillars of Sustainability: Social, Environmental and Economic, in: Molthan-Hill, P. (ed.) *The Business Student's Guide to Sustainable Management*, Sheffield: Greenleaf Publishing, 2nd ed., pp. 85–128.

Epstein, M. J. (2008) *Making Sustainability Work: Best Practices in Managing and Measuring Corporate Social, Environmental, and Economic Impacts*, Sheffield: Greenleaf Publishing.

Esty, D. C. and Winston, A. S. (2009) *Green to Gold: How Smart Companies Use Environmental Strategy to Innovate, Create Value and Build Competitive Advantage*, Hoboken, NJ: Wiley.

European Commission (2019) Eco-Management and Audit Scheme, webpages, available at http://ec.europa.eu/environment/emas/index_en.htm (Retrieved, 01.07.2019).

Figge, F., Hahn, T., Schaltegger, S. and Wagner, M., (2002) The Sustainability Balanced Scorecard – Linking Sustainability Management to Business Strategy, in: *Business Strategy and the Environment*, 11(5): 269–284.

Fiksel, J. (2012) *Design for Environment: A Guide to Sustainable Product Development*, New York: McGrawHill, 2nd ed.

Friedman, F. B. (2011) *Practical Guide to Environmental Management*, Washington, DC: Environmental Law Institute, 11th ed.

GEMI (1993) Total Quality Environmental Management: The Primer, Global Environmental Management Initiative, available at www.gemi.org/resources/TQE_101.pdf (Retrieved, 26.06.2019).

Goldman, D. (2016) Microsoft Just Put a Data Center Under Water, Cable News Network (CNN), 01.02.2016, available at https://money.cnn.com/2016/02/01/technology/microsoft-underwater-data-center/ (Retrieved, 17.07.2019).

Gond, J.-P., Grubnic, S., Herzig, C. and Moon, J. (2012) Configuring Management Control Systems: Theorizing the Integration of Strategy and Sustainability, in: *Management Accounting Research*, 23(3): 205–223.

Grant, D. B., Trautrims, A. and Wong, C. Y. (2017) *Sustainable Logistics and Supply Chain Management: Principles and Practices for Sustainable Operations and Management*, London: Kogan, 2nd ed.

Griffin, A. (2019) How Using the Internet Damages the Environment, *The Independent*, 12.07.2019, available at www.independent.co.uk/environment/internet-environment-data-climate-change-global-warming-servers-centre-a8968296.html (Retrieved, 23.07.2019).

Guenther, E., Endrikat, J. and Guenther, T. W. (2016) Environmental Management Control Systems: A Conceptualization and a Review of the Empirical Evidence, in: *Journal of Cleaner Production*, 136: 147–171.

Hart, S. L. (2010) *Capitalism at the Crossroads: Next Generation Business Strategies for a Post-Crisis World*, Upper Saddle River, NJ: Pearson, 3rd ed.

Henri, J. and Journeault, M. (2010) Eco-control: The Influence of Management Control Systems on Environmental and Economic Performance, in: *Accounting, Organizations and Society*, 35(1): 63–80.

Hoffman, A. J. and Bansal, P. (2012) Retrospective, Perspective, and Prospective: Introduction to the Oxford Handbook of Business and the Natural Environment, in: Bansal, P. and Hoffman, A. J. (eds.) *The Oxford Handbook of Business and the Natural Environment*, Oxford: Oxford University Press, pp. 3–25.

IEA (2009) *Gadgets and Gigawatts. Policies for Energy Efficient Electronics*, International Energy Agency, Paris.

IEA (2017a) *Energy Efficiency Indicators, Highlights 2017*, International Energy Agency, Paris.

IEA (2017b) *Digitalization and Energy*, International Energy Agency, Paris.

IEA (2018) *Key World Energy Statistics 2018*, International Energy Agency, Paris.

ISO (2019) ISO 14000 Family – Environmental Management, International Standards Organisation, webpages, available at www.iso.org/

iso-14001-environmental-management.html (Retrieved, 27.06.2019).

Kaplan, R. S. and Norton, D. S. (1996) Using the Balanced Scorecard as a Strategic Management System, in: *Harvard Business Review*, 74(1): 75–85.

Knight, D. (2006) The SIGMA Management Model, in: Jonker, J. and de Witte, M. (eds.) *Management Models for Corporate Social Responsibility*, Berlin: Springer, pp. 11–18.

Leipziger, D. (2003) *The Corporate Responsibility Code Book*, London: Routledge.

Linger, H., Fisher, J., Barnden, A. Barry, C., Lang, M. and Schneider, C. (2013) *Building Sustainable Information Systems*, New York: Springer.

Liu, J. (2011) Information Technology, in: Mulvaney, D. (ed.) *Green Technology: An A-to-Z Guide*, Los Angeles: Sage Publications, pp. 255–260.

Lueg, R. and Radlach, R. (2016) Managing Sustainable Development with Management Control Systems: A Literature Review, in: *European Management Journal*, 34(1): 158–171.

Marsden, J. (2011) Auditing, in: Brady, J., Ebbage, A. and Lunn, R. (eds.) *Environmental Management in Organizations, The IEMA Handbook*, London: Earthscan/Routledge, 2nd ed., pp. 399–417.

Meadowcroft, J. (2018) Green Technology, in: Castree, N., Hulme, M. and Proctor, J. D. (eds.) *Companion to Environmental Studies*, London: Routledge, pp. 694–697.

Melville, N. P. (2012) Information Systems, Business, and the Natural Environment: Can Digital Business Transform Environmental Sustainability? in: Bansal, P. and Hoffman, A. J. (eds.) *The Oxford Handbook of Business and the Natural Environment*, Oxford: Oxford University Press, pp. 327–344.

Nemetz, P. N. (2013) *Business and the Sustainability Challenge: An Integrated Perspective*, London: Routledge.

NPR (2019) How Virtual Reality Can Help People Better Understand Climate Change, *National Public Radio*, 18.11.2019, available at www.npr.org/transcripts/780373013 (Retrieved, 29.11.2019).

Piasecki, B. W., Fletcher, K. A. and Mendelson, F. J. (1999) *Environmental Management and Business Strategy: Leadership Skills for the 21st Century*, New York: John Wiley & Sons.

Pondeville, S., Swaen, V. and De Rongé, Y. (2013) Environmental Management Control Systems: The Role of Contextual and Strategic Factors, in: *Management Accounting Research*, 24(4): 317–332.

Reuters (2011) Thai Floods Batter Global Electronics, Auto Supply Chains, *Reuters World News*, 28.10.2011, available at www.reuters.com/article/us-thai-floods/thai-floods-batter-global-electronics-auto-supply-chains-idUSTRE79R0QR20111028 (Retrieved, 20.06.2019).

Riffle, C. (2017) What Artificial Intelligence Means for Sustainability, *GreenBiz*, 19.07.2017, available at www.greenbiz.com/article/what-artificial-intelligence-means-sustainability (Retrieved, 17.07.2019).

Roach, J. (2018) Under the Sea, Microsoft Tests a Datacenter that's Quick to Deploy, could Provide Internet Connectivity for Years, *Microsoft News*, 05.06.2018, available at https://news.microsoft.com/features/under-the-sea-microsoft-tests-a-datacenter-thats-quick-to-deploy-could-provide-internet-connectivity-for-years/ (Retrieved, 17.07.2019).

Robertson, M. (2017a) *Dictionary of Sustainability*, London: Routledge.

Robertson, M. (2017b) *Sustainability Principles and Practice*, London: Routledge, 2nd ed.

Schaltegger, S., Burritt, R. and Peterson, H. (2003) *An Introduction to Corporate Environmental Management*, Sheffield: Greenleaf Publishing.

Schaltegger, S., Burritt, R., Zvezdov, D., Hörisch, J. and Tingez-Holyoak, J. (2015) Management Roles and Sustainability Information. Exploring Corporate Practice, in: *Australian Accounting Review*, 75(25): 328–345.

Schaltegger, S. and Sturm, A. (1998) *Eco-efficiency through Eco-control*, Zürich: VDF.

Schmelzer, R. (2019) AI and Climate Change: The Mixed Impact of Machine Learning, *SearchEnterpriseAI*, 15.11.2019, available at https://searchenterpriseai.techtarget.com/feature/AI-and-climate-change-The-mixed-impact-of-machine-learning (Retrieved, 29.11.2019).

Sheldon, C. (2011) Getting Beyond Integrated Thinking and Governance, in: Brady, J., Ebbage, A. and Lunn, R. (eds.) *Environmental Management in Organizations, The IEMA Handbook*, London: Earthscan/Routledge, 2nd ed., pp. 533–558.

Simons, R. (1995) *Levers of Control: How Managers use Innovative Control Systems to Drive Strategic Renewal*, Boston: Harvard Business School Press.

vom Brocke, J., Watson, R. T., Dwyer, C., Elliot, S. and Melville, N. (2013) Green Information Systems: Directives for the IS Discipline, in: *Communications of the Association for Information Systems (CAIS)*, 33(30): 509–520.

Weybrecht, G. (2014) *The Sustainable MBA: A Business Guide to Sustainability*, Chichester: Wiley, 2nd ed.

CHAPTER 12

Accounting
Sustainability reporting, accounting and finance

INTRODUCTION

The previous chapter on information technology and management systems focused on the role of information and how it is used in moving companies towards greater environmental sustainability. This chapter follows on from that by looking at how accounting provides and makes use of information internally to aid decision-making – what is generally known as *management accounting* – and externally to report on corporate activities to stakeholders – generally known as *financial accounting* (or cost accounting) and involving financial information. When moving accounting into a sustainability context, such information and reporting extends from the financial to the non-financial, from the monetary to the physical. In the realm of management accounting, accounting supports environmental management, also known as environmental management accounting, whereas in the realm of financial accounting we move to environmental reporting, or external environmental accounting (Gray 2011; Schaltegger et al. 2003).

When introducing environmental sustainability to this corporate function, conventional accounting faces a challenge as it often uses narrow (financial) performance measures, and sustainability involves interconnected events for which companies are variously liable or by which they are affected (Blowfield 2013). Sustainability accounting and reporting can help to replace business-as-usual attitudes with a more balanced focus on the triple bottom line and not just on profitability as a success measure for business (Lodhia 2013). As many corporate initiatives never take off without accounting's involvement, its role in greening an organisation is vital (Weybrecht 2014). Robertson (2017: 308) also emphasises that measurement – a key purpose of accounting – is "the key to getting sustainability work done". Similarly, a company's finance department can drive responsible and green business (Laasch and Conaway 2015).

Accounting is very much related to accountability and discharging it, as it can help "make things account-*able*" whilst also providing mechanisms of control (Brown and Fraser 2006: 107). Accountability is the principle of providing information to society, in particular information that society has a right to know (Gray and Milne 2004). It can also be defined as the readiness of an organisation to explain and justify its actions and decisions as well as omissions (Laasch and Conaway 2015). Accounting and reporting is then about discharging that accountability duty. We can see the link here to the previous two chapters – where Chapter 11 very much dealt with information and how it is used for environmental management in a systematic way, whilst in Chapter 10 we looked at marketing and communication. Here in the chapter on accounting and reporting, we focus on information that can be used for internal decision-making and on information that is used to report on a company's environmental performance to external stakeholders.

Whilst we emphasise accounting and reporting in the chapter, we include finance where relevant. As accounting and reporting as well as finance generally, and more specifically in environmental sustainability terms, are complex and far-reaching topics, the chapter can only highlight some of the tools, concepts and standards used in environmental accounting. Going into greater detail is beyond the scope of this chapter, so the interested reader is pointed to other sources.

In the following, we provide a brief outline of accounting, reporting and finance before introducing the business case for environmental accounting. We proceed to discuss environmental accounting (or accounting for sustainability) and then introduce a number of standards, tools and concepts that companies can apply to accounting and finance to make their activities more sustainable.

Although sustainability accounting and reporting concerns environmental, social and economic activities i.e. the triple bottom line, we use the terms interchangeably with environmental accounting and reporting on the understanding that as with the rest of the book we focus on environmental sustainability. This does of course not diminish the importance of social accounting and reporting, which takes into account social impacts using tools such as for example Social Return on Investment.

ACCOUNTING, REPORTING AND FINANCE

Overview

The key role of accounting and accountants is in the measurement, provision and communication of information that is used in decision-making, both internally and externally (Weybrecht 2014). Inside companies, the accounting function broadly fulfils three roles – support for strategic decisions, control of costs and cash flows, and reporting required information (Piasecki et al. 1999).

Within accounting, we can differentiate between financial accounting and management accounting. Broadly speaking, management accounting is the financial system that

> identifies, tracks and records the detailed internal costs of the organisation and . . . offers management a range of control and planning alternatives through processes such as budgeting performance appraisal and investment appraisal.
>
> (Gray 2011: 435)

On the other hand, financial accounting serves to generate financial information as a basis for stakeholder relationships, founded on transparency and accountability. The information pertains to a company's financial position and performance, and changes to them (Schaltegger et al. 2003). As a key role of accounting is in the provision of information, there is also an element of choice – subjective and objective – of what kind of information is provided.

Both accounting and auditing are subject to financial reporting standards, as many organisations regularly need to issue some form of financial statements, which include a profit and loss account (also called income statement) and a balance sheet, normally published as part of an annual report (Gray 2011). These financial statements, issued annually (or quarterly), are subject to a process of auditing or assurance, which serves to verify accuracy and credibility (Weybrecht

2014). Such external independent verification and endorsement of accounting processes and outcomes ensure reliability and compliance with rules and regulations, in turn leading to environmental improvements (Laasch and Conaway 2015).

In a broader context, finance provides "grease to the economy" (Scholtens 2006: 19). Similarly, in a corporate context, financial management and finances are central to any company and its processes. Financial management concerns how an organisation's financial activities are planned, organised, directed, controlled and governed. It generally serves three functions – financing (the external procurement of funds), budgeting (internal investment decisions), and distribution of financial results to shareholder (pay-outs or dividends) (Laasch and Conaway 2015). Finance is closely interlinked with corporate governance, which involves making sure that managers lead companies in the best interest of owners.

Environmental impact of accounting and reporting

In itself, the accounting function could be argued to have a small direct environmental impact, compared to for example supply chain management and production. Its environmental impact is through the use of resources, for example for paper-based reports, and the use of energy for data analysis and report production.

Nevertheless, it has an indirect environmental impact: through its emphasis on the economic bottom line and the maximisation of profit, accounting – as it has been conventionally carried out – has mostly ignored the environmental impacts of organisations and could thus be said to have encouraged unsustainable operations at worst, or hiding these at best (Bebbington et al. 2014) – or provided distorted information to managers on which to base their decisions (Burritt and Schaltegger 2010). Financial management has similarly been perceived as driving unsustainable short-term and profit-driven corporate behaviour (Laasch and Conaway 2015).

Furthermore, as accounting has a key involvement in the two issues of organisational decision-making and external reporting, it makes this function particularly well-placed to ensure the integration of sustainability in everyday management (Weybrecht 2014).

Impact of environment on accounting and reporting

Traditionally, companies have not been required to acknowledge environmental issues in their financial statements as there is no intrinsic interest in environmental matters as such (Gray 2011). Environmental issues are only recognised if there is a material financial impact on an organisation. A material impact can be in the form of impaired assets and environmental liabilities, for example as a result of breaching legal or regulatory compliance. Another area where environmental topics impact on accounting and reporting is in the trading and offsets from carbon accounting, which attaches monetary values to carbon emissions, the practice of carbon pricing.

As companies increasingly recognise the risk from the climate emergency, this has had an influence on the accounting function as data collection is no longer limited to financial information but also non-financial information. Financial institutions, including central banks, and financial market participants such as investors are becoming more concerned about the resilience of the financial system in light of environmental challenges (Scholtens 2017). Particularly insurance companies, threatened by weather-related losses, were early voices in calling for corporate action on climate change and are in increasing numbers pulling out of fossil fuel investments (mostly coal) due to financial risks as well as refusing to provide insurance to new fossil fuel projects.

Business case for sustainability accounting and reporting, and sustainable finance

The preceding two sections have indicated the need for integrating environmental sustainability considerations into accounting and reporting practices. In addition, there are further drivers for environmental accounting.

Reporting on environmental performance can lead to costs savings and productivity improvements for companies on the basis of better resource use and management, waste reduction, energy and transport management, risk management and lower insurance premiums (Newbold 2011). Compliance is also supported by reporting as it makes breaching regulations and costly fines less likely. Environmental accounting as the systematic analysis of environmental costs provides benefits in the form of sound business decision-making (Piasecki et al. 1999). In a similar line, environmental metrics in the form of data and indicators are key to fact-based decision-making and good environmental management as they drive continuous improvement and allow companies to measure progress (Esty and Winston 2009).

Reporting accurately and fully about environmental and social performance is a foundation for stakeholder engagement and social legitimacy (Stead and Stead 2014). Brown and Fraser (2006) also emphasise that environmental reporting can help manage threats to organisational legitimacy and maintain (or protect) the license to operate. In addition, sustainability reporting can enhance corporate reputation and increase credibility (Fogel 2016). Beyond stakeholder communication, measuring and reporting shows progress in environmental performance and helps motivate employees towards improving such performance further (Robertson 2017). It can also improve brand loyalty (Weybrecht 2014) and builds trust with all stakeholders (Esty and Winston 2009).

One particular stakeholder community for which environmental reporting is important is the investment community – according to Newbold (2011), such reporting makes companies more attractive as environmental risks can influence investment decisions. In most financial models, such environmental risks are not included and the financial system itself is at risk if it continues to ignore these, in particular climate change (Shrivastava et al. 2019). Institutional investors, in particular, have been pushing for stronger environmental performance and governance, forcing firms to rethink (Dycket al. 2019).

As emphasised throughout the book, sustainability necessitates taking long-term perspectives, for which traditional financial accounting has little to offer as it was not designed with environmental data in mind (Stead and Stead 2014; Weybrecht 2014). In response, full-cost accounting systems have been developed which can account for both the short-term and the long-term, taking into account not just economic costs but also the social and environmental costs of business operations.

Other business benefits from accounting and reporting for sustainability that can drive companies towards such practices include value creation, risk awareness, innovation and continuous improvement as well as attracting long-term capital (Brown and Fraser 2006, drawing on WBCSD 2003, see also WBCSD 2017).

ENVIRONMENTAL ACCOUNTING AND REPORTING – ACCOUNTING FOR SUSTAINABILITY

Overview

Broadly speaking, environmental accounting applies the established tools of accounting (such as information management, analysis and communication) to environmental management

(Schaltegger et al. 2003). In order to be able to integrate sustainability into corporate strategy, relevant environmental data needs to be in a form that is recognised as well as usable by key corporate decision-makers (Nemetz 2013).

The field of environmental accounting and reporting is a complex one, with different strands and terminologies (Gray 2011). The 1970s saw a rise in social reporting, particularly in Europe, generally seen as the first instances of non-financial corporate reporting. Environmental reporting gained importance in the 1980s and 1990s, especially in more polluting industries, whereas the 2000s saw a move towards sustainability reporting in an effort to integrate both environmental and social issues in corporate reporting (Herzig and Ghosh 2017). Japan has played a significant role in promoting environmental auditing, accounting and reporting through issuing guidelines and advice on quantifying environmental costs and benefits (Visser 2012). Other governments have played a role in environmental reporting by making it compulsory through new laws, though most environmental reporting is still voluntary (Herzig and Ghosh 2017). Increasingly, global standards are being developed for companies to use.

These historical developments are reflected in the various terms used for non-financial reports: depending on industry and company context, names range from corporate responsibility, corporate social responsibility, and social report to corporate citizenship, sustainability or environmental report, and more recently integrated reports and extended external reports. Gray and Herremans (2012) argue that the distinctions between the various types of accounting and reporting – triple bottom line reporting, sustainability reporting, environmental and social reporting are not necessarily helpful. Here we focus on environmental accounting and reporting but acknowledge that the terms sustainability and TBL reporting can apply to our context, too, and thus these are used as well.

As conventional accounting differentiates between (internal) management accounting and (external) financial accounting, we can differentiate between environmental management accounting and external environmental accounting (Schaltegger et al. 2003). Both types of environmental accounting record monetary as well as physical information, a crucial difference from conventional accounting. For Gray (2011), the equivalent of financial accounting is about the environmental impact on financial statements and auditing, whilst the equivalent of management accounting is accounting in support of environmental management (Gray 2011). Burritt and Schaltegger (2010) differentiate between the *outside-in approach* of sustainability accounting, which includes stakeholder dialogue and producing information for external parties, and the *inside-out approach* focused on internal decision-making for strategy implementation on the basis of measuring, managing and reporting sustainability performance.

In environmental management accounting, the financial as well as non-financial information can be used by management for decision-making and control (Buhr and Gray 2012; Guenther et al. 2016). It thus addresses a shortcoming of conventional accounting where environment-related costs can often not be found in records, or information is hidden in overhead accounts (ibid.). For Schaltegger and colleagues (2003), environmental accounting is the most important approach to managing environmental information.

The communication to external stakeholders is in the form of environmental reporting (or triple bottom line or sustainability reporting). We can think of environmental reporting as

> communicating an accurate, although much simplified, overview of the organization's environmental interactions to the stakeholders of the organization through the annual report, in a stand-alone report, on the company's website or by a combination of these methods
>
> (Gray 2011: 437)

Historically, such reports have been stand-alone ones, but increasingly they are published together with annual reports – integrated reporting will be further detailed in the following section.

Environmental reporting provides a window of corporate performance to the public at a particular point in time (Piasecki et al. 1999). As this requires information that is credible and useful, this encourages the development of environmental performance metrics internally. This in turn can drive environmental performance (ibid.). A good environmental report should include key aspects of a company's environmental footprint, using quantitative metrics and covering core issues which will differ from company to company (Esty and Winston 2009).

Whilst traditional accounting is focused on the financial flows of an organisation, primarily of interest to shareholders and managers, sustainability accounting also takes into account external flows and impacts – whether positive or negative – that concern other stakeholders (Richardson 2004). External impacts are also known as externalities, which were introduced in Chapters 2 and 3. Examples of external costs and benefits include the costs or benefits for consumers in using and disposing of products, the environmental impacts linked to producing purchased goods and services in terms of suppliers, environmental risks or benefits linked to the workplace for employees, and the emissions, effluents and waste to air, land and water for the community as a stakeholder, as well as the risks to investors if a company has a poor reputation for its environmental management (Richardson 2004). Accountants can play a key role with regard to environmental performance as they can help organisations understand what the true environmental costs are. In the growing field of true cost or full-cost accounting, such external environmental costs are internalised and are included in company reports (Blowfield 2013). It is the basic idea behind environmental management accounting (Lodhia 2013).

Visser (2012) points out that accounting for physical impacts is easier as well as less risky and controversial than financially quantifying external costs – this has resulted in measuring and reporting that is a perhaps more superficial representation of the environmental impacts of business compared to e.g. full-cost accounting. Monetising key environmental values sends clear economic signals to internal decision-makers and external stakeholders (Nemetz 2013). Nevertheless, reliable and effective quantification and monetisation of externalities that are commensurate (comparable) and enable financial internalisation remain a challenge (Unerman et al. 2018).

Environmental management accounting techniques can fall into three broad categories: operational and capital budgeting, costing and measuring, and monitoring (Buhr and Gray 2012). In budgeting, the environment often fails to feature, but can be incorporated through, environmental criteria. There are a number of techniques for costing and measuring, including activity-based costing, quality costing, product/service costing, physical flow/eco-balance analysis and eco-efficiency indicators. Other techniques include full-cost accounting, life cycle (cradle to grave) costing and cost-benefit analysis (ibid.). For monitoring, environmental auditing can be used to assess environmental performance and a sustainability balanced scorecard for internal reporting.

Lamberton (2005, based on Gray 1993) identifies sustainable cost, natural capital inventory accounting, and input-output analysis as methods of sustainability accounting, to which he adds full-cost accounting and triple bottom line (TBL) accounting. Fogel (2016) establishes three types of information that are key to nonfinancial reports: ESG (environmental, social and governance), intangible assets and key performance indicators (KPI). We return to some of these later in the chapter.

Auditing for sustainable management involves the evaluation of a company's economic, social and environmental performance which provides useful data for a firm wishing to close its sustainability performance gaps (Stead and Stead 2014). Environmental audits began as simple management tools but have evolved to detailed procedures to check compliance and examine management systems (Friedman 2011).

Environmental auditing comprises a range of investigative and checking activities against established criteria on the basis of a common methodology and other protocols, resulting in multiple methodologies. It serves to support decision-making and management processes with regard to compliance, effectiveness and environmental performance (Marsden 2011). Environmental auditing plays a key role in environmental management and can be done internally or externally. Internal auditing is particularly used to check whether environmental management systems (EMS) meet certain criteria, such as the ISO 14001 specifications or the EMAS (Eco-Management and Audit Scheme) requirements, and provides internal (management) assurance. There is thus a close connection between EMS and auditing, highlighting the link between this and the previous chapter. In contrast, audits intended for external consumption (also called second- and third-party audits) provide external (stakeholder) assurance and can include certification[1] (Marsden 2011). Whilst audits are usually focused on compliance – particularly in a conventional accounting context – they should go beyond this in a sustainability context in order to provide information about operational effectiveness and current best practice to inform managers strategically for better environmental management (Weybrecht 2014).

For investors, ESG issues (environment, social, governance) can strongly influence such intangible assets as reputation and brand, and thus feed into the total market value of listed companies (Weybrecht 2014). As shareholders become more attuned to the potential financial risks of environmental issues, companies can send signals to investors if they are included in ratings, indices or rankings that assess organisations not only by their financial performance but also by their social and environmental one, such as the Dow Jones Sustainability Index (DJSI) or FTSE4Good (Laasch and Conaway 2015). Similarly, sustainable investors send signals to companies by ruling out certain unsustainable investments ("negative screening") which can lead to divestment, or by deliberately making sustainable investment decisions using positive and norms-based screening (Weybrecht 2014). Such financing and investment decisions can be made on the basis of the environmental reports issued by companies.

It is important to note than an underlying assumption in the rise of environmental accounting is that providing relevant information to stakeholders can alter corporate impacts on the environment (Lamberton 2005). This may, however, not necessarily be the case, and often sustainability reporting is used to camouflage poor sustainability practices. It is therefore imperative to briefly shed a more critical light on environmental accounting practices, particularly where there is a so-called reporting-performance portrayal gap, i.e. a disconnect between performance and disclosure (Herzig and Ghosh 2017, based on Adams 2004). This makes companies susceptible to accusations of greenwashing, which we covered in Chapter 10 on marketing and communication. Similarly, Lamberton (2005) acknowledges that the reporting of sustainability accounting information is vulnerable to being manipulated by vested interests.

In order to counter such developments and to challenge corporate reporting that falls below stakeholder expectations, shadow reporting (or counter-accounting, Gray and Herremans 2012) has been put forward. Shadow accounting uses independent external sources (e.g. newspaper articles, worker testaments, NGO reports) to highlight contradictions between disclosure and actual performance (Herzig and Ghosh 2017). It can be seen as a supplement to company reports or even as an alternative (Weybrecht 2014). Such adversarial accounting and externally prepared "anti-reports" can challenge the status quo and give voice to alternative worldviews (Brown and Fraser 2006). It is also broadly known as external accounting, which comprises silent, shadow and counter accounting, recasting figures to provide alternative accounts.

Whilst the number and the quality of environmental reports has generally improved, there are still concerns about whether such reports simply focus on measuring internal performance (e.g. amount of emissions or waste) and leave out substantial external impacts of activities that are more material to stakeholders (Herzig and Ghosh 2017).

The visual imagery of reports needs to be mentioned in this context, as the power and ubiquity of pictorial and graphic forms are used by accounting for impression management, and by companies in their communication more generally to produce desired messages about their performance (Davison 2015; Greenwood et al. 2019). Hrasky (2012) found that less sustainable companies use powerful visual rhetoric, particularly photographs, to gain pragmatic legitimacy through symbolism rather than action, and are less likely to use quantitative data and graphs that more accurately reflect operations and their impacts. This is in contrast to more sustainable companies which use a combination of graphs and photographs to gain moral rather than only symbolic legitimacy. The interplay between images, numbers and the narrative thus influences what kind of sustainability impression is created.

There is also a tension between reporting guidelines used as a legitimating exercise and showing accountability for a broader range of stakeholders including their views and needs (Herzig and Ghosh 2017). Similarly, Gray (2011) acknowledges the tension that sits between the voluntary nature of most environmental reporting and the various audience needs for reports that are not the result of greenwashing and PR exercises. This often results in reports that are honest and substantial but also selective and partial.

To counter this companies can use a number of standards and tools that have been developed for use in environmental accounting and reporting.

A selection of standards, tools, concepts, principles and indicators

Introduction

As sustainability is a complex and multi-dimensional challenge for companies, choosing what to account for and report on is not easy, particularly in light of various environmental challenges, whether it is climate change, pollution or waste (or any other challenge not covered in the book, see also introduction to part two). It is important to not only measure and track data that are readily available and avoid those that are not easily measurable, but to ensure indicators are relevant to a firm's context and chosen sustainability goals, thus setting an appropriate framework[2] (Robertson 2017). As Piasecki and colleagues (1999: 67) put it:

> Selectivity is the biggest problem in the use of performance indicators.

The authors call for realistic cautions and a qualifying statement to be attached to any measurement results of corporate environmental performance in order to support corporate decision-making processes. In addition, Weybrecht (2014) argues that companies should choose metrics that drive performance rather than simply measure it, and use metrics that are forward-looking, in line with a company's vision. Esty and Winston (2009) suggest three broad guidelines for collecting environmental data – tracking relative and absolute metrics, capturing data at multiple levels within an organisation, and collecting the same information across the supply chain. As firms are looking to integrate environmental strategies, they might need unique approaches to measurement which combine financial and non-financial data (Fogel 2016, see also Maas et al. 2016a). Maas and colleagues (2016b) also stress that there is no one-size-fits-all approach, and firms need to be creative as well as strategic and targeted in integrating tools of accounting and reporting in order to become transition leaders towards sustainability.

Regardless of approach, if companies adhere to the principle of *materiality*, they can address the issue of selectivity and the challenge of facing a choice of too many variables to report on, as no company can be expected to report on all sustainability issues (Weybrecht 2014). Materiality reflects the shared importance and relevance of a particular issue to both the company and stakeholders, and how this is identified and prioritised (Laasch and Conaway 2015; AccountAbility 2019). In order to determine material issues, environmental impacts (direct or indirect) can be ranked according to influence of (or interest of) and importance for stakeholders. Dialogue and engagement with stakeholders is crucial for determining which sustainability issues are material to companies (GRI 2019); the result of such stakeholder inclusion can then be mapped for example in a materiality matrix or calculated in a materiality score.[3]

There are numerous efforts and initiatives that use accounting, auditing and reporting tools, techniques and formats to incorporate sustainability into corporate decision-making and external communication. We can but offer only a snapshot in the following in terms of the standards and tools available. It is useful here to distinguish amongst different approaches to standards. Oakley and Buckland (2004) classify corporate sustainability standards based on principles, performance and process, and provide examples for all, some of which are covered in the following:

- *Principle-based standards* set out broader principles of behaviour: UN Global Compact, Caux Roundtable Principles for Business, OECD Guidelines for Multinational Enterprises.
- *Process-based standards* outline processes companies can follow to improve their performance: ISO 14000, ISO 9000.
- *Performance standards* help measure actual achievements of companies using indicators: Global Reporting Initiative (GRI), SA8000.[4]
- *Hybrid standards* combine elements of the approaches mentioned earlier. Examples include FTSE4Good and SIGMA, which was covered in Chapter 11.

Oakley and Buckland (2004) stress as well that a single approach is not sufficient and that a combination is needed that provides a framework of principles, practicable guidance on what to do and making performance assessable. With this in mind, we now turn to briefly outlining some of the most used and relevant tools and standards, taking into account that these instruments and guidelines differ both quantitatively and qualitatively in how sustainability performance is assessed, whether it is through outputs, processes or impacts (Maas et al. 2016a).

The global reporting initiative

The Global Reporting Initiative (GRI) is behind one of the world's most widely used frameworks for sustainability reporting (Fogel 2016). The framework was and is developed through a multi-stakeholder process and sets out principles and performance indicators for measuring and reporting a company's economic, social and environmental performances i.e. the triple bottom line. It defines the *content* of a sustainability report and prescribes a *standard quality* for sustainability reporting (Herzig and Ghosh 2017).

The initiative was formed in 1997 by the Coalition for Environmentally Responsible Economies (CERES) and the United Nations Environmental Program (UNEP), and issued its first guidelines (G1) in 2000. Second and third generations of the guidelines (G2 and G3) followed in the 2000s, whilst the most recent guidelines (G4) were issued in 2013, which were subsequently developed and refined into the first global standard for sustainability reporting in 2016 (GRI 2019). The GRI standards as issued by the independent standard-setting body Global

TABLE 12.1

GRI reporting principles

Reporting principles for defining report *content*	Brief explanation
Stakeholder inclusiveness	Stakeholders should be identified and the reporting entity should explain its response to stakeholders' expectations and interests.
Sustainability context	This refers to the wider context of the organisation in terms of industry or sector, and local, regional or global level trends in sustainable development.
Materiality	The shared importance of a particular topic to both company and stakeholders. Reports should reflect significant (environmental and other sustainability) impacts for such topics and how these influence decisions and assessments by stakeholders. Materiality maps can help visualise and rank impacts by importance and influence.
Completeness	Reports should cover material topics within the reporting timeframe and for the agreed entity according to topic boundaries
Reporting principles for defining report *quality*	**Brief explanation**
Accuracy	The principle looks for detail correctness, ensuring a low margin of error in the reporting information.
Balance	Both positive and negative aspects should be reflected to allow a reasoned assessment of performance by stakeholders.
Clarity	Information in the report should be accessible and easy to understand.
Comparability	Comparisons over time and across different companies should be made possible by using consistent data and information in reports.
Reliability	The information that is collected, recorded, compiled, analysed and reported should be examined in order to ascertain quality and materiality.
Timeliness	Reports should be published regularly to allow stakeholders to make informed decisions.

Sources: GRI (2019), GRI (2016), Laasch and Conaway (2015).

Sustainability Standards Board (GSSB) are modular, divided into universal standards (foundation, general disclosures and management approach) and topic-specific standards, organised into economic, environmental and social series. Companies (and other organisations) choose and report only according to relevant topic-specific standards that are material to them.

The principles the GRI standards set out in order to ensure high-quality sustainability reporting cover both report content and report quality, which are detailed in Table 12.1.

GRI has been criticised in various respects; for example its ability to promote sustainability has been questioned in terms of allowing window dressing, where reporting might hide issues, whilst other topics might be under-reported and over-claimed. Other criticisms include the lack of audits and external verification for many reports and the subsequent consequences for internal management (particularly as report preparation takes up many resources), as well as a lack of adoption by the investment community (see e.g. Fogel 2016). Nevertheless, GRI has led do a remarkable uptake of sustainability reporting, unthinkable 20 years ago (see e.g. Buhr et al. 2014).

Beyond the reporting standards, the GRI organisation also provides guidance on how they link to other norms, such as for example the EU directive on non-financial reporting or ISO standards. It is part of the Corporate Reporting Dialogue which brings together major reporting frameworks to work towards transparency and accountability.[5] One organisation it works closely with is the International Integrated Reporting Council (IIRC).

Integrated Reporting

The International Integrated Reporting Council, which provides guidelines for integrated reporting, was co-founded in 2010 by a group of international partners, such as regulators, NGOs, investors, companies and standard setters, such as the GRI and the Prince's Accounting for Sustainability Project (Herzig and Ghosh 2017; IIRC 2019). As such, it adopts a stakeholder approach that focuses on integrated thinking about value creation over time that also takes in the context of the external environment of a company on the basis of its strategy, governance, performance and prospects (IIRC 2013).

Whilst sustainability reporting has grown phenomenally in recent years, this development has mostly been isolated from financial reporting developments. Integrated reporting addresses this by bringing together material information of a financial and non-financial nature to enable stakeholders to understand the underlying health of a company (Lodhia 2013). It is centred around

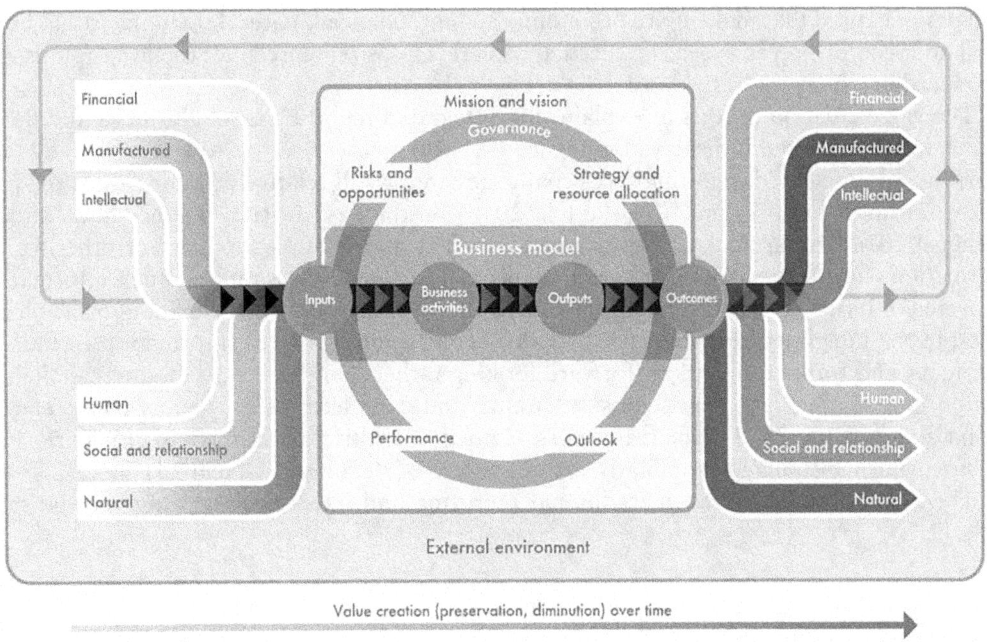

FIGURE 12.1
The value creation process

Source: IIRC, used with permission.

TABLE 12.2

Key reporting differences

	From traditional approach	To integrated approach
Thinking	Isolated, in silos	Integrated, full complexity of value creation process
Stewardship	Financial capital	All forms of capital and their interdependent contribution to success
Focus	Past, financial, annual	Past and future, connected, strategic
Timeframe	Short term	Short, medium and long term
Trust	Narrow disclosures	Greater transparency, broader disclosure of positive and negative
Adaptive	Rule bound, compliance	Responsive to individual circumstances, principles-based, material factors
Concise	Long and complex	Concise and material
Technology enabled	Paper based	Technology enabled

Sources: IIRC (2011).

six types of capital that make up the economic system: financial, natural, manufactured, human, social and relationship as well as intellectual capital. This is reminiscent of the five capitals model introduced in Chapter 2 of the book.

The value creation process as explained in integrated reporting through stocks and flows of various forms of capital is displayed in Figure 12.1.

Although integrated reporting faces challenges and is still relatively young, it has the potential to transform business practices (Lodhia 2013) on the basis of transparency and connectivity (Fogel 2016). In particular, an integrated report is more than a summary of other forms of information (such as a financial or sustainability report) but focuses on how such information is connected in terms of value creation. Similar to the GRI standard, other guiding principles that underpin the integrated reporting framework beyond connectivity of information include strategic focus and future orientation (forward looking rather than backward looking), stakeholder relationships, materiality, conciseness, reliability and completeness as well as consistency and comparability (IIRC 2013). The framework is not prescriptive about methods for performance measurement, which may be qualitative or quantitative (Herzig and Ghosh 2017).

The key differences between traditional reporting and integrated reporting can be seen in Table 12.2.

Auditing and assurance: ISAE 3000 and AA1000AS

As the transparency, reliability and credibility of environmental (or integrated) reports are important to stakeholders (whether internal or external), assurance and auditing can be used to enhance these characteristics. Whilst this can be done internally, external assurance provides more independence and objectivity (Rezaee 2015). Although efforts are made on the environmental reporting front, there is no agreed international set of principles or standard for assuring and verifying such non-financial reports, though the accounting standards ISAE 3000 and 3410 offer a starting point alongside AA1000AS, a multi-stakeholder initiative (Weybrecht 2014).

The International Standards on Assurance Engagements[6] (ISAE) 3000 is an international assurance standard for non-financial information, issued by the International Auditing and Assurance Standards Board (IAASB), which is the issuing agency of the International Federation of Accountants (IFAC) (Laasch and Conaway 2015). It is more generic in nature as it is not limited to sustainability, as it also covers outsourcing audits and compliance. ISAE 3410 concerns carbon emissions in particular. ISAE 3000 provides detailed guidance and is considered the most influential of accountancy-based standards (Cooper and Owen 2014). ISAE 3000 is more concerned with data quality and integrity as well as procedures around reporting, controlling and evidence gathering (ISAE 3000 2019).

In contrast, the AccountAbility[7] series of standards holds stakeholder engagement central, according to the principles of inclusivity (or completeness), materiality and responsiveness (Cooper and Owen 2014; ISAE 3000 2019). In particular, the AA1000AS (2008) assurance standard requires auditors to asses adherence to principles and the quality of information in disclosing sustainability performance. AccountAbility issued stakeholder engagement standards in 2015 (AA1000SES) and updated its assurance standards in 2018 according to its AccountAbility Principles (AA1000AP), which give companies an internationally accepted framework for identifying, prioritising and responding to sustainability challenges and improvement of long-term performance (AccountAbility 2019).

Since 1993, accounting and consultancy firm KPMG has published biennial reports on how firms report on sustainability and corporate responsibility. Whilst by far not every environmental report is assured, the share of the world's 250 largest firms (by revenue) that invest in third-party assurance of their non-financial report has consistently risen, with over two-thirds (67%) at the latest count (KPMG 2017). For the 100 largest companies in each of the 49 countries surveyed, the share of firms seeking assurance was much lower, at 45%, though with a rising trend that is expected to pass the 50% threshold within the next five years.

Finance for sustainability

Financing and investing for environmental sustainability could cover a whole textbook, so the following is a series of snapshots of initiatives and tools, and by necessity short, as a comprehensive discussion is not possible within the limitations of this chapter. As a short reminder, financial institutions contribute to environmental problems as in most cases they do not take into account ecological information (such as for example climate change risks or biodiversity loss) in their decision-making, resulting in corporate strategies and activities based on incomplete information and ill-designed financial models (Scholtens 2017; Shrivastava et al. 2019). As information is key to finance, measuring and reporting on environmental impact is therefore crucial.

So-called green finance encompasses both "greening finance" and "financing green" (DBEIS 2019), i.e. integrating environmental factors into financial decision-making as well as investing for example in clean technology, environmental services and green products. Other elements include the strengthening of environmental disclosure and developing green dialogue (SSE initiative 2019). Green financial instruments include for example green bonds and commodity derivatives. Whilst green finance has been a niche so far, it is certainly a growing area, as exemplified by the international Task Force on Climate-related Financial Disclosures (TCFD) and the United Kingdom's launch of a Green Finance Strategy and Green Finance Institute in 2019 (DBEIS 2019).

Increasingly, financial actors and institutions are using screening and various principles to change corporate behaviour (Scholtens 2017).

We can differentiate between negative and positive screening. Funds and other financial institutions can use negative screening to rule out investments that do not meet certain environmental criteria, or investment in whole industries, so-called sin stocks, which might lead to divestment where stocks are taken off portfolios (Laasch and Conaway 2015). In contrast, in positive screening, investors identify exemplary companies and invest accordingly.[8]

On a global level, the United Nations has encouraged the development of principles of more sustainable finance, providing two partners (the UN Environment Programme Finance Initiative and UN Global Compact) for the Principles for Responsible Investment (PRI), which sets out six principles towards developing a more sustainable global financial system (UNPRI 2019). Importantly, the principles were developed by investors and encourage the integration of ESG issues into investment decisions as these are often not priced into markets with resultant risks, particularly in the case of climate change and mortgages and insurance for property (Tett 2019). The UN Environment Programme Finance Initiative also launched the Principles for Sustainable Insurance (PSI), which is a framework for the insurance industry to address environmental risks and opportunities. Similarly, the Sustainable Stock Exchanges Initiative (SSE) was set up by UN bodies and the PRI to enable stock exchanges to stimulate sustainable investment and advance ESG issues (SSE initiative 2019). These initiatives are informed by the UN Sustainable Development Goals (SDGs).

On a regional level, the EU has developed a Sustainable Finance Taxonomy as part of its Action Plan on Sustainable Finance which helps investors assess how environmentally sustainable an economic activity is by providing a common language between policymakers, investors and issuers (UNPRI 2019). The taxonomy sets screening criteria in relation to environmental objectives. It does not constitute a rating or ranking of environmentally good or bad companies (European Commission 2019).

Rating schemes that rank companies according to their sustainability performance include the Dow Jones Sustainability Indexes and the FTSE4Good Index Series (Fogel 2016). Dow Jones Sustainability Indexes were among the first global sustainability benchmarks, launched in 1999, whilst the FTSE4Good opened in 2001. Both use inclusion as well as exclusion criteria to help asset owners, fund managers, brokers and stock exchanges and so on to screen investments (Weybrecht 2014). In addition to being used for investment decisions, environmental sustainability corporate rankings can also be used to identify good practices (Fogel 2016). Other rating schemes include for example Bloomberg ESG Disclosure Scores, Sustainalytics, Newsweek Green Rankings and Oekom Corporate Ratings, amongst others (Fogel 2016). Other indexes and ratings work at the national, regional or industry level, whilst other are consumer, product or topic focused (Weybrecht 2014). Increasingly, credit rating agencies use ESG factors to identify sectors where environmental risks threaten a company's credit ratings (e.g. mining, oil and gas, chemicals, utilities, logistics and automobile producers) (Nauman and Gross 2019).

CERES (formerly the Coalition for Environmentally Responsible Economies) was set up in response to the Exxon Valdez oil spill in 1989 by investors and environmentalists. Whilst the CERES principles[9] are more general in nature, the Ceres Investor Network on Climate Risk and Sustainability helps investors with managing environmental risks and increasing environmental investments by developing their expertise and by pressuring regulators and stock exchanges for greater environmental (risk) disclosures (CERES 2019).

The Equator Principles for project finance set out a framework for managing environmental risks. The framework can be adopted by financial institutions as good practice when providing finance for projects. The principles set a minimum standard for decision-making in terms of due diligence and monitoring (Equator Principles 2019). They are used by many banks, which also have the Principles for Responsible Banking as a framework on which to draw to integrate sustainability into their operations. Like the PRI, the Principles for Responsible Banking were initiated in 2019 by the UN Environment Programme Finance Initiative.

Furthermore, the Green Loan Principles (GLP) were launched in 2018 by the regional loan market associations for Asia Pacific, Europe, Middle East and Africa, which build on the Green Bond Principles, supported by the Global Green Finance Council. The new Green Loan Principles establish a framework for green lending which standardises environmentally friendly lending (Allen 2018).

On a national level, The Prince's Accounting for Sustainability Project (A4S) established by The Prince of Wales in 2004 works with the finance and accounting community in the UK and beyond to encourage a shift towards more resilient and sustainable business models that integrate environmental and social factors into financial decision-making.

Scholtens (2017) highlights the use of discount rates in aligning ecology and finance. Discount rates – which were introduced in Chapter 3 (particularly Box 3.5) – are also referred to as the rate of time preference and are used to calculate net present value and to discount future costs and benefits (Anderson 2014). Using typical market discount rates can introduce biases against preventive action, and, as a result, investment decisions by financial markets tend to be biased towards projects generating current rather than future benefits (Hackett 2011). As environmental sustainability requires a long-term view, such market myopia when choosing discount rates that do not value future benefits sufficiently has significant environmental implications. By amending (lowering) discount rates through taking into account environmental factors, expected rates of returns from investment projects are adjusted and better decisions may be made.

In the context of finance, corporate governance should briefly be mentioned as the two topics are closely connected in their focus on how a company and its owners – often shareholders – relate (Laasch and Conaway 2015). Corporate governance can be narrowly defined as "the ways in which suppliers of finance to corporations assure themselves of getting a return on their investment" and thus deals with the separation of management and finance, also known as the principal-agent problem (Shleifer and Vishny 1997: 737). It plays a significant role with regard to the quality of financial reports and how efficiently financial markets work by promoting accountability and improving financial performance (Rezaee 2015). Reforms in recent years, particularly following a number of corporate scandals and the global financial crisis, have broadened the scope of corporate governance beyond shareholder returns to encourage for example board diversity, incorporation of ethics and the inclusion of stakeholders (Blowfield 2013; Crane and Matten 2016). Investors are increasingly looking for good governance and may use sustainability as a proxy for good management (Weybrecht 2014). In particular, environmental governance can be defined as the set of incentives and pressures motivating companies to improve their environmental performance, which may include markets for green products and investments, and relationships with regulators and other stakeholders (Baron and Lyon 2012; for environmental risk governance see Pollard 2011). We return to corporate governance and issues of compensation in the following chapter on human resource management.

Tools and principles for measuring, accounting and reporting for sustainability

As sustainability is both complex and multidimensional, firms have a number of tools and indicators at their disposal to measure and account for their environmental impact in order to then report and manage accordingly. Indicators are thus not ends in themselves but should be used as tools for learning and improvement as well as guidance for decision-making (Robertson 2017).

We introduced the concept of ecological footprint in the context of the earth's carrying capacity in Chapter 2. Footprint analysis allows the calculation of the total impact on the environment by a company or a product or an individual person (Fogel 2016, see also Wackernagel and Rees 1996). In particular, it allows comparisons between different entities and enables application to activities across the life cycle, from manufacturing of a product to its use phase, and can thus cover waste, pollution and climate change (and beyond) in terms of environmental challenges. Ecological footprints are based on rigorous calculations yet are also easily understandable and therefore a good tool for communicating, thus combining the two important functions of accounting and reporting. Other footprint methodologies include water and carbon footprints, materials footprints as well as ecological rucksacks which measure the weight of material that is "carried" by a product, including the hidden material flows needed to produce a good, thus indicating the material intensity (Robertson 2017). Related methodologies include life cycle analysis (or assessment, LCA), covered in previous chapters, and input-output analysis. Input-out analysis provides a disciplined approach to providing environmental information and "accounts for the physical flow of materials and energy inputs and product and waste outputs in physical units"; it has roots in materials accounting and is based on the principle *what goes in must come out* (Lamberton 2005: 10).

The measurement of greenhouse gas emissions on the basis of the Greenhouse Gas (GHG) Protocol, developed by the World Resources Institute (WRI) and the World Business Council for Sustainable Development (WBCSD), was mentioned in Chapter 5 on climate change. Here we highlight other frameworks to calculate carbon footprints. This is also known as carbon accounting which, as a subset of sustainability accounting,[10] focuses on the management and reporting of carbon emissions (Lodhia 2013). Other carbon accounting and reporting frameworks beyond the GHG Protocol that help companies monitor, control, measure and report their GHG emissions include the following:

- the CDP (formerly the Carbon Disclosure Project), which engages investors, companies and policymakers in better disclosure of environmental information to enable better decision-making on climate action,
- the CDSB (Climate Disclosure Standards Board) which works with the Task Force on Climate-related Financial Disclosures (TCFD) to move climate disclosure into the mainstream and is committed to putting natural capital on a level footing with financial capital when it comes to corporate reporting,
- the ISO 14064 from the ISO 14000 series for carbon analysis which specifies principles and requirements for quantifying and removing GHG emissions.

Industry-specific frameworks and guidelines for environmental accounting and reporting include for example: IPIECA, the global oil and gas industry association for environmental and social issues and its climate change reporting framework; IOGP, the International Association of Oil & Gas producers, and; STARS, the Sustainability Tracking, Assessment and Rating System developed by the Association for the Advancement of Sustainability in Higher Education (AASHE) (Robertson 2017). On a national level, the SASB – Sustainability Accounting Standards Board – provides reporting guidance for publicly traded companies in the US (Fogel 2016). On a global level, the United Nations Global Compact initiative encourages companies to adopt sustainability policies and to report on their implementation and performance accordingly in the areas of human rights, labour, environment and anticorruption (Rezaee 2015). The UN Global Compact's strategy is based on ten principles and supports the UN's Sustainable Development Goals (SDGs) (UN Global Compact 2019).

Environmental (cost) accounting methodologies

In this section we provide a brief overview of the main methodologies of environmental accounting that companies have started using (Laasch and Conaway 2015). For a more detailed overview and background, see Schaltegger and Burritt (2000) and Schaltegger et al. (2003).

Full or true cost accounting helps to identify the environmental costs not identified or measured in conventional accounting which results in an incomplete or inaccurate understanding of the true costs of providing goods or services (Weybrecht 2014). In attempting to allocate all direct and indirect costs of a company's economic activities, i.e. the total cost[11] in financial terms, full-cost accounting serves to counter the misinformation of market prices that do not account for the environmental costs (Lamberton 2005) – as a reminder, environmental externalities represent a market failure that is not financially internalised in company accounts (Unerman et al. 2018). One form of full-cost accounting can be found in the accounting technology Sustainability Assessment Model SAM, which incorporates sustainability thinking into corporate decision-making processes (Bebbington and Fraser 2014).

In a similar line to full-cost accounting, sustainable cost accounting looks at "the (hypothetical) cost of restoring the earth to the state it was in prior to an organisation's impact" (Lamberton 2005: 8). Sustainable cost draws on the accounting concept of capital maintenance and applies it to natural rather than financial capital. It also draws on the sustainability concept of maintaining this capital for the use of future generations (Gray 1994).

Whereas sustainable cost is a financial measure and deducted from accounting profit to calculate a notional or hypothetical level of sustainable profit or loss, natural capital inventory accounting is concerned with the non-financial recording (and reporting) of natural capital stocks over time (Laasch and Conaway 2015). Changes in these stock levels are a proxy indicator for the quality of the natural environment. The different types of natural capital can be categorised according to how critical, non-renewable/non-substitutable, non-renewable/substitutable and renewable they are (Gray 1994). The Natural Capital Protocol by the Natural Capital Coalition provides a framework on how to produce reliable data to inform corporate decision-making based on an organisation's direct and indirect impacts as well as dependencies on natural capital (NCC 2019).

Input-output analysis and life cycle assessment were mentioned in an earlier section. To this we can add life cycle costing, which assesses project costs from inception to retirement, whilst total cost assessment is a form of capital budgeting that attempts to calculate not only conventional costs but also the costs associated with environmental impacts (Lifset and Boons 2012). Here we find that industrial ecology overlaps with and complements accounting and finance. For example, material flow analysis is used as a key tool in industrial ecology and is similar to input-output analysis, whilst material flow cost accounting combines monetary units and physical flows, with roots in energy and material efficiency analysis, and management as well as environmental accounting. Material flow cost accounting is a tool that supports decisions for improved economic and environmental performance by revealing material efficiency and cost-saving opportunities (Christ and Burritt 2015; Guenther et al. 2015).

A related methodology that is material-flow oriented is activity-based costing (ABC) (Nemetz 2013). Activity-based costing assumes that any activities related to goods, services and customers lead to costs (Laasch and Conaway 2015). Using activity-based costing (or accounting) to assess environmental costs allows the integration of environmental cost accounting into strategic management processes (Schaltegger et al. 2003). As ABC identifies cost drivers, it results in more meaningful information about costs and a clearer identification of the environmental issues driving those costs (Buhr and Gray 2012; Weybrecht 2014). One form of ABC is life cycle accounting, highlighting the links and overlaps between various methods of environmental accounting, which can also include cost-benefit analysis (Fiksel 2012).

An environmental profit and loss (EP&L) statement provides another opportunity to put a monetary value on a company's environmental impacts, whether direct or indirect, through its use of ecosystem services (Weybrecht 2014). The methodology was developed by PriceWaterhouseCoopers and Trucost, which were commissioned by sportswear company Puma (Nemetz 2013). The innovative methodology has given the company insights into its impact on the supply chain and how to structure its activities in a more sustainable way. As the EP&L methodology is an open source approach, Puma (and its subsequent parent companies) have shared it with other companies to use as a model for greater transparency and improved performance that takes into account the costs of using natural resources (Nemetz 2013).

Sustainable balanced scorecards were mentioned in the previous chapter on environmental management and information systems. As they encompass financial and non-financial performance indicators, they can also be used for accounting and reporting purposes as well as for controlling, where we find a significant overlap between environmental management control systems ("eco-control") and environmental accounting as also highlighted in the previous chapter (Nemetz 2013; Figge et al. 2002; Guenther et al. 2016).

Accounting and reporting sustainably

We have focused on the role of accounting and reporting in making companies more environmentally sustainable on the basis of better measuring and reporting of physical, non-financial data. When it comes to making the function itself more sustainable, this addresses the environmental impact referred to at the beginning of the chapter, which arguably is small compared to other corporate functions.

Companies are increasingly moving away from printed reports and their limitations towards web-based reporting on the internet, making use of possibilities and advantages of providing more information and increased accessibility (e.g. social media) and comprehensibility. Although this saves paper, energy is used for data analysis and internet use. By moving energy supplies to renewable sources, the environmental impact can be further reduced.

Spotlight on . . . examples and practices of accounting, finance and reporting for sustainability

In the following we briefly highlight some examples of accounting, finance and reporting for environmental sustainability. According to the most recent KPMG survey on corporate responsibility reporting, over three-quarters of the world's 250 largest firms now include corporate responsibility information in their annual (financial) reports, which is a significant increase from 44% in 2011 (KPMG 2017). The most commonly used framework for reporting is the GRI, whilst integrated reporting is on the rise, particularly in Mexico, Brazil and Japan. The IIRC provides a database of examples of companies that use the emerging practice of integrated reporting.

Many companies reference the UN's SDGs in their corporate responsibility reporting, which – as highlighted previously – inform many of the principles and guidelines for sustainable reporting and finance. Whilst the majority of companies surveyed by KPMG report on non-financial matters, very few companies acknowledge the financial risks posed by climate change, which highlights the need for the work of the Task Force on Climate-related Financial Disclosures (TCFD) (KPMG 2017).

In addition to the rise in environmental reporting, the launch of the so-called Moral Money newsletter by the *Financial Times* in June 2019, which features news, analysis and opinion, is a further sign of how seriously the mainstream finance and accounting world now takes ESG issues, sustainable finance, impact investing and the UN's SDGs – all topics featuring in the newsletter (FT 2019). This interest is reflected in the record number of green bonds being issued in 2019 and is driven also by the above-average performance of sustainable funds (Thompson 2019).

In contrast to the increase in sustainable investing, insurance companies were the first organisations alongside universities to divest from fossil fuels, selling off holdings in coal, oil and gas companies over climate change fears and financial risks (Watts 2017). They are increasingly followed by other institutional investors such as pension funds, key medical organisations, major cities such as London and New York, and countries, such as for example Ireland. Calculations estimate that these divestments have amounted to over $6 trillion (Carrington 2018). Sustainable investments amount to over $22 trillion in total, accounting for over a quarter of professionally managed assets in 2017 (Bernow et al. 2017). Negative screening is the most widely used strategy to identify sustainable investments, whilst positive screening and proactive engagement are also used to encourage companies to engage in ESG issues.

In order to help its clients improve their environmental performance, Dutch-based global bank ING offered the world's first sustainability improvement derivative in 2019, which links the price of the derivative to the sustainability performance alongside traditional measures such as trading risk, profit and capital requirements. In 2018, the bank launched a sustainability improvement loan, also a world first (ING 2019). Visser (2012) highlights the role of the Netherlands as a leader in sustainable finance and socially responsible investing.

Puma, which pioneered environmental profit and loss accounts by valuing and monetising the ecosystem services it uses, has been followed by other companies in doing so. Examples include Vodafone, Novo Nordisk, Arla Foods, Philips, and Puma's parent company Kering and thus comprise companies from a variety of sectors. Whilst Trucost helped Puma develop its EP&L methodology, another company that assesses the environmental footprints is Sustainalytics, which has also supported ING (Fogel 2016; ING 2019). Other organisations in this space include SustainAbility and the Sustainable Consortium.

SUMMARY

As environmental sustainability is increasingly integrated into decision-making processes in companies, environmental accounting offers a powerful tool to facilitate this (Piasecki et al. 1999). There is value in collecting environmental performance data for reporting purposes if companies use it not just for external communication purposes but also for internal performance improvement and management control (Maas et al. 2016a). This links environmental accounting with the other corporate functions covered in the book, particularly production and operations and the wider supply chain where companies' environmental impact is mostly found, but also links in with marketing and management systems and how information is used and communicated as accountants are also providers and preparers of information, not just users of it.

Similar to these other corporate functions and the wider case for sustainability, there are a number of drivers for integrating sustainability into accounting and reporting that we highlighted in the chapter. These include performance improvements, cost savings, risk management and waste reduction as well as stakeholder management and social legitimacy. Esty and Winston (2009) encourage companies to be upfront about the bad news as well and not just the good news, as transparency is a powerful driver for corporate environmental leaders. As emphasised before, accounting is also about the discharging of accountability.

Having made the case for accounting for sustainability, we introduced environmental accounting and reporting in an overview, including an element of finance. As there is no one-size-fits-all standardised approach, the instruments and standards then introduced in the chapter here can be thought of as a "toolbox" servicing different needs, purposes and stakeholders (Maas et al. 2016a: 241). This should involve both leading and lagging indicators as sustainability is a long-term business, and thus environmental accounting and reporting should combine the conventional rear-view mirror of financial reporting into past performance and the important view ahead of future-oriented challenges and topics.

Regardless of which framework companies use, good communication is key in order to implement sustainability, particularly as teams work together (Robertson 2017). Ultimately perhaps the goal is to create one report that combines financial with nonfinancial reporting (Fogel 2016). As sustainability draws on a variety of disciplines and backgrounds, the role of human resource management in enabling people to work together within and across functions and departments is critical. As accounting and reporting uses performance indicators, it can inform changes to performance appraisal systems to move the behaviour of managers and employees away from environmentally destructive (yet economically attractive) activities to environmentally desirable ones (Gray 2011). It is therefore to HRM that we now turn to in Chapter 13 as the final chapter in part three on corporate functions' contribution to corporate environmental sustainability.

SUGGESTED QUESTIONS AND ACTIVITIES

- Check out the database of examples by the IIRC to look at companies engaging in integrated reporting: http://integratedreporting.org/resource/emerging-integrated-reporting-database/
- Look up some of the corporate examples and practices mentioned in the chapter, and research them in more detail.
- Select a company of your choice and look up its sustainability or environmental report (terms vary). Who are the stakeholders identified by the company and what do you consider to be the company's material impacts? Does the company report on the identified material issues? Does the company use one of the frameworks introduced in the chapter, according to which it measures and reports its environmental impact?
- For more in-depth background material on environmental and sustainability accounting and reporting, check out the following books and websites:
  - Bebbington, J., Unerman, J. and O'Dwyer, B. (2014) (eds.) *Sustainability Accounting and Accountability*, London: Routledge, 2nd ed.
  - Schaltegger, S. and Burritt, R. (2000) *Contemporary Environmental Accounting: Issues, Concepts and Practice*, Sheffield: Greenleaf Publishing.
  - The Centre for Social and Environmental Research (CSEAR) and its The Social and Environmental Accountability Journal – available at www.st-andrews.ac.uk/csear/.
  - The Global Reporting Initiative: www.globalreporting.org.

- The Principles for Responsible Investment: www.unpri.org.
- Carbon reporting frameworks: www.cdp.net, www.cdsb.net/, https://ghgprotocol.org/.
- *Financial Times* Moral Money newsletter: www.ft.com/moral-money.

NOTES

1 Friedman (2011) uses the term assessment for internal environmental audits as they are less formal than e.g. a financial audit but stresses that for external, independent audits, the term audit is appropriate.

2 For environmental accounting considerations at a national level, see Pearce et al. (1999). See also Adams and Frost (2008) for developing key performance indicators for measuring sustainability performance in a bid to integrate it into management practices.

3 See Schaltegger and colleagues (2003) for issues of complexity when applying the materiality principle to environmental and social reporting compared to using it in traditional financial reporting.

4 SA8000® is a social certification standard established by Social Accountability International (www.sa-intl.org).

5 The Global Reporting Dialogue was convened by the International Integrated Reporting Council, bringing together financial and non-financial reporting frameworks and standards. Participants beyond the GRI and IIRC include the CDP (Carbon Disclosure Project), the Climate Disclosure Standards Board, International Accounting Standards Board, the International Organization for Standardization (ISO) and the Sustainability Accounting Standards Board, plus the Financial Accounting Standards Board as observer.

6 The full name is "International Standard on Assurance Engagements Other than Audits or Reviews of Historical Financial Information".

7 AccountAbility is a global think tank and consultancy. It was previously known as the Institute of Social and Ethical Accountability (ISEA, not to be confused with ISAE) (Laasch and Conaway, 2015).

8 Krosinsky and Purdom (2017) differentiate between seven distinct types of sustainable investing (see also Shrivastava et al. 2019).

9 The Ceres Principles were originally called the Valdez Principles (Visser 2012).

10 Nemetz (2013: 347) calls carbon accounting (or carbon footprinting) a "hybrid of life-cycle analysis and the ecological footprint".

11 See also total impact accounting (Matthews 1993).

SOURCES AND FURTHER READING

AccountAbility (2019) AccountAbility, webpages, available at www.accountability.org (Retrieved, 28.08.2019).

Adams, C. A. (2004) The Ethical, Social and Environmental Reporting-Performance Portrayal Gap, in: *Accounting, Auditing & Accountability Journal*, 17(5): 731–757.

Adams, C. A. and Frost, G. R. (2008) Integrating Sustainability Reporting into Management Practices, in: *Accounting Forum*, 32(4): 288–302.

Allen, K. (2018) Loan Market Association Sets Out Green Guidelines, *Financial Times*, 21.03.2018, available at www.ft.com/content/053061ee-2cf0-11e8-a34a-7e7563b0b0f4 (Retrieved, 29.08.2019).

Anderson, D. A. (2014) *Environmental Economics and Natural Resource Management*, London: Routledge, 4th ed.

Baron, D. P. and Lyon, T. P. (2012) Environmental Governance, in: Bansal, P. and Hoffman, A. J. (eds.) *The Oxford Handbook of Business and the Natural Environment*, Oxford: Oxford University Press, pp. 122–139.

Bebbington, J. and Fraser, M. (2014) Organizational Change and Sustainability Accounting, in: Bebbington, J., Unerman, J. and O'Dwyer, B. (eds.) *Sustainability Accounting and Accountability*, London: Routledge, 2nd ed., pp. 141–153.

Bebbington, J., Unerman, J. and O'Dwyer, B. (2014) Drawing to a Close and Future Horizons, in:

Bebbington, J., Unerman, J. and O'Dwyer, B. (eds.) *Sustainability Accounting and Accountability*, London: Routledge, 2nd ed., pp. 286–289.

Bernow, S., Klempner, B. and Magnin, C. (2017) *From 'why' to 'why not': Sustainable Investing as the New Normal*, McKinsey & Company, available at www.mckinsey.com/industries/private-equity-and-principal-investors/our-insights/from-why-to-why-not-sustainable-investing-as-the-new-normal (Retrieved, 02.09.2019).

Blowfield, M. (2013) *Business and Sustainability*, Oxford: Oxford University Press.

Brown, J. and Fraser, M. (2006) Approaches and Perspectives in Social and Environmental Accounting: An Overview of the Conceptual Landscape, in: *Business Strategy and the Environment*, 15(2): 103–117.

Buhr, N. and Gray, R. (2012) Environmental Management, Measurement, and Accounting: Information for Decision and Control? in: Bansal, P. and Hoffman, A. J. (eds.) *The Oxford Handbook of Business and the Natural Environment*, Oxford: Oxford University Press, pp. 425–443.

Buhr, N., Gray, R. and Milne, M. J. (2014) Histories, Rationales, Voluntary Standards and Future Prospects for Sustainability Reporting: SCR, GRI, IIRC and Beyond, in: Bebbington, J., Unerman, J. and O'Dwyer, B. (eds.) *Sustainability Accounting and Accountability*, London: Routledge, 2nd ed., pp. 51–71.

Burritt, R. L. and Schaltegger, S. (2010) Sustainability Accounting and Reporting: Fad or Trend? in: *Accounting, Auditing & Accountability Journal*, 23(7): 829–846.

Carrington, D. (2018) Fossil Fuel Divestment Funds Rise to $6tn, *The Guardian*, 10.09.2018, available at www.theguardian.com/environment/2018/sep/10/fossil-fuel-divestment-funds-rise-to-6tn (Retrieved, 02.09.2019).

CERES (2019) Coalition for Environmentally Responsible Economies, webpages, available at www.ceres.org (Retrieved, 29.08.2019).

Christ, K. L. and Burritt, R. L. (2015) Material Flow Cost Accounting: A Review and Agenda for Future Research, in: *Journal of Cleaner Production*, 108: 1378–1389.

Cooper, S. and Owen, D. (2014) Independent Assurance of Sustainability Reports, in: Bebbington, J., Unerman, J. and O'Dwyer, B. (eds.) *Sustainability Accounting and Accountability*, London: Routledge, 2nd ed., pp. 72–85.

Crane, A. and Matten, D. (2016) *Business Ethics*, Oxford: Oxford University Press, 4th ed.

Davison, J. (2015) Visualising Accounting: An Interdisciplinary Review and Synthesis, in: *Accounting and Business Research*, 45(2): 121–165.

DBEIS (2019) *Green Finance: Guidance, Department for Business*, Energy & Industrial Strategy, available at www.gov.uk/guidance/green-finance (Retrieved, 29.08.2019).

Dyck, A., Lins, K. V., Roth, L. and Wagner, H. F. (2019) Do Institutional Investors Drive Corporate Social Responsibility? International Evidence, in: *Journal of Financial Economic*, 31(3): 693–714.

Esty, D. C. and Winston, A. S. (2009) *Green to Gold: How Smart Companies Use Environmental Strategy to Innovate, Create Value and Build Competitive Advantage*, Hoboken, NJ: Wiley.

Equator Principles (2019) Equator Principles, webpages, available at https://equator-principles.com (Retrieved, 29.08.2019).

European Commission (2019) Financing a Sustainable European Economy: Using the Taxonomy, Supplementary Report 2019 by the Technical Expert Group on Sustainable Finance, available at https://ec.europa.eu/info/sites/info/files/business_economy_euro/banking_and_finance/documents/190618-sustainable-finance-teg-report-using-the-taxonomy_en.pdf (Retrieved, 29.08.2019).

Figge, F., Hahn, T., Schaltegger, S. and Wagner, M. (2002) The Sustainability Balanced Scorecard – Linking Sustainability Management to Business Strategy, in: *Business Strategy and the Environment*, 11(5): 269–284.

Fiksel, J. (2012) *Design for Environment: A Guide to Sustainable Product Development*, New York: McGrawHill, 2nd ed.

Fogel, D. S. (2016) *Strategic Sustainability: A Natural Environmental Lens on Organizations and Management*, London: Routledge.

Friedman, F. B. (2011) *Practical Guide to Environmental Management*, Washington, DC: Environmental Law Institute, 11th ed.

FT (2019) Moral Money Newsletter, *Financial Times*, available at www.ft.com/moral-money (Retrieved, 30.08.2019).

Gray, R. (1993) *Accounting for the Environment*, London: Paul Chapman.

Gray, R. (1994) Corporate Reporting for Sustainable Development: Accounting for Sustainability in 2000AD, in: *Environmental Values*, 3(1): 17–45.

Gray, R. (2011) Reporting and Accounting, in: Brady, J., Ebbage, A. and Lunn, R. (eds.) *Environmental Management in Organizations, The IEMA Handbook*, London: Earthscan/Routledge, 2nd ed., pp. 433–443.

Gray, R. and Herremans, I. M. (2012) Sustainability and Social Responsibility Reporting and the Emergence of the External Social Audits: The Struggle for Accountability, in: Bansal, P. and Hoffman, A. J. (eds.) *The Oxford Handbook of Business and the Natural Environment*, Oxford: Oxford University Press, pp. 405–424.

Gray, R. and Milne, M. (2004) Towards Reporting on the Triple Bottom Line: Mirages, Methods and Myths, in: Henriques, A. and Richardson, J. (eds.) *The Triple Bottom Line: Does it all add up? Assessing the Sustainability of Business and CSR*, London: Earthscan, pp. 70–80.

Greenwood, M., Jack, G. and Haylock, B. (2019) Toward a Methodology for Analyzing Visual Rhetoric in Corporate Reports, in: *Organizational Research Methods*, 22(3): 798–827.

GRI (2016) Global Reporting Initiative, GRI 101: Foundation 2016, available at www.globalreporting.org/standards/media/1036/gri-101-foundation-2016.pdf (Retrieved, 16.08.2019).

GRI (2019) Global Reporting Initiative, webpages, available at www.globalreporting.org (Retrieved, 16.08.2019).

Guenther, E., Endrikat, J. and Guenther, T. W. (2016) Environmental Management Control Systems: A Conceptualization and a Review of the Empirical Evidence, in: *Journal of Cleaner Production*, 136: 147–171.

Guenther, E., Jasch, C., Schmidt, M., Wagner, B. and Ilg, P. (2015) Material Flow Cost Accounting – Looking Back and Ahead, in: *Journal of Cleaner Production*, 108: 1249–1254

Hackett, C. (2011) *Environmental and Natural Resource Economics*, London: Sharpe, 4th ed.

Herzig, C. and Ghosh, B. (2017) Sustainability Reporting, in: Molthan-Hill, P. (ed.) *The Business Student's Guide to Sustainable Management*, Sheffield: Greenleaf Publishing, 2nd ed., pp. 160–195.

Hrasky, S. (2012) Visual Disclosure Strategies Adopted by More and Less Sustainability-Driven Companies, in: *Accounting Forum*, 36(3): 154–165.

IIRC (2011) Towards Integrated Reporting: Communicating Value in the 21st Century, Discussion Paper, International Integrated Reporting Council, available at http://integratedreporting.org/wp-content/uploads/2011/09/IR-Discussion-Paper-2011_spreads.pdf (Retrieved, 28.08.2019).

IIRC (2013) The International<IR> Framework, International Integrated Reporting Council, available at http://integratedreporting.org/wp-content/uploads/2015/03/13-12-08-THE-INTERNATIONAL-IR-FRAMEWORK-2-1.pdf (Retrieved, 28.08.2019).

IIRC (2019) International Integrated Reporting Council, webpages, available at http://integratedreporting.org (Retrieved, 28.08.2019).

ING (2019) Introducing the World's First Sustainability Improvement Derivative, *ING News*, 13.08.2019, available at www.ing.com/Newsroom/All-news/Introducing-the-worlds-first-sustainability-improvement-derivative.htm (Retrieved, 02.09.2019).

ISAE 3000 (2019) International Standard on Assurance Engagements 3000 Standard, webpages, available at https://isae3000.com/ (Retrieved, 28.08.2019).

KPMG (2017) The Road Ahead: The KPMG Survey of Corporate Responsibility Reporting 2017, available at https://assets.kpmg/content/dam/kpmg/xx/pdf/2017/10/kpmg-survey-of-corporate-responsibility-reporting-2017.pdf (Retrieved, 28.08.2019).

Krosinsky, C. and Purdom, S. (2017) *Sustainable Investing: Revolutions in Theory and Practice*, London: Routledge.

Laasch, O. and Conaway, R. (2015) *Principles of Responsible Management: Glocal Sustainability, Responsibility and Ethics*, Stamford, CT: Cengage Learning.

Lamberton, G. (2005) Sustainability Accounting – A Brief History and Conceptual Framework, in: *Accounting Forum*, 29: 7–26.

Lifset, R. and Boons, F. (2012) Industrial Ecology: Business Management in a Material World, in: Bansal, P. and Hoffman, A. J. (eds.) *The Oxford Handbook of Business and the Natural Environment*, Oxford: Oxford University Press, pp. 310–326.

Lodhia, S. (2013) Sustainability Accounting and Reporting: An Overview, Contemporary Developments and Research Possibilities, in: Wells, G. (ed.) *Sustainable Business: Theory and Practice of Business under Sustainability Principles*, Cheltenham: Edward Elgar, pp. 73–88.

Maas, K., Schaltegger, S. and Crutzen, N. (2016a) Integrating Corporate Sustainability Assessment, Management Accounting, Control, and Reporting, in: *Journal of Cleaner Production*, 136: 237–248.

Maas, K., Schaltegger, S. and Crutzen, N. (2016b) Advancing the Integration of Corporate Sustainability Measurement, Management and Reporting, in: *Journal of Cleaner Production*, 133: 859–862.

Marsden, J. (2011) Auditing, in: Brady, J., Ebbage, A. and Lunn, R. (eds.) *Environmental Management in Organizations, The IEMA Handbook*, London: Earthscan/ Routledge, 2nd ed., pp. 399–417.

Matthews, M. R. (1993) *Socially Responsible Accounting*. London: Chapman & Hall.

Nauman, B. and Gross, A. (2019) Credit Rating Agencies Focus on Rising Green Risks, *Financial Times*, 27.11.2019, available at www.ft.com/content/45d721ee-1036-11ea-a7e6-62bf4f9e548a (Retrieved, 02.12.2019).

NCC (2019) Natural Capital Protocol, Natural Capital Coalition, available at https://naturalcapitalcoalition.org/natural-capital-protocol/ (Retrieved, 02.12.2019).

Nemetz, P. N. (2013) *Business and the Sustainability Challenge: An Integrated Perspective*, London: Routledge.

Newbold, J. (2011) Environmental Performance Indicators, in: Brady, J., Ebbage, A. and Lunn, R. (eds.) *Environmental Management in Organizations, The IEMA Handbook*, London: Earthscan/ Routledge, 2nd ed., pp. 419–432.

Oakley, R. and Buckland, I. (2004) What if Business as Usual Won't Work, in: Henriques, A. and Richardson, J. (eds.) *The Triple Bottom Line: Does it all add up? Assessing the Sustainability of Business and CSR*, London: Earthscan, pp. 131–141.

Pearce, D., Markandya, A. and Barbier, E. (1999) *Blueprint for a Green Economy*, London: Earthscan (first published 1989).

Piasecki, B. W., Fletcher, K. A. and Mendelson, F. J. (1999) *Environmental Management and Business Strategy: Leadership Skills for the 21st Century*, New York: John Wiley & Sons.

Pollard, S. (2011) Environmental Risk Governance, in: Brady, J., Ebbage, A. and Lunn, R. (eds.) *Environmental Management in Organizations, The IEMA Handbook*, London: Earthscan/Routledge, 2nd ed., pp. 495–509.

Rezaee, Z. (2015) *Business Sustainability: Performance, Compliance, Accountability and Integrated Reporting*, Sheffield: Greenleaf Publishing.

Richardson, J. (2004) Accounting for Sustainability: Measuring Quantities or Enhancing Qualities, in: Henriques, A. and Richardson, J. (eds.) *The Triple Bottom Line: Does it all add up? Assessing the Sustainability of Business and CSR*, London: Earthscan, pp. 34–44.

Robertson, M. (2017) *Sustainability Principles and Practice*, London: Routledge, 2nd ed.

Schaltegger, S. and Burritt, R. (2000) *Contemporary Environmental Accounting: Issues, Concepts and Practice*, Sheffield: Greenleaf Publishing.

Schaltegger, S., Burritt, R. and Peterson, H. (2003) *An Introduction to Corporate Environmental Management*, Sheffield: Greenleaf Publishing.

Scholtens, B. (2006) Finance as a Driver of Corporate Social Responsibility, in: *Journal of Business Ethics*, 68(1): 19–33.

Scholtens, B. (2017) Why Finance Should Care about Ecology, in: *Trends in Ecology & Evolution*, 32(7): 500–505.

Shleifer, A. and Vishny, R. W. (1997) A Survey of Corporate Governance, in: *Journal of Finance*, 52(2): 737–783.

Shrivastava, P. Zsolnai, L., Wasieleski, D., Stafford-Smith, M., Walker, T., Weber, O., Krosinsky, C. and Oram, D. (2019) Finance and Management for the Anthropocene, in: *Organization & Environment*, 32(1): 26–40.

SSE initiative (2019) The Sustainable Stock Exchanges (SSE) Initiative, webpages, available at https://sseinitiative.org (Retrieved, 29.08.2019).

Stead, J. G. and Stead, W. E. (2014) *Sustainable Strategic Management*, Sheffield: Greenleaf Publishing, 2nd ed.

Tett, G. (2019) Climate Change could Cause a New Mortgage Default Crisis, *Financial Times*, 26.09.2019, available at www.ft.com/content/7ec25f94-e04f-11e9-9743-db5a370481bc (Retrieved, 04.12.2019).

Thompson, J. (2019) Sustainable Funds More Likely to be Top Performers, Study Shows, *Financial Times*, 12.08.2019, available at www.ft.com/content/9e71cf86-ba2d-345c-bb3b-0d5887abbc6a (Retrieved, 04.12.2019).

Unerman, J., Bebbington, J. and O'Dwyer, B. (2018) Corporate Reporting and Accounting for Externalities, in: *Accounting and Business Research*, 48(5): 497–522.

UN Global Compact (2019) United Nations Global Compact, webpages, available at www.unglobalcompact.org (Retrieved, 30.08.2019).

UNPRI (2019) Principles for Responsible Investment, webpages, available at www.unpri.org (Retrieved, 29.08.2019).

Visser, W. (2012) *The Quest for Sustainable Business: An Epic Journey in Search of Corporate Responsibility*, Sheffield: Greenleaf Publishing.

Wackernagel, M. and Rees, W. (1996) *Our Ecological Footprint: Reducing Human Impact on the Earth*, Gabriola Island: New Society Publishers.

Watts, J. (2017) Growing Number of Global Insurance Firms Divesting from Fossil Fuels, *The Guardian*, 15.11.2017, available at www.theguardian.com/environment/2017/nov/15/growing-number-of-global-insurance-firms-divesting-from-fossil-fuels (Retrieved, 02.09.2019).

WBCSD (2003) Sustainable Development Reporting: Striking the Balance, World Business Council for Sustainable Development, available at https://docs.wbcsd.org/2002/12/SustainableDevReporting-StrikingTheBalance.pdf (Retrieved, 13.08.2019).

WBCSD (2017) Reporting Matters: Striking a Balance Between Disclosure and Engagement, 2017 Report, World Business Council for Sustainable Development, available at https://docs.wbcsd.org/2017/10/WBCSD_Reporting_matters_2017_interactive.pdf (Retrieved, 13.08.2019).

Weybrecht, G. (2014) *The Sustainable MBA: A Business Guide to Sustainability*, Chichester: Wiley, 2nd ed.

CHAPTER 13

Human resource management for environmental sustainability

INTRODUCTION

Human resource management has a core role to play when it comes to both finding and developing the people who contribute to an organisation's success and who are the primary sources of competitive advantage (Daft et al. 2010). As people are therefore also "key to success" in any corporate environmental management programme (Friedman 2011: 135), a chapter on human resource management is a fitting topic to end this part on corporate functions. In a sense human resources are both the starting and end point to any corporate endeavour, and none of the functions covered so far can succeed without their input, skills and experience. Addressing how they are managed and led is therefore crucial. As sustainability in most cases requires organisations to change how they do things, change management, good leadership and governance are crucial. Whilst we covered change and leadership in Chapter 4 in the context of where companies are in their journey towards sustainability and in terms of the business models available, here we highlight their role in the context of human resource management. We set this against the broader background of organisational structure and culture as well as corporate governance before highlighting a number of tools and concepts that can be used in HRM processes to make organisations more sustainable as well as to make the function itself more sustainable. As with previous chapters, we start with a brief overview of the function, its impact on the environment and vice versa, and the business case for sustainable HRM, which we then introduce.

HUMAN RESOURCE MANAGEMENT

Overview

In short, human resource management (HRM) consists of activities that attract, develop and maintain a workforce that is effective (Daft et al. 2010). It generally refers to all activities associated with managing the people working for an organisation (Mariappanadar 2019). In the broader context of the core management of planning, organising, leading and controlling, HRM encompasses both organising and leading.

A key function of HRM is managing recruitment and selection processes to identify the most suitable employees for an organisation based on their skills, knowledge and attitudes. Other core processes include training and development, performance management and employee communications (Laasch and Conaway 2015). For Daft et al. (2010) finding the right people includes HRM planning, job analysis and forecasting as well as recruiting and selecting, whilst managing and developing talent includes training, development and appraisal; and maintaining an effective workforce concerns wages and salary, benefits, labour relations and terminations. In short, HRM comprises systems, practices and policies intended to influence the behaviour, attitudes and performance of employees in pursuit of organisational goals and strategies (Noe et al. 2015).

Environmental impact of HRM

Similar to the other supporting functions of accounting and information systems covered in the previous two chapters, we could argue that the HRM function in itself has limited environmental impact. However, regardless of the size of a company, its employees and their activities will have an environmental impact, whether positive or negative (Baxter 2011). For example, inadequately trained or uninformed employees could fail to ensure regulatory compliance, which could lead to costly fines for a company (Yoxon 2011). Uninformed and unaware employees contribute to an organisation's environmental footprint e.g. in terms of saving or wasting energy and recycling or disposal of waste. Travelling to and from work, and business travel for meetings coupled with a reluctance to adopt virtual meeting tools or to consider alternative modes of transport further increase the environmental impact of human resources (Laasch and Conaway 2015). Although it is difficult to say to what extent all of this is the responsibility of the HR department, the development of employees through training, communication and awareness raising is part of HRM.

Looking at HRM processes more specifically, we find environmental impacts in the recruitment process, for example how candidates are travelling to interview locations, particularly if flying in from abroad or for international positions. Similarly, if the HRM function is only located in a multinational's headquarters and a subsidiary is recruiting, this requires international travel. The organisational structure might have an impact if large distances between corporate locations require large amounts of travelling by road or air transport. Where training – internal or external – takes place might have similar effects. Travel impact largely falls under pollution and climate change. Other impacts might be in the form of waste from training materials and employee communications provided in paper or even glossy format rather than electronically, and perhaps in the form of (cheap) motivational branded gifts for employees produced in an unsustainable manner.

Impact of environment on HRM

As with the previous chapters on marketing and accounting, it is less clear what the environmental impact on the human resource (HR) function is.[1] However, if we think of the HRM function as providing the working environment for a firm's staff, environmental factors may have an impact for example in how air pollution affects human health, in particular lung and heart functions. As outlined in Chapter 6 on pollution, productivity may be affected as a result of employees being ill and less productive at work, or ill and absent from work (Birchby et al. 2014). Businesses are thus affected both financially and socially as staff absence puts additional pressures on other employees. When it comes to climate change, employees may be affected by more frequent and more intense heatwaves and droughts, as well as floods, which may negatively impact health, well-being and productivity (e.g. through infectious diseases, weather-related mortality and malnutrition). This naturally depends on the location of the organisation. Furthermore, if employees for example live in areas prone to frequent flooding, this will also have an impact on their ability to work and contribute productively. Even if the sites of a company might be flood-proof, if its employees are personally affected, this will affect the organisation's ability to operate. With regard to waste, exposure to toxic waste can be dangerous to human health and life, and affect a company's ability to be staffed appropriately.

In short, environmental factors can impact the HRM function in terms of lost work capacity and reduced labour productivity.

Business case for sustainable HRM

Although the sections presented earlier might indicate that the business case for making HRM more sustainable is less clear-cut than for other corporate functions, it is clear that companies affect the environment through their HRM and are indeed affected, financially and otherwise. In particular, employees are the stakeholder group that most influences the results of a company and thus are a significant factor in advancing sustainability within organisations (Cohen and King 2017). Esty and Winston (2009) argue that they are possibly the most powerful player as they can influence the success or failure of an initiative with implications for the entire company. This has ramifications both inside and outside the company – for a company wanting to market its green (or greener) products, employees who are competent and capable in engaging customers are needed (Laasch and Conaway 2015). Similarly, without employees supporting for example waste and energy reduction initiatives, companies will not be able to achieve their sustainability goals in this area.

Environmental initiatives are thus positively influenced by human resource management in terms of success (Daddi et al. 2015). A company can have the best senior managers or the most sophisticated system of management controls, but without appropriate staff, its environmental mission cannot be achieved (Friedman 2011). As new graduates increasingly look for employment in companies that take ethics and sustainability seriously, employers need to be aware of their reputation when it comes to environmental issues if they want to attract the best suited candidates and motivate their existing workforce (Brady 2011).

Furthermore, if employees are not engaged, educated and informed, a company risks being perceived as greenwashing (Weybrecht 2014). In contrast, when employees are actively involved in environmental management systems, opportunities for environmental improvement are increased (Daddi et al. 2015). Employees are often internal drivers for environmental improvements as they can spot opportunities for energy savings and cost reductions more easily. Making this possible and easier can be facilitated by the HR function, thus contributing to both the economic and environmental bottom line (and the social bottom line in terms of employee satisfaction).

A strong environmental reputation can be a draw for superior talent (Russo and Fouts 1997) and can differentiate companies from their competitors when it comes to hiring (Robertson 2017). This makes a strong case for companies in terms of the recruitment aspect of HR. In other elements of HR, linking compensation to environmental stewardship can lead to an improvement in environmental performance as incentive systems can change behaviour and help achieve environmental objectives (Russo and Minto 2012; Daddi et al. 2015).

Similar to other chapters and the wider case for sustainability, the business case is thus also made for the human resource function, and we can proceed to introducing sustainable HRM or HRM for sustainability.

HRM FOR SUSTAINABILITY – GREEN(ING) HRM

Overview

Sustainable HRM as a field developed in recognition that HR is a key element for achieving sustainability outcomes as part of broader corporate strategies (Mariappanadar 2019). As there is often a tension and paradox between achieving the financial and economic, social and environmental outcomes of sustainability – as also highlighted in other parts of the book – sustainable HRM is tasked with synthesising the achievement of organisational performance outcomes and the reduction of unsustainable impacts on stakeholders (employees) and on the natural environment (ibid.).

In a report for the Society for Human Resource Management (SHRM), Cohen and colleagues (2011: 1) define sustainable HRM as "using the tools of HR to create a workforce that has the trust, values, skills and motivation to achieve a profitable triple bottom line". For the environmental aspect of the triple bottom line, the term "green HRM" has emerged as a description for HR practices that tap into employees' engagement, creativity and motivation in order to reduce material inputs, design eco-friendly goods and services, and to improve product quality (ibid.). Green HRM is a subset of sustainable HRM (Mariappanadar 2019) and can be defined as the HRM aspects of environmental management (Renwick et al. 2013).

Sustainable HR works by "leveraging HR tools and processes to support sustainable business" (Cohen and King 2017: 265). Senior management commitment, an appropriate organisational structure and suitable staffing provide a solid foundation for corporate environmental sustainability (Friedman 2011).

In order to integrate sustainability into the way a company operates, this requires not only training and development but also a shift in mind-set and wider organisational cultures (Cohen and King 2017). Mariappanadar (2019) emphasises the need for a bundle of internally consistent sustainable HRM practices to engage employees in achieving sustainability outcomes. Similarly, Forbes and Jermier (2012) highlight that HRM's role is both strategic as a functional tool in greening companies as well as symbolic when it comes to culture, which can either support or undermine sustainability efforts. In the following we cover both aspects – the strategic and functional, and the symbolic and cultural – as we look at organisational structure as well as culture, at corporate governance and leadership, at change management and the various tools and concepts for making HRM, its elements and processes, more environmentally sustainable.

Organisational structure and corporate governance

Governance is about how an organisation is run, and for sustainability programmes it is a major factor in whether programmes are successful or not (Robertson 2017). Corporate governance in particular concerns how businesses are directed and controlled in terms of structure, objectives, responsibilities and performance (Weybrecht 2014). There are some overlaps between corporate governance and organisational structure (e.g. how boards are organised or how board members are compensated).

Organisational structure is the framework that defines how an organisation divides tasks, deploys resources and coordinates departments (Daft et al. 2010). Approaches to organisational structure – which largely follows strategy – traditionally encompass functional and divisional approaches with more recent developments also including matrix structures, team approaches and virtual networks, which all come with respective advantages and disadvantages and differing needs for coordination[2] (Daft et al. 2010).

Historically, corporate governance and organisational structure have moved from including health and safety in their policies and positions to health, safety and environment (HSE), elevating the status of the natural environment. Changing the organisational structure to include roles, positions or departments that are dedicated to environmental sustainability has significant symbolic significance, which is further underlined if the broader structural arrangement includes a formal environmental management system (Forbes and Jermier 2012).

Integrating sustainability into the composition, structure and capacity of a company's board through recruitment and training of new and existing members is a measure that companies can take to make corporate governance more effective (Kakabadse and Grayson 2012). As directors and board members often have legal responsibilities and could potentially be liable for penalties for non-compliance with environmental regulations (Dunphy et al. 2007), using corporate governance for sustainability is therefore also about risk management and corporate accountability.

Regardless of whether the organisational structure is centralised or decentralised, it should be designed for effective implementation of environmental policies and strategies, and supported by senior management as that sends the strongest message (Friedman 2011). This also means that potential conflicts or turf problems between different divisions are minimised (ibid.). An effective structure alongside finding the right people is the first step in sustainability initiatives (Robertson 2017).

Furthermore, sustainability starts with where a company is and then takes deliberate steps in the direction it wants to move towards – a vision (Robertson 2017). This is followed by measurement, action, monitoring, evaluation and adaptation and a return to the vision – the cyclical steps for continuous improvement (Robertson 2017), which we also covered in Chapter 11 on environmental management systems. When it comes to a clear vision and purpose for corporate sustainability, leadership and senior level support underline a company's commitment (Ferguson and Clarke 2012). It is to leadership we now turn.

Leadership, organisational culture and change management

Leadership is a key variable in developing environmentally sustainable organisations, as demonstrated in many studies (Russo and Minto 2012). Leadership, as shown in Figure 1.2 in the book's introductory chapter, brings together the various levels of sustainability – societal, organisational and individual – across all three elements of the triple bottom line: people, planet and profit (Cavagnaro and Curiel 2012). Although leadership is defined and theorised in a number of ways, it is commonly understood as the process of influencing or the ability to influence others to attain goals based on a shared vision – this process involves creating direction and alignment, and maintaining commitment (Laasch and Conaway 2015).

Leadership has an impact on organisational behaviour, which is about the dynamics and interplay between employees, leaders, teams and the organisation itself (Slocum and Hellriegel 2011). This also links in with culture, the importance of which for sustainability is stressed by Esty and Winston (2009). Although culture is an important foundation and driver for sustainability initiatives, it can also be an inhibitor, and therefore an understanding of how it impacts on employees' behaviour and actions is important in order to transform business practice (Howard-Grenville and Bertels 2012).

Every company comes with its own history, issues, constraints and organisation, which feed into the organisational culture (Friedman 2011). Beyond that, culture in companies consists of values, beliefs and underlying assumptions as well as attitudes and behaviours shared by employees, which therefore means that systems and processes are not enough to implement sustainability in an organisation (Weybrecht 2014). This is particularly important to remember as sustainability in one form or another is about change – whether it is changing how things work, changing people's thinking or their actions and behaviours (ibid.). HRM support can be key in enabling and driving a culture of environmental efficiency (Laasch and Conaway 2015).

Cavagnaro and Curiel (2012: 107) emphasise that "change will not come from the outside" and therefore change is an inside-out process. Although the external environment and the regulatory framework can drive companies towards sustainability, and there is indeed a role for external stakeholders (Howard-Grenville and Bertels 2012), it is leadership inside companies that will set change in motion and guide it by grounding it strategically in values, vision and mission first and then taking steps to implement it operationally and culturally.

Here we can differentiate between *management* of the more operational aspects and *leadership* of the more strategic aspects and culture[3] (Cavagnaro and Curiel 2012). Where leadership develops vision and ideas, and influences others to follow, management is concerned with coordinating, directing and controlling (Slocum and Hellriegel 2011). Despite this differentiation, there are significant overlaps, and effective implementation of sustainability requires leaders to incorporate a number of managerial roles, whether of an informational (monitor, disseminate, spokesperson), interpersonal (figurehead, leading, liaising) or decisional (entrepreneurial, handling disturbances, allocating resources, negotiating) nature (Daft et al. 2010). Although it is clear that we cannot rely on individuals alone when it comes to leading and managing for sustainability, the role of change agents in leading transformational change towards greening an organisation is crucial, alongside wider buy-in.

Leadership for sustainability thus requires bold as well as collaborative leadership, given the size and timescale of the challenges companies face (Visser 2012). The role of a sustainability leader is therefore often to steer her or his company through a process of transformation, which often involves helping employees make sense of the changes needed and thus building commitment (Blowfield 2013). Piasecki and colleagues (1999: 3) emphasise the comprehensive and diverse set of skills uniquely needed by "environmental path-breakers". Other terms we can use here beyond leadership for sustainability include environmental leadership, green leadership or eco-leadership.

The leadership literature is full of the different traits, skills, styles and knowledge that leaders should have, based on a number of different theories (Blowfield 2013; Stead and Stead 2014). In his research, Visser (2012) concludes that there are seven key elements that differentiate leadership for (environmental) sustainability: systemic understanding, strong values, emotional intelligence, vision, inclusion, innovation and long-term perspective. For Stead and Stead (2014) servant leadership is an important foundation for sustainable management, as it is based on (environmental) stewardship as a core value and uses trust to build social capital. This links in with the guiding principle of product stewardship for e.g. strategy and design, as highlighted in Chapters 4 and 9.

As a single leader is unlikely to fully reflect Visser's (2012) key elements, collaborative work is all the more important. According to Robertson (2017), teams are at the core of any sustainability change initiative, whether they are transition teams for a time or permanent elements of the organisational structure. Looking then at the role of change agents more closely, we can set this in the context of change management, which is the process of transforming a company towards a shared goal, which involves both elements of facilitating and driving change as well as overcoming barriers to it (Laasch and Conaway 2015). It is important to acknowledge the latter, as change is hard, particularly when the broader worldview and fundamental assumptions of an economy based on a take-make-waste model run counter to the change needed (Amel et al. 2017). Millar and colleagues (2012) also acknowledge a number of tensions and dilemmas when it comes to organisational change and transformational vision for sustainability – for example balancing short-term priorities and day-to-day implementation with a long-term vision and strategic goals, change with stability, international with domestic responsibilities. Slawinski and colleagues (2017) emphasise the danger of short-termism and uncertainty avoidance that could lead to overall inaction, particularly in the area of climate change.

According to Visser (2012), there are four types of change agents – the expert, the facilitator, the catalyst and the activist. Although every change agent may draw on all four types, in reality each will have a preferred mode of operating. One particular role of change agents may be in being pioneers and innovators for finding entrepreneurial solutions to sustainability challenges in organisations. If working inside companies, such change agents are also called intrapreneurs who are tasked with introducing innovation to a business (Blowfield 2013). Other terms found in this context are sustainability intrapreneurs or entrepreneurs, environmental entrepreneurs or ecopreneurs. These can lead change without having a formal power base or leadership role.

Change can thus be led from the top or be initiated from the bottom up (Robertson 2017). When a top-down approach is used, change agents are crucial first-movers who should be nurtured, guided and supported in order to be successful in their roles (Ferguson and Clarke 2012). Training is one of the most used approaches when it comes to changing behaviour and culture (Daft et al. 2010), which will be covered in the next section.

As culture, change management and leadership cannot be covered comprehensively, we refer readers back to Chapter 4, in particular Box 4.10 and Table 4.3, for more on leadership and how change can be incremental or transformational. Furthermore, Weybrecht (2014), and Laasch and Conaway (2015) highlight the process of managing change with regard to steps for success and basic principles for avoiding errors and overcoming obstacles. For Elkington (1997) the sustainability transition (or revolution in his words) requires changes in a number of areas and as this happens both top-down and bottom-up, leadership is key.

Tools and concepts in HRM processes

In order to get employees, and (top) managers in particular, to focus on environmental sustainability and make it part of their job, appropriate signals need to be sent, whether they are in the form of including environmental goals in job descriptions and linking them to bonuses and awards or making environmental metrics part of a company's key performance indicators against which staff are assessed, appraised and promoted, all of which serve to embed environmental performance in company culture (Esty and Winston 2009). In the following we cover the key elements of the HRM process and the opportunity to embed sustainability in each.

Recruitment and selection

As environmental sustainability is complex and multidimensional (as frequently highlighted throughout the book), human resources need to come from a range of backgrounds and disciplines (Friedman 2011). When embarking on a journey towards greater sustainability, a starting point from an HRM perspective can be the review of job descriptions when recruiting (Friedman 2011). This could include the identification of the skills, knowledge and experience needed, which then needs to be communicated accordingly in order to find the right people via recruitment goals and strategies (Weybrecht 2014; Forbes and Jermier 2012). Making a move towards sustainability could also involve the creation of new roles related to the environment (Howard-Grenville and Bertels 2012). How resources are allocated reinforces the commitment of a company's leadership and management team and sends a signal to emphasise how important sustainability is to the organisation (ibid.).

This also means for example that interviewers should be familiar with the company's sustainability priorities, thus aligning recruitment strategy with sustainability strategy, regardless of which function or position is recruited for (Weybrecht 2014). The need for recruiters to be informed is confirmed by Klingenberg and Kochanowski (2015) as HR and thus recruitment is key to providing an organisation with appropriately trained and prepared staff when journeying towards sustainability. In their study they find that there is often a lack of alignment and connection between recruitment activities and a company's sustainability efforts, and they call for clearer recruitment policies that communicate the needs of companies and for educational activities for recruiters to ensure these needs are fulfilled in hiring.

In addition to recruiting for sustainability and offering green jobs, the process of hiring itself can be made more sustainable as well. The already widely implemented practice of online recruiting (or e-cruiting, Daft et al. 2010) cuts down on paper use whilst at the same time extending the recruitment reach of an organisation, which can result in a wider pool of candidates at reduced costs. It also allows recruiters to provide more information on a company's environmental activities (Renwick et al. 2013). A wider reach can, however, also lead to a greater geographical spread of potential hires. Companies therefore often use air transport to conveniently fly in candidates when train or road travel is not possible. Here, companies can reduce their environmental impact by replacing face-to-face interviews with virtual meetings, supported by communications and video conferencing software such as for example Skype and Zoom.

Whilst sustainability efforts can lead to marketing advantages more generally – as explored in Chapter 10, it can also be a factor in attracting more qualified (and motivated) staff and the ability to retain those already in the organisation (Robertson 2017; Renwick et al. 2013). For some potential employees a pro-environmental stance might be more important than financial remuneration (Cohen et al. 2011). This is supported by surveys of graduates when asked about their values and employment expectations (Esty and Winston 2009). Increasingly, graduates search for jobs that are in line with their values and ethics, looking for companies that make a positive impact (Jenkin 2015).

How a company contributes to sustainability can thus be a distinct factor in making a recruitment drive successful or not (Laszlo and Zhexembayeva 2011). And it does not stop there – once recruited, the commitment to sustainability continues to matter and contributes to employee engagement.

Working practices, employee retention and engagement

As working practices are also included in the other corporate functions in this part of the book, we cover them briefly and generally here.

In addition to reviewing the job description for new posts when recruiting, examining and rewriting existing job descriptions can support the integration of environmental goals in HRM (Friedman 2011) and thus change working practices.

Earlier in the chapter, we highlighted how commuting and business travel to meetings is one area where human resources impact on the environment. There are a number of ways in which companies can use HRM to reduce the environmental impact of their staff, for example in requiring employees to live within a certain radius, encouraging more home-based and flexible working, as well as telecommuting and virtual teams. Telecommuting involves using electronic equipment to do work that does not involve travelling to an office, whilst virtual teams are composed of geographically or organisationally dispersed members who hardly meet face-to-face but use electronic means to work together (Daft et al. 2010). In addition to reducing commuting through e-meetings, companies can encourage more environment-friendly means of transport for employees through cycle-to-work schemes, season ticket loan schemes, car-sharing schemes, providing facilities to park and change for cyclists, providing charging points for electric cars and charging for parking according to vehicles' CO_2 emissions. These can all serve to engage employees in environmental initiatives (Renwick et al. 2013, see also Weybrecht 2014).

As people are key to success in any corporate environmental management programme (Friedman 2011), they are particularly important when setting up programmes for greening offices and buildings, which not only concern the physical infrastructure but also how they are used in terms of energy, water, waste and recycling, electronics (Weybrecht 2014).

One particular important tool for companies to implement sustainability throughout an organisation is environmental management systems (EMSs), which were addressed in Chapter 11. EMSs rely on staff participating and being involved. This is particularly stressed by the EU's Eco-Management and Audit Scheme (EMAS) as it creates a sense of responsibility for the environment (Daddi et al. 2015). At a minimal level, this includes informing employees, though active participation and making use of staff expertise will yield better results, particularly if there is support from senior management (ibid.). Green HRM facilitates such employee participation and engagement and is therefore key in implementing EMSs in a bid to improve environmental performance (Mariappanadar 2019).

For Esty and Winston (2009), giving employees permission to question any sustainability initiative is an engagement tool that can lead to environmental improvements. Furthermore, enabling employees to contribute suggestions for how to make such improvements and linking them to awards can change the culture and influence behaviour. This may start with identifying problems which may lead to suggestions or how their own jobs could be done better (Weybrecht 2014). Here, having a whistleblowing system in place, where employees who witness unsustainable behaviour can report this, can help the company in its sustainability efforts (Weybrecht 2014). This should be confidential and accompanied by training and communication as well as support by senior management.

Another starting point is the formation of so-called green teams, which are employee-driven initiatives to generate awareness and enhance practices around energy use, waste reduction and recycling, paper use for printing and so on (Cohen and King 2017; Laasch and Conaway 2015). This can lead to cost savings and reduced environmental impacts as well as to employees taking personal action to reduce their environmental footprints at home. From an HR perspective, there are additional benefits in terms of retention and motivation that participants derive from being ambassadors for sustainability. Further advantages come in the form of strengthened retention in that recruitment and training costs are lower as well as job satisfaction and commitment higher in companies with HRM policies on sustainability (Cohen et al. 2011).

Whilst green teams and building coalitions are important in advancing sustainability, environmental (or eco) champions can also be good starting points for bringing such issues on to the agenda of a company, acting both as models for behaviour and ambassadors for values

(Howard-Grenville and Bertels 2012). Corporate policies can make such behaviour explicit and acceptable, and promote shared sustainability values (ibid., Forbes and Jermier 2012).

Engaged employees can also help recruitment through recommendations (Weybrecht 2014). Other benefits of employee engagement with corporate sustainability include improved productivity and resource efficiency, reduced environmental impact, a more committed and loyal workforce that is retained and a stronger identity and brand (Ferguson and Clarke 2012). Engaging employees requires both rational and emotional aspects and recognising that top-down is not the only approach. Furthermore, employee involvement can be formal or informal (ibid.)

In order to empower and engage employees, employee involvement practices can include suggestions schemes, newsletters, problem-solving groups, green (action) teams, green or low-carbon champions, which should all be supported by managerial support and leading by example as part of a wider pro-environment supportive organisational culture (Renwick et al. 2013). In addition to providing such green opportunities it is also crucial to develop green abilities beyond recruitment and selection through training and development.

Training and organisational development

> Sustainability is a learning process.
>
> (Visser 2012: 207)

Training and development for environmental sustainability ranges from raising awareness to building concrete skills and knowledge, and practical training. As the sustainability profession is still an evolving one, the role of training and development is particularly important in light of the need for multiple skills (e.g. project management), diversity and integrated thinking across and within systems.

It starts with employee orientation for new recruits and continues with training for existing employees. Renwick and colleagues (2013) stress that training is seen as a key intervention in Green HRM as it can make staff aware of the environmental impact of a company's activities, equip them with skills and raise the level of expertise and eco-literacy, whether it is regarding waste and recycling, energy and transport, or procurement and sourcing supplies.

Esty and Winston (2009) differentiate between specific training on particular topics such as eco-efficiency or regulatory compliance, general education on sustainability to raise knowledge and awareness and to help employees engage with sustainability in their lives as well as work, and high-level big-picture programmes for executives. Yoxon (2011) distinguishes between strategic, functional and individual training needs which feed into the training cycle of initial analysis and design, delivery and evaluation of effectiveness. Staff development is a supporting factor for all corporate functions covered e.g. Seuring and Müller (2008) emphasise the role of training employees as well as suppliers in making supply chain management more sustainable.

Training might need to start with educating the HR department or function so it can fully support a company's sustainability efforts (Friedman 2011). Weybrecht (2014) identifies leadership and influencing, communicating, knowledge and awareness as well as hard and soft skills as competences needed for sustainability.

EMSs can be a key mechanism for building awareness and providing training to employees (Grant et al. 2017). They allow companies the identification of training needs within the organisational structure defined for the EMS (Daddi et al. 2015). Roles and responsibilities may need to be defined and the structure adapted by the HR function in order to achieve the goals set. Only if staff are appropriately informed and trained can an EMS be effective (ibid.). Such training can consist of education ("know who you are"), information ("know-how") and practical training ("how to"), all geared towards promoting learning, and changes in behaviour and organisational culture (Daddi et al. 2015: 93). Failure to train employees appropriately could lead to costly errors for organisations, for example when environmental regulations are breached and thus a culture of learning should be encouraged as implementing legislation is non-negotiable (Yoxon 2011). Successful training should change how people behave and could thus also contribute to a change in corporate culture.

Training can take a number of forms and shapes, depending on resources available and corporate needs. Weybrecht (2014) differentiates between in-class and web-based training, whilst Yoxon (2011) distinguishes between group and individual training methods. Training can be offered internally as well as externally. In the latter category, we find professional bodies and chartered institutes for various professions, which can offer qualifications, accreditation and recognition of use to employees beyond their immediate employers. Table 13.1 lists some of the methods available.

Regardless of which method is used, post-training follow-up can be as important as the training itself to ensure the learning is then applied at the workplace, and to measure its effectiveness

TABLE 13.1

Training methods for green HRM

Training method	Characteristic
Lectures	Low cost but also low effectiveness. Good for general overviews
Group discussions	Better participation compared to lectures
Role playing	Simulates real world and can thus be a powerful technique to bridge theory and practice
Audio-visual	Videos and audio-visual presentations can break up training sessions, but effectiveness depends how well aligned they are with organisational needs
Case studies	Can relate specifically to organisational needs and works with other methods
Games (board or computer based)	Effective if integrated with overall training strategy
Outdoor training	Good for team building and team skills, higher costs
Open/distance learning	High flexibility but requires strong self-motivation. Best when combining knowledge-based activities with work-based practical activity.
IT-based or web-based learning	Easy to track progress. Can fully engage learner at own pace.
Coaching and mentoring	More experienced staff can provide support; complements other trainings interventions
On-the-job-training	Combines knowledge, skills, and techniques. Depends highly on instructor
Other	Secondments, job rotation, placements, community engagement

Sources: Yoxon 2011; Weybrecht 2014.

(Webybrecht 2014; Yoxon 2011). Furthermore, Renwick and colleagues (2013) stress the importance for training to create an emotional involvement in environmental matters which can result in greater ownership of the issue by employees.

Development of green abilities initially through recruitment and then training, and providing green opportunities for employee engagement provides a solid foundation for green HRM practices. In addition, it is important to motivate green employees through performance management and appraisal, and through pay and reward systems (Renwick et al. 2013).

Performance management: incentives, rewards, motivation, compensation and benefits

Performance management concerns the process of monitoring employees' performance and its improvement, aligned with a company's goals (Laasch and Conaway 2015).

Reflecting environmental objectives and priorities in bonuses and performance review systems makes employees more accountable and enables them to take ownership of environmental success, thus advancing the corporate environmental strategy (Esty and Winston 2009). Howard-Grenville and Bertels (2012) also highlight that when individual efforts are rewarded, commitment to environmental objectives is increased. Whether it is pollutant emissions or nonproduct output (waste), tying bonuses and other rewards to environmental performance indicators is a starting point (Piasecki et al. 1999). This is where we see the link between the chapter on accounting and reporting and this chapter, as the indicators and measurements used in accounting are not ends in themselves but can be used to foster learning and adaptation in behaviour (Robertson 2017). If incorporated in performance appraisal, organisational sustainability goals and targets can be aligned with personal objectives and thus ensure both are met (Weybrecht 2014). A helpful tool here can be the sustainability balanced scorecard, which we introduced in Chapters 11 and 12, which combines financial and non-financial measures of performance and allows for strategic alignment of economic and sustainable measures. Another link to accounting and reporting can be found in the Global Reporting Initiative (GRI), where half of all indicators are considered to relate to HR (Weybrecht 2014).

A further link to the previous chapter on accounting can also be made in terms of eco-controlling, which is the "systematic process of managing and achieving environmental goals" (Schaltegger et al. 2003: 178). Part of eco-control is the allocation of resources and the setting of incentive systems designed to manage and achieve such environmental goals set by an organisation.

When environmental goals are not met, a performance management system can also include "dis-benefits", in the form of negative reinforcements such as criticism, warnings and even suspensions, though there is the danger that this does not necessarily educate employees in good practice and that environmental problems risk not being reported as a result of self-protective behaviour (Renwick et al. 2013: 9).

Adding another element to performance management, not only bonuses but also promotions can be linked to achieving sustainability objectives, particularly if integrated in job descriptions (Weybrecht 2014). Objectives should be measurable with the help of KPIs (key performance indicators) (Ferguson and Clarke 2012).

An alternative perspective is provided by Francoeur and colleagues (2017) who find that when it comes to designing compensation for CEOs, CEOs do not necessarily act opportunistically as agency theory would suggest, being motivated largely by monetary compensation, but may be rewarded by taking on a stewardship role for the natural environment as a major organisational stakeholder. Compensation committees therefore need to consider not just financial but also psychological and institutional factors when it comes to compensating and motivating CEOs (Francoeur et al. 2017). The importance of intrinsic motivation and moral duties in pursuing environmental goals as part of wider corporate strategies applies beyond the executive level. We already highlighted in the recruitment section that a firm's environmental stance can be more important than financial remuneration for some job candidates (Cohen et al. 2011).

As well as for senior management, it is often non-monetary recognition that is given to reward environmental performance achieved by employees (Renwick et al. 2013). This might include individual or team awards for most waste reduced or most energy saved. Such rewards can significantly influence employee willingness to produce eco-initiatives.

Other financial incentive alternatives to manage and motivate staff for environmental sustainability include tax exemptions for example for bicycle loans, or cash incentives for buying or using lower-polluting cars. Alternatively, green benefits rather than pay benefits can be offered (Renwick et al. 2013).

Employee relations and communications

Employee (or employment) relations concerns the relationship between employers and employees and is often associated with freedom of association and unionisation (Laasch and Conaway 2015). It is also called industrial relations, which concerns the various employment relationships between not just employers and employees, but also employees and their trade or labour unions, and employers and their associations as well as government as it regulates employment (Mariappanadar 2019). As trade unions have an established history of taking action on environmental issues, for example in workplace health and safety, they should be mentioned in the context of HRM as well, since they can be important partners for joint training programmes for environmental management, for example (Renwick et al. 2013). Furthermore, through collective bargaining – a basic labour right (Laasch and Conaway 2015) – employees can jointly press for better working conditions that protect them from environmental harm (e.g. when working with hazardous chemicals or toxic waste).

In order for a corporate environmental strategy to be successful, it needs to be communicated, and opportunities have often been missed as a result of failed communication of change (Piasecki et al. 1999). Employees need to hear about positive examples of the change initiated and will be in a better position to make decisions when they are well informed (Robertson 2017). Sustainability communication can be both top-down or bottom-up, although traditionally such internal communication has been one-way from the top (Weybrecht 2014; Laasch and Conaway 2015). The importance of communication is underlined by Yoxon (2011), who warns that failures may result in costly errors, loss of trust and cooperation between managers and staff, and may affect motivation. Communication (alongside training) can help equip employees with skills, knowledge and motivation in facilitating change. If done openly, collaboratively and effectively, two-way communication can turn employees into ambassadors for a company's sustainability approach, in turn building value and reputation for it (Laasch and Conaway 2015).

Spotlight on . . . examples and practices of HRM for environmental sustainability

In the following, a few examples and practices of HRM for environmental sustainability and green(ing) HRM will be highlighted.

When Walmart asked its top-tier suppliers in 2010 to assess their environmental practices and informed them that future supplier selection would include environmental criteria, the company contributed to a hiring drive of environmental specialists (Laasch and Conaway 2015). The company has also built employee buy-in and employee engagement through launching a Personal Sustainability Project (PSP) programme, where employees could voluntarily integrate sustainability into their daily lives (Laszlo and Zhexembayeva 2011, see also Esty and Winston 2009).

Telecommunications provider ECI Telecom has been running internal competitions to engage employees in environmental awareness (Cohen and King 2017). Similarly, SAP has used internal comparisons to change organisational and individual behaviour by making use of information systems, which shows the link between this chapter and Chapter 11 on environmental management systems. By using key environmental performance indicators that are broken down on an individual basis, employees receive information about their footprint and how much smaller or bigger it is in comparison to other employees. According to the company, this empowers employees to make more informed decisions (e.g. about travel options) and to positively change behaviour. Furthermore, sustainability targets and goals are set to increase awareness and encourage personal accountability of employees for their environmental actions (PWC 2011).

In terms of addressing the environmental impact of staff commutes, several universities in the UK have started charging their employees according to the CO_2 emissions of their vehicles, with vehicles emitting less than 100 or 120 g CO_2/km being charged a significantly reduced fee (e.g. Nottingham University) or no fee at all (e.g. Loughborough University), and more carbon-intensive cars being charged more, resulting in five or six different fee bands. In order to address concerns over social and economic impacts, charging is often differentiated according to pay grade, thus not disadvantaging staff on lower incomes disproportionately.

In addition to setting public goals for reducing carbon emissions not just from its own operations but also in its supply chain,[4] BHP, one of the world's largest mining companies, has started linking executive pay and remuneration plans with the achievement of environmental targets (Hume 2019).

The importance of leadership in making organisations more sustainable is demonstrated by the values-based leadership of Paul Polman, former CEO of Unilever, who abandoned quarterly financial statements to focus on sustainable long-term perspectives. In its Sustainable Living Plan launched in 2010, the company ambitiously announced doubling revenues, whilst halving its environmental footprint[5] (Visser 2012; Blowfield 2013). Polman's leadership extends beyond Unilever, as he persuaded other business leaders to support the 2015 Paris climate accord (Gelles 2019). Similar transformational leadership was exhibited by Lee Scott, who ran Walmart from 2000 to 2009, and started the company's sustainability programme – his influence can be seen in the company's continued efforts, including the Sustainable Consortium which helps a range of commercial and non-profit stakeholders to collaborate on sustainability (Spicer and Hyatt 2018).

SUMMARY

The success of sustainability strategies and programmes depends on their integration into day-to-day operations and a company's culture (Webyrecht 2014). They cannot be done without employees and thus the human resource management function, as people are key to success when it comes to environmental management (Friedman 2011). As with previous chapters in part three, we introduced the function with an overview and highlighted its environmental impact and how it is impacted by the environment, thus making the business case for greening HRM. Since employees are also part of the environmental impacts of all the other functions, this puts HRM in a significant leadership role for making whole organisations more sustainable. The effort includes how organisations are structured and governed. As already emphasised in Chapter 4, leadership, organisational culture and change management are central to HRM's role in sustainability. There is a place for entrepreneurial initiatives in the form of ecopreneurship and environmental intrapreneurship, which recognise opportunities and drive innovation in this direction. In the final part of the chapter we looked at how each phase and element of the human resource management process, from recruitment and selection to training, from working practices to performance management as well as employee relations and communication can engage employees in a company's sustainability efforts through green HRM. Highlighting some selected examples of environmentally sustainable HRM practices underpinned this final part.

This chapter concludes part three on corporate functions, having looked at the main primary activities of the value chain as well as the main support activities.

In the following final part of the book, a chapter with case studies provides an opportunity to pull together the various parts of the book before we provide a conclusion and a look ahead to the future of environmental sustainability in the closing chapter of the book.

SUGGESTED QUESTIONS AND ACTIVITIES

- Explore recruitment agencies and online job boards that focus on sustainability and the environment.
- Choose one of the environmental challenges covered in part two and explore in detail how HRM can contribute to addressing it.
- Look up some of the corporate examples and practices mentioned in the chapter, and research them in more detail.
- What other examples of green HRM (green recruitment, green performance management, green training and development and so on) can you find?
- Check out the Institute for Environmental Management and Assessment (www.iema.net) – what training courses are available and which one(s) would you consider taking? What do other professional bodies and chartered institutes offer in terms of environmental management training and greening HRM?
- For more in-depth background material, check out the following books and websites:
  - Dunphy, D., Griffiths, A. and Benn, S. (2007) *Organizational Change for Corporate Sustainability: A Guide for Leaders and Change Agents of the Future*, London: Routledge, 2nd ed.
  - Special issue of *Journal of Organizational Change Management*, Volume 25, Issue 4 in 2012 on sustainability and the need for change in organisations: www.emerald.com/insight/publication/issn/0953-4814/vol/25/iss/4.
  - Benn, S., Dunphy, D. and Griffiths, A. (2014) *Organizational Change for Corporate Sustainability*, London: Routledge, 3rd ed.
  - Mariappanadar, S. (2019) *Sustainable Human Resource Management: Strategies, Practices and Challenges*, London: Red Globe Press.

NOTES

1 HR and HRM are used interchangeably.

2 For more on organisational structure see Daft et al. 2010.

3 Much has been written about the difference and overlap between management and leadership, which we cannot cover in greater detail here (see Daft et al. 2010 and Slocum and Hellriegel 2011 for more). Sustainability certainly needs both the vision, courage and creativity of leadership as well as the structured approach, attention to detail and problem-solving attitude of management.

4 Scope 3 emissions in addition to Scope 1 and 2, see Chapter 5 on climate change for more details.

5 This is reminiscent of *Factor Four* by Weizsäcker and colleagues (1998), which is about doubling wealth and halving resource use.

SOURCES AND FURTHER READING

Amel, E., Manning, C., Scott, B. and Koger, S. (2017) Beyond the Roots of Human Inaction: Fostering Collective Effort Toward Ecosystem Conservation, in: *Science*, 356(6335): 275–279.

Baxter, M. (2011) Environmental Management Systems, in: Brady, J., Ebbage, A. and Lunn, R. (eds.) *Environmental Management in Organizations, The IEMA Handbook*, London: Earthscan/Routledge, 2nd ed., pp. 261–273.

Birchby, D., Cooke, S., Stedman, J., Hurley, F., Searl, A. and Hunt, A. (2014) Valuing the Impacts of Air Quality on Productivity, Report for the Department for Environment, Food and Rural Affairs, Ricardo-AEA/R/3417, available at https://uk-air.defra.gov.uk/assets/documents/reports/cat19/1511251135_140610_Valuing_the_impacts_of_air_quality_on_productivity_Final_Report_3_0.pdf (Retrieved, 28.10.2019).

Blowfield, M. (2013) *Business and Sustainability*, Oxford: Oxford University Press.

Brady, J. (2011) The Response of Organizations, in: Brady, J., Ebbage, A. and Lunn, R. (eds.) *Environmental Management in Organizations, The IEMA Handbook*, London: Earthscan/Routledge, 2nd ed., pp. 251–260.

Cavagnaro, E. and Curiel, G. (2012) *The Three Levels of Sustainability*, Sheffield: Greenleaf Publishing.

Cohen, E. and King, D. (2017) Human Resources Management: Developing Sustainability Mindsets, in: Molthan-Hill, P. (ed.) *The Business Student's Guide to Sustainable Management*, Sheffield: Greenleaf Publishing, 2nd ed., pp. 263–289.

Cohen, E., Taylor, S. and Muller-Camen, M. (2011) HR's Role in Corporate Social Responsibility and Sustainability, SHRM Foundation Executive Briefing, Society for Human Resource Management, available at www.shrm.org/foundation/ourwork/initiatives/building-an-inclusive-culture/Documents/HR%27s%20Role%20in%20Corporate%20Social%20Responsibility.pdf (Retrieved, 16.09.2019).

Daddi, T., Iraldo, F. and Testa, F. (2015) *Environmental Certification for Organisations and Products*, London: Routledge.

Daft, R. L., Kendrick, M. and Vershinina, N. (2010) *Management: International Edition*, Andover: Cengage Learning EMEA.

Elkington, J. (1997) *Cannibals with Forks: The Triple Bottom Line of 21st-Century Business*, Oxford: Capstone.

Esty, D. C. and Winston, A. S. (2009) *Green to Gold: How Smart Companies Use Environmental Strategy to Innovate, Create Value and Build Competitive Advantage*, Hoboken, NJ: Wiley.

Ferguson, D. and Clarke, M. (2012) Enabling the Change: Corporate Sustainability and Employee Engagement, in: Grayson, D. and Exter, N. (eds.) *Cranfield on Corporate Sustainability*, Sheffield: Greenleaf Publishing, pp. 167–181.

Forbes, L. D. and Jermier, J. M. (2012) The New Corporate Environmentalism and the Symbolic Management of Organizational Culture, in: Bansal, P. and Hoffman, A. J. (eds.) *The Oxford Handbook of Business and the Natural Environment*, Oxford: Oxford University Press, pp. 556–571.

Francoeur, C., Melis, A., Gaia, S. and Aresu, S. (2017) Green or Greed? An Alternative Look at CEO Compensation and Corporate Environmental Commitment, in: *Journal of Business Ethics*, 140(3): 439–453.

Friedman, F. B. (2011) *Practical Guide to Environmental Management*, Washington, DC: Environmental Law Institute, 11th ed.

Gelles, D. (2019) He Ran an Empire of Soap and Mayonnaise. Now He Wants to Reinvent Capitalism, *New York Times*, 29.08.2019, available at www.nytimes.com/2019/08/29/business/paul-polman-unilever-corner-office.html (Retrieved, 04.10.2019).

Grant, D. B., Trautrims, A. and Wong, C. Y. (2017) *Sustainable Logistics and Supply Chain Management: Principles and Practices for Sustainable Operations and Management*, London: Kogan, 2nd ed.

Howard-Grenville, J. and Bertels, S. (2012) Organizational Culture and Environmental Action, in: Bansal, P. and Hoffman, A. J. (eds.) *The Oxford Handbook of Business and the Natural Environment*, Oxford: Oxford University Press, pp. 194–210.

Hume, N. (2019) BHP to Set Targets for Reducing Customers' Carbon Emissions, *Financial Times*, 23.07.2019, available at www.ft.com/content/90b8fdd0-ac87-11e9-8030-530adfa8

79c2?emailId=5d378119a2f8970004ba7b3a (Retrieved, 04.10.2019).

Jenkin, M. (2015) Millennials Want to Work for Employers Committed to Values and Ethics, *The Guardian*, 05.05.2015, available at www.theguardian.com/sustainable-business/2015/may/05/millennials-employment-employers-values-ethics-jobs (Retrieved, 16.09.2019).

Kakabadse, A. and Grayson, D. (2012) Embedding the Governance of Responsibility in the Business of the Board, in: Grayson, D. and Exter, N. (eds.) *Cranfield on Corporate Sustainability*, Sheffield: Greenleaf Publishing, pp. 64–79.

Klingenberg, B. and Kochanowski, S. M. (2015) Hiring for the Green Economy: Employer Perspectives on Sustainability in the Business Curriculum, in: *Journal of Management Development*, 34(8): 987–1003.

Laasch, O. and Conaway, R. (2015) *Principles of Responsible Management: Glocal Sustainability, Responsibility and Ethics*, Stamford, CT: Cengage Learning.

Laszlo, C. and Zhexembayeva, N. (2011) *Embedded Sustainability: The Next Big Competitive Advantage*, Sheffield: Greenleaf Publishing.

Mariappanadar, S. (2019) *Sustainable Human Resource Management: Strategies, Practices and Challenges*, London: Red Globe Press.

Millar, C., Hind, P. and Magala, S. (2012) Sustainability and the Need for Change: Organisational Change and Transformational Vision, in: *Journal of Organizational Change Management*, 25(4): 489–500.

Noe, R. A., Hollenbeck, J. R., Gerhart, B. and Wright, P. M. (2015) *Human Resource Management: Gaining a Competitive Advantage*, Maidenhead: McGraw-Hill Education, 9th ed.

Piasecki, B. W., Fletcher, K. A. and Mendelson, F. J. (1999) *Environmental Management and Business Strategy: Leadership Skills for the 21st Century*, New York: John Wiley & Sons.

PWC (2011) Building Sustainable Companies, in: *PwC Technology Forecast*, 2011(4), available at www.pwc.pt/pt/technology/images/pwc-building-sustainable-companies.pdf (Retrieved, 04.10.2019).

Renwick, D. W. S., Redman, T. and Maguire, S. (2013) Green Human Resource Management: A Review and Research Agenda, in: *International Journal of Management Reviews*, 15(1): 1–14.

Russo, M. V. and Fouts, P. A. (1997) A Resource-Based Perspective on Corporate Environmental Performance and Profitability, in: *Academy of Management Journal*, 40(3): 534–559.

Russo, M. V. and Minto, A. (2012) Competitive Strategy and the Environment: A Field of Inquiry Emerges, in: Bansal, P. and Hoffman, A. J. (eds.) *The Oxford Handbook of Business and the Natural Environment*, Oxford: Oxford University Press, pp. 29–49.

Schaltegger, S., Burritt, R. and Peterson, H. (2003) *An Introduction to Corporate Environmental Management*, Sheffield: Greenleaf Publishing.

Seuring, S. and Müller, M. (2008) From a Literature Review to a Conceptual Framework for Sustainable Supply Chain Management, in: *Journal of Cleaner Production*, 16(15): 1699–1710.

Slawinski, N., Pinkse, J., Busch, T. and Banerjee, S. B. (2017) The Role of Short-Termism and Uncertainty Avoidance in Organizational Inaction on Climate Change: A Multi-Level Framework, in: *Business and Society*, 56(2): 253–282.

Slocum, J. W. Jr. and Hellriegel, D. (2011) *Principles of Organizational Behavior*, Andover: Cengage Learning, 13th ed.

Spicer, A. and Hyatt, D. G. (2018) Walmart Tried to Make Sustainability Affordable. Here's what Happened, The Conversation, 13.08.2018, available at https://theconversation.com/walmart-tried-to-make-sustainability-affordable-heres-what-happened-76771 (Retrieved, 04.10.2019).

Stead, J. G. and Stead, W. E. (2014) *Sustainable Strategic Management*, Sheffield: Greenleaf Publishing, 2nd ed.

Visser, W. (2012) *The Quest for Sustainable Business: An Epic Journey in Search of Corporate Responsibility*, Sheffield: Greenleaf Publishing.

von Weizsäcker, E., Lovins, A. B. and Lovins, L. H. (1998) *Factor Four: Doubling Wealth, Halving Resource Use*, London: Earthscan.

Weybrecht, G. (2014) *The Sustainable MBA: A Business Guide to Sustainability*, Chichester: Wiley, 2nd ed.

Yoxon, M. (2011) Engaging with People in Your Organization, in: Brady, J., Ebbage, A. and Lunn, R. (eds.) *Environmental Management in Organizations, The IEMA Handbook*, London: Earthscan/ Routledge, 2nd ed., pp. 333–353.

Conclusion

In this concluding part of the book we draw the main three parts of the book on foundations, challenges and functions together, firstly, in Chapter 14 through integrative case studies allowing the application of the material covered so far. Secondly, in Chapter 15 we conclude the book by providing an outlook into the future for the environmental challenges and the corporate functions covered, drawing together the introductory chapter on the past and present with reflections on what the future might hold for environmental sustainability in business.

CHAPTER 14

Integrative case studies

INTRODUCTION

In the following, three integrative case studies provide readers with an opportunity to put the material covered in the book so far to action in terms of foundations, challenges and functions. Students and readers are encouraged to think, plan and act systematically, strategically and holistically when applying sustainability in practice. The case studies are of different lengths and cover different sectors to ensure both depth and breadth. To start with, we provide a brief overview of sectors and their classifications and then a brief introduction to the three cases. The cases are then presented, all including discussion questions pertaining to the three main parts of the book.

Classification of industries and sectors

It is useful to differentiate between different types of businesses and industries as the environmental challenges they face can vastly differ as well as be shared to a large extent. The following thus serves as a background for applying the book material to cases and companies. It is also accompanied by some general questions to compare and contrast different sectors of the economy from an environmental perspective.

Industries and sectors can be classified in various ways. Broadly, we can differentiate between three sectors – the primary (raw material extraction – e.g. agriculture, mining), the secondary (manufacturing) and the tertiary (services) sectors, to which some also add the quaternary sector of knowledge-based industries.

International institutions as well as governments (e.g. through national statistics offices) use various systems to classify industries and sectors, often based on the Standard Industrial Classification (SIC) system for purposes of data provision, analysis and comparison. At the national level, examples include the United Kingdom Standards Industrial Classification of Economic Activities (UKSIC) and the Singapore Standard Industrial Classification (SSIC), whilst at the regional level we find The North American Industry Classification System (NAICS) and the European Union's classification system (NACE[1]). The latter corresponds to the United Nations' industry classification system, the so-called International Standard Industrial Classification of All Economic Activities (ISIC), which is the international reference classification of productive activities. These various systems are mainly based on activities and production but can also be based on products or occupations, and often countries adopt similar systems (UN Statistics Division 2019).

TABLE 14.1

United Nations International Standard Industrial Classification of All Economic Activities (ISIC)

A	Agriculture, forestry and fishing
B	Mining and quarrying
C	Manufacturing
D	Electricity, gas, steam and air conditioning supply
E	Water supply; sewerage, waste management and remediation activities
F	Construction
G	Wholesale and retail trade; repair of motor vehicles and motorcycles
H	Transportation and storage
I	Accommodation and food service activities
J	Information and communication
K	Financial and insurance activities
L	Real estate activities
M	Professional, scientific and technical activities
N	Administrative and support service activities
O	Public administration and defence; compulsory social security
P	Education
Q	Human health and social work activities
R	Arts, entertainment and recreation
S	Other service activities
T	Activities of households as employers; undifferentiated goods- and services-producing activities of households for own use
U	Activities of extraterritorial organizations and bodies

Source: based on UN Statistics Division 2008.

Table 14.1 shows the latest revision to the UN's ISIC.

Many national systems, such as the UKSIC system, use the same headlines, which are then further broken down into (industry) groups, industries or subsectors and classes. In its sector analysis, the UK government furthermore differentiates between *low-med tech* (e.g. shipbuilding, metal products, food, beverages and tobacco) and *med-high tech* (e.g. pharmaceuticals, automotive and electrical equipment) in manufacturing, and between *knowledge services* (e.g. education, communications, financial and digital services) and *other services* (e.g. retail, real estate, and transport), whilst the term *other production* includes agriculture, mining, utilities and construction (BIS 2012), i.e. activities in the primary sector.

Other, non-governmental classifications are based on markets and companies rather than production and establishment and include for example the Global Industry Classification Standard (GICS), developed by Morgan Stanley Capital International (MSCI) and Standard & Poor's, and Thomson Reuters Business Classification (TRBC) as well as Bloomberg's.

The TRBC classifies companies according to five (hierarchical) levels: economic sectors, business sectors, industry groups, industries and activities. The ten economic sectors are energy, basic materials, industrials, consumer cyclicals, consumer non-cyclicals, financials, healthcare, telecommunications services, and utilities (Refinitiv 2019). A similar taxonomy is used by the GICS, which differentiates between consumer discretionary and consumer staples, and adds real estate to its list (GICS 2018).

At this point, it is pertinent and appropriate to acknowledge the limitations of such classifications – as we have highlighted in various parts of the book (e.g. Chapter 4 and Chapter 9), the move to servicisation (or servitization) where products are substituted by services blurs the boundaries between manufacturing and service activities. It represents a shift in thinking towards a service and flow economy where companies keep ownership of products through their life cycle, embodying product stewardship by delivering functionality and a service rather than ownership and a product. Chemical leasing and pay-per-use schemes are examples in this respect.[2]

SUGGESTED QUESTIONS AND ACTIVITIES

- How useful is the classification of the economy into sectors and industries?
- Choose one sector or industry and explore the state of sustainability thinking in the sector. A starting point for sources of information here might be the industry association.
- What kind of environmental challenges is the sector faced with (e.g. in terms of cause and effect)? Climate challenge, pollution, waste, others?
- What ecosystem services does the sector/ industry use?
- What's the ecological footprint of the sector/ industry? (carbon, water, materials and so on)
- Is there an industry or sector code of conduct? Or a set of guidelines and principles?
- What role do standards, labels and/or certificates play for the sector (e.g. ISO 14000 series)?
- Find examples of Life Cycle Analysis (LCA), Design for Environment (DFE), Industrial Ecology used in the sector.
- Are any of the functions covered in part three of the book of particular relevance for the sector in question?
- If you need a suggestion for a sector to research, use the fashion industry, in particular fast fashion, and explore how sustainable fashion (also called ethical fashion) is addressing the environmental impact of the sector.

Overview of case studies

The first case study is a longer, three-part case study on manufacturing, whilst the second case study is of medium length and on services. The third case study is a short scenario set in the primary sector of farming. Here, we briefly introduce each.

Case study 1: Start-ups and coffee cups – manufacturing sector

The first case study is a three-part case on the decisions a young student start-up venture faces when trying to launch a sustainability product – reusable coffee cups. This involves marketing and pricing decisions, production and material decisions, negotiations with suppliers and so on. Whilst the first part (A) considers the broader remit of sustainability and making the business case for sustainability as well as starting a business, the second part (B) concerns the decisions around material choice for production, location choice and marketing in the context of various environmental concerns. The third part (C) narrows down the decisions around material choice and considers strategies for pricing, marketing and growth.

Part (A) could for example be used after part one of the book, and part (B) after part two, whilst part (C) could be used after part three. Alternatively, the entire case study can be used at the end of a course on business and environmental sustainability to apply the knowledge gained and learning undertaken. The main environmental challenge the case addresses is waste, but students are encouraged to consider others as well.

This case is an intermediate experience case and has been inspired by the author's different experiences of sustainability while living and studying in Germany and then living, studying and working in the UK, complemented by experiences of travelling in other countries.

Case study 2: Greenbelt Festival – services sector

The second case study is about how the Greenbelt Festival, a festival for arts, faith and activism, navigated the environmental challenges of changing location and sites from a predetermined environment at a race course to a greenfield site near a stately home. The festival attracts upwards of 10,000 visitors every year on the August bank holiday weekend in the UK. The event combines "artistry and activism, spirituality and politics, faith and justice" and includes many known and lesser known musicians, comedians, politicians, writers, academics, poets, artists and so on. (Greenbelt 2019).

The festival's main environmental impact is in the form of waste and particularly of energy use, and thus climate change. Other challenges are faced as well, e.g. depending on the type of energy used (e.g. diesel generators contributing to noise and air pollution).

The case study looks at the general environmental impact of the festival and how it uses its progress to share and spread good practice. In 2018, the founder of the world's largest festival – Glastonbury – Michael Eavis, visited Greenbelt. The following year, Glastonbury followed Greenbelt's lead in banning the sale of plastic drink bottles on site, demonstrating a multiplication of Greenbelt's environmental efforts as Glastonbury attracts over 130,000 visitors a year compared to the 10,000 to 20,000 visitors Greenbelt attracts. The festival has announced plans to get rid of single-use plastic by 2020. The move is reflective of wider discussions in the festival and music industry with bands (e.g. Massive Attack and Coldplay) and singers increasingly acknowledging their environmental impact, particularly of touring and trying to address it, supported by e.g. the Green Touring Network, Julie's Bicycle and climate action group Music Declares Emergency (Holden 2018; Ellcock 2019).

The case is based on an interview with the Event Director of Greenbelt, Mary Corfield, in 2018, and the author's experience of attending the festival.

The discussion questions encourage students to apply and integrate the various parts of the book.

Case study 3: Established soya farmer – primary sector

The final case is a short scenario inspired by Ray Anderson's story, who radically changed his company Interface having read *The Ecology of Commerce* by Paul Hawken (1993). It centres on the decisions of a soya farmer in Latin America who decides to stop supplying beef farmers and to start growing soy sustainably for human consumption, shifting from conventional to sustainable production. This follows a change of heart of the farm owner, who had a lightbulb moment reading *Eating Animals* and *Farmageddon*.

The case is set within an emerging market, marking a contrast to the previous two cases. All three parts of the book are reflected in the case and material from all chapters may be applied. Whilst climate change is not explicitly mentioned in the case, this is an important issue to consider. Other issues and topics that the case addresses include genetically modified crops, growing crops for animals versus for humans, support of buyers and customers for a change in direction, food security, the rise of the market for vegan and vegetarian products. More implicitly, the case also allows students to explore the importance of environmental regulations and lobbying, the feasibility for vertical integration and downstream diversification, responsible resource use, EMS certification, and the importance of culture, leadership and values when taking possibly controversial decisions.

CASE STUDY 1: START-UPS AND COFFEE CUPS – MANUFACTURING SECTOR

Start-ups and coffee cups: frustration and first ideas (A)

Laura sat in the new university café, staring at the frothy remains of her cappuccino at the bottom of the paper cup. She felt frustrated – not just with herself for breaking her supposedly unbreakable coffee cup in a rush to get out of the house for her first lecture this morning but also with her university's catering services for not offering drinks in reusable mugs as standard, as they did in the staff canteen. They were offering reusable coffee cups for purchase, but money was tight this month and she had not budgeted for this extra expense. Her best friend Alex – possibly anticipating a different culture of recycling and waste? – had given her a coffee cup made from sustainably grown bamboo as a goodbye present before she left Denmark to start a Masters in International Business in the UK following her Bachelor's in Economics and Political Science. Although the cup had appeared sturdy, it did not survive the drop on the hard kitchen floor in her student dig this morning and bounced back with a big crack (Figure 14.1).

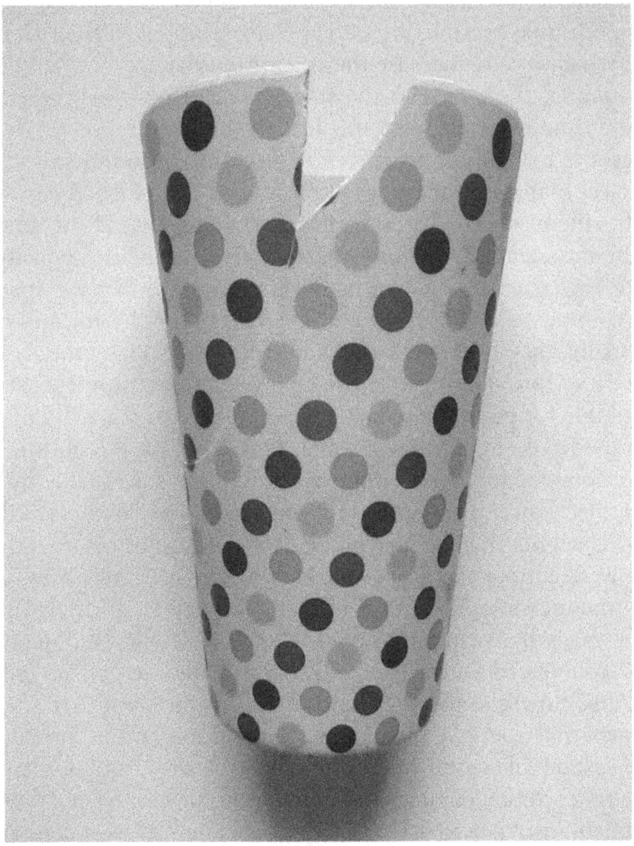

FIGURE 14.1
Broken bamboo cup.

In her university back home, the campus cafeteria would sell coffee only in ceramic mugs and factor in lost, broken or otherwise appropriated cups with a slightly higher (though still subsidised) price of hot drinks. Why could the university she had chosen for her postgraduate studies in the UK not do the same? They were supposedly big into sustainability, though she was starting to doubt that commitment. She was determined to find out more and tried to arrange a meeting with the service responsible for catering on campus. This was harder than expected as responsibilities for sustainability seemed to be across different campus services and posts: in the end she ended up with two meetings. The first one was with a manager of campus catering outlets who brought along a sustainability manager, and the second one with the senior project manager of the estates department alongside a junior project manager who had been involved in building the new catering outlet. Although the importance of sustainability was acknowledged in both meetings, operational constraints were emphasised and Laura couldn't help but think that more could be done to emphasise and implement the university's sustainability pledges.

In the first meeting, the catering staff were knowledgeable about sustainability concerns in their area and talked extensively about life cycle analysis, which in their area meant that the use of ceramic mugs and dishwashers were not necessarily more sustainable than paper cups for hot drinks, as water usage, detergent use and electricity to run dishwashers all needed to be part of the equation and calculation. The main constraint for them was, however, the lack of space in most catering outlets that had been designed for serving drinks in single-use cups, leaving no space for storing and cleaning reusable ceramic mugs. They also pointed to the risk of attrition because of breakage and theft, which would increase the cost of provision.

In the second meeting, she met with the staff who had been instrumental in getting the new university café up and running, which had involved reusing and recycling furniture and equipment from other buildings and installing passive ventilation and a ground source heat pump. This had been part of the project management team's effort to attain BREEAM certification, a scheme that independently certifies the sustainability performance of buildings. The team's project brief had only included the actual premises, and operational matters were not their responsibility and had therefore not been considered. Nevertheless, the two managers estimated that the space, energy and staffing needed to operate a commercial dishwasher, and the costs of purchasing and storing crockery would be significant. Informally, they told Laura that whilst they personally totally supported the concept of reusable mugs, whether provided by the café or brought by customers themselves – highlighting the KeepCups® available for purchase which gave students and staff 10p off hot drinks – they had the impression that the head of catering was mainly focused on the bottom line and cost-efficient purchasing, and therefore not so interested in exploring more sustainable catering.

The two meetings left Laura wondering – the university had clearly signed up to becoming more sustainable, that was clear, but there seemed to be a lack of resources and it didn't look like all parts of the organisation were supporting this effort or working together. Why wasn't there a sustainability champion in the university executive and at other levels, and why was there no joined-up approach or if there was, why wasn't it more visible? What could be done beyond that? Although the government had announced that it planned to ban single-use plastics (such as for example the lids for disposable coffee cups) and to introduce a so-called "latte levy", this was nowhere near being discussed in parliament and a long way from implementation and becoming a reality. Was there a different solution to sustainable coffee catering if the university could not live up to its promises?

It was a lecture on entrepreneurship and innovation on her master's course that finally made her realise that she might just have to do it herself. Launching her own company was not something she had ever contemplated, but her dissatisfaction with her experience showed how strongly she felt about this and gave her the motivation she needed. It shouldn't be this hard to get coffee in non-disposable mugs! After discussing her intention with her lecturer, she joined the entrepreneurship society on campus and tested her ideas for an affordable and sustainable coffee cup in the dragons' den style competition the club ran every term for students with proposals for their own venture. Laura was subjected to some tough questions, but she used the feedback she received to refine her concepts and plans, and entered the Entrepreneurship Awards of the university's Enterprise Hub.

While she appreciated the availability of KeepCups® for purchase on campus, they were still made of either plastic or glass – implying limited recyclability and increased breakability – and Laura wanted to address both with her product. Aware that she was inspired by KeepCup®, she thought long and hard about how she could acknowledge the inspiration without infringing any copyrights or trademarks. The design would definitely be different, but what about the name? She wanted to put more emphasis on the owner of the cups, so the name *KeepersRGood* (or *KeepersЯGood*) came about. Laura thought this could be abbreviated to KRG, KReG or possibly KeG. To make it visually more intriguing, she played around with K=G and Kr=G. With the name pretty much settled, she turned her attention to production and pricing.

DISCUSSION QUESTIONS

1 Why is it difficult for organisations to integrate sustainability concerns more broadly? How could sustainability be embedded not only into building projects but into operations as well? What could for example help estates and campus services work better together? How would you make the business case for sustainability in this context?

2 How do you evaluate Laura's name for her product? Which one would you choose? Could you come up with a better one? Explain your rationale.

3 Who are Laura's potential competitors other than KeepCup®?

4 How do you assess the effectiveness of campus outlets offering discounts of 10p for drinks bought with reusable cups? Compare and contrast with major coffee shop chains – how many of them offer discounts or rewards for bringing your own cup?

5 Life cycle analysis – are reusable mugs and disposable mugs comparable in life cycle analysis? What parameters need to be taken into account? What weighting should be attached to the different considerations in light of different environmental concerns?

6 The new university café was built to BREEAM certification. What other schemes and accreditation programmes are available for organisations to demonstrate their commitment to sustainability?

7 What other solutions could you come up with to ensure sustainable coffee catering?

8 What would your recommendation be regarding Laura's next steps?

9 What kind of strategy and business model for sustainability should Laura aim for? Go back to Chapter 4 for some ideas.

Start-ups and coffee cups: fitting fabric (B)

Having finally made up her mind about her new company's name, Laura enjoyed her favourite coffee on a break from her efforts to get this new venture off the ground. As usual, she'd get the best ideas when on a caffeine break. She was pleased with how far her plans had advanced since the initial idea for reusable coffee cups had formed, but there was still much to do.

The decision on where and how to produce was a difficult one. She knew she wanted to offer a product that made sustainable living easier, but she wasn't sure how she could ensure that it was also produced in an environmentally sustainable manner. The conventionally most cost-effective options had been ruled out on principle – plastics and glass, as they were either difficult to recycle or easy to break. She stared at the possible materials she had listed with some initial thoughts and points:

- *Plastics:* difficult to recycle, reliance on fossil fuels; light weight, not easily breakable.
- *Glass:* breaks too easily, heavy (!), but recyclable, too hot to handle?
- *Recycled plastic:* better than virgin plastics, but would it meet food safety standards? (will the coffee taste of plastic?!).
- *Bamboo:* used by other coffee cup companies, but issues with quality (mine broke!), can it be grown commercially in Europe?
- *Rice husks:* used by other coffee cup sellers, but would it imply sourcing from Asia rather than more locally or regionally (e.g. neighbouring countries in continental Europe)?
- *Seaweed:* Stavros Tsompanidis – his company Phee was one of the many start-ups founded by students (yay!) – it collects seaweed and processes it into a natural surface from which recyclable products were made. No food or drinks containers so far, but would this be possible? (have to find out!!!).
- *Bioplastics:* One of the lecturers who was one of the judges at the entrepreneurship award asked about it. Started looking at bioplastics, but one of the materials I came across, PLA (Poly Lactic Acid), a plant starch extract, is not leak-proof for hot drinks when the lid is closed.

Looking at the list and deliberating how to decide, Laura remembered that when presenting her initial plans to the dragons' den style competition the student entrepreneurship society had organised, one of the tough questions she had faced during the competition was whether she was considering additive manufacturing, to which she had had to admit (politely) that she didn't have a clue. Clearly, more detailed research was needed! Following her mid-morning coffee break, she decided to spend the rest of the day in the library with research.

Additive manufacturing involved digital design and producing objects layer by layer. Although three-dimensional (3D) printing was often used as a synonym for additive manufacturing, she learned, there were in fact a number of different processes involved in layer manufacturing, subject to which material and which technology was used. Unlike much of conventional production and operations, which involves removing material through machining, carving and shaping (also called subtractive manufacturing) and is therefore often wasteful, additive manufacturing carefully adds material when creating objects and is considered more sustainable as a result. 3D printing also enabled more localised production, which appealed to Laura as it eliminated the need for long-distance transportation. But could the technology be used for drinks containers? Would it meet food safety standards?

After a day of research in the library and exploring options online, Laura had a better grasp of additive manufacturing. She loved the idea of customers being able to print their own products at home, though from her research she understood that there was still a need for plastics in the process and there were issues with emissions from the process. It might be suitable for rapid prototyping, but there were additional concerns of food safety with 3D printing: there was the risk of bacteria building up in the design and the most commonly used materials for additive manufacturing were not suitable, whilst the process also emitted ultra-fine particles which might stick to the surface of the printed product. Laura concluded that this type of production might be something to consider in the long term, once the technology had developed further, but for now it wasn't an option.

At the Entrepreneurship Awards of the university's Enterprise Hub she had entered, Laura had been told by one of the potential investors that her requirements were too difficult to meet – the cups needed to be near unbreakable, leak-proof, whilst easy to clean and not leave any permanent coffee (or tea!) stains. So what would be a material that could satisfy these requirements, be environmentally sustainable to produce and economically viable? And how could she find out? She decided to ask at tonight's social of the sustainability society on campus, which she had joined in addition to becoming an active member of the entrepreneurship society.

There were students from most other departments at the university in the society, ranging from geography, history and management to economics, political science, drama, medical and physics students. Over a pint of local ale, Laura explained her situation and asked what she should do next. Jay, who was studying for a postgraduate degree in management and sustainability asked whether she had conducted a life cycle analysis of all the potential materials to help her make a decision. Laura knew this was probably a good idea but confessed that she didn't know where and how to start or carry out one. Fortunately for her, Ellen and Marcus, who both majored in engineering as part of their environmental science degree, offered to conduct a life cycle analysis for her, comparing the environmental merits of two materials of her choice from cradle to grave. She promised she'd look at her list and provide her choice the following day.

When Laura got back to her student dig, she couldn't sleep, she was so excited. In the end, she decided to give in and get up to look at her list of materials again. She had already ruled out glass and plastics, so now it was a matter of whittling the others down to two. Recycled plastics was still plastics, so that came off the list. The choice between bamboo and rice husks was made easier after her Italian flatmate Angelo had pointed out that rice was grown in Italy (yes, that delicious risotto he always made) as well as in Spain (Paella, anyone?) and that farmers often didn't know what to do with the rice husks left over after the harvest. And what about the other two, seaweed and bioplastics? As Phee was still a young enterprise, she decided to email the founder to enquire as to whether his processed seaweed could be used as a drinks container, and in particular for hot drinks. She drafted an email and decided to wait until the morning before sending it.

Over the following days, as she was waiting for a reply from Phee, she turned her attention to possible production and sourcing locations. She loved the idea of buying and producing locally, and from a financial point of view sourcing within the UK could be more economical, given the development of the pound's exchange rate, yet she was also somewhat wary about "buying British" in the current political climate (she was a European citizen of the world after all!) . . . focus, Laura, focus! She reminded herself – she was determined that this shouldn't prejudice her search for the most environmentally sustainable option.

When she hadn't heard back from Phee for three weeks, she decided this was not worth pursuing, particularly after considering that the material only seemed to produce flat surfaces, making it unsuitable for coffee cups. That left rice husks and bioplastics. But she had second thoughts . . . should she leave out bamboo just because of her own experience? Laura decided to discuss her dilemma with Ellen and Marcus in person.

DISCUSSION QUESTIONS

1 Critically evaluate Laura's conclusion that 3D printing (or more correctly, additive manufacturing) was not suitable for her venture.
2 What kind of production would you recommend instead?
3 Which materials would you choose for a life cycle analysis and why?
4 What kind of life cycle analysis would you conduct – a conceptual one or a detailed one? Discuss what is involved in either. What other types of life cycle analysis are you aware of?
5 What results would you expect for the life cycle analysis of your chosen materials? Which one would be more environmentally sustainable and why?
6 Where should production be located? What factors play into the decision, generally and specifically (e.g. financially and environmentally)?
7 Where and how should Laura sell her coffee cups? How should she market them?
8 What kind of environmental challenges is Laura likely to face in her decision-making process and which challenges could she actively address with her product beyond the obvious issue of waste?

Start-ups and coffee cups: fluke and finance – fighting to get off the ground (C)

Following the surprisingly inconclusive results of the life cycle analysis that Ellen and Marcus had conducted for her, Laura needed to decide sooner rather than later which material to use if her reusable coffee cup venture was to stay on track. Ellen and Marcus had warned her that the results might not necessarily be clear-cut – something that was evident in an earlier study they had come across as part of their environmental science degree where paper cups, Styrofoam cups and ceramic mugs had been compared. That must have been the study referred to by the catering staff Laura met when she first questioned the campus catering outlets' lack of reusable mugs. With both rice and bamboo as possible materials, the question turned to availability and price. She knew that rice was grown commercially in Italy and Spain, but what about bamboo? It certainly grew as a decorative house or garden plant in European climes, but what about larger scale cultivation?

As the costs and distances of transport hadn't been part of the life cycle analysis due to Marcus and Ellen's time constraints, Laura spent another day in the university library, which almost started to feel like home. . . . Bamboo turned out to be a more versatile material than expected – Laura knew it was used for clothing (she loved her soft bamboo socks!) and garden fences, but she didn't know that it could be used to build bicycle frames and to make toilet paper as well as kitchen utensils and musical instruments. What settled the matter for her, though, was the fact that for now, bamboo couldn't be grown commercially in Europe. It needed a tropical or subtropical climate which even with climate change wasn't (yet) the case for Europe. She decided to call it a day and join Angelo, her flatmate, for supper at home – he was the one who had originally pointed her to the rice grown in Europe; perhaps he might have some more ideas to help her?

As Laura left the library, she ran into Martina, who was studying on the same international business master's course as her. Martina was struggling with job applications, so in the spur of the moment, Laura invited her for supper – knowing her Italian flatmate Angelo's cooking, she knew there'd be enough to eat for everyone and a good laugh was always included – Martina certainly looked like she could do with some cheering up. As usual, supper was delicious and the conversation flowed easily, which is why Laura decided to ask Angelo and Martina to help her brainstorm more ideas on how to move forward with her chosen production material of rice husks. She confessed that with Martina being from Spain and Angelo from Italy she was (stereotypically!) hoping they could shed some more light on rice cultivation in the two countries.

Drawing on one of their previous chats, Laura asked Angelo how he knew that rice farmers often had rice husks left over after harvesting which they needed to dispose of. Angelo admitted that what he hadn't told her back then was that two of his cousins were running a rice farm in Northern Italy, producing Arborio rice for risotto that they were selling mainly to Northern European customers. "What??!" Laura exclaimed. "Why didn't you tell me before?" – "I don't talk about my farming background very often . . . during my first degree, people used to make fun of me". Angelo sheepishly responded. He explained that his cousins were always complaining about the cost of disposing of rice husks since their local council had raised waste collection fees and they didn't have space on their farm to deal with it themselves. Could Laura's coffee cup venture be the solution? "Well . . ." Martina said, "As you know from our entrepreneurship lectures, luck sometimes plays a role as well. . . . Why don't we go and visit? I'd like to find out, too!". "Luck indeed", Laura said, "I can't believe it!"

With the question of raw materials in the process of being answered and a trip to Northern Italy in planning, Laura moved ahead with the next challenges – what to do about making a prototype and how to go about the question of finance? She decided to meet up with Karen to see whether that would help make up her mind. Karen had started several companies, successfully and unsuccessfully, some of which she had sold and were still an ongoing concern, and some in which she was still involved. It was clearly something she enjoyed doing, Laura could tell, and she had been offering her advice through the local start-up hub's mentoring scheme, which is where they had met. Through her various ventures Karen had a good network in the local area and connected Laura with a small manufacturing business that she thought might be able to offer a prototype. It was a local firm producing catering equipment that was looking for further opportunities in order to grow its business.

Although Laura had not been successful at either the dragons' den style competition organised by the entrepreneurship society on campus, nor the Entrepreneurship Awards of the university's Enterprise Hub, the questions potential would-be investors had asked at both events had helped her refine her pitch and her calculations of the finance needed. Karen had usefully reminded Laura that although start-ups founded by women often received less investment initially and were often subject to challenges and questions, in the long run they outperformed young companies founded by men (BCG 2018). The two worked on Laura's presentation of her pitch, making sure she could answer tricky questions and deliver confidently without underselling her proposals.

With the advice of Karen and the support of Martina – who had started to take a keen interest in the venture – she was thus well prepared for meeting Luke, the managing director of the local manufacturing firm, to negotiate the production of a prototype and explore the potential for larger-scale manufacturing.

At the same time as considering production, Laura knew she needed to secure finance. She was very aware that as a student it would be difficult to get a loan, so with her improved pitch and the prospect of a manufacturing partner, she entered the Santander Universities Enterprise Awards, an annual business pitching competition. She was eligible to enter as her university was one of the bank's UK partner universities. It was a tough competition, but she managed to hold her nerve, present her pitch perfectly and answer all questions and challenges calmly and confidently – Karen's coaching had paid off, and she secured funding for £10,000! She reckoned that this would be sufficient to finance the making of the prototype, produce some advertising, transport rice husk bales from Italy, and manufacture a first run of 1,000 mugs.

Another decision Laura had to take was at which price to market and sell the cups. The standard size KeepCups® were selling for £7.95 on campus . . . could she price hers cheekily at £7.90 or possibly £6.95? She decided to sit down with Martina and crunch the numbers.

Three months later. . .

Laura was chuffed. The student union (SU) on campus had agreed not only to stock her product alongside KeepCups® (giving customers a choice!) but also to trial a different pricing policy. Whilst all catering outlets on site offered discounts for customers bringing their own cup, the prices displayed were for single-use cups. Laura's suggestion of displaying prices for drinks bought in reusable cups wouldn't change the price of coffee but would add a charge for disposable cups rather than offering the discount. Instead of displaying £2.00 for a cappuccino and customers paying £1.80 if they brought their own mug, the price on the board would be £1.80 and customers wishing to have a coffee to go in a disposable cup would have to pay an additional £0.20. In the talks with the student union she referred to the nudge theory from her undergraduate economics studies. Perhaps she was lucky that one of the student union employees had just come across the theory in their part-time studies so was willing to give it a try and use the two SU-run outlets as an experiment where one would run the new pricing policy and the other the conventional one.

The only thing left for Laura to decide, now that her product was on the market (albeit only the local one), was whether she could pursue a sustainable pathway to growth or whether a conventional path was the route to scaling up her endeavours. There were no easy answers as most of the teaching she had encountered so far paid little attention to sustainability. Could her success at the Santander Universities Enterprise Awards give her some leverage for more funding to grow her venture as responsibly and sustainably as possible, and even provide some employment not just for herself but possibly for Martina and for Luke's staff?

DISCUSSION QUESTIONS

1 To what extent do you agree with Laura's decision to make her coffee cups from rice husks. Why? Why not? What other materials could she have used and why?
2 What other sources of funding could Laura have sought?
3 How realistic are Laura's numbers? What pricing strategy would you recommend for Laura?
4 Based on average coffee consumption, how quickly could consumers recoup the cost of Laura's cups?
5 What could Laura do to change the wider pricing policy for coffee shops on campus (beyond the student union) from offering discounts on drinks bought with reusable cups to charging extra for drinks bought with disposable cups?
6 Discuss how the new pricing policy in the student union might be received by consumers on campus – students, staff and visitors.
7 Find out more about nudge strategy – how could it be used beyond pricing for sustainability purposes?
8 What kind of strategy for growth and scaling up should Laura pursue? Is sustainable growth possible?
9 What are the sustainability implications of growth for each of the corporate functions covered in part three of the book?

REFERENCE

BCG (2018) Why Women-Owned Startups Are a Better Bet, Boston Consulting Group, available at www.bcg.com/publications/2018/why-women-owned-startups-are-better-bet.aspx (Retrieved, 07.10.2019).

CASE STUDY 2: GREENBELT FESTIVAL – SERVICES SECTOR

Mary Corfield, the festival's event director, had been employed for less than three months when she delivered her first festival in 2013 – held at Cheltenham Race Course. The previous year's festival in 2012 had been infamous for being the wettest festival in years, with vehicles being banned from driving on site to protect the flooded and muddy grass areas. The location had meant a lot of existing infrastructure was in place for the festival to work with, but when the racecourse was about to undergo substantial remodelling with the usual festival areas no longer available, Greenbelt had to look for an alternative. When the festival moved to the new site at Boughton House in 2014, the organisers had to get their head around building everything from scratch, but this also provided new opportunities as more control over decisions was possible. The two biggest environmental issues for the festival are energy usage and waste, and the festival hopes to be single-use plastic-free by 2020.

The following is Mary's account of how the festival engages its stakeholders with environmental sustainability, including crew, traders and caterers, performers and artists, and festival-goers.

"We've decided on a strategy of picking a few things each year, and try and improve those. As we're only running the event for four days, we can't do it as if we were doing something week in, week out, you have to do change in a slightly different way. So we looked at what we build, how we build it, how we're powering everything, what we waste. One of the things I chose to focus on to start with was, where all the things we bring to site end up? That then impacts the choices you make, where you get them from, what they're made out of, and what the options of disposal are at the end.

"For instance, we took the decision to require all of our caterers and traders to use only fully compostable packaging. Having spoken with our waste contractors and made sure that there is a local commercial composting facility which could process that. This varies a lot in the UK, depending on where you are. There's no point in getting everyone to use compostable cups if there is nowhere you can compost them. That was part of the starting point of making things more visible in the public-facing aspect of our operations rather than the things we do behind the scenes. It was part of the deal – if you want to trade here, your packaging has got to be compostable, and actually all caterers were great about that, we didn't get any arguing, although there were some questions over it, so we offered advice and communicated clearly in order to explain to them why we're doing it. We had worked out that essentially, particularly inside the festival village, the bulk of the waste being created by our festival-goers is food and food containers. So if we could ensure that all of that was going to be repurposed, that is a huge achievement.

"The next thing, this year with food and drinks, we took the decision that nothing could be purchased anywhere on site that was in plastic bottles so all the caterers had to have drinks canned and our bars have re-usable cups and use a deposit scheme. Even the tea bags on site are plastic-free – not many people know that these often contain micro-plastics. For a lot of drinks, avoiding plastic is not a problem, Coke, Fanta etc., they can easily get them in cans instead of a bottle. Water was the one thing the traders were concerned about because normally, particularly if it's been a hot summer, fizzy water can be a quite popular sale for them and none of them knew how they could sell fizzy water without plastic bottles. We did some research and found a company that sells canned fizzy water and we brought fizzy water onto the site, a pallet of it, which the caterers could then buy off us at cost price so that they could meet their clients' demand. Because it's such a new product, it's not easily available in small quantities but we took the hit and bought a chunk. As it wasn't a very hot festival weekend, we haven't sold very much, but even just helping the traders to understand that these products are actually available means that when they go to other events they now know about products and packaging options they didn't know about before.

"With regard to engaging our artists and performance with environmental sustainability, we've bought lots of metal bottles so all of our artists on the main stages have been provided with one. We obviously need to provide them with water and what we did is give them a bottle they could take away with them, and we're hoping they would use them all weekend here, get into the habit, keep them and then use them at all the other venues where they've got gigs. There they could say, actually I don't need bottled water, I've got this. Similarly, all our crew, our electricians, our plumbers, they were given a metal water bottle with their name on. They were told, here's a tap, here's your bottle, we're not having plastic on site. And that's been really well received.

"It's not always been easy, some of the festival-goers complain about having their choice taken away saying 'what if buy my child a fizzy drink and they only drink half of it?'. We've done some messaging around bringing your own bottles and containers, using the many water taps on site, and there are silicon lids you can put on top of a can, which are not plastic but washable and reusable. We also didn't want Tetra Paks as they've often got plastic straws on them and they're lined with plastic. In the end we found a supplier of canned orange juice, but we had lots and lots of conversations with drinks companies, particularly with Fair Trade drinks companies – who you think would be more aware of this – saying to them 'look we can't buy your products unless you give it to us in a container that isn't plastic', and increasing numbers of them are saying, 'we've had lots of phone calls about this'. When we ask 'That's great, but what are you actually doing?' the answer to that is, 'well we don't know'. So . . . for us it's important to model what we're doing and we know that a few other events have also been doing this, which is making companies stop and think about what they're doing, and about the choices they're making. So even if we don't always succeed in getting the products that we're after, the pressure we're applying means hopefully by future years' festivals those products can be accessed.

"One product we've been looking at replacing are our festival wristbands. We're aiming to phase out single use plastic by 2020 and for example with the drinks, it's not been easy, but reasonably straight-forward and fairly visible as part of the front-facing things we do. It's the back-of-house things like hazard tapes and cable ties that are more problematic, though we've managed to replace plastic tape with coloured ropes to mark out the campsite and parking. Wristbands might seem like a small thing, but to the best of our knowledge there are no plastic-free ones available; even the fabric ones we're using in 2018 have plastic in them. Less than a solid plastic one, but still. . . . So we were so thrilled when after talking to many people, we finally found a Finnish company that could supply us with plastic-free wristbands for the first time for 2019 made of soft bamboo which is completely recyclable. It's taken us a long time, but challenging the status quo isn't always quick and easy and we're glad we stuck with our search for it".

"In terms of energy, when we first moved to the site at Boughton House here, everything we were doing was diesel generators. We tried quite hard to bring in the only firm of solar hybrid generators, who were new in the UK, but we couldn't get them as they were busy with long-term contracts and low on stock. Instead, we worked really hard with our power team and some volunteers to monitor in great detail how much power we use on the whole site. This enables us to know what we're putting in, how much power the lights and the PA use, but also all the other people on site, such as caterers and traders. They tend to overspec [overspecify] it when they tell us what they want, which is understandable as they want to make sure they've got enough, but there is actually quite a big disparity between what they ask for and what they use in reality. We've got a guy with us who's been doing lots of real-time monitoring for us in terms of the

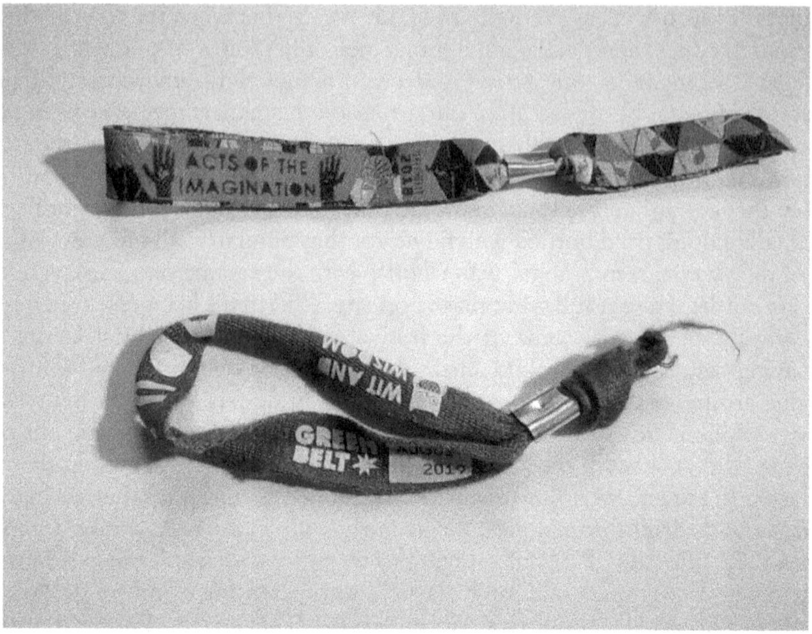

FIGURE 14.2A
Participant wristbands at Greenbelt Festival in 2018 (top) and 2019 (bottom)

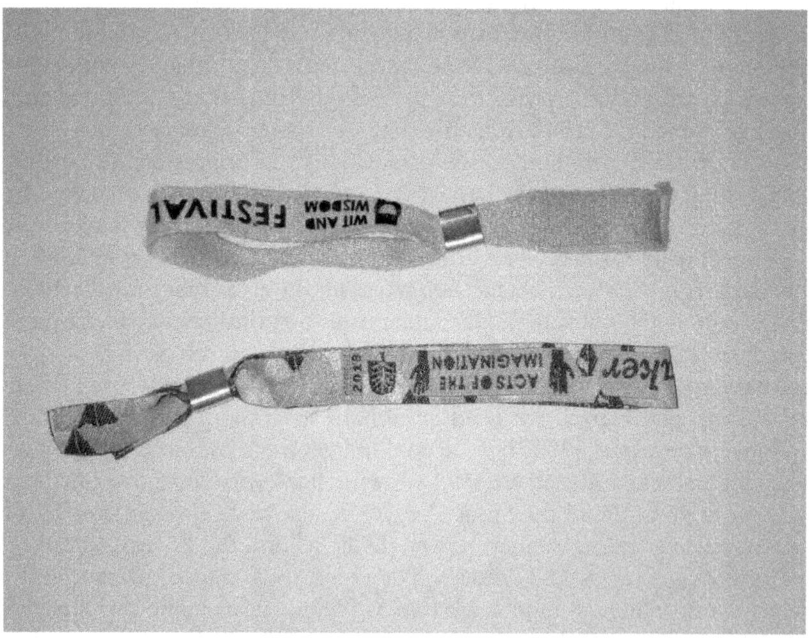

FIGURE 14.2B
Volunteer wristbands at Greenbelt Festival in 2018 (top) and 2019 (bottom)

power we're using, where and when, quiet periods, busier periods, all the different areas, what is generated in the different zones, what is the reality there.

"The first year that we went through in such great detail to analyse what people want and what they need, we managed to reduce energy usage by 15%. Last year we reduced it by another 15%. This year, our knowledge is now so detailed, with the kit that is now starting to be available in the UK, we were able to be confident enough to be able to switch a lot of our power over to hybrid. We still have areas with diesel-based generators, smaller ones than in previous years based on the data, but we've now got some solar-powered flood lights on the campsite and in the car park. They're a new thing. We've been working with the guy who makes them for building works at railway tracks, so we know they work, but the columns are so huge they're very difficult to transport as they don't fold down, so we've worked with him, trying to persuade him that if he can make them fold, then the events industry as a whole would love them. That would also be a whole festival business for him. We could all be solar powering our lighting, car parks, walk ways, which would just be better for everybody. So yes, we've been strongly encouraging him to work through that, we think that would be a great thing for the industry as a whole to have people becoming more sustainable.

"So all in all for energy, we've now got a variety of solar-powered flood lights, a combination of generators, hybrid generators, and some really good battery systems, where we draw on the power of solar rays and store it, and then we can use it when we need it. That's pretty new as well, and it's been challenging to get companies to provide it, the UK is a bit behind the curve in that respect compared to places like Germany. We decided not to go down the biodiesel route because there are so many complications with that. How can you be completely sure that your fuel has come from a source that's not contributing to food crops not being grown? for example. I'm really thrilled that we've managed to substantially reduce our fuel usage each year, and it's getting less and less, despite increasing numbers of visitors coming on site.

"Furthermore, where we can, we try to use local suppliers to reduce the miles our supplies travel. That also included where we sourced our metal bottles from – we hunted all over the place trying to find a company that could provide them, but the only ones we could find were in China, which would be counterproductive, sourcing water bottles to reduce our environmental impact in terms of waste and plastic, but then having them travel all the way from China. In the end we managed to work with the Raw Foundation, whose mission is to educate teenagers about the environment. They set up a trading arm selling metal water bottles to events, with profits going back into educating people, and importantly for us, the bottles are made in Britain, which is great. Unfortunately, there have been issues in the UK plants, so whilst we have reduced our waste in this area, the bottles are currently being made in China again – that's one of the challenges of going green, trying to keep on top of the constant changes and constraints, and the RAW foundation are working towards bringing production back to the UK.

"We also rent a barn on site year-round so we can keep our storage really close to the festival. That's different from many other festivals that keep things in central storage compounds. For us it also means that we only get the equipment out that we really need.

"Speaking of other festivals, we work quite hard to talk about what we do, we talk to trade publications, because we want to inspire other people at events to realise that if Greenbelt can do it – who are not a big spender in festival terms – then other festivals that have a much higher spend can do it, too. We're part of the Greener Festivals system, they send an assessor each year and grade us accordingly. It provides us with advice and knowledge, but also gives us an opportunity to talk about what we're doing. It doesn't sell tickets, but it encourages people in the events industry. I'm also an ambassador for Positive Impact which is an industry environmental body. Things like the Showman's Show which happens in October every year and is sort of an annual trade show for the outdoor events industry, is also an opportunity where everybody in the events industry goes and we can talk about what we're doing.

"This work also links in with the visit of Michael Eavis from Glastonbury to Greenbelt. He's legendary in the outdoor events world. We had him at the festival to speak and show him around. We gave him one of our bottles and he said, 'oh well, they're nice, I guess you get them from China?' He was genuinely surprised to hear that they weren't and asked for the details of the supplier. He thought modelling those at Glastonbury would have a huge impact. So yes, we try to connect lots of different people where we can and try and make an impact. I also get quite a lot of university students who're doing some kind of project on sustainability in the events industry phoning up and saying can I talk to you, can I ask a question about my dissertation, because they see us listed in Greener Festival awards, so they think we're interesting people to talk to. And that interest has definitely increased compared to five years ago, and it's no longer a weird add on at Greenbelt but lots of people in the industry asking, what should we be doing?

"You asked about the traditional argument that there's a trade-off between environmental objectives versus financial performance. It's been mixed, to be honest. With energy I would say, it's reasonably cost-neutral – the kit costs more, but then you're not buying the fuel in the same quantities because you're using solar and hybrid. It roughly balances. Getting rid of the plastic bottles was more expensive, not in terms of the catering aspect of it, but buying the metal bottles for our crew and artists. But we felt that was the right thing to do. Using upcycling and designing the signs for example in a sensible way so you can reuse them every year has been a great way of avoiding costs and saving money when decorating the site for the festival. Ultimately, we've found that if you approach everything with the question "what do we need and where will it end up?" you can actually be much more cost-effective about it. All the guys, all our crew on the team are briefed: 'We are going to be keeping this, so put it up in a way that we can take it down without it being damaged.' Having that in people's heads affects how they put things together and makes the whole retrieval much easier. And that doesn't cost anything, it's just the time of thinking it through and making sure you brief your people".

DISCUSSION QUESTIONS

1 Find out more about the festival by visiting its website www.greenbelt.org.uk/ and various social media feeds. To get a feel for the Greenbelt festival, watch footage of the festival's most recent events: https://youtu.be/A193nEgLpu8 and https://youtu.be/vgpg0nUs3iE.

2 How do you evaluate Greenbelt's strategic decision to change and improve a few things each year to become more sustainable? Would the relocation to a new site have been an opportunity to go for transformational change rather than incremental change? What would have been the risks involved? Return to Chapter 4 to revisit the different types of change.

3 According to Mary, energy and waste are the two biggest environmental challenges for festivals. What other environmental issues might be a concern as well that are not covered in the case study? Look to the introduction to part two and its constituent chapters (5 to 7) for more.

4 Critically evaluate how Greenbelt has addressed the traditionally perceived trade-off between environmental objectives and financial performance. Return to Chapter 3 for a discussion of trade-offs in environmental sustainability.

5 How does Greenbelt communicate its sustainability commitment with its users and stakeholders? In particular, consider suppliers (e.g. caterers and traders) and

festival-goers and consumers. What role do collaboration and partnerships play? To what extent does the festival engage in choice editing for sustainable consumption? Revisit Chapters 8 and 10 for more.

6 Compare and contrast the public information available on the web for the Greenbelt festival and the Glastonbury festival regarding environmental sustainability: browse www.greenbelt.org.uk/ and www.glastonburyfestivals.co.uk/ as a starting point. Can you find other sources of information?

7 What role does sharing of good practice play in the events industry? Evaluate Greenbelt's efforts in this respect.

8 Using Chapter 11 on Environmental Management Systems and Chapter 12 on Accounting and Reporting, discuss the importance of measuring and monitoring for the festival, and the relevance of certification and labelling. Visit www.agreenerfestival.com for more information and draw on other certification and labelling schemes that might be applicable to this and other festivals.

9 To what extent can you apply concepts from Chapter 13 on human resource management to the case study?

10 What further steps could the festival take to become more environmentally sustainable? Is a Mission Zero (see Chapter 3) possible? Use the knowledge you have gained throughout the various parts of the book.

CASE STUDY 3: ESTABLISHED SOYA FARMER – PRIMARY SECTOR

Fernando put the book down. He was bewildered. How could he not have known this? He had always been an earnest farmer, wanting to do right by his family and his employees, but according to the book he had received from his niece Maria, who was studying in London, the way he was farming was damaging the environment and put food security at risk. The book was called *Farmageddon* and documented the cost of current food production practices – exactly the industry in which he was invested. It didn't make for encouraging reading.

He had inherited his father's crop farm in the 1980s and had increasingly shifted to growing soya and more lately genetically modified soya as that was more herbicide-resistant. He had managed to expand the farm since taking over, and business was good as beef farmers across the world were buying soya from Latin America. Nevertheless, Fernando was in a thoughtful mood – he loved his niece and was in awe of her willingness to take risks and learn – one reason why she had gone to the UK for further studies in biodiversity. Her brother Pablo was similarly eager to learn and had managed to secure a scholarship to study agricultural technology in the US. As close-knit relatives, they all often talked on the phone. Fernando and his wife Olivia couldn't have children of their own, so their nephew and niece were all the more important to them, and they were hoping that one (or both) of them could take on the farming business one day.

Further fodder to this reflective mood was added when a week later a parcel from Pablo in California arrived. Pablo was not one for slow mail; he preferred being in touch via e-mail and mostly via his latest smart phone apps. Fernando couldn't even remember the last time he'd received something handwritten from him other than a co-signed birthday card. Yet, there it was, a letter from Pablo accompanying the content of the parcel – a book called *Eating Animals*. In his letter, Pablo confirmed his intention to get involved in the family farm on his return from the US, but he needed Fernando to read this book, as he was concerned about the future viability and sustainability of their established way of doing things. Curious and intrigued, Fernando got to reading. He finished it over the weekend.

The book got to him. Together with the effect Maria's book had had on him, Fernando realised that something had to change if the farm were to have a long-term future in the family. A light had switched on in his head and in his heart, and he was determined to change his company's business model for the better to ensure its future. This could not go on. But he knew this wouldn't be easy – the farming company's culture and relationships with buyers and suppliers were well-established, and it would be a challenge to move his soya production from supplying beef farmers to start growing soy in a more sustainable way destined for human consumption rather than cattle consumption. But it needed to happen; that's what he had decided, and it was not just a gut feeling.

He decided to give the books to his chief marketer and his head accountant before taking any further steps. They had been in the business for long enough to know what was possible, but they were also young enough to be open for change – they had told him their teenage children had participated in the climate youth strikes against the advice of their school friends. And they knew that Fernando was considering who was going to take over from him before too long. What would be their reaction?

DISCUSSION QUESTIONS

1 Find out what the environmental issues are with growing soya as animal feed. The story mentions food security; what about climate change, waste and pollution as covered in part two of the book?

2 Are there environmental challenges associated with growing soya for human consumption? What is the share of global soya cultivation and what role does the increasing uptake of flexitarian diets play in this regard and the rise in vegan products, in which soy has a major part to play?

3 Debate the merits and de-merits of GM crops – in particular genetically modified soy.

4 Find out more about the two books Fernando read – are they still relevant? How trustworthy are they and how were they received by the public and industry and farming communities?

5 Suggest steps Fernando could take to transform his agricultural business in the long term. What should he do within the first year? Outline further steps for three, five and ten years. Can he scale the "seven fronts" mount sustainability? Refer to Box 4.6 for a reminder on mount sustainability.

6 In Chapter 4, we talked about the different kinds of change, transformational and incremental. What type of change would be most appropriate according to your answers to the previous question? Where is Fernando on the trajectory to sustainability, using the phases and stages from the same chapter?

7 In Chapter 4 we also encountered business models for sustainability. Can you apply any of them to Fernando's business?

8 Consider Fernando's relationships with his buyers – how can he and his team build support for his (possibly controversial) decision? What resistance are they likely to face?

9 Discuss the costs and benefits of diversification and vertical integration as possibilities for the farm.

10 Use the functions covered in part three of the book to explore respectively what options Fernando and his team have. Choose from e.g. supply chain management, production and resource use, EMS certification, marketing and accounting, HRM.

11 How controversial might Fernando's decision be for his employees and his senior team? Include the importance of culture, leadership and values in your considerations.

12 What role do environmental regulations play for agriculture? How significant is lobbying in this respect? Return to Chapter 3 for more material.

REFERENCES

Lymbery, P. and Oakeshott, I. (2014) *Farmageddon: The True Cost of Cheap Meat*, London: Bloomsbury.

Saffran Foer, J. (2009) *Eating Animals*, New York: Back Bay Books.

NOTES

1 NACE is the acronym for Nomenclature statistique des Activités Économiques dans la Communauté Européenne, French for Nomenclature of Economic Activities in the European Communities.

2 See Figure 4.3 for more types of delivering functionality rather than ownership, as well as Chapter 9 for examples.

SOURCES AND FURTHER READING

BCG (2018) Why Women-Owned Startups Are a Better Bet, Boston Consulting Group, available at www.bcg.com/publications/2018/why-women-owned-startups-are-better-bet.aspx (Retrieved, 07.10.2019).

BIS (2012) Industrial Strategy: UK Sector Analysis, BIS Economics Paper No. 18, Department for Business, Innovation and Skills, available at https://assets.publishing.service.gov.uk/government/uploads/system/uploads/attachment_data/file/34607/12-1140-industrial-strategy-uk-sector-analysis.pdf (Retrieved, 17.10.2019).

Burrett, C. (2018) What's the Return on Investing in a Reusable Coffee Cup? *Financial Times*, 08.02.2018, available at www.ft.com/content/edddb47c-0b22-11e8-839d-41ca06376bf2 (Retrieved, 07.10.2019).

Defra (2008) A Framework for Pro-environmental Behaviours, Department for Environment, Food and Rural Affairs, available at https://assets.publishing.service.gov.uk/government/uploads/system/uploads/attachment_data/file/69277/pb13574-behaviours-report-080110.pdf (Retrieved, 17.10.2019).

Ellcock, L. (2019) 'I Made Excuses': Music Industry Frets Over Becoming Carbon Neutral, *The Guardian*, 29.11.2019, available at www.theguardian.com/music/2019/nov/29/i-made-excuses-music-industry-frets-over-becoming-carbon-neutral (Retrieved, 02.12.2019).

GICS (2018) Global Industry Classification Standard, S&P Global and MSCI, available at www.spglobal.com/marketintelligence/en/documents/112727-gics-mapbook_2018_v3_letter_digitalspreads.pdf (Retrieved, 17.10.2019).

Greenbelt (2019) Greenbelt, webpages, available at www.greenbelt.org.uk/ (Retrieved, 21.10.2019).

Hawken, P. (1993) *Ecology of Commerce: A Declaration of Sustainability*, New York: Harper Business.

Hocking, M. B. (1991) Paper Versus Polystyrene: A Complex Choice, in: *Science*, 251(4993): 504–505.

Holden, S. (2018) The Festivals Aiming to Get Rid of Single-Use Plastic by 2021, *BBC Newsbeat*, 19.04.2018, available at www.bbc.co.uk/news/newsbeat-43809182 (Retrieved, 02.12.2019).

Kellens, K., Baumers, M., Gutowski, T. G., Flanagan, W., Lifset, R. and Duflou, J. R. (2017) Environmental Dimensions of Additive Manufacturing: Mapping Application Domains and Their Environmental Implications, in: *Journal of Industrial Ecology*, 21(S1): 49–68.

Lymbery, P. and Oakeshott, I. (2014) *Farmageddon: The True Cost of Cheap Meat*, London: Bloomsbury.

Middlemiss, L. (2018) *Sustainable Consumption: Key Issues*, London: Routledge.

Refinitiv (2019) TRBC Sector Classification, Thomson Reuters Business Classification, available at www.refinitiv.com/en/financial-data/indices/trbc-business-classification (Retrieved, 17.10.2019).

Saffran Foer, J. (2009) *Eating Animals*, New York: Back Bay Books.

Thaler, R. H. and Sunstein, C. R. (2008) *Nudge: Improving Decisions About Health, Wealth and Happiness*, London: Penguin Books.

UN Statistics Division (2008) International Standard Industrial Classification of All Economic Activities: Revision 4, available at https://unstats.un.org/unsd/classifications/Econ/Download/In%20Text/ISIC_Rev_4_publication_English.pdf (Retrieved, 17.10.2019).

UN Statistics Division (2019) National Classifications, United National Statistics Division, https://unstats.un.org/unsd/classifications/nationalclassifications/ (Retrieved, 17.10.2019).

CHAPTER 15

Conclusion
The future

INTRODUCTION

We started our introductory chapter with a look at the past and the present. In concluding the book, we look back at, and reflect on, what we have covered, and we look ahead to what the future of environmentally sustainable business might look like.

The chapter begins with a brief review and summary of the three main parts of the book and their constituent chapters, before looking ahead at what the future of sustainability tools and concepts, the regulatory framework, the challenges they are intended to address, and of what corporate functions might look like – despite the difficulties in predicting or forecasting the future we can look at current trends and use conjecture to draw some inferred conclusions. We wrap up this part with a look at sustainability as a vocation and the importance of education before concluding with an emphasis on the importance of integration in achieving sustainability.

LOOKING BACK – FROM FOUNDATIONS OF PRINCIPLES AND MODELS (PART I) TO ENVIRONMENTAL CHALLENGES (PART II) AND CORPORATE FUNCTIONS FOR ENVIRONMENTAL SUSTAINABILITY (PART III)

We started the journey in Chapter 1 by looking at the origins of sustainability, which, depending on where we anchor our starting point, go back to the 1960s and the environmental movement inspired by Rachel Carson's *Silent Spring*, to the 18th century and forest management in Germany or even to unsustainable agriculture and deforestation in Babylonian times 4,000 years ago. This historical background was set alongside the political background, where the 1970s and 1980s saw policymakers react to the combined pressures from grassroots movements, scientific evidence and corporate environmental disasters by introducing environmental regulations.

In part one of the book, we unpacked this historical and political setting further by looking at the various contextual foundations, principles and models that underpin corporate environmental sustainability. We set the scene in Chapter 2 by looking behind those scenes. By using the five capitals model of natural and man-made capital, we combined science and economics to see how systemically intertwined nature and the economy are. Businesses use ecosystem services of the earth system for free and produce negative externalities in their activities – in both cases no price is set for the goods used and the costly environmental damage done. This environmental impact of industry and humanity on the planet can be measured through ecological footprints which are then contrasted with the carrying capacity of the earth. It became clear that planetary boundaries are crossed and limits breached as the ecological footprint exceeds the carrying capacity, resulting in overshoot. The concepts of weak and strong sustainability were introduced to provide guidance in how to allocate resources on the basis of substitutability, particularly in the case of natural capital. The tensions and trade-offs in sustainability, such as between natural and human-made capital and between economic growth and environmental protection, and the role of technology, were further explored in Chapter 3 before turning our attention to how the regulatory framework can help address these. Principles such as the polluter pays, prevention and the precautionary principle underpin regulations, which can be direct, in the form of command and control, indirect in the form of market-based incentives, or voluntary. Feeding into both the legislative framework and the corporate context are a number of guiding tools that were then introduced, from environmental impact assessment, including risk assessment and cost-benefit analysis, and life cycle analysis, to eco-efficiency, the triple bottom line, The Natural Step, and mission zero, as the culmination of regulatory and corporate sustainability efforts aiming at a footprint of zero – zero emissions, zero waste and zero pollution. In Chapter 4 we looked at how the economic and scientific foundations of Chapter 2, and the principles, regulations and guiding tools from Chapter 3 could translate into tangible strategies and models for business to respond to the sustainability challenge. As a first step this requires companies to move from seeing sustainability as cost to seeing it as an opportunity for value creation. In this, they can use conventional strategy frameworks, such as generic strategies and the resource-based view, or sustainability-focused ones, of which we covered The Natural Step, natural capitalism as well as eco-effectiveness and cradle-to-cradle. Commonalities were found in e.g. the importance of natural capital, re-thinking business models from offering product ownership to providing functionality and service, mimicking nature and introducing circularity and life cycle thinking. Although the move towards sustainability can involve short-term wins, it is a long, and at times difficult, journey requiring sustained efforts. We therefore also included the (incremental and transformational) change and leadership that are needed to move through the various phases, waves and stages of sustainability, which can help signpost and map the road towards corporate environmental sustainability.

The foundations, principles and models of part one and the corporate functions of part three together underpin and guide how companies can address the sustainability challenges they face. In order to make this link more tangible, we covered three selected environmental challenges – climate change, pollution and waste – in part two that are illustrative of the broader, interconnected challenges that society and industry face in terms of inefficient use and depletion of resources, and disruptive ecosystem change. For all three challenges, we stressed the systemic nature of environmental sustainability and the interlinkages and interdependencies of the various issues. Each challenge was defined and explained in terms of why it is a concern, before the business impact in terms of cause and effect was covered and how regulators respond to the challenges – mitigating, adapting, avoiding and restoring (EEA 2016). The focus then shifted to how businesses respond to each challenge, first internally to minimise their contribution to the challenge and then externally in terms of capitalising on opportunities to offer solutions to other companies and to consumers. Climate change (Chapter 5) is arguably the biggest of the three challenges, requiring not only mitigation but also adaptation to a significantly changed context in which businesses operate. Solutions are not only technological ones, such as clean technologies and geoengineering (which often come with their own problems[1]) but require also behavioural change of companies and consumers. Pollution of air, water and land (Chapter 6) is the challenge that has been known the longest and is addressed through regulations as well as remediation, abatement, control and prevention. Waste (Chapter 7) is an all-pervasive problem of the *throwaway society* centred around planned and built-in obsolescence that can be addressed through reducing, reusing and recycling, but above all through re-thinking how we do business and moving from a linear cradle-to-grave to a circular cradle-to-cradle model. Common to the challenges discussed is that the planet's natural capacity to absorb and treat excessive greenhouse gas emissions, pollution and waste is exceeded. Underpinning corporate responses to the challenges is the practice of measuring – knowing what and how much we emit, pollute and waste – which then enables control and prevention, reduction and minimisation. The broader context of addressing these challenges is extended producer responsibility and circular thinking which emphasise that companies are responsible for the environmental impact of their goods and services across the whole life cycle beyond the factory gate.

How the challenges covered in part two can be addressed not just generally by companies, but by each corporate function specifically, was covered in part three of the book. We looked at how the individual management tasks of planning, organising, leading and controlling and the activities in the value chain (primary and supporting ones) are affected by and impact on the environmental challenges, making the business case for corporate environmental sustainability in the area. We introduced each corporate function generally as well as specifically under a sustainability umbrella, providing an overview as well as covering elements that make up the environmentally sustainable function and the tools that companies can use to green the role, before highlighting selected examples and practices of companies. Chapter 8 provided the initial foundation, covering the whole supply chain of activities from raw materials and transformation through to transport and delivery to the end user. Tools introduced here extend to other functions as well, such as circular economy principles and practices of closing the loop, industrial ecology, cradle-to-cradle, life cycle analysis, and certification and labels, whilst reverse logistics, transport and collaboration with suppliers are more specific to supply chain management. Building on that, Chapter 9 on production and operations focused on extended producer responsibility and product stewardship which start with eco-design where environmental considerations are systemically included from the start of a product's life cycle. This can then be used to inform how operations and production are made leaner, cleaner and greener with the help of technology and innovation. Chapter 10 on marketing and communication also looked at the other side of production, i.e. consumption, at consumer behaviour and how it is encouraged by marketing and contributes to society's environmental problems. Solutions offered therefore also include choice editing in addition to looking at how green products are marketed and marketing itself is made more sustainable. Credible marketing with the help of eco-labelling and the avoidance of greenwashing are part of addressing the tension between consumers' willingness to go green and their willingness to pay and between marketing's encouragement of consumption and sustainability's need to reduce it. From the primary activities in the value chain we then moved on to support activities, starting in Chapter 11 with information technology and management systems where data and information is used to improve performance, often in the iterative Deming cycle of plan-do-check-act. Emphasis was placed on the ISO 14000 series of environmental management systems and on EMAS, the Eco-Management and Audit Scheme, both of which provide standards for sustainability management and reporting. These naturally fed into Chapter 12 on accounting, reporting and finance for sustainability, with an emphasis on environmental accounting and how information from measuring can be used and provided to improve decision-making and performance. Various standards and tools were introduced, such as the Global Reporting Initiative and Integrating Reporting as well as initiatives for auditing and finance for sustainability. Chapter 13 on human resource management concluded part three, emphasising that all functions in essence rely on people and their skills and experience to implement sustainability across an organisation. The broader context of organisational structure and culture, corporate governance, leadership and change management was set before highlighting various tools and concepts that can be used to make the HRM process more sustainable, from recruitment and training, to motivation, benefits, employee relations and communication.

Having laid the groundwork, Chapter 14 then provided an opportunity to put into practice the materials of parts one, two and three, with the help of three case studies on reusable coffee cups, a festival, and soya farming, covering the broad primary, secondary and tertiary sectors of an economy.

LOOKING AHEAD: THE FUTURE OF BUSINESS AND ENVIRONMENTAL SUSTAINABILITY

Environmental challenges – can we address them or is time running out?

We covered three challenges that are increasingly interconnected with each other but also with other issues, as emphasised in the introduction to part two. This interdependence sets the tone and frames the context for businesses in terms of the reality they and society need to countenance, now and in the future.

The year 2014 was the first year where greenhouse gas emissions fell while the economy grew. This decoupling of the environmental impact and the economy is needed if the environmental challenges society faces are to be addressed substantially and sufficiently, particularly when confronted with potential tipping points. Such points of no return are reached when thresholds effects occur, resulting in irreversibility of the environmental impact on natural systems. This relates particularly to climate change but has implications for pollution and waste, too. These three challenges are part of a wider picture of the sustainability challenge where the planet's biodiversity and ecosystem (which provides services to business and society alike) as a whole are threatened at an accelerating pace – 1 million plant and animal species currently face extinction (IPBES 2019).

For climate change, it is very clear that mitigation alone is no longer sufficient and that in addition adaptation is necessary, as avoiding climate change altogether is no longer a possibility. A landmark IPCC report released in 2018 warns that time as well as carbon budgets are running out for limiting global warming to the 1.5°C temperature rise to prevent further climate instability in the form of more extreme weather (floods, droughts, extreme heat), rising sea levels, reduced arctic sea ice and disrupted ecosystems.[2] The year 2018 was a record-breaking year with global CO_2 concentrations reaching levels last experienced 3–5 million years ago when the temperate was 2 to 3°C warmer and sea levels 10 to 20 meters higher (WMO 2019a). Limiting global warming to 1.5°C would require CO_2 levels to drop by 45% by 2030 compared to 2010, resulting in net zero emissions by around 2050 (IPCC 2018). A net zero carbon target has been adopted by many governments as well as businesses in response, some of which have gone even further, setting or demanding the stretching target of net zero carbon by 2030. Increasingly, the terms global heating and climate crisis, breakdown or catastrophe are used to signal the urgency and seriousness of this challenge which is no longer a future challenge – climate emergency was declared word of the year in 2019 by Oxford Dictionary, the same year when scientists reached 100% consensus on anthropogenic global warming at the end of the hottest decade on record (Powell 2019; WMO 2019b). The ten years left until 2030 are paramount and present a true fork in the road – can humanity as a whole through civil society, the market and the state transition rapidly to a carbon-neutral way of operating, or will the inability and inconvenience of turning from business as usual risk overshoot and interference of natural systems at such a scale that these ecosystems are irreversibly damaged, with a risk of 250,000 additional deaths from malaria, diarrhoea, malnutrition and heat stress per year as the World Health Organization (WHO 2018) warns?

The first choice would require the equivalent of a supertanker U-turn, reversing the growth in emissions whilst also allowing an expanding economy, particularly in developing countries where the population and their needs are still growing (Economist 2019). The second choice, we could argue, is simply not an option if we want future generations to meet their needs as this generation does. This has been particularly highlighted by the youth climate strikes that started in Stockholm in 2018 with Greta Thunberg and then spread across the world in 2019, leading Collins Dictionary to declare climate strike word of the year. With increased incidences of floods (e.g. in the UK or Venice, Italy) and wildfires (e.g. California and Australia), and national governments, cities and other public institutions (such as universities) declaring a climate emergency, public awareness seems to be reaching a tipping point[3] which might give hope to reversing the supertanker and help avoid tipping greenhouse gas emissions beyond the point of no return – there is a narrow window of opportunity to allow for an ecological "soft landing" (King 2011). On the one hand, this means not waiting for peak oil but keeping fossil fuels in the ground through a variety of means (using clean energy alternatives, reforming fossil fuel subsidies, divestments) as well as taking out CO_2 from the air (e.g. using carbon capture, afforestation, rewilding). On the other hand, it requires rethinking and transforming business as usual and coming up with imaginative, creative and restorative ways of doing business, of consuming, of doing life differently. Despite the continued heavy reliance on fossil fuels for meeting over 80% of global energy needs, there are some encouraging signs – e.g. in 2017 the UK experienced its first coal-free day of power generation since the industrialisation, which extended to an entire fortnight in 2019 when coal's share in electricity fell to a record low of less than 1%.

Keeping fossil fuels in the ground also helps address air pollution, which is in large part due to fossil fuel burning (see Table 6.1 for a reminder). Although both challenges – climate change and air pollution – come largely from the same source, the effects are very different. Whereas climate change has global effects, air pollution can be more concentrated, as the effects are direct and immediate. It is often the main environmental threat in urban areas, with substantial effects on human health that are hard to escape, whether it is in the form of mental or physical health, impaired cognitive performance or illness – with potentially 7 million deaths each year linked to air pollution and its effects, it is the major environmental killer (Hook et al. 2019). With rapid urbanisation, it is expected that by 2050 over two-thirds of the world's population will be urban compared with just over half in 2018 (UN 2019). If no action is taken, air pollution, alongside water and ground pollution, will continue to have poisonous effects on ecosystems and human activities,[4] in turn constraining industry, agriculture and commerce. In response to this "airpocalypse", China declared a "war against pollution" (Smedley 2017) – illustrating the need for urgent action, similar to the climate emergency. Addressing the problem requires both the mitigation of already existing pollution and restoration of air, water and ground quality as well as the avoidance of further pollutants being emitted. The latter is – as with climate change – feasible through phasing out fossil fuels and finding alternative sources of energy and cleaner fuels, whereas the former has been trialled through clean technology such as the smog free tower, the world's first smog vacuum cleaner, building surfaces that convert harmful substances into harmless ones with the help of sunlight (Smedley 2017). In line with the increased urbanisation, better urban planning is vital for providing sustainable solutions to pollution. Much has already been done to counter pollution, but much more needs to happen to not only mitigate the effects but to prevent it as much as possible.

Whilst some air and water pollution are very visible (e.g. the smog in Delhi in 2019, oil spills), such pollution is often invisible, as is global warming and thus the immediate link to their effects might not always be evident. This is very different from our final challenge covered, which in most cases is very visible and a symbol of the so-called throwaway culture in which many societies live – waste. Its generation is directly linked to economic development, population growth and urbanisation. With the global population expected to grow from just under 8 billion in 2020 to about 10 billion in 2050, the earlier-mentioned urbanisation and the economic growth in lower middle-income countries, global waste is expected to increase by 70% over the next three decades (World Bank 2018). In particular, plastic waste that ends up polluting the ocean – both less visible microplastics and larger items – is expected to outweigh fish by weight by 2050 in a business-as-usual scenario, i.e. if we produce and manage waste as we currently do, following a tenfold increase in plastic pollution since 1980 (EMF 2016; IPBES 2019). In addition to managing the vast physical quantities of waste, the environmental impact extends to air, water and soil which waste (and its treatment) pollutes, and to climate change, to which waste contributes in the form of methane released from e.g. landfill and food waste. Minimising waste is therefore an important starting point that addresses not just the issue of waste but can support a holistic and integrated approach to creating a more sustainable society, economy and healthier environment. Waste minimisation requires moving from the planned obsolescence of take, make, waste (or take, make, use, lose) to reduce, reuse, recycle, from rethinking the linear way of business as usual to a circular economy where waste is never just waste but input (or food) for another process or product.

In brief, decarbonisation, pollution prevention and control, as well as dematerialisation are key to addressing all three challenges covered, with the ultimate aim of zero emissions, zero pollution and zero waste. This requires sustainability as well as resilience – sustainability could be argued to control and limit change based on a model of conservation and maintaining, whereas resilience focuses more on coping and managing impact in dynamic systems that are subject to continuous change (Nieuwenhuis and Touboulic 2017).

These challenges are complex and interrelated issues, often presenting as wicked problems dependent on many factors, making them difficult to tackle. As such, they represent not just a technological and engineering problem on the basis of science and innovation, but they also require financial support and increasing recognition that these are social, cultural and behavioural challenges as well, and one could argue a democratic one, too, as the topics require collective decisions that take into account current as well as future generations and that provide an appropriate legislative framework with clear regulatory and market signals. Robertson (2017) argues that policy and legislation are crucial for successfully dealing with the systemic global problems facing humankind, whilst Mason (2011) cautions that whilst there is general recognition of the sustainability challenge, there is disagreement about how to address it effectively. So what does the future for environmental regulations hold, and what makes them fit for purpose?

Environmental regulations – is the legislative framework fit for purpose?

Environmental regulations provide the legislative context within which companies operate. Legislation addresses the reality of environmental challenges which neither the market nor society on their own can meet as the environmental costs – also known as externalities – are not factored into conventional market and price mechanisms of demand and supply. Whether it is consumer convenience or the expediency of business as usual and the resulting inertia of both, behavioural change does not happen automatically, and therefore the state needs to legislate to correct these market failures and to mitigate environmental challenges by providing regulations and incentives that frame the operating context for companies.[5] Such regulations need to be both socially and politically acceptable, technologically realistic and ambitious, economically efficient, easily enforceable and, above all, not only mitigate against and adapt to the environmental challenge but also avoid future impacts and restore the environment by remediating environmental degradation.

Although at a global level the Paris Climate Accord agreed in 2015 represents a powerful agreement by over 190 countries to limit global warming to 2°C, with a further ambitious target of 1.5°C, the slowdown or decline in GHG needed for them to peak by 2020 has not been achieved and even a peak by 2030 is becoming less realistic given current climate policies and ambitions within the nationally determined contributions (NDCs) (WMO 2019a). Fundamental changes are needed to close the gap between the targets agreed and the obvious inadequacies of existing regulations, and the gap between policymakers who need to be (re-)elected in the short term (every four to five years) and the lasting efforts required to address the climate breakdown that are in the public interest in the long term, but probably quite painful and costly in the short term. Carbon taxes and carbon pricing offer market-based approaches to speed up the transition to carbon neutrality. However, as neither business as usual nor politics as usual appear to be able to fix the climate crisis, citizens' assemblies have been suggested as alternatives, as these are in a better position to take far-sighted and far-reaching decisions for the long term, beyond election cycles (Farrell 2019). Whilst the public might – in principle – support net zero carbon emissions by 2050, the implications are probably not fully understood and require sustained efforts of not only regulations but also education and information at global, national, regional, and local levels as well as innovative technologies. Arguably though, only a 2030 goal can realistically set us on the path to carbon neutrality,[6] which will require adaptation and mitigation, and possibly geoengineering, and above all, political will, during a decade of delivery in the 2020s. At the time of writing, 15 of the 20 wealthiest countries do not have a timeline for such a net zero target at a time when faster and deeper cuts to GHG are needed (BBC 2019).

Tackling climate change is particularly dependent on action and regulations at a global level, which can also help address pollution. Nevertheless, as pollution generally tends to be geographically more confined, local and national regulatory efforts can be especially powerful. This does not diminish the importance of transboundary action, especially when considering e.g. pollutants that are persistent, bio-accumulate and subject to long-range atmospheric transport or pollution of international waters by ships. Experience e.g. from China shows that wide-ranging clean air policies that are effectively implemented with foresight for the long term can indeed bring air pollution under control (Cross 2019). From the point of view of companies, pollution regulations should be aligned across all these levels, whether it is urban planning at a local level that improves transport infrastructures, national regulations on cleaner fuels and quality of air, water and land, regional legislation on permitted pollutant emissions, or international conventions on hazardous chemicals or toxic waste at a global level.

The same applies to waste, where companies are often faced with an array of different regulations at local, national and regional levels in terms of volume and weight of waste to be recycled or recovered, types of waste (and their long-range transport) being phased out or banned outright, and limits to what can be disposed of in landfill. Increasingly, extended producer responsibility is used by policymakers to tackle the built-in or planned obsolescence that is emblematic of the throwaway society. Such legislation has for example led to take-back programmes where companies have to accept the return of products at the end of their life. Whilst significant progress has been made in this regard, efforts need to go further for companies to fully take extended responsibility for their products beyond the factory gates both before and after production in terms of the supply chain and consumer behaviour and product disposal. Moving from the linear cradle-to-grave to a more circular cradle-to-cradle model will require a more concerted effort across the board, at local, national and transnational levels, closing legal loopholes that previously might have allowed companies to exploit weaker regulations in other sovereignties.

This points to two wider issues that regulations concerning all three challenges need to address in future. Legislation needs to be aligned and consistent across all geographical levels, all the way from local, regional and national levels to the transnational regional and global levels. Furthermore, legislation needs to be aligned and consistent across the different environmental challenges. In order to avoid contradictory requirements for companies, policymakers need to take into consideration whole life cycles of products and services and their various environmental impacts, which requires a more holistic approach. Such an approach should be cost-effective, based on science and informed by social and environmental values, allowing the integration of the three triple bottom line objectives of the environment, economy and society rather than playing them against each other (see also Wagner 2011). The resulting regulations should provide clarity, a level playing field and a predictable planning horizon for companies, which would lead to a resilient as well as a sustainable economy, society and environment.

The 17 United Nations Sustainable Development Goals (UN SDGs), adopted by all member states in 2015, are a recognition that meeting today's interconnected challenges – whether it is the environmental ones such as climate crisis, clean water, clean energy, sustainable cities and responsible consumption and production, or social and economic ones – cannot happen in isolation. The emphasis is on integration and on a systems approach as the different goals and actions to achieve them all affect each other. In order to move from commitment to action, partnerships between governments, civil society, citizens and the private sector, i.e. industry, are needed. What reinforces this work is designing policy for the future, access to and raising of finances, robust data as well as innovation and learning, all in partnership. To support such learning, the United Nations launched the Principles for Responsible Management Education (PRME) initiative to equip today's business students with the knowledge and ability for delivering the change needed to realise the SDGs.

Regulating for the sustainability challenge requires a combination of empirical assessment and normative claims as a structural realignment is needed to decouple economic development from unsustainable energy- and material-intensive processes. This is a form of ecological modernisation which together with economic dematerialisation and decarbonisation can deliver environmental sustainability (Mason 2011). Future environmental regulations need to be problem-led, creative and flexible, moving away from reactive, incremental policymaking of the past towards anticipatory, integrated approaches that use all available instruments (direct, indirect, voluntary) in a strategic and selective way (ibid.), clearly communicated to provide business with legislative certainty in an uncertain world.

What tools then can businesses build on in future to contribute to achieving the SDGs, in particular environmental sustainability and responsible production?

Sustainability tools and concepts – from the triple bottom line to the triple top line?

Reflecting the SDGs, the concept of doughnut economics illustrates both the social foundations and the planetary boundaries between which the safe operating space for humanity is placed, meeting life's basics and not overshooting beyond the ecological ceiling (Raworth 2017). Although the image of a doughnut is used, it could as easily be an inflatable swim ring – in some ways this is more helpful as it illustrates the importance of resilience, which is the ability of a system to handle disruptions and reorganise while retaining in essence the same structure, function, identify and feedbacks, based on the system's capacity to adapt and transform (Walker et al. 2004).

Resilience is key to biomimicry, where companies copy nature's economics, using nature as a model, measure and mentor. Biomimicry has influenced the circular economy concept which is also part of doughnut economics. They all have in common a more cyclical approach to industry and commerce where we move from the conventional linear take, make, consume, use and lose (or waste) cradle-to-grave model to a regenerative and restorative repair, reuse, refurbish, recycle cradle-to-cradle model, where value is captured from waste – which can be seen as "lost profit" (Nemetz 2013: 495) – in each stage of decomposition. The global growth opportunity from a transition towards a circular economy is estimated to represent $4.5 trillion by 2030 (WEF 2019).

Feeding into the circular economy and its life cycle view of production and consumption are various concepts covered throughout the book. Extended producer responsibility which puts product stewardship at the centre is underpinned by life cycle analysis and industrial ecology, where the waste product of one process becomes the input (or food) for another, by closed-loop systems and reverse logistics. A key starting point is design for environment (or eco-design) where environmental considerations are designed into a product from the onset. To ensure credibility and eco-effectiveness, cradle to cradle certification can be obtained, and consumers as well as buyers and suppliers can use this programme, other eco-labels and environmental management systems to ascertain how sustainable an organisation and its products is. Further tools that feed into this include ecological footprinting and environmental accounting, providing data for more informed decision-making based on measuring and monitoring environmental performance. On the distribution and marketing side, such decisions can also include choice editing, i.e. removing environmentally harmful products from the market altogether, giving buyers and consumers only a choice of e.g. green(er) products or of services instead of products as part of a move to servicisation and product-service systems. This builds reuse and upcycling into consumption and allows companies to keep ownership of a product while providing a service to buyers, thus practicing true product stewardship.

Other tools, frameworks and strategies covered include The Natural Step framework, natural capitalism, eco-efficiency, many of which are summarised in the sustainable business model archetypes in Figure 4.3. In particular, The Natural Step framework allows strategic planning from a visionary future point of success, the road to which is then operationalised through backcasting.

All of these help companies to integrate and embed sustainability thinking from the start, rather than treating it as a bolt-on to add to existing practices once they have been developed. Whilst the triple bottom line is a helpful concept, it measures corporate performance across the three dimensions of economy, equity and environment (or ecology) at the end; if companies want to incorporate sustainability from the beginning, they need to start using a triple top line approach, which puts sustainability considerations at the beginning of any value creation endeavour.

In order to adopt such a triple top line approach and to become truly sustainable in the future, companies need to take a strong sustainability stance that is regenerative as it protects and restores natural capital rather than a weak sustainability standpoint, where natural capital is only one of several types of capital that can be substituted with other types, whether constructed, social, manufactured, cultural or financial.[7] The difference between the two types of sustainability is perhaps best expressed in the Orwellian-style phrase, "all capitals are created equal except natural capital, which is more equal than others" (Sheldon 2011: 538).

Such a transition from weak to strong sustainability, from the triple bottom to the triple top line will require a major shift and radical as well as rapid action for businesses to rethink and transform business as usual, to redesign and reimagine ways of doing things in their ambition to scale mount sustainability and achieve mission zero.

Box 15.1 provides some examples of what the future could like for companies on the transition path to sustainability.

BOX 15.1 THE FUTURE OF AGRICULTURE, MANUFACTURING AND SERVICES – EXAMPLES

To complement the outlook on challenges, regulations, functions and tools and concepts, we provide some examples from various sectors of the economy that indicate how the future might look like.

In agriculture, vertical farming gives a glimpse of the future of the sector where crops grown in multi-storey high-tech containers can help address some of the world's big challenges, from loss of biodiversity to climate change. Vertical farming gets around the problem of space as food is grown upwards in stacked containers, and enables production of local food that does not need much shipping. The food is organic as it is grown in a pest- and pesticide-free environment (Whipple 2019). Another development in agriculture is the growing of compostable drink containers, including coffee cups, through a fast-growing fruit in the pumpkin family, gourds, that can be shaped through moulds and then dried before being used to drink from (*The Telegraph* 2019). Edible drink and dessert cups made of vanilla-flavoured biscuit are produced by *twiice* in New Zealand (see https://twiice.co.nz), while various companies are developing edible cutlery. Edible and biodegradable packaging for sauces and drinks are made from seaweed and plants by Notpla – its Ooho! is an edible water and an alternative to plastic and bottled water, for example at sports events (Platt Leonard 2017, see also www.notpla.com). Hot drink cups are also made by Gumdrop Ltd, which recycles and processes chewing gum into a sustainable compound called Gum-tec® for the plastic and rubber industry, having collected them in dedicated Gumdrop bins, with other products including guitar picks, lunch boxes, and door stops also made by the company (see http://gumdropltd.com for more).

In manufacturing as well as services, there is an increasing move to a subscription economy. Whether it is software that is bought as a regular subscription rather than a one-off, or music and film streaming services, or physical products such as automobiles or razors, food or recipe boxes, the subscription economy frees consumers from the burden of ownership and of responsibility for disposal, and moves it onto companies who can extend their producer responsibility by adopting product stewardship.

Other developments in manufacturing include transparent solar panels that could be applied to windows as a large source of untapped energy, thus using surfaces in the existing infrastructure of the built environment rather than building new solar power parks (Traverse et al. 2017). In Israel, UBQ materials takes mixed household rubbish and upcycles it by converting it into a pseudo-plastic material which the company calls "thermoplastic, composite, bio-based, sustainable, climate-positive" that can be made into multiple other products, helping address both climate change and waste (Morrison and Kordova 2019). The Smog Free Project run by Daan Roosegaarde in the Netherlands is a campaign for clean air

BOX 15.1 (CONTINUED)

using urban innovations that provide clean air locally in public spaces. Products include the smog free tower and the smog free bicycle, which both inhale polluted air and release cleaned air back into the environment, but also a smog free ring made from compressed smog particles (see https://www.studioroosegaarde.net for more). A similar approach to climate change is taken by *made of air*, which is a biochar-based material made of 90% atmospheric carbon which can for example be used in construction, interiors and for other purposes, capturing carbon and putting it to use (see http://www.madeofair.com/). It is one of many projects and companies that mine carbon from the air that are part of the *air miners* collective of researchers, entrepreneurs and funders exploring such opportunities whilst Impossible Labs is an innovation consultancy focusing on new opportunities for business from climate change (see http://www.airminers.org and https://impossiblelabs.io/).

There are many more examples too numerous to list that provide reason to be hopeful, whether it is regarding plastic waste, clean energy, climate change, electrification of transport or the circular economy (see Makower 2018 for more instances).

How do companies then address the environmental challenges that are framed through environmental regulations beyond the general guiding frameworks of tools and concepts? How can companies apply these in each corporate function in the future?

Corporate functions for environmental sustainability – from functional silo mind-sets to cross-functional teams

In the third part of the book we covered the three functions of primary value chain activities – supply chain management, production and operations (including design), and marketing (including consumer behaviour), and the three functions of supporting activities – information technology and management systems, accounting and finance, and human resource management. Whilst economics was covered in Chapter 2, entrepreneurship was referred to in Chapter 13 as well as in Chapter 4 where we looked at strategy and business models for sustainability. Although we covered each corporate function that was chosen for a chapter separately and individually, we stressed throughout the book that an integrated approach to sustainability is necessary, as there are overlaps, interconnections and interdependencies. Going forward, this is even more important for businesses if they are to manage the transition from reactive and compliance-oriented modes of operating to proactive, regenerative and sustaining ones successfully. Sheldon (2011) comments that the approach to corporate functions and management systems has traditionally been less than coordinated. The need for better integration and coordination is also stressed by professor Bhattacharya and former Unilever CEO Polman (2017) who identify six points companies need to consider when integrating the environment into business decisions: sustainability is more than simply a change initiative; it requires the whole value chain and a redefinition of competitive space by collaborating, support from boards as a priority as well as buy-in from the undecided, and ensuring that sustainability is part of every employee's job, whether procurement, marketing, development, finance, regardless of geographic location or business unit.

To illustrate: if production and operations do not incorporate environmental design principles from the start, then the marketing function will struggle to promote products as environmentally friendly and risk being accused of greenwashing. Robust data (e.g. of life cycle analysis or environmental footprints) collected with the help of information technology and broader environmental management systems can inform design decisions and become a basis for obtaining certification and eco-labels. The data can furthermore be used to inform a company's accounting and reporting division, which communicates environmental performance to external stakeholders as well as to internal stakeholders and managers for decision-making. In turn, this can feed into integrated marketing communications. If employees are not appropriately trained and engaged, all other functions will struggle to move towards environmental sustainability. Communication and integration are therefore key, particularly for the supply chain management function which works across departments within a company and across companies within the supply chain, whether upstream or downstream.

As with environmental regulations, corporate functions cannot operate in isolation to achieve corporate environmental sustainability; they need to work in partnership and be integrated and coordinated within the wider organisational structure and corporate strategy. This will enable sustainability to be embedded systematically throughout an organisation, underpinned by communication and strong values that drive the company's mission and vision.

Moving from compartmentalised functions to integrated cross-functional teams has implications for how managers and employees are trained and developed, as in particular tertiary education is traditionally based on separate distinct disciplines rather than interdisciplinarity.

Sustainability as a vocation and profession – educating and managing for sustainability

Sustainability requires employees and managers who appreciate the complexity and systemic nature of the sustainability challenge. The very act of managing necessitates an understanding of the different disciplines and corporate functions (Sheldon 2011). This is a shift away from the conventional assumption of managers that employees only need to understand their immediate tasks rather than see the organisation's work as a whole – these traditional structures facilitate the linear take-make-waste approach that is based on organisational silos, whereas what is needed is a borrow-use-return system that operates cyclically. In such an organisation employees are engaged, well-informed and collectively contribute to the development of vision and strategies, supported by learning and innovation (Robertson 2017).

It is important to remember that in terms of education and learning, this entails learning from good (or even best) practice as well as learning from failure, which Bhattacharya and colleagues (2018) emphasise as key to creating sustainable businesses. This relates equally to policymakers who might need to experiment with policies to manage the sustainability transition and its complex demands (Mason 2011). The importance of learning applies across an organisation; in particular, educating boards of directors, investors and senior management are needed if a company is to commit to take the environment seriously (King 2011). This includes learning by doing or experiential learning as an essential element of sustainability education – such active learning is more likely to result in changing thinking and practice (Robertson 2017; Visser 2016). Whether the learning happens in the workplace, in schools or universities, sustainability education needs to be embedded, not separate, multidisciplinary and connected, as it is likely to be part of every job and every business in the future (Weybrecht 2017). Such an approach to learning and education for sustainability is also supported by the UN's PRME initiative. Going forward, it is not just about teaching sustainability; it is also about making education and research activities more sustainable themselves (see e.g. Jäckle 2019 or Goodall and Hill 2019).

Much progress has been made since the 1970s to move the work of environmental managers and sustainability professionals from focusing on operational compliance and control to more strategic cost-saving and value-creating activities (King 2011). Whilst compliance activities are important starting points, it is crucial to avoid the impression that companies have "done" the environment once they have implemented environmental management systems, achieved ISO 14001 certification or conducted environmental impact assessments (Sheldon 2011). To become truly sustainable, companies need to go beyond mechanical schemes and formal systems – it requires a vision as well of what and where a business wants to become and go to. A positive vision is key for moving away from "should" tenets and for focusing on opportunity rather than threat (although important, too). Industry can use the experience of transition towns – which focus on hope and action – as an example (Robertson 2017).

The future of the environmental profession thus encompasses a range of levels, from the tactical, operational, short term to the strategic, visionary and long term, which all contribute to the transition to sustainability across an organisation based on being able to make sense of these different levels in terms of tasks, responsibilities, work involved, time horizons – we outlined these seven levels based on King (2011) in Chapter 4 on change and leadership for sustainability. What this requires from managers is the capability to recognise complex system elements as interdependent where changing one part alters the whole. It requires working at multiple scales and across boundaries equipping not just businesses but wider society and humanity for the sustainability transition[8] (Robertson 2017).

This is also emphasised by the long-term, dynamic and integrated thinking inherent in systems thinking (Valero-Silva 2017). Systems thinking is a key sustainability literacy skill as applied to business and management alongside values reflection and philosophical assumptions, as highlighted previously in Chapter 4. Such systems thinking is complemented by subject-specific knowledge (e.g. in marketing), specialised knowledge such as in LCA or EMS, general sustainability knowledge on climate change or pollution, and strategic decision-making rooted in practice and supported by key stakeholders (Molthan-Hill et al. 2017).

Systems thinking powerfully counters the reductionism and silo mentality of which business education is often accused (Valero-Silva 2017). It underpins strong sustainability by challenging the view of humans being superior to other species and by highlighting the value of natural resources and the limits to infinite growth. Systems thinking also reminds us that sustainability is inherently integrated and holistic as it encompasses environment, ecology and economy, as depicted in Figure 1.2, reproduced here.

Working towards sustainability thus also requires the ability to resolve competing demands of economy, society and environment (King 2011) and to reconcile the threats of environmental challenges and the opportunities of corporate decision-making for sustainability[9] inherent in the tensions and trade-offs encountered in corporate sustainability (Haffar and Searcy 2017, see also Chapter 3, Hahn et al. 2015, 2018). A new legal form for companies encourages just such reconciliation – the benefit corporation, or B Corp, that balances profit and purpose, where companies use financial profits for environmental and social benefits (Robertson 2017) is part of a wider movement for purpose-led business, as for example demonstrated by The British Academy's Principles for Purposeful Business, and the World Economic Forum's Davos Manifesto 2020, both put forward in 2019. Purpose also ties in with the concept of vocation, where a job provides meaning beyond pay, which sustainability work often does.

Sustainability professionals and managers of the future need to have a variety of skills, both hard and technical as well as soft ones, based on evolving job descriptions involving multi-dimensional and changing tasks (Robertson 2017). Pioneers and innovators who understand the

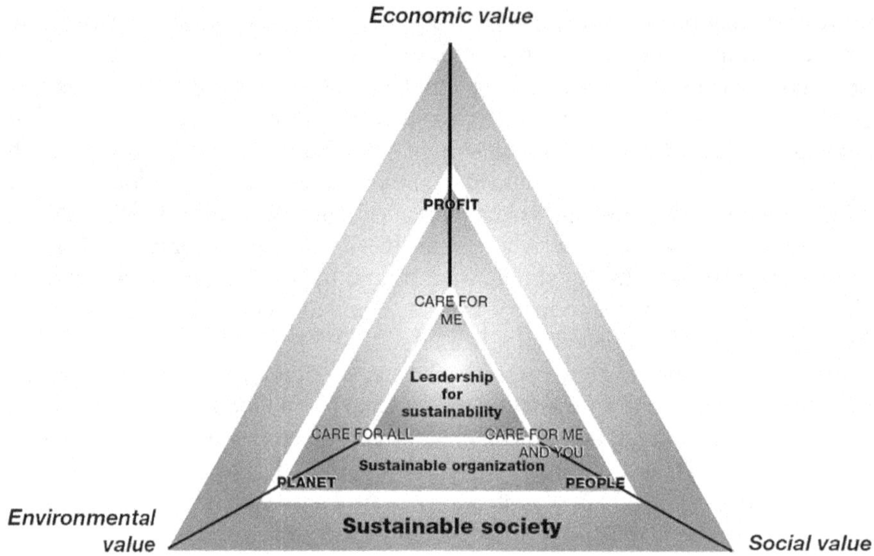

FIGURE 15.1
Three levels of sustainability

importance of culture, adaptive leadership and governance have a key role to play, as emphasised in the following quotation:

> the essence of how environmentalism changes the corporate enterprise lies not primarily in its technical adjustments but more important, in its structural, strategic and cultural transformation.
>
> (Hoffman 2001: 21)

In order to enable such transformation that firstly reduces unsustainability and secondly creates sustainability and that is embedded into organisational culture, companies need to foster commitment, build momentum for change and enable capacity for such change, which in the case of individuals is more lasting when incremental and in the case of companies more likely when not disruptive (Bertels et al. 2011; Robertson 2017; Dowell and Muthulingam 2017; Hoffman 2018). Although each organisation will have its own particular place-based set of challenges and opportunities to which it needs to tailor its own strategies, general principles and tools can be adjusted to any kind of organisation (Robertson 2017) – such as the ones provided in the various parts in the book. Whichever framework is used, it is important to communicate well and clearly in order to ensure commitment throughout the organisation for implementing sustainability, which in all cases requires a team effort (ibid.), even more so in the future. Such a team will consist of individuals for whom sustainability is a conviction based on moral duty and a vocation to which they feel called based on the values they hold, individuals for whom sustainability is a profession based on the knowledge they have gained through studying and learning, and individuals who come to sustainability with initial scepticism based perhaps on ignorance but persuaded by the business case and the scientific evidence.

INTEGRATION – SUSTAINABILITY AS A TEAM EFFORT

> Sustainability in an organization is a collaborative enterprise.
>
> (Robertson 2017: 314)

We conclude with the theme we have probably stressed most throughout the book – interdependence and interconnectedness, requiring a holistic and integrated approach to managing sustainability, particularly if we look at all aspects of sustainability, the triple bottom (or top) line of people, planet and profit, or equity, environment and economy. The dynamic complexity of multiplicity, ambiguity, interdependence and flux that characterises the task alongside the uncertainty, irreversibility, scale and discontinuities accompanying the consequences and possible threshold effects as environmental degradation reaches points of no return calls for an integrated rather than a fragmented approach, for a global perspective that is implemented locally as well as nationally. This applies across the regulatory framework provided by governments as well as the corporate functions within a company.

The team effort needs to go beyond companies alone – it requires a joint effort of the societal triangle of state, the market and civil society (see Figure 1.1), as well as humanity working in partnership with nature, which humanity can use as mentor and model as in biomimicry – in an effort to emulate nature, working as an interconnected ecosystem, using and reusing natural as well as waste material.

Such a team effort and integration does not imply uniformity – it is diversity (as we find it in nature) that underpins resilient systems (Robertson 2017). Therefore, it is a diversity of skills, backgrounds, disciplines, organisations – whether for-profit or not-for-profit, whether governmental or non-governmental – industries, regulations and countries across all levels that is needed to address the array of environmental challenges we face.

From a corporate as well as an academic point of view, an interdisciplinary approach where corporate functions and levels collaborate within and across each other as well as with other organisations is needed, encompassing local as well as global levels.

As Robertson (2017: 335) emphasises, systemic problems necessitate systemic solutions. This is where we need to remind ourselves again that environmental sustainability is only one element of sustainability. Social sustainability goes hand in hand with it, as the environmental damage done is often borne by those not yet born or by those not responsible for the damage. This is recognised by the growing global movement calling for a so-called Green New Deal which combines decarbonisation with creating jobs and transforming the economy for the benefit of people and planet. It is illustrated in Figure 15.1 and for example by the concept of planetary health which emphasises that everything is connected – human health is not separate from the planet's health as humanity is impacted by how human-caused disruptions of the Earth's natural systems affect the planet's well-being (PHA 2019).

Bringing it all together we can perhaps attempt to summarise the book in the following matrix (Table 15.1), integrating environmental challenges, regulatory and corporate levels as well as corporate functions. X denotes weight of environmental regulations at each level, O denotes importance of corporate response to environmental challenge at each level, whereas the abbreviations indicate at which level which corporate function should act. It is not entirely comprehensive or conclusive, as the picture changes according to size and scale of companies (not all companies operate at an international, let alone global level), for example, and not all meso-levels of government (both types of regional) will legislate directly and instead draw on other policy instruments such as voluntary agreements or indirect tools. The matrix also does not account for the work of

TABLE 15.1

Matrix of integration

Environmental regulations (X)	Environmental challenges			Corporate levels (0) (time spans increase L1 to L7)
	Climate change	Pollution	Waste	
LOCAL	X 0 A&F P&0 EMS HRM	X 0 P&0 EMS A&F HRM	X 0 P&0 EMS A&F HRM	LEVEL 1: task focus LEVEL 2: first line supervision LEVEL 3: department and system management
REGIONAL (SUB-NATIONAL)	X 0 Mkt SCM HRM	X 0 Mkt P&0 SCM HRM	X Mkt SCM HRM	LEVEL 4: general management of division or function
NATIONAL	X 0 A&F HRM Mkt SCM P&0 Strat	X 0 A&F Mkt HRM SCM P&0 DfE Strat	X 0 HRM A&F Mkt SCM DfE Strat	LEVEL 5: business unit, corporate division management
REGIONAL (SUPRA-NATIONAL)	X 0 Strat SCM	X 0 Strat	X 0 Strat SCM	LEVEL 6: international level
GLOBAL	X 0 A&F EMS Strat SCM DfE P&0	X 0 A&F EMS Strat SCM DfE P&0	X EMS Strat DfE SCM P&0	LEVEL 7: global level

Source: compiled by author based on chapters 1-14, using King (2011) for corporate levels. Abbreviations: SCM - Supply Chain Management, P&0 - Production & Operations, DfE - Design for Environment (includes research and development), EMS - environmental management systems (includes information technology), Mkt - Marketing, A&F - Accounting and Finance, HRM - Human Resource Management, Strat - Strategy. Font size – weighting.

non-governmental organisations from civil society or the market that provide e.g. standards or codes of conducts and is not able to accommodate the various tools and concepts that can be used by companies. Furthermore, it is limited in comprising only the three environmental challenges selected for the book rather than the broader set of challenges of which they are a part. Nevertheless, it provides a tentative integrative picture of the scale, complexity and interdependencies, and possibilities inherent in tackling the corporate environmental sustainability challenge, which truly is a team effort.

SUMMARISING CONCLUSION

We have taken quite a journey in this book, from the past to the present to the future, from foundations and principles to tools and concepts for putting the principles into action, from overarching corporate strategies and business models for sustainability to individual corporate functions, and from surveying environmental challenges in general to covering climate change, pollution and waste in detail. Despite all the wide-ranging complexities and challenges that characterise the interdependencies and interconnections in environmental sustainability for business and corporate functions, Everard (2011: 39) encourages us with a reminder that sustainability is "really quite a simple concept" as it "literally . . . relates to the capacity to continue indefinitely".

Whether you're a student, an academic, a manager or simply an interested layperson, whether you have dipped in and out of the various parts, or whether you have studiously read the chapters in sequence, I hope the book has provided insight into the reality and scale of the environmental challenges business and society face – which are ground for deep concern and pessimism – as well as insight to a box of tools and concepts for how to address them, both strategically and functionally, which can give us hope, optimism and guidance to put business and industry on a more sustainable path to a more resilient future. What role will you play in this transformative work?

SUGGESTED QUESTIONS AND ACTIVITIES

- Learn more about the UN's SDGs and check out the work on integration: https://sustainabledevelopment.un.org/ and https://sdgintegration.undp.org/. For a critique of the contradiction between growth and ecology in the SDGs, see Hickel (2019).
- The UN's PRME (Principles for Responsible Management Education) initiative incorporates the UN SDGs into management education. Find out which education institutions in your country participate in the initiative: www.unprme.org. Get inspired by student-written stories about innovative, profitable and positive businesses supporting the UN SDGs: https://aim2flourish.com/innovations.
- Check out the coalition advocating for leaving fossil fuels in the ground: Leave it in the Ground Initiative (LINGO) www.leave-it-in-the-ground.org/ – explore their suggestions of how we tackle climate change and evaluate the options.
- Check out the post carbon institute's work on building sustainable and resilient communities: www.resilience.org/ and www.postcarbon.org/.
- Find out more about B Corps: https://bcorporation.net/ is a starting point.
- Test your sustainability literacy skills: www.sulitest.org/en/index.html.
- Prepare for a debate for one of the following motions:
  - A truly environmentally sustainable company is impossible.
- What are key elements in making organisations more sustainable in the future?
- What will the sustainable company look like in the future?
- Check out some of the companies and examples mentioned in Box 15.1. Can you find other innovative and creative ventures that rethink business as usual to tackle environmental challenges?

NOTES

1 E.g. artificial intelligence and machine learning can greatly contribute to studying and addressing environmental impact with the use of big and open data but have themselves high energy requirements.

2 The biophysical system of the Arctic is already moving away from its relatively stable 20th-century state into an unprecedented state that will have implications beyond the Arctic (Box et al. 2019).

3 The shift in public awareness is reflected in the other shortlisted terms for Oxford Dictionary's word of the year: eco-anxiety, flight shame, extinction, ecocide, climate crisis and plant-based (Kahn 2019).

4 The poisonous effect is not just on health and premature deaths – increases in air pollution are also linked to higher levels of crime and poorer school and university performance (Hogenboom 2019).

5 We should add here that beyond the state and civil society, it is also the market in the form of industry context that drives corporate sustainability (see e.g. Chen et al. 2017).

6 Climate neutrality equates to net zero carbon emissions. Activities are *climate positive* if they go beyond net zero emission and generate environmental benefits (e.g. by removing CO_2 from the atmosphere). An alternative term for this is *carbon negative*, although some companies (e.g. Unilever) confusingly use the term carbon positive to indicate their efforts to eliminate fossil fuels and to e.g. generate more renewable energy than they consume (Anzilotti 2018, Unilever website: www.unilever.com/sustainable-living/reducing-environmental-impact/greenhouse-gases/how-were-becoming-carbon-positive-in-our-operations/).

7 A proverbial critique of weak sustainability and how it is often narrowly understood is best captured in a quote ascribed to Alanis Obomsawin, a North American native, who said in the early 1970s about Canada, a most affluent country, that it "operates on a depletion economy which leaves destruction in its wake. Your people are driven by a terrible sense of deficiency. When the last tree is cut, the last fish is caught, and the last river is polluted; when to breathe the air is sickening, you will realize, too late, that wealth is not in bank accounts and that you can't eat money" (Poole 1972: 43). The quote is now widely used in the form, "When the last tree is cut down, the last fish eaten and the last stream poisoned, you will realize that you cannot eat money" and generally attributed to a Native American source.

8 Turnheim and Nykvist (2019) explore the potential and feasibility of sustainability transition pathways and the conditions of their realisation.

9 For Nemetz (2013) corporate strategic decision-making sits at the nexus of the potential threats from sustainability (based on the scientific evidence and effects of environmental degradation which influence media attention, regulatory responses, civil society and other stakeholder responses, and international agreements) and its opportunities (such as lower costs, liabilities and risks linked to legislation, turning waste into profits, improved reputation and access to markets, improved efficiency and profitability as well as competitive advantage).

SOURCES AND FURTHER READING

Anzilotti, E. (2018) Climate Positive, Carbon Neutral, Carbon Negative: What do They Mean? Fast Company, 06.03.2018, available at www.fastcompany.com/40583176/climate-positive-carbon-neutral-carbon-negative-what-do-they-mean (Retrieved, 25.11.2019).

BBC (2019) Climate Change: 'Bleak' Outlook as Carbon Emissions Gap Grows, *BBC*, 26.11.2019, available at www.bbc.co.uk/news/science-environment-50547073 (Retrieved, 26.11.2019).

Bertels, S., Papania, L. and Papania, D. (2011) Embedding Sustainability in Organizational Culture: A Systematic Review of the Body of Knowledge, Network for Business Sustainability, available at www.nbs.net/articles/systematic-review-organizational-culture (Retrieved, 27.11.2019).

Bhattacharya, C. B., Ciorra, E. and Radeke, J. (2018) Tolerating Failure: A Key to Creating Sustainable Business, *The European Business Review*, 17.01.2018, available at www.europeanbusinessreview.com/tolerating-failure-a-key-to-creating-sustainable-business/ (Retrieved, 27.11.2019).

Bhattacharya, C. B. and Polman, P. (2017) Sustainability Lessons from the Front Lines, in: *MIT Sloan Management Review*, 58(2): 71–78.

Box, J. E., Colgan, W. T., Røjle Christensen, T., Schmidt, N. M., Lund, M., Parmentier, F.-J. W., Brown, R., Bhatt, U. S., Euskirchen, E. S., Romanovsky, V. E., Walsh, J. E., Overland, J. E., Wang, M., Corell, R. W. Meier, W. N., Wouters, B., Mernild, S., Mård, J., Pawlak, J. and Skovgård Olsen, M. (2019) Key Indicators of Arctic Climate Change: 1971–2017, in: *Environmental Research Letters*, 14(4), https://doi.org/10.1088/1748-9326/aafc1b.

Chen, H., Zeng, S., Lin, H. and Ma, H. (2017) Munificence, Dynamism, and Complexity: How Industry Context Drives Corporate Sustainability, in: *Business Strategy and the Environment*, 26(2): 125–141.

Cross, D. T. (2019) Air Pollution is Reaching Endemic Proportions, *Sustainability Times*, 22.11.19, available at www.sustainability-times.com/environmental-protection/air-pollution-is-reaching-endemic-proportions/ (Retrieved, 25.11.2019).

Dowell, G. W. S. and Muthulingam, S. (2017) Will Firms Go Green if it Pays? The Impact of Disruption, Cost, and External Factors on the Adoption of Environmental Initiatives, in: *Strategic Management Journal*, 38(6): 1287–1304.

Economist (2019) Global Warming 101 – The Past, Present and Future of Climate Change, Briefing, 21.09.2019, available at www.economist.com/briefing/2019/09/21/the-past-present-and-future-of-climate-change (Retrieved, 21.11.2019).

EEA (2016) *The European Environment – State and Outlook 2015 Report*, European Environment Agency, available at www.eea.europa.eu/soer (Retrieved, 15.11.2019).

EMF (2016) The New Plastics Economy – Rethinking the Future of Plastics, Report Produced by World Economic Forum, Ellen MacArthur Foundation and McKinsey & Company, available at www.ellenmacarthurfoundation.org/assets/downloads/EllenMacArthurFoundation_TheNewPlasticsEconomy_Pages.pdf (Retrieved, 25.11.2019).

Everard, M. (2011) Sustainability and Sustainable Development, in: Brady, J., Ebbage, A. and Lunn, R. (eds.) *Environmental Management in Organizations, The IEMA Handbook*, London: Earthscan/Routledge, 2nd ed., pp. 39–61.

Farrell, D. (2019) Politics-as-usual Can't Fix the Climate Crisis. Maybe it's Time to Try a Citizens' Assembly, *The Guardian*, 28.08.2019, available at www.theguardian.com/commentisfree/2019/aug/28/climate-crisis-citizens-assembly-extinction-rebellion (Retrieved, 25.11.2019).

Goodall, A. and Hill, S. (2019) No More Business: Scholars Need Support to Move Teaching and Research on Climate Change into the Mainstream, *Times Higher Education*, print edition, 14.11.2019.

Haffar, M. and Searcy, C. J (2017) Classification of Trade-offs Encountered in the Practice of Corporate Sustainability, in: *Journal of Business Ethics*, 140(3): 495–522.

Hahn, T., Figge, F., Pinkse, J. and Preuss, L. (2018) A Paradox Perspective on Corporate Sustainability: Descriptive, Instrumental, and Normative Aspects, in: *Journal of Business Ethics*, 148(2): 235–248.

Hahn, T., Pinkse, J., Preuss, L. and Figge, F. (2015) Tensions in Corporate Sustainability: Towards an Integrative Framework, in: *Journal of Business Ethics*, 127(2): 297–316.

Hickel, J. (2019) The Contradiction of the Sustainable Development Goals: Growth Versus Ecology on a Finite Planet, in: *Sustainable Development*, 27(5): 873–884.

Hoffman, A. J. (2001) *From Heresy to Dogma: An Institutional History of Corporate Environmentalism*, Stanford, CA: Stanford University Press.

Hoffman, A. J. (2018) The Next Phase of Business Sustainability, in: *Stanford Social Innovation Review*, Spring: 35–39.

Hogenboom, M. (2019) How Air Pollution is Doing more than Killing us, *BBC Future*, 16.04.2019, available at www.bbc.com/future/article/20190415-how-air-pollution-is-doing-more-than-killing-us (Retrieved, 22.11.2019).

Hook, L., Munshi, N., Hornby, L., Kuchler, H., Schipani, A., Bernard, S. and Harlow, M. (2019) How Safe is the Air we Breathe? *Financial Times*, 05.09.2019, available at www.ft.com/content/7d54cfb8-cea5-11e9-b018-ca4456540ea6 (Retrieved, 04.12.2019).

IPBES (2019) Summary for Policymakers of the Global Assessment Report on Biodiversity and Ecosystem Services, Intergovernmental Science-Policy Platform on Biodiversity and Ecosystem Services, available at https://ipbes.net/global-assessment-summary-policymakers-final-version-now-available (Retrieved, 29.11.2019).

IPCC (2018) Summary for Policymakers of IPCC Special Report on Global Warming of 1.5°C Approved by Governments, Intergovernmental Panel on Climate Change, Press Release 08.10.2018, available at www.ipcc.ch/2018/10/08/summary-for-policymakers-of-ipcc-special-report-on-global-warming-of-1-5c-approved-by-governments/ (Retrieved, 21.11.2019).

Jäckle, S. (2019) WE have to Change! The Carbon Footprint of ECPR General Conferences and Ways to Reduce it, in: *European Political Science*, 18(4): 630–650.

Kahn, B. (2019) Why Oxford Dictionary's 'Word of the Year' Matters This Time, *Gizmodo Earther*, 21.11.2019, available at https://earther.gizmodo.com/why-oxford-dictionarys-word-of-the-year-matters-this-ti-1839970234 (Retrieved, 21.11.2019).

King, R. (2011) The Future of the Environmental Profession, in: Brady, J., Ebbage, A. and Lunn, R. (eds.) *Environmental Management in Organizations, The IEMA Handbook*, London: Earthscan/Routledge, 2nd ed., pp. 543–558.

Makower, J. (2018) Two Steps Forward – 2018 was the Year that. . ., *GreenBiz*, 31.12.2018, available at www.greenbiz.com/article/2018-was-year (Retrieved, 03.12.2019).

Mason, M. (2011) The Sustainability Challenge, in: Brady, J., Ebbage, A. and Lunn, R. (eds.) *Environmental Management in Organizations, The*

IEMA Handbook, London: Earthscan/ Routledge, 2nd ed., pp. 525–532.

Molthan-Hill, P., Hill, S. and Parkes, C. (2017) A New Framework for Embedding Sustainability into the Business School Curriculum, in: Molthan-Hill, P. (ed.) *The Business Student's Guide to Sustainable Management*, Sheffield: Greenleaf Publishing, 2nd ed., pp. 13–51.

Morrison, J. and Kordova, S. (2019) Revolutionary Recycling? A New Technology Turns Everyday Trash into Plastic Treasure, *The Washington Post*, 18.11.2019, available at www.washingtonpost.com/graphics/2019/climate-solutions/israeli-startup-ubq-turning-trash-into-plastic-products/ (Retrieved, 04.12.2019).

Nemetz, P. N. (2013) *Business and the Sustainability Challenge: An Integrated Perspective*, London: Routledge.

Nieuwenhuis, P. and Touboulic, H. (2017) Is SSCM Sustainable? – The Case for 'Ecological Supply Chain Management', Presentation at International Symposium, Greening of Industry Network, 27.05.2017.

PHA (2019) Planetary Health Alliance, webpages, available at https://planetaryhealthalliance.org/ (Retrieved, 22.11.2019).

Platt Leonard, J. (2017) Edible Water: How Eating Little Balls of H2O could be the Answer to the World's Plastic Pollution, *The Independent*, 14.04.2017, available at www.independent.co.uk/life-style/food-and-drink/edible-water-eating-ooho-skipping-rocks-lab-no-packaging-plastic-pollution-world-h20-a7682711.html (Retrieved, 03.12.2019).

Poole, T. (1972) Conversations with North American Indians, in: Osborne, R. (ed.) *Who is the Chairman of This Meeting? A Collection of Essays*, Toronto: Neewin Publishing Co., pp. 39ff.

Powell, J. (2019) Scientists Reach 100% Consensus on Anthropogenic Global Warming, in: *Bulletin of Science, Technology & Society*, 1–2, https://doi.org/10.1177/0270467619886266.

Raworth, K. (2017) *Doughnut Economics: Seven Ways to Think Like a 21st-Century Economist*, London: Random House Business Books.

Robertson, M. (2017) *Sustainability Principles and Practice*, London: Routledge, 2nd ed.

Sheldon, C. (2011) Getting Beyond Integrated Thinking and Governance, in: Brady, J., Ebbage, A. and Lunn, R. (eds.) *Environmental Management in Organizations, The IEMA Handbook*, London: Earthscan/ Routledge, 2nd ed., pp. 533–542.

Smedley, T. (2017) How the World's Biggest Cities are Fighting Smog, *BBC Future*, 12.01.2017, available at www.bbc.com/future/article/20170112-how-the-worlds-biggest-cities-are-fighting-smog (Retrieved, 22.11.2019).

The Telegraph (2019) Biodegradable Coffee Cups Being 'Grown' in Bid to Cut Down on Plastic Waste, *The Telegraph*, 17.03.2019, available at www.telegraph.co.uk/news/2019/03/17/biodegradable-coffee-cups-grown-bid-cut-plastic-waste/ (Retrieved, 03.12.2019).

Traverse, C. J., Pandey, R., Barr, M. C. and Lunt, R. R. (2017) Emergence of Highly Transparent Photovoltaics for Distributed Applications, in: *Nature Energy*, 2(11): 849–860.

Turnheim, B. and Nykvist, B. (2019) Opening up the Feasibility of Sustainability Transitions Pathways (STPs): Representations, Potentials, and Conditions, in: *Research Policy*, 48(3): 775–788.

UN (2019) World Urbanization Prospects 2018: Highlights, United Nations, Department of Economic and Social Affairs, Population Division, ST/ESA/SER.A/421, available at https://population.un.org/wup/Publications/Files/WUP2018-Highlights.pdf (Retrieved, 22.11.2019).

Valero-Silva, N. (2017) Systems Thinking and Sustainable Management, in: Molthan-Hill, P. (ed.) *The Business Student's Guide to Sustainable Management*, Sheffield: Greenleaf Publishing, 2nd ed., pp. 504–535.

Visser, W. (2016) Foreword, in: Sunley, R. and Leigh, J. (eds.) *Educating for Responsible Management: Putting Theory into Practice*, London: Greenleaf/Routledge.

Wagner, S. M. (2011) The Corporate Political Activities of Multinational Enterprises: The Automotive Industry and Environmental Regulations in the European Union, PhD Thesis, Loughborough University.

Walker, B., Holling, C. S., Carpenter, S. R. and Kinzig, A. (2004) Resilience, Adaptability and Transformability in Social – Ecological Systems, in: *Ecology and Society*, 9(2): 5, available at www.ecologyandsociety.org/vol9/iss2/art5/print.pdf (Retrieved, 25.11.2019).

WEF (2019) It's Time for the Circular Economy to go Global – and you can Help, World Economic Forum Annual Meeting, available at www.weforum.org/agenda/2019/01/its-time-for-the-circular-economy-to-go-global-and-you-can-help/ (Retrieved, 28.11.2019).

Weybrecht, G. (2017) *The Future MBA: 100 Ideas for Making Sustainability the Business of Business Education*, London: Routledge.

Whipple, T. (2019) Welcome to the Future of Farming, *The Times*, 07.06.2019, available at www.thetimes.co.uk/article/welcome-to-the-future-of-farming-c6lshbhsb (Retrieved, 03.12.2019).

WHO (2018) Climate Change and Health, World Health Organization, Fact Sheet, 01.02.2018, available at www.who.int/news-room/fact-sheets/detail/climate-change-and-health (Retrieved, 03.12.2019).

WMO (2019a) Greenhouse Gas Concentrations in Atmosphere Reach Yet Another High, World Meteorological Organization, Press Release, 25.11.2019, available at https://public.wmo.int/en/media/press-release/greenhouse-gas-concentrations-atmosphere-reach-yet-another-high (Retrieved, 25.11.2019).

WMO (2019b) 2019 Concludes a Decade of Exceptional Global Heat and High-Impact Weather, World Meteorological Organization, Press Release, 03.12.2019, available at https://public.wmo.int/en/media/press-release/2019-concludes-decade-of-exceptional-global-heat-and-high-impact-weather (Retrieved, 06.12.2019).

World Bank (2018) What a Waste: An Updated Look into the Future of Solid Waste Management, World Bank, available at www.worldbank.org/en/news/immersive-story/2018/09/20/what-a-waste-an-updated-look-into-the-future-of-solid-waste-management (Retrieved, 25.11.2019).

INDEX

For Product Safety Concerns and Information please contact our EU
representative GPSR@taylorandfrancis.com
Taylor & Francis Verlag GmbH, Kaufingerstraße 24, 80331 München, Germany

www.ingramcontent.com/pod-product-compliance
Lightning Source LLC
LaVergne TN
LVHW080121011026
849160LV00013B/1151

* 9 7 8 1 1 3 8 6 3 2 3 8 7 *